RF and Microwave Coupled-Line Circuits

For a complete listing of the *Artech House Microwave Library*, turn to the back of this book.

RF and Microwave Coupled-Line Circuits

Rajesh Mongia
Inder Bahl
Prakash Bhartia

Artech House
Boston • London

Library of Congress Cataloging-in-Publication Data
Mongia, Rajesh.
  RF and microwave coupled-line circuits / Rajesh Mongia, Inder J. Bahl, Prakash Bhartia.
    p.   cm.
  Includes bibliographical references and index.
  ISBN 0-89006-830-5 (alk. paper)
    1. Strip transmission lines. 2. Electric network analysis. 3. Microstrip antennaas. 4. Microwave circuits. I. Bahl, I. J. II. Bhartia, P. III. Title.
  TK7876.M634   1999
  621.381'32—dc21
                                                                99-21677
                                                                    CIP

British Library Cataloguing in Publication Data
Mongia, Rajesh
  RF and microwave coupled-line circuits. – (Artech House microwave library)
  1. Microwave circuits
  I. Title II. Bahl, I. J. (Inder Jit) III. Bhartia, P. (Prakash)
  621.3'813

  ISBN 0-89006-830-5

Cover design by Elaine Donnelly

© 1999 ARTECH HOUSE, INC.
685 Canton Street
Norwood, MA 02062

All rights reserved. Printed and bound in the United States of America. No part of this book may be reproduced or utilized in any form or by any means, electronic or mechanical, including photocopying, recording, or by any information storage and retrieval system, without permission in writing from the publisher.
  All terms mentioned in this book that are known to be trademarks or service marks have been appropriately capitalized. Artech House cannot attest to the accuracy of this information. Use of a term in this book should not be regarded as affecting the validity of any trademark or service mark.

International Standard Book Number: 0-89006-830-5
Library of Congress Catalog Card Number: 99-21677

10 9 8 7 6 5 4 3 2 1

Contents

	Foreword	**xiii**
	Preface	**xv**
1	**Introduction**	**1**
1.1	Coupled Structures	1
1.1.1	Types of Coupled Structures	4
1.1.2	Coupling Mechanism	6
1.2	Components Based on Coupled Structures	9
1.2.1	Directional Couplers	9
1.2.2	Filters	14
1.3	Applications	18
1.4	Scope of the Book	19
2	**Microwave Network Theory**	**23**
2.1	Actual and Equivalent Voltages and Currents	24
2.1.1	Normalized Voltages and Currents	24
2.1.2	Unnormalized Voltages and Currents	28
2.1.3	Reflection Coefficient, VSWR, and Input Impedance	29
2.1.4	Quantities Required to Describe the State of a Transmission Line	31
2.2	Impedance and Admittance Matrix Representation of a Network	33
2.2.1	Impedance Matrix	33
2.2.2	Admittance Matrix	34
2.2.3	Properties of Impedance and Admittance Parameters of a Passive Network	34

2.3	Scattering Matrix	35
2.3.1	Unitary Property	37
2.3.2	Transformation With Change in Position of Terminal Planes	38
2.3.3	Reciprocal Networks	39
2.3.4	Relationship Between Normalized and Unnormalized Matrices	40
2.4	Special Properties of Two-, Three-, and Four-Port Passive, Lossless Networks	40
2.4.1	Two-Port Networks	40
2.4.2	Three-Port Reciprocal Networks	42
2.4.3	Three-Port Nonreciprocal Networks	43
2.4.4	Four-Port Reciprocal Networks	43
2.5	Special Representation of Two-Port Networks	47
2.5.1	ABCD Parameters	47
2.5.2	Reflection and Transmission Coefficients in Terms of ABCD Parameters	49
2.5.3	Equivalent T and Π Networks of Two-Port Circuits	53
2.6	Conversion Relations	55
2.7	Scattering Matrix of Interconnected Networks	56
2.7.1	Scattering Parameters of Reduced Networks	58
2.7.2	Reduction of a Three-Port Network Into a Two-Port Network	59
2.7.3	Reduction of a Two-Port Network Into a One-Port Network	61
2.7.4	Reduction of a Four-Port Network Into a Two-Port Network	62
3	**Characteristics of Planar Transmission Lines**	**65**
3.1	General Characteristics of TEM and Quasi-TEM Modes	65
3.1.1	TEM Modes	70
3.1.2	Quasi-TEM Modes	71
3.1.3	Skin Depth and Surface Impedance of Imperfect Conductors	73
3.1.4	Conductor Loss of TEM and Quasi-TEM Modes	74

3.2	Representation of Capacitances of Coupled Lines	75
3.2.1	Even- and Odd-Mode Capacitances of Symmetrical Coupled Lines	76
3.2.2	Parallel-Plate and Fringing Capacitances of Single and Coupled Planar Transmission Lines	80
3.3	Characteristics of Single and Coupled Striplines	83
3.3.1	Single Stripline	84
3.3.2	Edge-Coupled Striplines	88
3.4	Characteristics of Single and Coupled Microstrip Lines	90
3.4.1	Single Microstrip	91
3.4.2	Coupled Microstrip Lines	98
3.5	Single and Coupled Coplanar Wave Guides	101
3.5.1	Coplanar Waveguide	102
3.5.2	Coplanar Waveguide With Upper Shielding	103
3.5.3	Conductor-Backed Coplanar Waveguide With Upper Shielding	104
3.5.4	Coupled Coplanar Waveguides	105
3.6	Suspended and Inverted Microstrip Lines	105
3.7	Broadside-Coupled Lines	109
3.7.1	Broadside-Coupled Striplines	109
3.7.2	Broadside-Coupled Suspended Microstrip Lines	112
3.7.3	Broadside-Coupled Offset Striplines	114
3.8	Slot-Coupled Microstrip Lines	117
4	**Analysis of Uniformly Coupled Lines**	**123**
4.1	Even- and Odd-Mode Analysis of Symmetrical Networks	124
4.1.1	Even-Mode Excitation	126
4.1.2	Odd-Mode Excitation	128
4.2	Directional Couplers Using Uniform Coupled Lines	130
4.2.1	Forward-Wave (or Codirectional) Directional Couplers	133
4.2.2	Backward-Wave Directional Couplers	136
4.3	Uniformly Coupled Asymmetrical Lines	140

4.3.1	Parameters of Asymmetrical Coupled Lines	140
4.3.2	Distributed Equivalent Circuit of Coupled Lines	145
4.3.3	Relation Between Normal Mode (c and π) and Distributed Line Parameters	150
4.3.4	Approximate Distributed Line or Normal-Mode Parameters of Asymmetrical Coupled Lines	152
4.4	Directional Couplers Using Asymmetrical Coupled Lines	154
4.4.1	Forward-Wave Directional Couplers	154
4.4.2	Backward-Wave Directional Couplers	157
5	**Broadband Forward-Wave Directional Couplers**	**161**
5.1	Forward-Wave Directional Couplers	162
5.1.1	3-dB Coupler Using Symmetrical Microstrip Lines	164
5.1.2	Design and Performance of 3-dB Asymmetrical Couplers	166
5.1.3	Ultra-Broadband Forward-Wave Directional Couplers	168
5.2	Coupled-Mode Theory	169
5.3	Coupled-Mode Theory for Weakly Coupled Resonators	177
6	**Parallel-Coupled TEM Directional Couplers**	**181**
6.1	Coupler Parameters	182
6.2	Single-Section Directional Coupler	183
6.2.1	Frequency Response	183
6.2.2	Design	185
6.2.3	Compact Couplers	192
6.2.4	Equivalent Circuit of a Quarter-Wave Coupler	192
6.3	Multisection Directional Couplers	193
6.3.1	Theory and Synthesis	193
6.3.2	Limitations of Multisection Couplers	208
6.4	Techniques to Improve Directivity of Microstrip Couplers	209
6.4.1	Lumped Compensation	209

6.4.2	Use of Dielectric Overlays	212
6.4.3	Use of Wiggly Lines	213

7	**Nonuniform Broadband TEM Directional Couplers**	**219**
7.1	Symmetrical Couplers	219
7.1.1	Coupling in Terms of Even-Mode Characteristic Impedance	221
7.1.2	Synthesis	224
7.1.3	Technique for Determining Weighting Functions	230
7.1.4	Electrical and Physical Length of a Coupler	233
7.1.5	Design Procedure	234
7.2	Asymmetrical Couplers	237

8	**Tight Couplers**	**243**
8.1	Introduction	243
8.2	Branch-Line Couplers	244
8.2.1	Modified Branch-Line Coupler	247
8.2.2	Reduced-Size Branch-Line Coupler	251
8.2.3	Lumped-Element Branch-Line Coupler	255
8.2.4	Broadband Branch-Line Coupler	260
8.3	Rat-Race Coupler	260
8.3.1	Modified Rat-Race Coupler	264
8.3.2	Reduced-Size Rat-Race Coupler	265
8.3.3	Lumped-Element Rat-Race Coupler	268
8.4	Multiconductor Directional Couplers	270
8.4.1	Theory of Interdigital Couplers	271
8.4.2	Design of Interdigital Couplers	274
8.5	Tandem Couplers	281
8.6	Multilayer Tight Couplers	284
8.6.1	Broadside Couplers	285
8.6.2	Embedded Microstrip Couplers	287
8.6.3	Re-Entrant Mode Couplers	294
8.7	Compact Couplers	296
8.7.1	Lumped-Element Couplers	298
8.7.2	Spiral Directional Couplers	299
8.7.3	Meander Line Directional Coupler	300
8.8	Other Tight Couplers	301

9	**Coupled-Line Filters**	**305**
9.1	Introduction	305
9.1.1	Types of Filters	307
9.1.2	Applications	308
9.2	Theory and Design of Filters	308
9.2.1	Maximally Flat or Butterworth Prototype	309
9.2.2	Chebyshev Response	310
9.2.3	Other Response-Type Filters	312
9.2.4	LC Filter Transformation	314
9.2.5	Filter Analysis and CAD Methods	319
9.2.6	Some Practical Considerations	320
9.3	Types of Coupled Filters	323
9.3.1	Parallel-Coupled Line Filters	323
9.3.2	Interdigital Filters	326
9.3.3	Combline Filters	328
9.3.4	The Hairpin-Line Filter	330
9.3.5	Parallel-Coupled Stepped-Impedance Filters	335
9.4	Miniature Filters	337
9.4.1	Lumped-Element-Based Miniaturization	337
9.4.2	Monolithic Microwave Integrated Circuit Filters	339
9.4.3	Miniaturization Using High-Dielectric Constant Ceramics	339
10	**Coupled-Line Circuit Components**	**353**
10.1	DC Blocks	353
10.1.1	Analysis	353
10.1.2	Broadband DC Block	357
10.1.3	Biasing Circuits	357
10.1.4	Millimeter-Wave DC Block	362
10.1.5	High-Voltage DC Block	362
10.2	Coupled-Line Transformers	364
10.3	Interdigital Capacitor	367
10.3.1	Approximate Analysis	370
10.3.2	Fullwave Analysis	372
10.4	Spiral Inductors	373
10.5	Spiral Transformers	383
11	**Baluns**	**391**
11.1	Introduction	391

11.2	Microstrip-to-Balanced Stripline Balun	395
11.3	Analysis of a Coupled-Line Balun	399
11.4	Planar Transmission Line Baluns	403
11.4.1	Analysis	405
11.4.2	Examples	408
11.5	Marchand Balun	411
11.5.1	Coaxial Marchand Balun	415
11.5.2	Synthesis of Marchand Balun	421
11.5.3	Examples of Marchand Baluns	426
11.6	Other Baluns	435
11.6.1	Coplanar Waveguide Baluns	435
11.6.2	Triformer Balun	438
11.6.3	Planar-Transformer Balun	441
12	**High-Speed Circuit Interconnects**	**447**
12.1	High-Speed Digital Circuit Interconnects	447
12.2	Analysis of Coupled Lossy Interconnect Lines	449
12.2.1	Frequency Dependence of Line Parameters	452
12.2.2	Results	456
13	**Multiconductor Transmission Lines**	**475**
13.1	Theory	477
13.1.1	Eigenvalues and Eigenvectors	479
13.1.2	Decomposition in Terms of Normal Modes	481
13.1.3	Admittance and Impedance Matrices	483
13.1.4	Eigenvalues and Eigenvectors of Lossy Lines	484
13.1.5	Equivalent Circuit of Multiconductor Transmission Line	485
13.2	Examples	488
13.2.1	Eigenvalues and Eigenvectors of Symmetrical Coupled Lines	488
13.2.2	Equivalent Circuit of Single Transmission Line Network	494
	Appendix	**499**
	Index	**503**

Foreword

It has been a privilege for me to go through the manuscript of the *RF and Microwave Coupled-Line Circuits*.

Sections of coupled transmission structures are critical components in distributed RF and microwave passive circuits. Significance of their role as basic building blocks is second only to the sections of single transmission structures. Their applications in design of directional couplers and filters are well known, but equally important is the role played by coupled-line sections in the design of baluns, capacitors, inductors, transformers and DC blocks. Availability of high dielectric constant materials has extended the usage of coupled-line sections to lower microwave and RF frequencies. Traditionally, coupled sections consisting of two single lines have been used extensively. However, as the circuit designers understand modeling and characterization of multiple coupled lines, we can look forward to significantly larger applications of multiple coupled transmission line structures. Three-line balun structures reported recently are a step in this direction. Also, as the multilayer RF and microwave circuits become more popular, couplings among the transmission lines at different levels of a multilayer structure become a critical design consideration. In some cases this multilayer-multiconductor coupling can be advantageous as a useful circuit component, while in other cases this coupling can become an undesirable effect that should be mitigated. Both of these situations need the modeling and characterization of multiconductor transmission line structures.

Recognizing the role of coupled lines, it is hard to comprehend why a comprehensive book on this topic has not been available so far. But then, someone has to be the leader. Bahl and Bhartia have a history of providing to the microwave design community several well-needed "firsts," and this book is their most recent contribution. Congratulations Rajesh, Inder, and Prakash on a book which I am confident will be very well received in the RF/microwave community.

K.C. Gupta
University of Colorado
at Boulder

Preface

There are a number of textbooks on microwave transmission lines. Recent ones include extensive information on the modern planar lines such as microstrip, slotlines, coplanar waveguides, and the like. At the next level of complexity are the various functional circuits such as couplers, hybrids, filters, and baluns, which use the elemental transmission line in different configurations to achieve the desired functionality and meet system performance requirements. Much of this functionality involves coupling between transmission lines, and extensive research has been conducted in the design and analysis of such structures. Initially, much of the literature was oriented to coaxial lines and waveguides. With the evolution and the popularity of planar transmission lines, however, it was felt desirable to put together all aspects of coupled circuits using these lines under one cover.

Most current texts, we found, contained perhaps a chapter or two on some specific components, especially couplers and filters. This text attempts to treat the topic in its entirety, starting with the fundamental theory of coupled structures and the application of this to the design and analysis of specific components such as couplers, filters, baluns, and so forth. This treatment emphasizes planar transmission lines, the CAD tools available for the design of these structures, use of full-wave analyses and accurate semiempirical equations for component design, novel structures and configurations, and new applications.

This book is primarily intended for design engineers and research and development specialists who are involved in the area of coupled-line circuit design, analysis, development, and fabrication. The layout of the book facilitates its use as a text for a graduate course and for short courses on specific component design.

The book is divided into 13 chapters. The first chapter introduces the reader to the topic and covers the nature of coupled structures, the importance of these structures in microwave circuits, and some applications. A good introduction to the principal components using coupled lines (i.e., directional couplers and filters) is also given.

The second chapter establishes the basic circuit parameters and representation of microwave networks. Fundamental network analysis tools such as impedance and scattering matrix techniques are introduced together with the properties of two-, three-, and four-port networks. Relationships between the commonly used matrix representation forms such as ABCD, scattering, and impedance are established to permit the researcher or designer to work in the system of his or her preference.

The fundamental building blocks for coupled-line circuits (i.e., transmission lines) are covered in Chapter 3. In particular, the characteristics of the commonly used planar lines, such as microstrip, coplanar, and striplines are covered in detail. In addition, the characteristics of coupled lines in these configurations under different conditions such as broadside coupling, edge coupling, or, in the case of coplanar waveguides, coupling with shields present are discussed. Whereas Chapter 3 concentrates on characteristics of physical transmission lines, Chapter 4 presents the general analysis of uniformly coupled asymmetrical lines, including forward and backward couplers. These fundamentals permit a more in-depth coverage of the coupling of uniform lines, which is covered in the next chapter. Even- and odd-mode analysis is covered together with an analysis of uniformly coupled asymmetrical lines. Forward and backward directional coupler design methods using the aforementioned techniques are also given in the chapter.

The next few chapters are devoted to the design of various types of directional couplers. Many directional couplers by their very nature and design have a narrowband performance. In a number of applications, broadband performance is essential. The design and performance of forward-directional couplers using asymmetrical coupled lines are the subject of Chapter 5. Coupled-mode theory, also discussed in this chapter, is very useful for the analysis of general weakly coupled systems.

Parallel-coupled backward TEM directional couplers using a single section or multisections are discussed in Chapter 6, together with limitations and methods for improving the directivity of such couplers. This permits the reader to have a good understanding of how these circuits work and the methods, including lumped-element compensation and dielectric overlay, that can be used to improve directivity. While we dealt with broadband couplers using multisection couplers in Chapter 6, one can also obtain this type of characteristic of performance using nonuniform lines. Additional flexibilities, and at the same time complexities, are introduced with this line, but in many cases it is essential to resort to this process because of physical or performance constraints imposed by the overall circuit design. In Chapter 7, the design and synthesis procedure for such couplers is outlined, together with some other techniques to obtain broadband performance.

Finally, the last type of coupler that requires special treatment is the tight coupler. Tight couplers, as described in Chapter 8, can be designed and fabricated in a number of configurations. Some of the most prevalent forms are the branch-line coupler, rat-race coupler, and lumped-element coupler. These are fully covered in this chapter, together with a large number of other layouts including the multiconductor couplers and tandem couplers. A number of novel designs are also discussed, including the interdigital Lange couplers and compact couplers for wireless applications. The material provided gives the designer a good grasp of the principles and techniques involved in the design of these coupler types, together with the advantages and disadvantages of the specific couplers. This information and understanding is critical to the designer in assisting him or her to choose the appropriate coupler type to meet not only the eletrical performance characteristics but also to meet any form, fit, and function requirements imposed.

Besides the directional couplers covered in the previous chapters, perhaps the most commonly used form of coupled line circuits is the filter. This is covered extensively in Chapter 9, starting with a definition of filter parameters and leading on to filter synthesis, design, and realization. Modern miniature filters are discussed, as they are critical to the wireless communications area, and an assessment of the capabilities of a number of software packages available for filter design is provided.

The next two chapters delve into a number of other commonly used coupled circuits. Chapter 10 covers the analysis and design of a number of dc blocks, coupled-line transformers, interdigital capacitors, spiral inductors, and transformers, while Chapter 11 treats the design and analysis of baluns. In particular, the Marchand balun is discussed in detail, together with other types of baluns such as the coplanar waveguide balun, triformer balun, and planar transformer balun.

Whereas the preceding chapters have used coupling as a means of achieving a specific function and performance, in many cases coupling is not desirable and can create problems. This is typically the electromagnetic compatability/electromagnetic interference problem that is encountered by any circuit designer. To try to cover the topic of coupled circuits in its entirety, we have included high-speed digital interconnects in Chapter 12 to bring about an awareness of the cross-talk problem and provide ways to mitigate this problem. Finally, many of the passive devices covered could use multiconductor lines for their design.

The literature on this topic is very dispersed. In Chapter 13, we have provided the essential information for the designer to permit the use of multiconductor lines as the building block for the type of coupled-line circuit one wished to design.

As with any comprehensive treatment of a topic, one must draw upon the works of a large number of researchers and authors. We have taken special care to reproduce equations and diagrams and believe that this text is a valuable addition to the microwave circuit designer's library.

The preparation of this text has depended on a number of very supportive individuals and organizations. Naturally, the time spent during evenings and weekends comes at the expense of time with our families. For their support and understanding we are eternally grateful. The organizations that we work for have also supported this project in many ways and we wish to express our thanks to them. While always dangerous to mention specific names because some others will feel left out, we have no hesitation in recognizing the contributions and acknowledging with our thanks the assistance of Josie Dunn for typing the manuscript and Bob Gervais of the Defence Research Establishment, Ottawa, who devoted large blocks of time in preparing the illustrations. Part of the manuscript has been handled efficiently by Tanya Morrison of ITT GaAsTEK, Roanoke. The Artech House team did an excellent job on the final book. We would like to thank Mark Walsh, Barbara Lovenvirth, Hilary Sardella, Judi Stone, Steve Cartisano, and Elaine Donnelly for their patience, support, and cooperation. Finally, we want to thank the reviewers for their thoroughness and excellent suggestions for improving the text.

R. Mongia
I. J. Bahl
P. Bhartia

1

Introduction

In microwave circuits, transmission lines are normally used in two ways: (1) to carry information or energy from one point to another; and (2) as circuit elements for passive circuits like impedance transformers, filters, couplers, delay lines, resonators, and baluns. Passive elements in conventional microwave circuits consist mainly of the distributed type and employ transmission line sections and waveguides in different configurations, thereby achieving the desired functionality and meeting performance specifications. This functionality is largely achieved by the use of coupled transmission lines. In this chapter, we briefly describe the various types of coupled-line structures, the coupling mechanism, and coupled-line components and their applications.

1.1 Coupled Structures

When two unshielded transmission lines (as shown in Figure 1.1) are placed in close proximity to each other, a fraction of the power present on the main line is coupled to the secondary line. The power coupled is a function of the physical dimensions of the structure, mode of propagation (TEM (transverse electromagnetic) or non-TEM), the frequency of operation, and the direction of propagation of the primary power. In these structures, there is a continuous coupling between the electromagnetic fields of the two lines. These parallel coupled lines are called *edge-coupled structures*. The structure shown in Figure 1.1(d) is an exception and is called a *broadside-coupled structure*.

Coupled lines can be of any form, depending on the application and generally consist of two transmission lines, but may contain more than two. The lines can be symmetrical (i.e., both conductors have the same dimensions)

2   RF and Microwave Coupled-Line Circuits

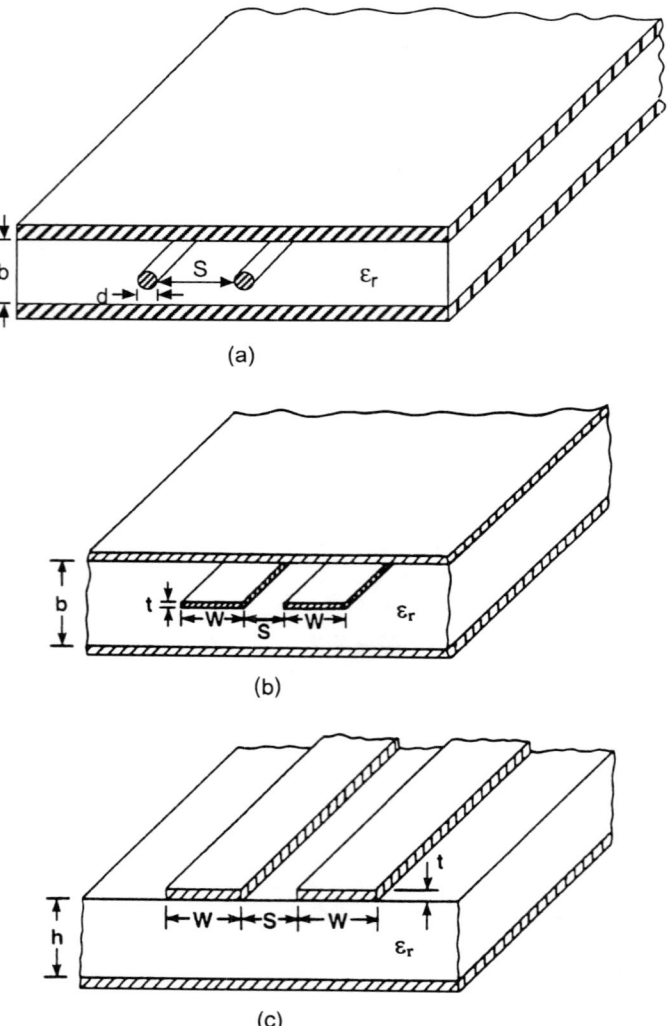

Figure 1.1 Coupled transmission lines: (a) coaxial lines, (b) striplines, (c) microstrip lines, (d) broadside striplines.

or asymmetrical. Both lines are placed in close proximity to each other so that the electromagnetic fields can interact. The separation between the lines may be either constant or variable. The closer the lines are placed together, the stronger the interaction that takes place. When one port is excited with a known signal, a part of this signal appears at other ports. This interaction effect known as *desirable*

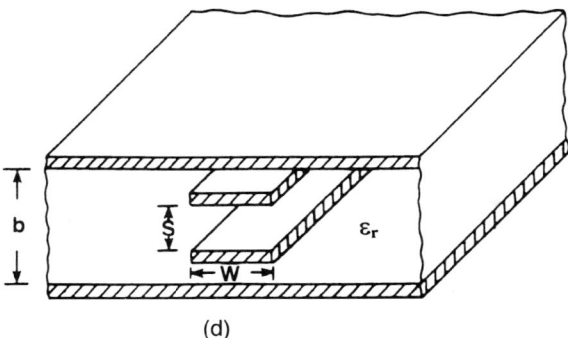

(d)

Figure 1.1 (*continued*)

coupling is used to advantage in realizing several important microwave circuit functions, such as directional couplers, filters, and baluns, with the coupled line length usually being approximately a quarter-wave long.

In addition, in closely packed hybrid and monolithic integrated circuits, *parasitic coupling* can take place between the distributed matching elements or closely spaced lumped elements, affecting the electrical performance of the circuit in several ways. It may change the frequency response in terms of frequency range and bandwidth and degrade the gain/insertion loss and its flatness, input and output voltage standing wave ratio (VSWR), and many other characteristics including output power, power-added efficiency, and noise figure, depending on the type of circuit. This coupling can also result in the instability of an amplifier circuit or create a feedback resulting in a peak or a dip in the measured gain response or a substantial change in a phase-shifter response. In general, this parasitic coupling is undesirable and an impediment in obtaining an optimal solution in a circuit design. However, this coupling effect can be taken into account in the design phase by using empirical equations and by performing electromagnetic (EM) simulations, or it can be reduced to an acceptable level by maintaining a large separation between the matching elements.

Multiconductor microstrip lines (Figure 1.2) are used in very-large-scale integrated (VLSI) chips for digital applications and three-dimensional microwave integrated circuits. Here, numerous closely spaced interconnection lines in different planes are used to integrate the components on a chip. The design of these interconnections is very important to satisfy the size, power consumption, clock frequency, and propagation delay requirements. Signal distortion, increase in background noise, and cross-talk between the lines from coupling are some of the undesirable characteristics. Proper design of these interconnects can reduce the distortion and cross-talk to acceptable levels and has played a significant role in the evolution of high-speed VLSI technology.

Figure 1.2 Cross-sectional view of a multiconductor and multilevel coupled-line configuration.

1.1.1 Types of Coupled Structures

Coupled-line structures are available for all forms and types of transmission lines/dielectric guides and waveguides. Striplines, microstrip lines, coplanar waveguides, image guides, and insular and inverted stripguides are the most popular planar forms. In Figure 1.3, cross sections of microstrip coupled lines and microstrip-like lines are shown. In these structures, practical spacing limitations between the lines limits the tight coupling achievable to about 8 dB over $\lambda/4$ sections. These configurations are edge-coupled structures. On the other hand, broadside-coupled lines (shown in Figure 1.4) are used extensively to realize tight couplings of the order of 2 to 3 dB. All three structures support TEM modes in the case of a homogeneous dielectric medium and quasi-TEM modes in the case of a nonhomogeneous media.

In Figure 1.5, we show coupled dielectric waveguides. They support non-TEM modes and forward-wave couplers are realized using these structures. These structures are commonly used at millimeter-wave frequencies, and continuous coupling occurs from one guide to another when they are placed in close proximity to each other.

The configurations shown in Figures 1.1, 1.3, 1.4, and 1.5 use equal widths for both conductors and guides and constant spacing between the conductors and guides. These structures are therefore called *symmetric and uniformly coupled*. Figure 1.6 shows an asymmetrically coupled microstrip line configuration with constant spacing between lines of unequal width. This structure is called a *uniformly coupled asymmetric line*. Figure 1.7 shows an example of a symmetric

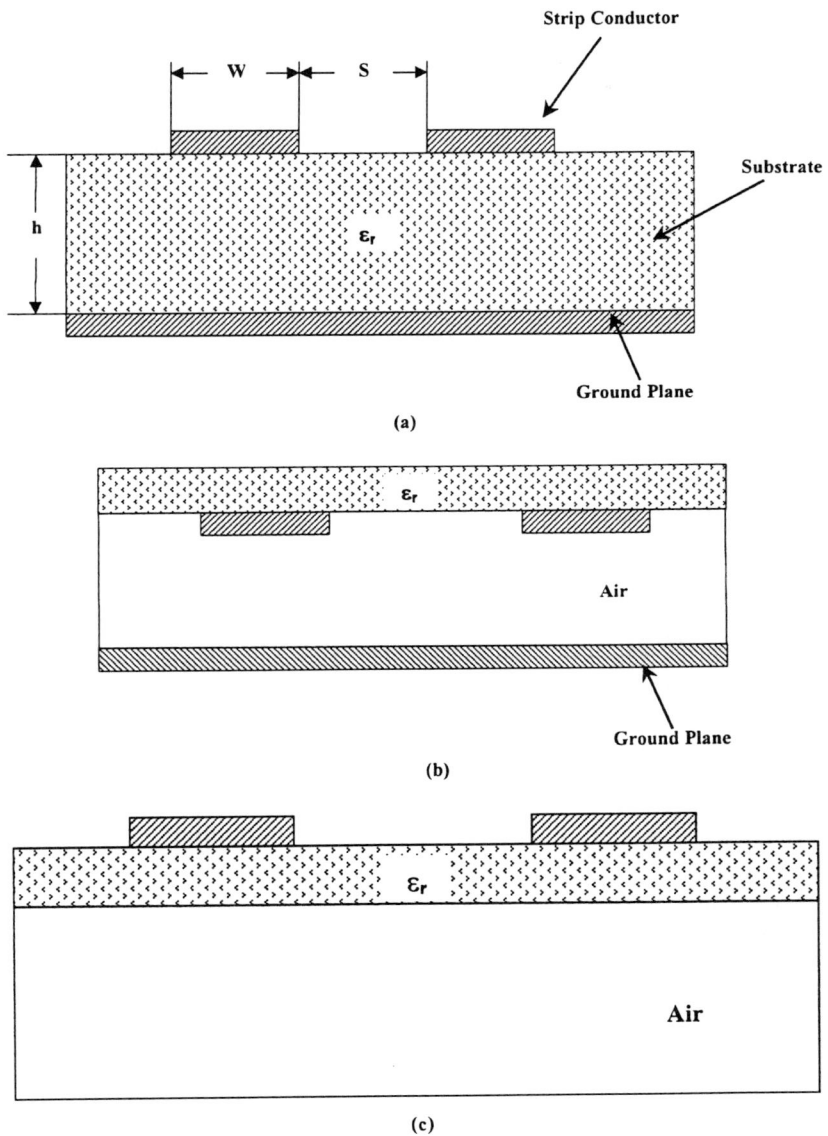

Figure 1.3 Coupled microstrip-like transmission lines: (a) microstrip lines, (b) inverted microstrip lines, (c) suspended microstrip lines, (d) coplanar waveguide.

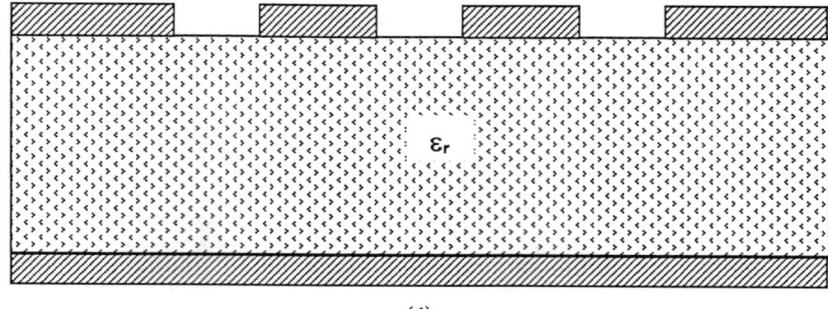

(d)

Figure 1.3 (*continued*)

coupled line with variable spacing between the microstrip conductors, called a *nonuniformly coupled symmetric line*.

1.1.2 Coupling Mechanism

The symmetric coupled-line structures, as shown in Figure 1.1, support two modes: even and odd. The interaction between these modes induces the coupling between the two transmission lines, and the properties of the symmetric coupled structures may be described in terms of a suitable linear combination of these modes. The field distributions for the even and odd modes on coupled microstrip lines are shown in Figure 1.8. In even-mode excitation, both microstrip conductors are at same potential while the odd mode delineates equal but of opposite polarity potentials. The even and odd modes have different characteristic impedances, and their values become equal when the separation between the conductors is very large (lines are uncoupled). The even-mode characteristic impedance (Z_{0e}) is the impedance from one line to the ground when both lines are driven in-phase from equal sources of equal impedances and voltages. The odd-mode characteristic impedance (Z_{0o}) is defined as the impedance from one line to the ground when both lines are driven out of phase from equal sources of equal impedances and voltages.

The velocities of propagation of the even and odd modes are equal when the lines are imbedded in a homogeneous dielectric medium (e.g., stripline). For transmission lines such as coupled microstrip lines, however, the dielectric medium is not homogeneous, and a part of the field extends into the air above the substrate, resulting in different propagation velocities for the two modes. Consequently, the effective dielectric constants (and the phase velocities) are different for the two modes. This nonsynchronous feature deteriorates the performance of circuits using these types of coupled lines. The voltage coupling coefficient of a coupling structure is generally expressed in terms of the

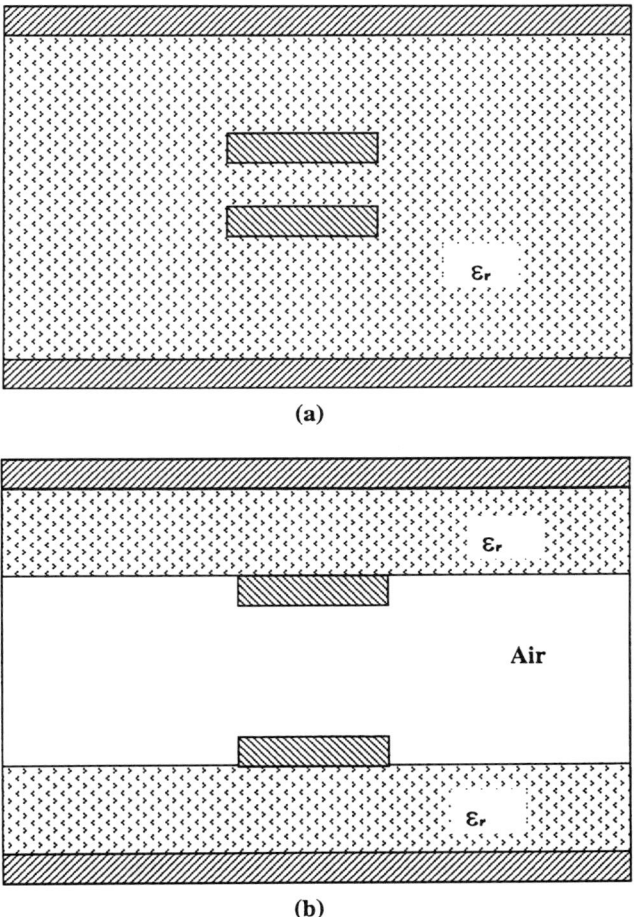

Figure 1.4 Coupled broadside transmission lines: (a) broadside-coupled striplines, (b) broadside-coupled inverted microstrip lines, (c) broadside-coupled suspended microstrip lines.

even- and odd-mode characteristic impedances, effective dielectric constants, and coupled structure line length. For a quarter-wave coupled section in a homogeneous dielectric medium, the coupling coefficient k is given by

$$k = \frac{Z_{0e} - Z_{0o}}{Z_{0e} + Z_{0o}} \quad (1.1)$$

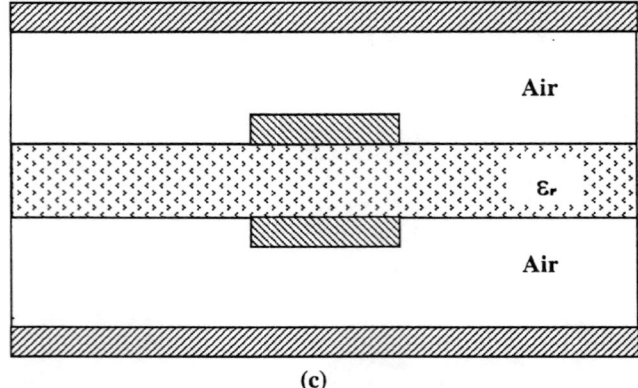

Figure 1.4 (*continued*)

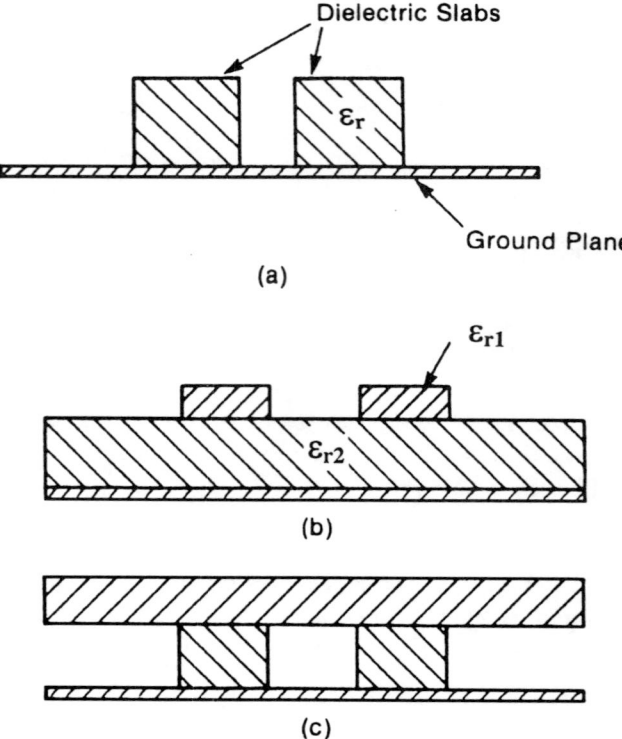

Figure 1.5 Coupled dielectric guides: (a) image, (b) insular, and (c) inverted strip.

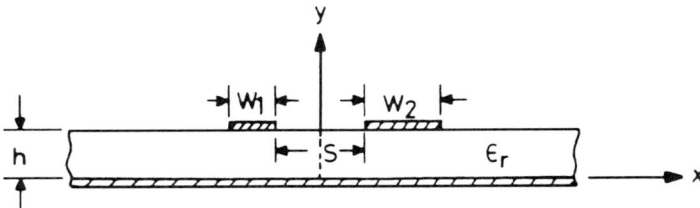

Figure 1.6 Coupled microstrip lines with unequal impedances (asymmetric lines).

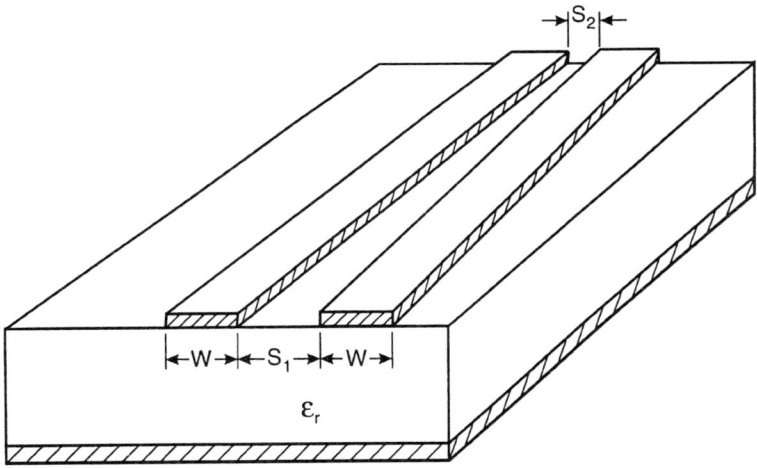

Figure 1.7 Nonuniformly coupled symmetric lines.

1.2 Components Based on Coupled Structures

There are numerous microwave passive components realized using coupled-line sections. They include directional couplers, filters, baluns, impedance transformer networks, resonators, inductors, interdigital capacitors, dc blocks, and others, of which directional couplers and filters are the most popular. A brief history of the latter is presented next.

1.2.1 Directional Couplers

Directional couplers perform numerous functions in microwave circuits and subsystems. They are used to sample power for temperature compensation and amplitude control and in power splitting and combining over the ultra-broadband frequency range. In balanced amplifiers they help obtain good input and output

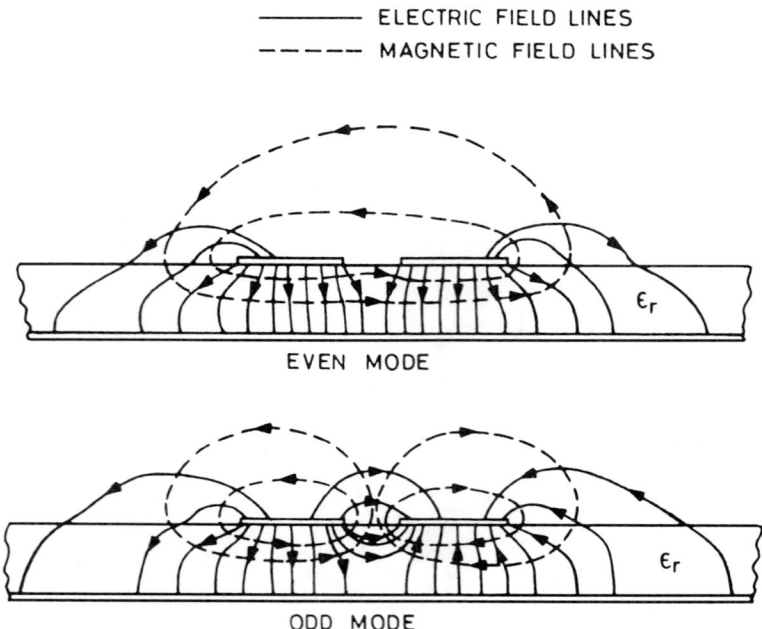

Figure 1.8 Even- and odd-mode field configurations in coupled microstrip lines.

VSWRs. In balanced mixers and microwave instruments (including network analyzers) they help sample incident and reflected signals. They have matched characteristics at all four ports, making them ideal for insertion in a circuit or subsystem.

A historical account of microwave directional couplers including an extensive reference list is given by Cohn and Levy [1]. The first directional coupler using a quarter-wave-long two-wire configuration was reported in 1922, and during the 1940s and 1950s, significant progress was made in waveguide couplers using apertures in the common wall. Directional couplers using planar TEM lines such as coupled striplines were developed in the mid-1950s. Numerous papers were published in the 1950s and 1960s describing the theory, design, fabrication, and measured data for the TEM-line edge-coupled directional couplers and significant contributions were made in the development of planar couplers. These couplers can provide coupling in the 8- to 40-dB range. Early work on these homogeneous couplers (single and multisections) can be found in references [1–5]. These couplers are also known as *backward-wave couplers* because the coupled wave on the secondary line travels in the opposite direction compared with the incident wave on the primary line when excited with a microwave signal.

In several applications, a tight coupler such as a 3-dB coupler is required and the cross-section shown in Figure 1.1(b) is difficult to realize as the very tight spacings required are limited by current photo-etch techniques. This problem is alleviated by using the three-dielectric-layer broadside-coupled striplines (including offset coupled strips [6]) and tandem-connection directional couplers [7] of Figure 1.9 or the vertically installed coupled-line configuration of Figure 1.10.

Multioctave bandwidth in the above-mentioned couplers is realized by using multisections of equal-length coupled sections. When these sections are joined, however, abrupt discontinuities in coupling and line widths occur resulting in coupling error and directivity degradation. Continuously tapered TEM couplers [8, 9] yield improved electrical performance including better bandwidth characteristics.

After widespread use of microstrip lines in microwave circuits, attention turned to microstrip line couplers [10–19]. One of the driving forces for the development of microstrip couplers was the higher level of integration of microwave circuits on a single substrate, including both active and passive components. Because a microstrip is inhomogeneous, the even- and odd-mode propagation

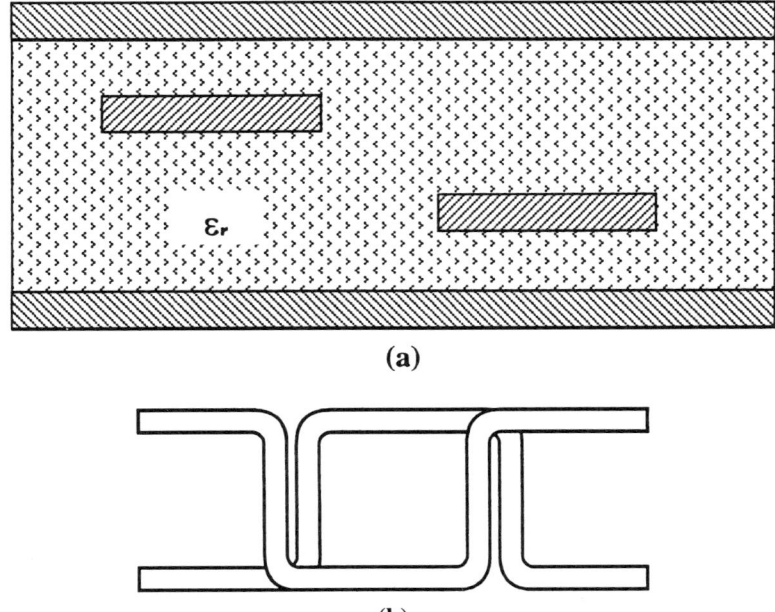

Figure 1.9 (a) Offset coupled striplines and (b) tandem-connection of directional couplers.

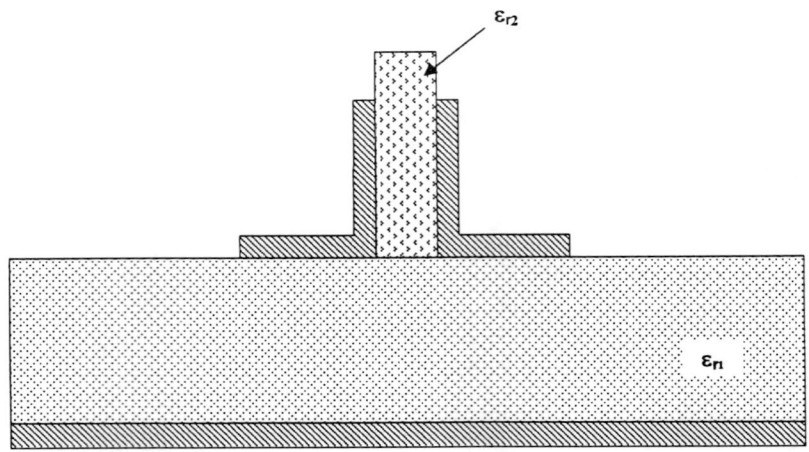

Figure 1.10 Vertically installed coupled-line coupler.

velocities for a coupled pair of microstrip lines are not equal resulting in poor directivity, which becomes worse as the coupling is decreased. For example, a 10% difference in phase velocities reduces the directivity of 10-, 15-, and 20-dB couplers to 13, 8, and 2 dB, respectively, from the theoretically infinite value with equal-phase velocities. Thus, the deterioration in directivity is higher for loose coupling and becomes worse with higher phase velocity differences. Therefore, couplers fabricated on low-dielectric constant substrates such as plastic ($\varepsilon_r = 2.5$), have better directivity performance than on alumina or GaAs. The directivity of stripline and homogeneous broadside directional couplers is much better than that of microstrip couplers.

Figure 1.11 depicts several techniques for equalizing or compensating for the difference in the modal phase velocities. Of these "wiggly lines" [10], dielectric overlays [11, 12] and capacitive compensation methods [13, 14, 17] are the most commonly used.

Tight microstrip couplers suffer from the same problem as their stripline counterparts; that is, requirements of impractically small spacing between the conductors. This problem was solved by Lange [20] in 1969 with his interdigital coupler (Figure 1.12) using four narrow strips. In this design, alternate strips in pairs are connected with wires or airbridges and the gaps, however small, are realizable. This coupler and its variations [20–25] are widely used in the microwave industry. Figure 1.13 shows other configurations for tight couplers, which include re-entrant structure, asymmetric broadside-coupled microstrip lines, and slot-coupled microstrip lines.

Introduction

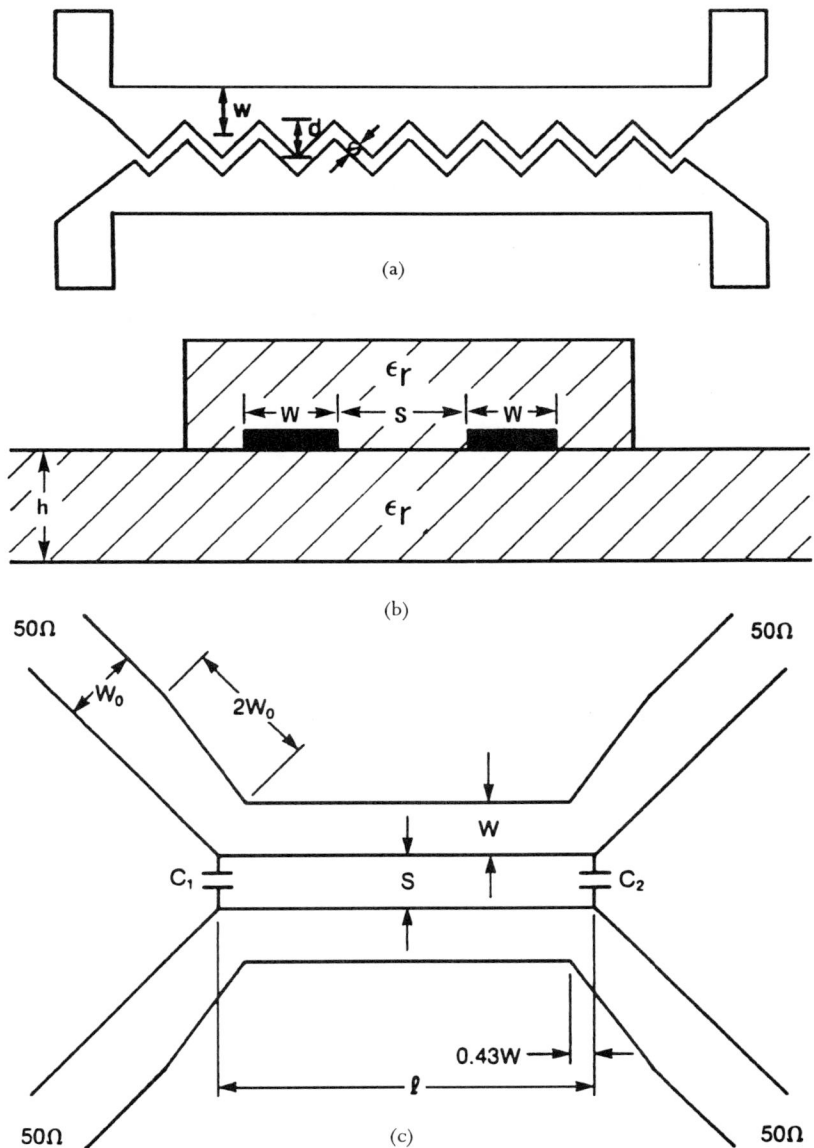

Figure 1.11 (a) Wiggly two-line coupler, (b) parallel coupled microstrip with overlay compensation, (c) lumped capacitor compensation of microstrip coupler.

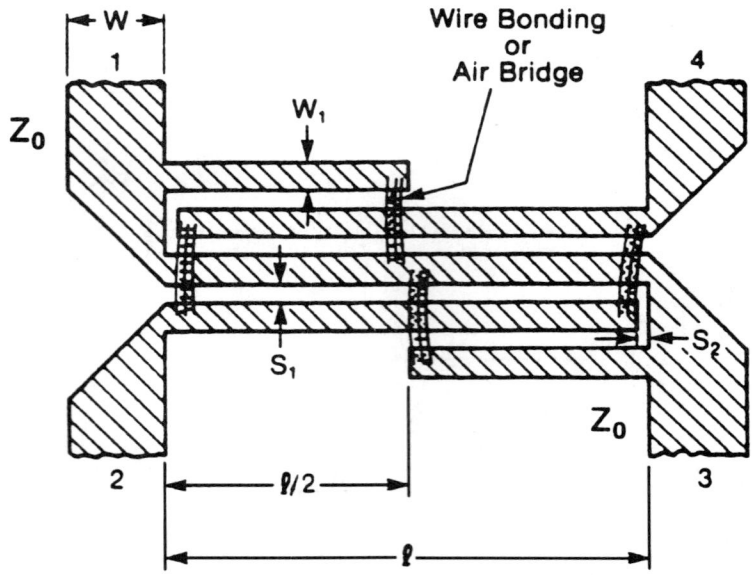

Figure 1.12 Ninety-degree hybrid coupler using interdigital Lange configuration.

1.2.2 Filters

Next to directional couplers, filters are the most important passive components used in microwave subsystems and instruments. Most microwave systems consist of many active and passive components that are difficult to design and manufacture with precise frequency characteristics. In contrast, microwave passive filters can be designed and manufactured with remarkably predictable performance. Consequently, microwave systems are usually designed so that all of the troublesome components are relatively wide in frequency response with filters being incorporated to obtain the precise system frequency response. Because filters are the narrowest bandwidth components in the system, it is usually the filters that limit such system parameters as gain and group delay flatness over frequency.

The first use of filters was reported in 1937. A historical survey of microwave filters including an extensive reference list is given by Levy and Cohn [26]. Pioneering work in coupled-line filters using TEM striplines was performed in the 1950s, 1960s, and 1970s. The most popular filter configurations are parallel-coupled-line [27, 28], interdigital [29, 30], combline [31, 32] and hairpin-line [33, 34]. All these filters are of the bandpass or bandstop type. Coaxial interdigital and combline configurations are shown in Figure 1.14, while stripline coupled-line configurations are shown in Figure 1.15.

Introduction 15

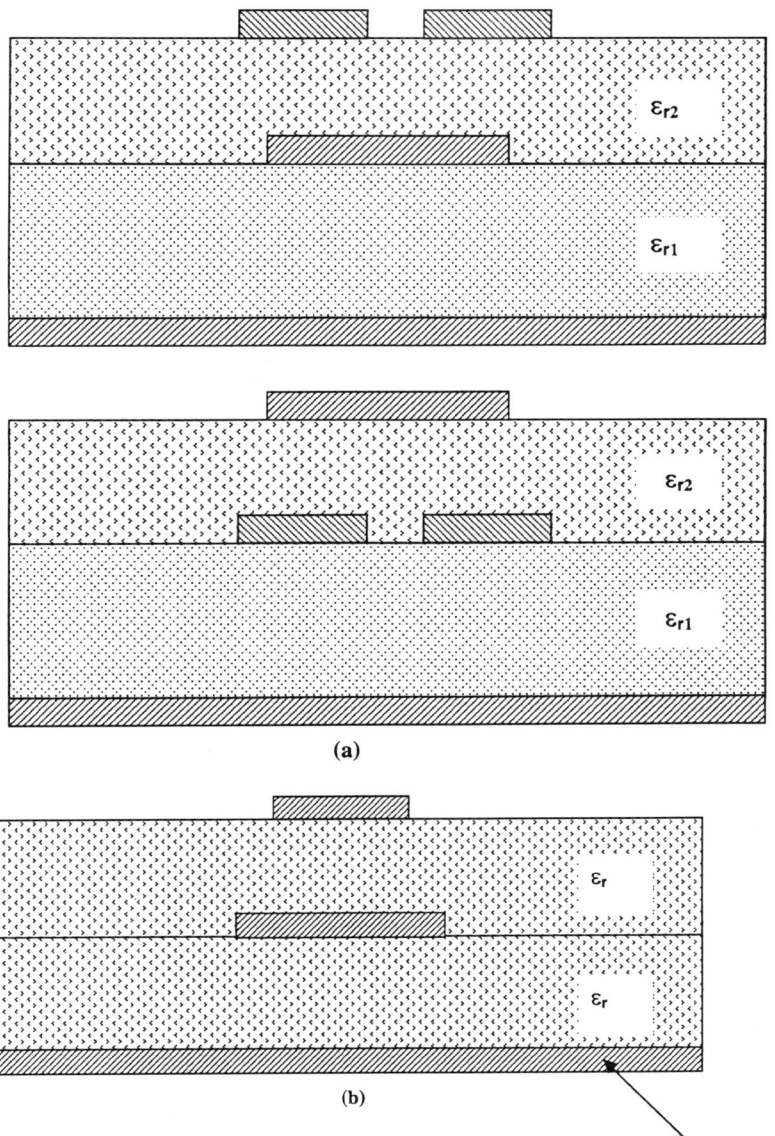

Figure 1.13 Other tight coupler configurations: (a) re-entrant structures, (b) asymmetric broadside coupled lines, (c) slot-coupled microstrip lines.

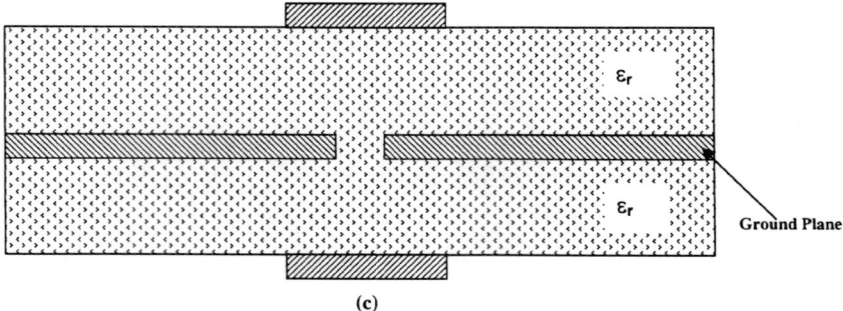

(c)

Figure 1.13 (*continued*)

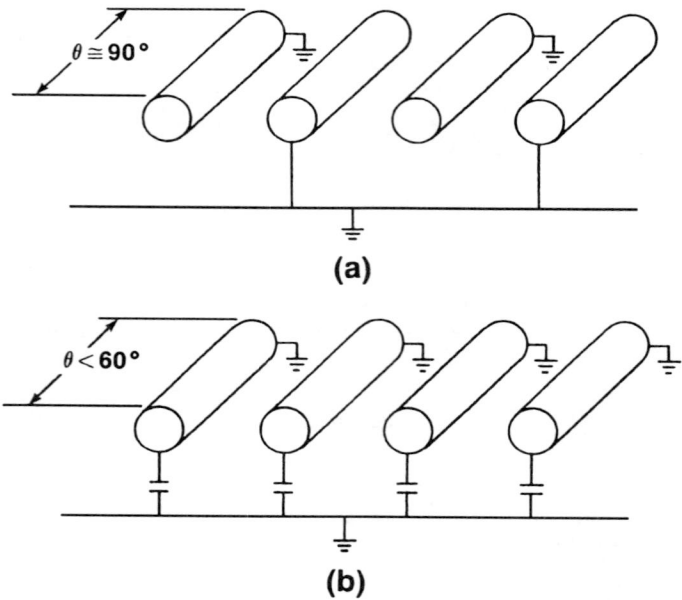

Figure 1.14 Coaxial-line filters: (a) interdigital, (b) combline.

Microstrip coupled line filters are similar to those shown in Figure 1.15. Because microstrip is a nonhomogeneous medium, the different even- and odd-mode phase velocities result in the filter having an asymmetrical passband response, deteriorating the upper stopband performance and moving the second passband (which is at about twice the center frequency) toward the center frequency. To overcome this problem, phase velocity equalization techniques

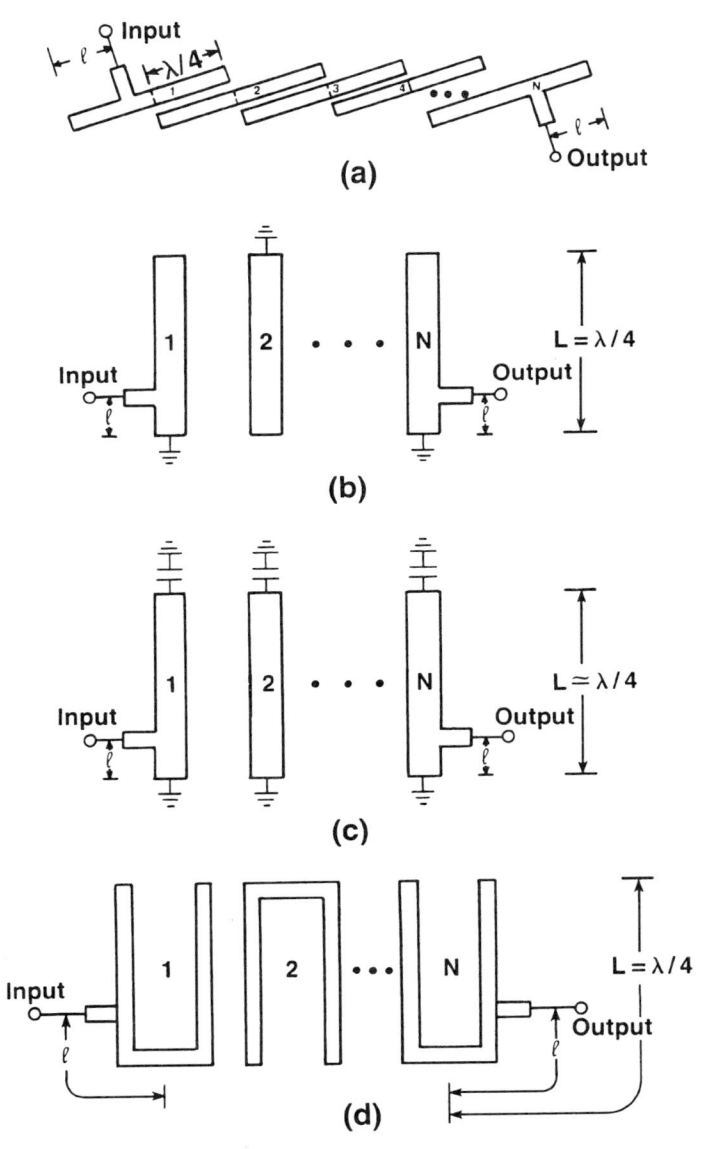

Figure 1.15 Typical coupled-line filter configurations: (a) parallel coupled, (b) interdigital, (c) combline, (d) hairpin-line.

similar to those employed for directional couplers can also be used [35]. Work on coupled line filters can also be found in the literature [2, 36, 37, 38].

In microstrip filters, temperature variation of ε_r, changes in ε_r from lot to lot, and substrate thickness variations usually mean that the bandwidth has to be wider than desired to accommodate manufacturing tolerances.

1.3 Applications

In the past decade, four major areas in the development of coupled-line components have been emphasized: (1) development of CAD tools, (2) full-wave analysis and accurate semiempirical expressions to enhance component designs, (3) the search for new structures and configurations, and (4) the search for new applications. Broader bandwidth, ease of fabrication and integration, compact size, and lower cost have been the driving factors. For example, in wireless applications, compact size and lower cost requirements triggered investigation of new configurations and the transformation of existing structures into new layouts such as meander line and spiral to realize compact components.

Other applications of coupled-line sections are in baluns, impedance transformers, dc blocks, interdigital capacitors, and spiral inductors. The spiral inductor is the most popular and is used extensively in hybrid and monolithic microwave integrated circuits (MICs). In particular, the compact size and low-cost circuits used for wireless applications in the L- and S-bands are based on inductors as matching elements.

Over the past two decades, because of the rapidly growing use of MICs in radar, satellite and mobile communications, electronic warfare (EW) and missiles, couplers and filter technologies have undergone a substantial change in terms of bandwidth, size, and cost. For example, in wireless applications, a 90-degree hybrid/coupler (whose output ports have signals of equal magnitude but with 90-degree phase difference) is needed to determine the phase error of the transmitter using the $\pi/4$ quadrature phase-shift keying (QPSK) modulation scheme common to digital cellular radio systems. Basic requirements for this coupler are small size, low cost, and tight amplitude balance and quadrature phase between the output ports. This was met by the coupled microstrip line couplers using the meander-line approach [39, 40, 41] and spiral configuration [41] on high-dielectric constant substrate compatible with MIC, and meander configuration [39] on a GaAs substrate compatible with monolithic microwave integrated circuits (MMICs). These couplers have potential to meet the \$2 to \$5 price goals when housed in plastic packages and produced in large quantities.

Satellite, airborne communications, and EW systems require small size, lightweight, low-cost filters. Coupled-microstrip and stripline filters are very suitable for wideband applications where the demand on selectivity is not severe. Various kinds of filters, shown in Figure 1.15, can be realized using microstrip-type structures. For wireless applications, however, miniature versions of these filters are required because of space and cost constraints. Hairpin-line and combline filters using resonators on high-dielectric constant ($\varepsilon_r = 80$ and 90) substrate or embedded in dielectric cavities have been developed and can be designed using traditional methods and/or EM simulators. Each filter shown in Figure 1.15 has several other versions to make them compact, either by folding or modifying the layout to fit into a small size.

1.4 Scope of the Book

Microwave components based on coupled-line structures have been in use for over half a century. This text deals exclusively with these components. The purpose of this book is twofold; first, to help the reader understand the theory and working of coupled-line components, and second, to provide in-depth design information to supplement commercial CAD tools in the design of microwave integrated circuits. As far as possible, enough information has been included to permit the design of passive components for wireless applications covering radio frequency (RF), microwave, and millimeter wave frequencies.

References

[1] Cohn, S. B., and R. Levy, "History of Microwave Passive Components With Particular Attention to Directional Couplers," *IEEE Trans. Microwave Theory and Techniques*, Vol. MTT-32, Sept. 1984, pp. 1,046–1,054.

[2] Matthaei, G. L., L. Young, and E. M. T. Jones, *Microwave Filters, Impedance-Matching Networks and Coupling Structures*, McGraw-Hill, N.Y., 1964 (reprinted by Artech House, Norwood, MA, 1980).

[3] Levy, R., "Directional Couplers," in *Advances in Microwaves*, Vol. 1, Academic Press, New York, 1966, pp. 184–191.

[4] Howe, H., *Stripline Circuit Design*, Artech House, Norwood, MA, 1974.

[5] Malherbe, J. A. G., *Microwave Transmission Line Couplers*, Artech House, Norwood, MA, 1988.

[6] Shelton, J. P., Jr., "Impedances of Offset Parallel-Coupled Strip Transmission Lines," *IEEE Trans. Microwave Theory Tech.*, Vol. MTT-14, pp. 7–15, Jan. 1966, and correction *IEEE Trans. Microwave Theory Tech.*, Vol. MTT-14, May 1966, p. 249.

[7] Shelton, J. P., and J. A. Mosko, "Synthesis and Design of Wideband Equal-Ripple TEM Directional Couplers and Fixed Phase Shifters," *IEEE Trans. Microwave Theory Tech.*, Vol. MTT-14, Oct. 1966, pp. 462–473.

[8] Tresselt, C. P., "The Design and Construction of Broadband, High Directivity, 90-Degree Couplers, Using Nonuniform Line Techniques," *IEEE Trans. Microwave Theory Tech.*, Vol. MTT-14, Dec. 1966, pp. 647–657.

[9] Kammler, D. W., "The Design of Discrete N-Section and Continuously Tapered Symmetrical Microwave TEM Directional Couplers," *IEEE Trans. Microwave Theory Tech.*, Vol. MTT-17, Aug. 1969, pp. 577–590.

[10] Podell, A., "A High Directivity Microstrip Coupler Technique," *IEEE MTT-S Int. Microwave Symp. Dig.*, 1970, pp. 33–36.

[11] Sheleg, B., and B. E. Spielman, "Broadband Directional Couplers Using Microstrip With Dielectric Overlays," *IEEE Trans. Microwave Theory Tech.*, Vol. MTT-22, 1974, pp. 1,216–1,220.

[12] Paolino, D. D., "MIC Overlay Coupler Design Using Spectral Domain Techniques," *IEEE Trans. Microwave Theory Tech.*, Vol. MTT-26, 1978, pp. 646–649.

[13] Kajfez, D., "Raise Coupled Directivity With Lumped Components," *Microwaves*, Vol. 17, No. 3, Mar. 1978, pp. 64–70.

[14] March, S. L., "Phase Velocity Compensation in Parallel-Coupled Microstrip," *IEEE MTT-S Int. Microwave Symp. Digest*, 1982, pp. 410–412.

[15] Davis, W. A., *Microwave Semiconductor Circuit Design*, Van Nostrand, New York, 1983.

[16] Horno, M., and F. Medina, "Multilayer Planar Structures for High-Directivity Directional Coupler Design," *IEEE Trans. Microwave Theory Tech.*, Vol. MTT-34, Dec. 1986, pp. 1,442–1,449.

[17] Dydyk, M., "Accurate Design of Microstrip Directional Couplers With Capacitive Compensation," *IEEE MTT-S Int. Microwave Symposium*, digest of papers, 1990, pp. 581–584.

[18] Uysal, S., *Nonuniform Line Microstrip Directional Couplers and Filters*, Artech House, Norwood, MA, 1993.

[19] Gupta, K. C., R. Garg, I. J. Bahl, and P. Bhartia, *Microstrip Lines and Slot Lines*, Second Edition, Artech House, Norwood, MA, 1996.

[20] Lange, J., "Interdigitated Stripline Quadrature Hybrid," *IEEE Trans. Microwave Theory and Techniques*, Vol. MTT-17, Dec. 1969, pp. 1,150–1,151.

[21] Waugh, R., and D. LaCombe, "Unfolding the Lange Coupler," *IEEE Trans. Microwave Theory and Techniques*, Vol. MTT-20, Nov. 1972, pp. 777–779.

[22] Ou, W. P., "Design Equations for an Interdigitated Directional Coupler," *IEEE Trans. Microwave Theory and Techniques*," Vol. MTT-23, Feb. 1973, pp. 253–255.

[23] Paolino, D., "Design More Accurate Interdigitated Couplers," *Microwaves*, Vol. 15, May 1976, pp. 34–38.

[24] Presser, A., "Interdigited Microstrip Coupler Design," *IEEE Trans. Microwave Theory Tech.*, Vol. MTT-26, Oct. 1978, pp. 801–805.

[25] Bhartia P., and I. J. Bahl, *Millimeter Wave Engineering and Applications*, Wiley, New York, Ch. 7, 1984.

[26] Levy, R. and S. B. Cohn, "A History of Microwave Filter Research, Design and Development," *IEEE Trans. Microwave Theory and Techniques*, Vol. MTT-32, Sept. 1984, pp. 1,055–1,067.

[27] Cohn, S. B., "Parallel-Coupled Transmission-Line Resonator Filters," *IRE Trans. Microwave Theory Tech.*, Vol. MTT-6, Apr. 1958, pp. 223–231.

[28] Ozaki, H., and J. Ishii, "Synthesis of a Class of Stripline Filters," *IRE Trans. Circuit Theory*, Vol. CT-5, June 1958, pp. 104–109.

[29] Matthaei, G. L., "Interdigital Bandpass Filters," *IRE Trans. Microwave Theory Tech.*, Vol. MTT-10, Nov. 1962, pp. 479–491.

[30] Wenzel, R. J., "Exact Theory of Interdigital Bandpass Filters and Related Coupled Structures," *IEEE Trans. Microwave Theory Tech.*, Vol. MTT-13, Sept. 1965, pp. 559–575.

[31] Matthaei, G. L., "Comb-Line Bandpass Filters of Narrow or Moderate Bandwidth," *Microwave J.*, Vol. 6, Aug. 1963, pp. 82–91.

[32] Wenzel, R. J., "Synthesis of Comb-Line and Capacitively Loaded Interdigital Bandpass Filters of Arbitrary Bandwidth," *IEEE Trans. Microwave Theory Tech.*, Vol. MTT-19, Aug. 1971, pp. 678–686.

[33] Cristal, E. G., and S. Frankel, "Hairpin Line/Half-Wave Parallel-Coupled-Line Filters," *IEEE Trans. Microwave Theory Tech.*, Vol. MTT-20, Nov. 1972, pp. 719–728.

[34] Gysel, U. H., "New Theory and Design for Hairpin-Line Filters," *IEEE Trans. Microwave Theory Tech.*, Vol. MTT-22, May 1974, pp. 523–531..

[35] Bahl, I. J., "Capacitively Compensated High-Performance Parallel Coupled Microstrip Filters," *IEEE MTT-S Int. Microwave Symp. Dig.*, 1989, pp. 679–682.

[36] Malherbe, J. A. G., *Microwave Transmission Line Filters*, Artech House, Norwood, MA, 1979.

[37] Bahl, I. J., and P. Bhartia, *Microwave Solid-State Circuit Design*, Wiley, New York, 1988, Ch. 6.

[38] Sheinwald, J., "MMIC Compatible Bandpass Filter Design: A Survey of Applicable Techniques, *IEEE MTT-S Int. Microwave Symp. Dig.*, 1989, pp. 679–682.

[39] Arai, S., et al., "A 900-MHz Degree Hybrid for QPSK Modulator," *IEEE MTT-S Int. Microwave Symp. Dig.*, 1991, pp. 857–860.

[40] Tanaka, H., et al., "2-GHz One Octave-Band 90 Degree Hybrid Coupler Using Coupled Meandered Line Optimized by 3-D FEM," *IEEE MTT-S Int. Microwave Symp. Digest*, 1994, pp. 906–906.

[41] Tanaka, H., et al., "Miniaturized 90-Degree Hybrid Coupler Using High Dielectric Substrate for QPSK Modulator," *IEEE MTT-S Int. Microwave Symp. Dig.*, 1996, pp. 793–796.

2

Microwave Network Theory

Microwave coupled lines and components can be classified as N-port networks such as two-port, three-port, four-port, and so on. If the input-output parameters of an N-port network are known, its behavior under various conditions of excitation and termination can be determined. For example, if two ports of a four-port network are terminated in open circuit, the input-output parameters of the remaining two-port network can be determined from a knowledge of the parameters of the original four-port network. Further, in a microwave system or subsystem, many individual components are cascaded together and the input-output parameters of the cascaded network can be determined if those of the individual networks are known.

The input-output relationship of a linear microwave network can be described in many equivalent ways [1]. In this chapter, we discuss how an N-port network can be characterized by its impedance, admittance, or scattering matrix. Although any form of matrix can be used to describe a network, one form may be more suitable than another. In general, the scattering matrix representation has been the most popular way of describing the input-output relationship of a microwave network. Because a network is usually constructed to have specific reflection and transmission properties, one can directly express the desired response in terms of a scattering matrix. These quantities can also be easily measured using vector network analyzers. We discuss scattering matrices in more detail in this chapter together with the conditions imposed by the losslessness and reciprocity on the various representative matrices of a passive network. Some special properties of two-, three-, and four-port networks are described. The ABCD representation of two-port networks is then discussed. This representation is very useful when a number of two-port networks are cascaded to form a single two-port network. The relationship among various forms of matrices are also given.

If P ports of an N-port network are connected to P ports of another M-port network, a network with $M + N - 2P$ ports results. Given the scattering matrices of individual networks, the scattering matrix of the overall network can be determined. We give these relationships and apply them to some specific cases to find the modified scattering parameters of two-, three-, and four-port networks whose ports are not match-terminated.

2.1 Actual and Equivalent Voltages and Currents

For low-frequency networks, one can define (and measure) unique voltages and currents at various locations in the circuit. Unfortunately, the same is not true at microwave frequencies and beyond, where it is possible to define unique (actual) quantities only for transmission lines carrying power in TEM mode. Examples of transmission lines supporting the TEM mode of propagation are a coaxial line, stripline, microstrip line,[1] and so forth. Many other commonly used transmission lines such as hollow waveguides, dielectric guides, and fin lines do not support the TEM mode of propagation. Therefore, one resorts to the concept of *equivalent* voltages and currents, and this can be applied to both TEM- and non-TEM-mode transmission lines. Relationships involving equivalent voltages or currents lead to unique physical quantities such as reflection and transmission coefficients, *normalized* input impedance, and the like. Equivalent voltages and currents can be defined on a *normalized* or *unnormalized* basis. Because the representative matrix of a network may define a relationship between *normalized* or between *unnormalized* quantities, it is essential to understand their meaning.

2.1.1 Normalized Voltages and Currents

Figure 2.1(a) shows a two-port network. The power flows into and out of the network by means of transmission lines connected to the network.[2] Each transmission line may carry a wave propagating *toward* the network defined as the *incident wave* or *away* from the network defined as the *reflected wave*. If power is incident in the transmission line connected to port 1, the mode in which the power flows is a characteristic of the type of transmission line. Associated with a mode are unique electromagnetic fields. The transverse components of electric and magnetic fields (transverse to the direction of propagation) have a unique phase associated with them, which is the same for both fields. Further,

---

1. Microstrip line supports quasi-TEM mode.
2. The term *transmission line* is used in a general sense to denote any physical waveguide structure that can be a microstrip line, coaxial line, rectangular waveguide, optical waveguide and so on.

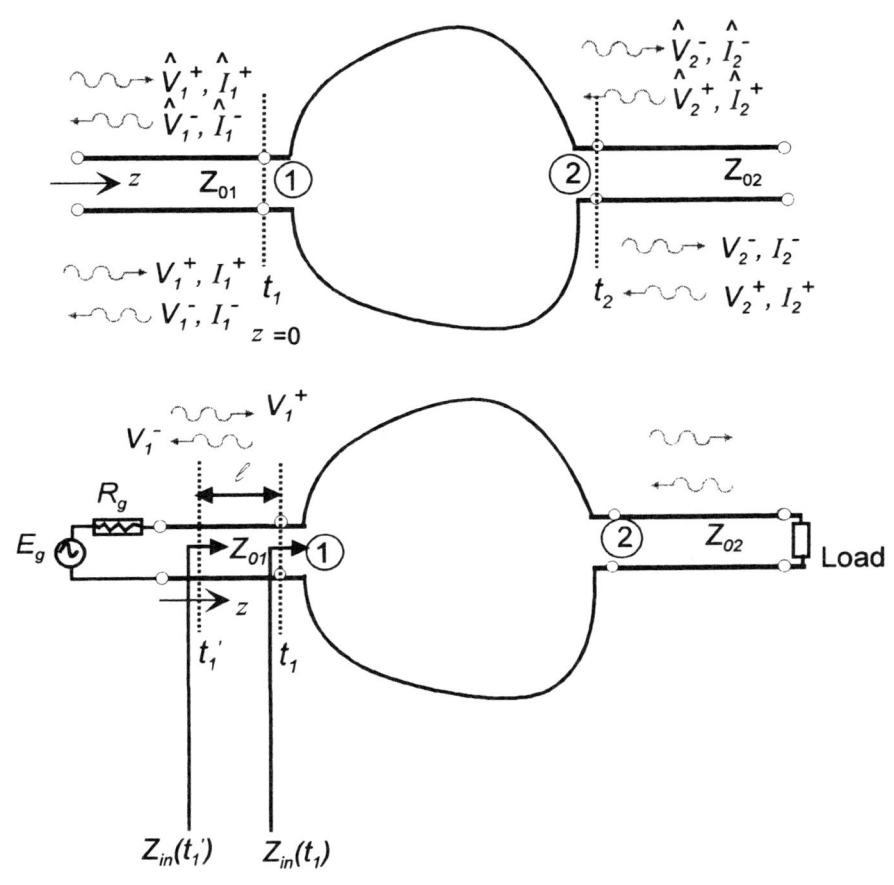

Figure 2.1 (a) Normalized and unnormalized voltage and current waves on transmission lines of a two-port network. (b) A two-port network connected to a source and load.

the z-variation of the incident electromagnetic wave (assuming that the power flow is in the positive z-direction) can be described by a simple factor $e^{-j\beta_1 z}$, where β_1 is a unique quantity and denotes the phase constant of the wave in the transmission line of port 1.

To determine the *normalized* voltage and current waves, we assume that the incident voltage and current waves have the same phase as that of the transverse electric and magnetic field components of the incident electromagnetic wave.[3]

---

3. The phase of an electromagnetic wave is unique and can be measured even if the wave is of a non-TEM type.

Further, the z-variation of voltage and current waves is also given by the same factor as that for the field components ($e^{-j\beta_1 z}$). Mathematically, the normalized incident voltage and current waves in the transmission line of port 1 can then be expressed as

$$\hat{V}_1^+(z) = \hat{V}_{10}^+ e^{-j\beta_1 z} = |\hat{V}_{10}^+| e^{j\psi_{i1}} e^{-j\beta_1 z} \qquad (2.1)$$

and

$$\hat{I}_1^+(z) = \hat{I}_{10}^+ e^{-j\beta_1 z} = |\hat{I}_{10}^+| e^{j\psi_{i1}} e^{-j\beta_1 z} \qquad (2.2)$$

where ψ_{i1} denotes the phase of the incident wave at $z = 0$, and $\hat{V}_{10}^+$ and $\hat{I}_{10}^+$ are the complex voltage and current, respectively, at the same terminal plane ($z = 0$). To reemphasize, the value of ψ_{i1} is the same as that of the transverse components of the electric and magnetic fields of the incident wave. The symbol "^" has been added to denote that the respective quantities are normalized.

When the characteristic impedance of a transmission line is real, the voltage and current waves can be expressed in terms of incident and reflected power.[4] At microwave frequencies, the characteristic impedances of practical transmission lines are generally real. To compute the values of $|\hat{V}_{10}^+|$ and $|\hat{I}_{10}^+|$, we force the condition that the average power flow is given by

$$|\hat{V}_{10}^+||\hat{I}_{10}^+| = P_1^+ \qquad (2.3)$$

where P_1^+ denotes the incident power, and $|\hat{V}_{10}^+|$ and $|\hat{I}_{10}^+|$ denote the *rms* quantities.

To determine $|\hat{V}_{10}^+|$ and $|\hat{I}_{10}^+|$, we need to have another relation between them. To define normalized quantities, we choose

$$\frac{|\hat{V}_{10}^+|}{|\hat{I}_{10}^+|} = 1 \qquad (2.4)$$

and hence from (2.3) and (2.4):

$$|\hat{V}_{10}^+| = |\hat{I}_{10}^+| = \sqrt{P_1^+} \qquad (2.5)$$

Substituting the values of $|\hat{V}_{10}^+|$ and $|\hat{I}_{10}^+|$ from the above equation in (2.1) and (2.2), we obtain

$$\hat{V}_1^+(z) = \hat{I}_1^+(z) = \sqrt{P_1^+} e^{j\psi_{i1}} e^{-j\beta_1 z} \qquad (2.6)$$

(2.6) is in a form that aids in understanding the physical meaning of normalized voltage and current waves.

---

4. This treatment is, however, not valid if the characteristic impedance is complex [2].

When the incident power reaches the network, a part of it (say, P_1^-) is reflected back. By analogy with incident voltage and current waves, the reflected waves can be expressed as

$$\hat{V}_1^-(z) = \hat{I}_1^-(z) = \sqrt{P_1^-} e^{j\psi_{r1}} e^{j\beta_1 z} \tag{2.7}$$

where the superscript " $-$ " is used to denote the reflected waves. ψ_{r1} is the phase of the transverse components of the electric and magnetic fields of the reflected wave at $z = 0$ and is a unique quantity. Because the reflected wave propagates in the negative z-direction, its z-dependence is described by the factor $e^{j\beta_1 z}$.

The total normalized voltage (at any value of z) in the transmission line of port 1 is thus given by,

$$\begin{aligned}\hat{V}_1(z) &= \hat{V}_1^+(z) + \hat{V}_1^-(z) \\ &= |\hat{V}_{10}^+| e^{j\psi_{i1}} e^{-j\beta_1 z} + |\hat{V}_{10}^-| e^{j\psi_{r1}} e^{j\beta_1 z} \\ &= \sqrt{P_1^+} e^{j\psi_{i1}} e^{-j\beta_1 z} + \sqrt{P_1^-} e^{j\psi_{r1}} e^{j\beta_1 z}\end{aligned} \tag{2.8}$$

On the other hand, the total current at any value of z is given by

$$\begin{aligned}\hat{I}_1(z) &= \hat{I}_1^+(z) - \hat{I}_1^-(z) = \hat{V}_1^+(z) - \hat{V}_1^-(z) \\ &= |\hat{I}_{10}^+| e^{j\psi_{i1}} e^{-j\beta_1 z} - |\hat{I}_{10}^-| e^{j\psi_{r1}} e^{j\beta_1 z} \\ &= \sqrt{P_1^+} e^{j\psi_{i1}} e^{-j\beta_1 z} - \sqrt{P_1^-} e^{j\psi_{r1}} e^{j\beta_1 z}\end{aligned} \tag{2.9}$$

Because the current flows in the axial direction, the net current is given by the difference of the currents flowing in the positive and negative z-directions. We now show that the net power flow (into the network) across any $z =$ constant plane in the transmission line of port 1 is given by the usual low-frequency relation, that is, $P = Re(vi^*)$, where v and i denote the total *rms* voltage and current, respectively, at the reference plane. In the present case, the term $\hat{V}_1 \hat{I}_1^*$ can be expressed as

$$\hat{V}_1 \hat{I}_1^* = (\hat{V}_1^+ + \hat{V}_1^-)(\hat{V}_1^+ - \hat{V}_1^-)^* \tag{2.10}$$

After substituting for $\hat{V}_1^+$ and $\hat{V}_1^-$ from (2.6) and (2.7) and noting that the conjugate of a complex number $(A + jB)$ is equal to $(A - jB)$, we obtain

$$\begin{aligned}\hat{V}_1 \hat{I}_1^* &= |\hat{V}_1^+|^2 - |\hat{V}_1^-|^2 + \text{imaginary term} \\ &= |\hat{V}_{10}^+|^2 - |\hat{V}_{10}^-|^2 + \text{imaginary term} \\ &= P_1^+ - P_1^- + \text{imaginary term}\end{aligned} \tag{2.11}$$

or

$$Re(\hat{V}_1 \hat{I}_1^*) = P_1^+ - P_1^- \qquad (2.12)$$

which is the desired result.

2.1.2 Unnormalized Voltages and Currents

If the ratio between the voltage and current of the incident wave (and the reflected wave) is chosen to be different from unity (2.4), the resulting quantities are called *unnormalized*. For TEM transmission lines, this ratio is generally chosen to be equal to the actual characteristic impedance of the line. In that case, the unnormalized voltages and currents reduce to actual quantities on the line. For non-TEM transmission lines, the characteristic impedance depends on the definition used. Referring to Figure 2.1(a), the unnormalized incident voltage and current waves in the transmission line of port 1 can be expressed as

$$V_1^+(z) = \sqrt{Z_{01} P_1^+} \, e^{j\psi_{i1}} e^{-j\beta_1 z} \qquad (2.13)$$

$$I_1^+(z) = \sqrt{\frac{P_1^+}{Z_{01}}} \, e^{j\psi_{i1}} e^{-j\beta_1 z} \qquad (2.14)$$

where Z_{01} denotes the characteristic impedance of the transmission line of port 1. In the above equations, the symbol "^" has been dropped to denote unnormalized quantities.

The unnormalized reflected voltage and current waves in the transmission line of port 1 can be expressed as

$$V_1^-(z) = \sqrt{Z_{01} P_1^-} \, e^{j\psi_{r1}} e^{j\beta_1 z} \qquad (2.15)$$

$$I_1^-(z) = \sqrt{\frac{P_1^-}{Z_{01}}} \, e^{j\psi_{r1}} e^{j\beta_1 z} \qquad (2.16)$$

Example 2.1

Consider the circuit shown in Figure 2.1(a). Assume that unit power is incident in the transmission line of port 1, of which a quarter is reflected back from the network and that the phase of the incident and reflected waves at $z = 0$ is equal to 0 and $\pi/6$ radians, respectively. From (2.6) and (2.7), the normalized voltage and current waves in the input transmission line are

$$\hat{V}_1^+(z) = \hat{I}_1^+(z) = \sqrt{P_1^+} e^{j\psi_{i1}} e^{-j\beta_1 z} = e^{-j\beta_1 z} \qquad (2.17)$$

$$\hat{V}_1^-(z) = \hat{I}_1^-(z) = \sqrt{P_1^-} e^{j\psi_{r1}} e^{j\beta_1 z} = 0.5 e^{j\pi/6} e^{j\beta_1 z} \qquad (2.18)$$

To determine the unnormalized voltage and current waves in the transmission line, we need to specify a value of the characteristic impedance. Let this value be 50 ohms. Using (2.13) and (2.14), the incident unnormalized voltage and current waves are then

$$V_1^+(z) = \sqrt{Z_{01} P_1^+} e^{j\psi_{i1}} e^{-j\beta_1 z} = \sqrt{50} e^{-j\beta_1 z} \qquad (2.19)$$

$$I_1^+(z) = \sqrt{\frac{P_1^+}{Z_{01}}} e^{j\psi_{i1}} e^{-j\beta_1 z} = \frac{1}{\sqrt{50}} e^{-j\beta_1 z} \qquad (2.20)$$

Using (2.13) and (2.14), the unnormalized reflected voltage and current waves in the transmission line can be expressed as

$$V_1^-(z) = \sqrt{Z_{01} P_1^-} e^{j\psi_{r1}} e^{j\beta_1 z} = \sqrt{12.5} e^{j\pi/6} e^{j\beta_1 z} \qquad (2.21)$$

$$I_1^-(z) = \sqrt{\frac{P_1^-}{Z_{01}}} e^{j\psi_{r1}} e^{j\beta_1 z} = \frac{0.5}{\sqrt{50}} e^{j\pi/6} e^{j\beta_1 z} \qquad (2.22)$$

2.1.3 Reflection Coefficient, VSWR, and Input Impedance

Referring to Figure 2.1(b), the voltage reflection coefficient Γ_1 in the transmission line of port 1 is defined as

$$\Gamma_1(z) = \frac{\hat{V}_1^-(z)}{\hat{V}_1^+(z)} = \frac{V_1^-(z)}{V_1^+(z)} \qquad (2.23)$$

Substituting values of $\hat{V}_1^-(z)$ and $\hat{V}_1^+(z)$ from (2.6) and (2.7) in (2.23), we have

$$\Gamma_1(z) = \frac{|\hat{V}_{10}^-| e^{j\psi_{1r} + j\beta_1 z}}{|\hat{V}_{10}^+| e^{j\psi_{1i} - j\beta_1 z}}$$

$$= \sqrt{P_1^- / P_1^+} e^{j(\psi_{1r} - \psi_{1i})} e^{2j\beta_1 z} \qquad (2.24)$$

We easily conclude from (2.24) that the reflection coefficient is a unique quantity, and the square of its modulus gives the fraction of the incident power that is reflected back. The ratio of the reflected-to-incident power is commonly referred to as *return loss*. The return loss (in decibels), which is a positive quantity, is given by

$$RL(\text{dB}) = -10 \log |\Gamma_1(z)|^2 = -20 \log |\Gamma_1(z)| \qquad (2.25)$$

The voltage standing wave ratio (VSWR) in the transmission line of port 1 is given by

$$\text{VSWR} = \frac{1 + |\Gamma_1(z)|}{1 - |\Gamma_1(z)|} \qquad (2.26)$$

More often, it is the practice to use the ratio of total voltage and current, which can be termed as input impedance. The ratio of total normalized voltage to current is defined as the normalized input impedance and is given by

$$\hat{Z}_{in}(z) = \frac{\hat{V}_1(z)}{\hat{I}_1(z)} = \frac{\hat{V}_1^+(z) + \hat{V}_1^-(z)}{\hat{V}_1^+(z) - \hat{V}_1^-(z)} = \frac{1 + \Gamma_1(z)}{1 - \Gamma_1(z)} \qquad (2.27)$$

Because $\Gamma_1(z)$ is a unique quantity, the normalized input impedance is also a unique quantity.

Similarly, the unnormalized input impedance Z_{in} is given by

$$Z_{in}(z) = Z_{01} \frac{1 + \Gamma_1(z)}{1 - \Gamma_1(z)} \qquad (2.28)$$

where Z_{01} denotes the characteristic impedance of the transmission line connected to port 1. Using (2.27), the reflection coefficient in terms of normalized input impedance is expressed as

$$\Gamma_1(z) = \frac{\hat{Z}_{in}(z) - 1}{\hat{Z}_{in}(z) + 1} \qquad (2.29)$$

and in terms of the unnormalized input impedance:

$$\Gamma_1(z) = \frac{\frac{Z_{in}(z)}{Z_{01}} - 1}{\frac{Z_{in}(z)}{Z_{01}} + 1} = \frac{Z_{in}(z) - Z_{01}}{Z_{in}(z) + Z_{01}} \qquad (2.30)$$

Transformation of Impedance

With $Z_{in}(t_1)$ as the input impedance at the terminal plane t_1 looking into the network, the input impedance at the terminal plane t_1' (which is closer to the generator compared with terminal plane t_1) is

$$Z_{in}(t_1') = Z_{01} \frac{Z_{in}(t_1) + jZ_{01} \tan \beta_1 l}{Z_{01} + jZ_{in}(t_1) \tan \beta_1 l} \quad (2.31)$$

where l is the distance between terminal planes t_1 and t_1'. If $Z_{in}(t_1)$ denotes the load impedance Z_L, the input impedance Z_{in} at distance l away (toward the generator) can be expressed as

$$Z_{in} = Z_{01} \frac{Z_L + jZ_{01} \tan \beta_1 l}{Z_{01} + jZ_L \tan \beta_1 l} \quad (2.32)$$

where $\beta_1 = 2\pi/\lambda_g$ denotes the propagation constant of the line, which is assumed to be *lossless*.

2.1.4 Quantities Required to Describe the State of a Transmission Line

Consider an N-port network as shown in Figure 2.2. The ports are numbered from $m = 1$ to $m = N$. The power is carried into and away from the network by means of transmission lines connected to each port. The characteristic impedance of the transmission line of the mth port is denoted by Z_{0m}. Because the voltages and currents vary along the length of the transmission line, fictitious terminal planes are located in each transmission line. Voltage or current at port m denotes the respective quantity at the specified terminal plane in the transmission line of port m. We use the notation of the sections above for normalized and unnormalized quantities and incident and reflected quantities.

Note that $\hat{V}_n^+$ and $\hat{I}_n^+$ (similarly, $\hat{V}_n^-$ and $\hat{I}_n^-$) are not independent quantities. The normalized quantities satisfy the following relationship:

$$\frac{\hat{V}_n^+}{\hat{I}_n^+} = \frac{\hat{V}_n^-}{\hat{I}_n^-} = 1 \quad (2.33)$$

If the two quantities $\hat{V}_n^+$ (or $\hat{I}_n^+$) and $\hat{V}_n^-$ (or $\hat{I}_n^-$) are known, the total voltage and current can be determined using

$$\hat{V}_n = \hat{V}_n^+ + \hat{V}_n^- = \hat{I}_n^+ + \hat{I}_n^- \quad (2.34)$$

and

$$\hat{I}_n = \hat{I}_n^+ - \hat{I}_n^- = \hat{V}_n^+ - \hat{V}_n^- \quad (2.35)$$

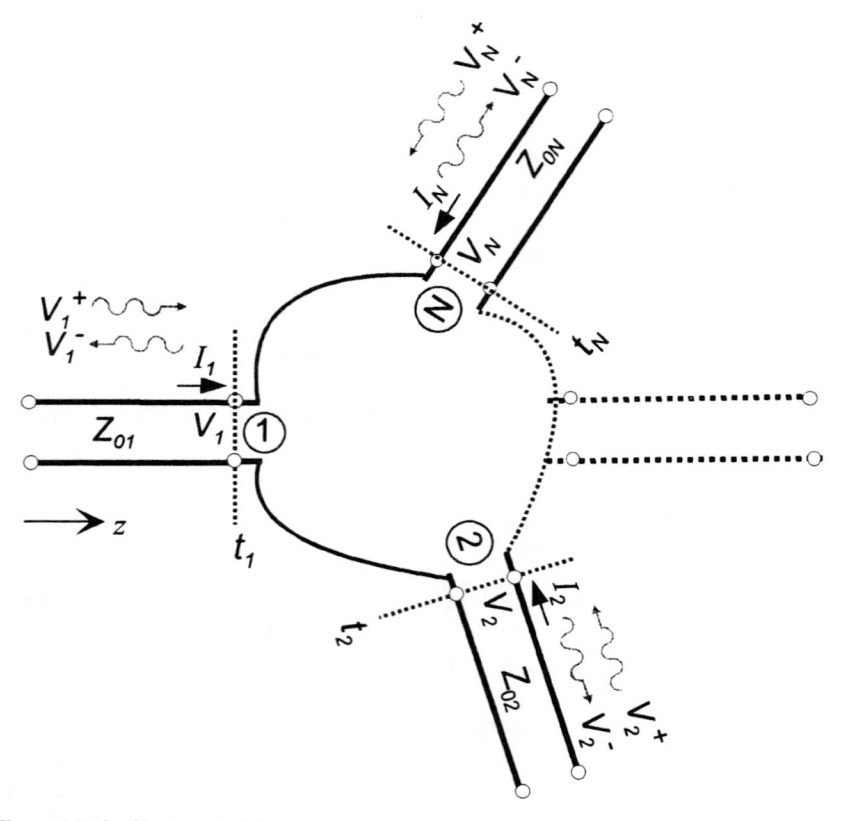

Figure 2.2 An N-port network.

Therefore, if any two of the four quantities $\hat{V}_n$, $\hat{I}_n$, $\hat{V}_n^+$ (or $\hat{I}_n^+$), and $\hat{V}_n^-$ (or $\hat{I}_n^-$) are known, all others can be determined. The same conclusion holds for unnormalized quantities if the characteristic impedances of all the transmission lines are known. For unnormalized quantities:

$$\frac{V_n^+}{I_n^+} = \frac{V_n^-}{I_n^-} = Z_{0n} \tag{2.36}$$

$$V_n = V_n^+ + V_n^- = Z_{0n}(I_n^+ + I_n^-) \tag{2.37}$$

$$I_n = I_n^+ - I_n^- = \frac{V_n^+ - V_n^-}{Z_{0n}} \tag{2.38}$$

Relationship Between Normalized and Unnormalized Quantities

The normalized and unnormalized quantities are related by

$$\hat{V}_n^\pm = \frac{V_n^\pm}{\sqrt{Z_{0n}}} \tag{2.39a}$$

$$\hat{V}_n = \frac{V_n}{\sqrt{Z_{0n}}} \tag{2.39b}$$

$$\hat{I}_n^\pm = \sqrt{Z_{0n}}\, I_n^\pm \tag{2.39c}$$

$$\hat{I}_n = \sqrt{Z_{0n}}\, I_n \tag{2.39d}$$

2.2 Impedance and Admittance Matrix Representation of a Network

2.2.1 Impedance Matrix

Consider the N-port network shown in Figure 2.2. In the impedance matrix representation, the voltage at each port is related to the currents at the different ports as follows:

$$\begin{aligned}
V_1 &= Z_{11}I_1 + Z_{12}I_2 + \cdots + Z_{1N}I_N \\
V_2 &= Z_{21}I_1 + Z_{22}I_2 + \cdots + Z_{2N}I_N \\
&\vdots \\
V_N &= Z_{N1}I_1 + Z_{N2}I_2 + \cdots + Z_{NN}I_N
\end{aligned} \tag{2.40}$$

In matrix notation, the above set of equations can be expressed as

$$[\mathbf{V}] = [\mathbf{Z}][\mathbf{I}] \tag{2.41}$$

where

$$[\mathbf{V}] = \begin{bmatrix} V_1 \\ \vdots \\ V_N \end{bmatrix} \tag{2.42}$$

$$[\mathbf{I}] = \begin{bmatrix} I_1 \\ \vdots \\ I_N \end{bmatrix} \tag{2.43}$$

and

$$[Z] = \begin{bmatrix} Z_{11} & Z_{12} & \cdots & Z_{1N} \\ Z_{21} & Z_{22} & \cdots & Z_{2N} \\ \vdots & \vdots & \vdots & \vdots \\ Z_{N1} & Z_{N2} & \cdots & Z_{NN} \end{bmatrix} \quad (2.44)$$

The impedance matrix [Z] is unnormalized because it relates unnormalized voltages and currents. The impedance matrix relating normalized voltages and currents is called normalized, and will be denoted as [$\hat{Z}$] with the "^" symbol. The normalized impedance matrix is denoted as

$$[\hat{Z}] = \begin{bmatrix} \hat{Z}_{11} & \hat{Z}_{12} & \cdots & \hat{Z}_{1N} \\ \hat{Z}_{21} & \hat{Z}_{22} & \cdots & \hat{Z}_{2N} \\ \vdots & \vdots & \vdots & \vdots \\ \hat{Z}_{N1} & \hat{Z}_{N2} & \cdots & \hat{Z}_{NN} \end{bmatrix} \quad (2.45)$$

2.2.2 Admittance Matrix

In the admittance matrix representation, the current at each port of the network as shown in Figure 2.2 is related to the voltages at the different ports as follows:

$$[I] = [Y][V] \quad (2.46)$$

where [V] and [I] are column vectors as defined by (2.42) and (2.43), respectively, and

$$[Y] = \begin{bmatrix} Y_{11} & Y_{12} & \cdots & Y_{1N} \\ Y_{21} & Y_{22} & \cdots & Y_{2N} \\ \vdots & \vdots & \vdots & \vdots \\ Y_{N1} & Y_{N2} & \cdots & Y_{NN} \end{bmatrix} \quad (2.47)$$

2.2.3 Properties of Impedance and Admittance Parameters of a Passive Network

For a network not containing any nonreciprocal media (ferrite, plasma, etc.),

$$\hat{Z}_{mn} = \hat{Z}_{nm} \quad (2.48)$$

and

$$\hat{Y}_{mn} = \hat{Y}_{nm} \quad (2.49)$$

Similar relationships are also satisfied by the elements of unnormalized impedance and admittance matrices that is:

$$Z_{mn} = Z_{nm} \tag{2.50}$$

and

$$Y_{mn} = Y_{nm} \tag{2.51}$$

Note that networks containing dielectrics and conductors are *reciprocal* and satisfy the above-mentioned properties.

For a lossless network, all the elements of an impedance or admittance matrix are imaginary. This is an expected result because any resistive element would imply loss.

The mn^{th} element of the unnormalized impedance matrix is related to the corresponding element of the normalized impedance matrix by

$$Z_{mn} = \hat{Z}_{mn}\sqrt{Z_{0m}Z_{0n}} \tag{2.52}$$

where Z_{0m} and Z_{0n} denote the characteristic impedances of transmission lines of ports m and n, respectively. Similarly, the mn^{th} element of the unnormalized admittance matrix is related to the corresponding element of the normalized admittance matrix by

$$Y_{mn} = \hat{Y}_{mn}\sqrt{Y_{0m}Y_{0n}} \tag{2.53}$$

where $Y_{0m} = 1/Z_{0m}$ and $Y_{0n} = 1/Z_{0n}$ denote the characteristic admittance of the transmission lines of ports m and n, respectively.

2.3 Scattering Matrix

A very popular method of representing microwave networks is by the scattering matrix. The scattering matrix is generally represented in a normalized form. In this representation, the normalized reflected voltage at each port of the network as shown in Figure 2.2 is related to the normalized incident voltages at the ports of the network as follows:

$$\begin{aligned}
\hat{V}_1^- &= \hat{S}_{11}\hat{V}_1^+ + \hat{S}_{12}\hat{V}_2^+ + \cdots + \hat{S}_{1N}\hat{V}_N^+ \\
\hat{V}_2^- &= \hat{S}_{21}\hat{V}_1^+ + \hat{S}_{22}\hat{V}_2^+ + \cdots + \hat{S}_{2N}\hat{V}_N^+ \\
&\vdots \\
\hat{V}_N^- &= \hat{S}_{N1}\hat{V}_1^+ + \hat{S}_{N2}\hat{V}_2^+ + \cdots + \hat{S}_{NN}\hat{V}_N^+
\end{aligned} \tag{2.54}$$

In matrix notation, the above set of equations can be expressed as

$$[\hat{\mathbf{V}}^-] = [\hat{\mathbf{S}}][\hat{\mathbf{V}}^+] \tag{2.55}$$

where

$$[\hat{V}^-] = \begin{bmatrix} \hat{V}_1^- \\ \vdots \\ \hat{V}_N^- \end{bmatrix} \qquad (2.56a)$$

$$[\hat{V}^+] = \begin{bmatrix} \hat{V}_1^+ \\ \vdots \\ \hat{V}_N^+ \end{bmatrix} \qquad (2.56b)$$

and

$$[\hat{S}] = \begin{bmatrix} \hat{S}_{11} & \hat{S}_{12} & \cdots & \hat{S}_{1N} \\ \hat{S}_{21} & \hat{S}_{22} & \cdots & \hat{S}_{2N} \\ \vdots & \vdots & \vdots & \vdots \\ \hat{S}_{N1} & \hat{S}_{N2} & \cdots & \hat{S}_{NN} \end{bmatrix} \qquad (2.57)$$

The scattering parameter $\hat{S}_{mn}$ is therefore given by

$$\hat{S}_{mn} = \left. \frac{\hat{V}_m^-}{\hat{V}_n^+} \right|_{\hat{V}_p^+ = 0 \text{ where } p=1,\ldots,N;\, p \neq n} \qquad (2.58)$$

In terms of the incident power P_n^+ in the nth transmission line, the amplitude of the normalized incident voltage wave at the nth port is given by

$$|\hat{V}_n^+| = \sqrt{P_n^+} \qquad (2.59)$$

Similarly, the amplitude of the normalized reflected voltage wave[5] at the mth port is given by

$$|\hat{V}_m^-| = \sqrt{P_m^-} \qquad (2.60)$$

where P_m^- denotes the reflected power at port m.

When the values of $|\hat{V}_n^+|$ and $|\hat{V}_m^-|$ from the last two equations are substituted in (2.58), we obtain

$$|\hat{S}_{mn}| = \frac{|\hat{V}_m^-|}{|\hat{V}_n^+|} = \sqrt{\frac{P_m^-}{P_n^+}} \qquad (2.61)$$

---

5. In the terminology used, any wave traveling toward the network is called the "incident" wave, and any wave traveling away from the network is called the "reflected" wave.

From the above equation, we see that $|\hat{S}_{mn}|^2$ denotes the ratio of power coupled from port n to port m when $\hat{V}_p^+ = 0$, where $p = 1, \ldots, N$; $p \neq n$. The condition $\hat{V}_p^+ = 0$, where $p = 1, \ldots, N$; $p \neq n$ is readily ensured by exciting only the nth port and terminating all the ports in matched loads. Similarly:

$$|\hat{S}_{nn}|^2 = |\Gamma_n|^2 = \frac{|\hat{V}_n^-|^2}{|\hat{V}_n^+|^2} = \frac{P_n^-}{P_n^+} \tag{2.62}$$

where Γ_n denotes the reflection coefficient at port n. Further, $|\hat{S}_{nn}|^2$ denotes the fraction of the incident power that is reflected back at port n.

The ports of a typical microwave network are usually match-terminated. Therefore, if some power is incident in one of the ports, the reflected power and the power coupled to the other ports of the network can be easily determined if the normalized scattering matrix is known.

2.3.1 Unitary Property

The elements of a normalized scattering matrix satisfy the following equation, which results from the Law of Conservation of Power:

$$\begin{bmatrix} \hat{S}_{11}^* & \hat{S}_{21}^* & \cdots & \hat{S}_{N1}^* \\ \hat{S}_{12}^* & \hat{S}_{22}^* & \cdots & \hat{S}_{N2}^* \\ \vdots & \vdots & \vdots & \vdots \\ \hat{S}_{1N}^* & \hat{S}_{2N}^* & \cdots & \hat{S}_{NN}^* \end{bmatrix} \begin{bmatrix} \hat{S}_{11} & \hat{S}_{12} & \cdots & \hat{S}_{1N} \\ \hat{S}_{21} & \hat{S}_{22} & \cdots & \hat{S}_{2N} \\ \vdots & \vdots & \vdots & \vdots \\ \hat{S}_{N1} & \hat{S}_{N2} & \cdots & \hat{S}_{NN} \end{bmatrix} = \begin{bmatrix} 1 & 0 & \cdots & 0 \\ 0 & 1 & \cdots & 0 \\ \vdots & \vdots & \vdots & \vdots \\ 0 & 0 & \cdots & 1 \end{bmatrix} \tag{2.63}$$

where the symbol * denotes the complex conjugate. Because the $[\hat{S}]$ matrix satisfies the above relationship, it is called a *unitary matrix*.

Power conservation is true for reciprocal as well as for nonreciprocal networks. Therefore, the normalized scattering matrix of any reciprocal or nonreciprocal lossless network is unitary.

In a compact form, (2.63) can be expressed as

$$[\hat{S}^*]_t [\hat{S}] = U \tag{2.64}$$

where $[\hat{S}^*]$ denotes the matrix formed by the conjugate of the elements of the $[\hat{S}]$ matrix and $[\hat{S}^*]_t$ denotes the transpose of matrix $[\hat{S}^*]$. U is a unit matrix of order N. All the diagonal elements of a unit matrix are 1, while all its nondiagonal elements are 0.

From (2.63), it is seen that if the nth row of the $[\hat{\mathbf{S}}^*]_t$ matrix is multiplied with the nth column of the $[\hat{\mathbf{S}}]$ matrix, the following equation results:

$$\hat{S}_{1n}\hat{S}_{1n}^* + \hat{S}_{2n}\hat{S}_{2n}^* + \ldots + \hat{S}_{Nn}\hat{S}_{Nn}^* = 1$$

or

$$|\hat{S}_{1n}|^2 + |\hat{S}_{2n}|^2 + \ldots + |\hat{S}_{Nn}|^2 = 1 \tag{2.65}$$

where $n = 1, 2, \ldots, N$.

On the other hand, if the nth row of the $[\hat{\mathbf{S}}^*]_t$ matrix is multiplied with the mth column of the $[\hat{\mathbf{S}}]$ matrix with $m \neq n$, then the following equation results:

$$\hat{S}_{1n}^*\hat{S}_{1m} + \hat{S}_{2n}^*\hat{S}_{2m} + \ldots + \hat{S}_{Nn}^*\hat{S}_{Nm} = 0 \tag{2.66}$$

where $m = 1, \ldots, N$, $n = 1, \ldots, N$, and $m \neq n$. The unitary properties of the scattering matrix as given by (2.65) and (2.66) lead to very useful predictions about the properties of a lossless network. For example, the unitary property of a scattering matrix leads to the result that it is impossible to match a lossless, three-port reciprocal network at all its ports simultaneously. We state some special properties of two-, three-, and four-port lossless networks derived using the unitary property of the scattering matrix later in this chapter.

2.3.2 Transformation With Change in Position of Terminal Planes

Assume that the scattering matrix of the network with the location of terminal planes denoted by t_p, $p = 1, \ldots N$ as shown in Figure 2.3 is given by (2.57). If the terminal plane in each transmission line is moved to new locations denoted by t'_p where $p = 1, \ldots N$, we define the scattering matrix with the location of ports denoted by t'_p as $[\hat{\mathbf{S}}']$, which is related to the scattering matrix $[\hat{\mathbf{S}}]$ as

$$[\hat{\mathbf{S}}'] = \begin{bmatrix} e^{-j\beta_1 l_1} & 0 & \cdots & 0 \\ 0 & e^{-j\beta_2 l_2} & \cdots & 0 \\ \vdots & \vdots & \vdots & \vdots \\ 0 & 0 & \cdots & e^{-j\beta_N l_N} \end{bmatrix} \begin{bmatrix} \hat{S}_{11} & \hat{S}_{12} & \cdots & \hat{S}_{1N} \\ \hat{S}_{21} & \hat{S}_{22} & \cdots & \hat{S}_{2N} \\ \vdots & \vdots & \vdots & \vdots \\ \hat{S}_{N1} & \hat{S}_{N2} & \cdots & \hat{S}_{NN} \end{bmatrix}$$

$$\times \begin{bmatrix} e^{-j\beta_1 l_1} & 0 & \cdots & 0 \\ 0 & e^{-j\beta_2 l_2} & \cdots & 0 \\ \vdots & \vdots & \vdots & \vdots \\ 0 & 0 & \cdots & e^{-j\beta_N l_N} \end{bmatrix} \tag{2.67}$$

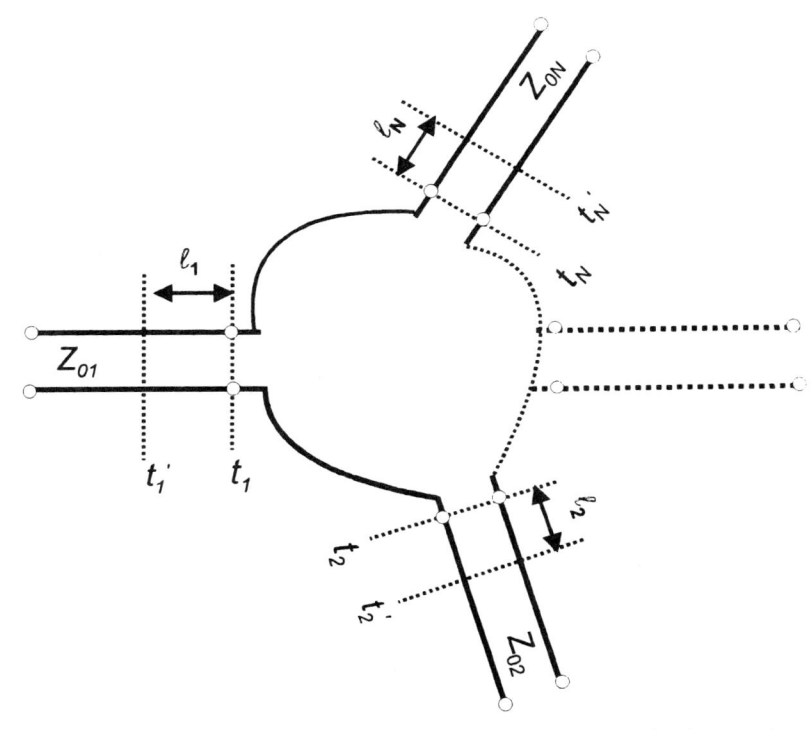

Figure 2.3 An N-port network. The elements of a representative matrix change when the location of ports (terminal planes) is changed.

where β_p ($p = 1, \ldots, N$) denotes the phase constant of the wave in the transmission line of the pth port. From (2.67), it follows that the mn^{th} element of the modified scattering matrix is related to the corresponding element of the original scattering matrix by

$$\hat{S}'_{mn} = \hat{S}_{mn} e^{-j\beta_m l_m - j\beta_n l_n} \tag{2.68}$$

2.3.3 Reciprocal Networks

If a network does not contain any nonreciprocal media (e.g., ferrite, plasma), then the following relation holds for the elements of its normalized scattering matrix:

$$\hat{S}_{mn} = \hat{S}_{nm} \tag{2.69}$$

In matrix notation, the condition of reciprocity is stated as

$$[\hat{S}] = [\hat{S}]_t \tag{2.70}$$

2.3.4 Relationship Between Normalized and Unnormalized Matrices

The mn^{th} elements of the unnormalized and normalized scattering matrices are related by

$$S_{mn} = \hat{S}_{mn}\sqrt{\frac{Z_{0m}}{Z_{0n}}} \qquad (2.71)$$

where Z_{0m} and Z_{0n} denote the characteristic impedances of the transmission lines at ports m and n. The above equation leads to an important conclusion that if all the ports of a network have the same characteristic impedance, then the unnormalized scattering matrix of the network is the same as the normalized matrix. In microwave networks, all the ports usually have the same characteristic impedance. Therefore, in these cases it is not *necessary* to specify whether the scattering matrix is normalized or unnormalized. However, if the characteristic impedances of all the ports of a network are not the same, such as in the case of asymmetrical coupled lines, impedance-transforming couplers, baluns, and so forth it should be specified whether the scattering matrix is normalized or unnormalized.

2.4 Special Properties of Two-, Three- and Four-Port Passive, Lossless Networks

All the elements of a network matrix of a passive lossless network cannot be chosen independently. For example, if a network does not contain any non-reciprocal media (e.g., ferrite, plasma), then $\hat{S}_{mn} = \hat{S}_{nm}$. Therefore, one cannot design a network containing a passive reciprocal medium having different values of $\hat{S}_{mn}$ and $\hat{S}_{nm}$. As another example, if a network has a plane of symmetry, then $\hat{S}_{mm} = \hat{S}_{nn}$ where m and n are symmetrical ports. The preceding properties described are true in general for any N-port network. There are, however, some special properties of a network depending on the number of ports it has. In the following, we discuss some special properties of passive lossless, two-, three-, and four-port networks.

2.4.1 Two-Port Networks

Figure 2.4 shows a two-port passive and lossless network. The normalized scattering matrix of a two-port network can be written as

$$[\hat{\mathbf{S}}] = \begin{bmatrix} \hat{S}_{11} & \hat{S}_{12} \\ \hat{S}_{21} & \hat{S}_{22} \end{bmatrix} = \begin{bmatrix} |\hat{S}_{11}|e^{j\theta_{11}} & |\hat{S}_{12}|e^{j\theta_{12}} \\ |\hat{S}_{21}|e^{j\theta_{21}} & |\hat{S}_{22}|e^{j\theta_{22}} \end{bmatrix} \qquad (2.72)$$

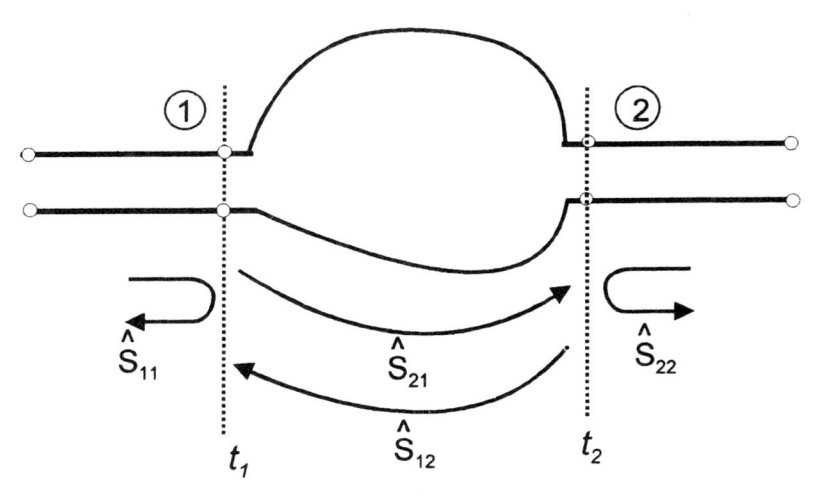

Figure 2.4 A passive, lossless two-port network.

The unitary property of the normalized scattering matrix as given by (2.65) and (2.66) leads to the following relations:

$$|\hat{S}_{11}|^2 + |\hat{S}_{21}|^2 = 1 \tag{2.73a}$$

$$|\hat{S}_{12}|^2 + |\hat{S}_{22}|^2 = 1 \tag{2.73b}$$

$$\hat{S}_{11}^*\hat{S}_{12} + \hat{S}_{21}^*\hat{S}_{22} = 0 \tag{2.73c}$$

The solution of the above set of equations leads to the following conclusions:

$$|\hat{S}_{11}| = |\hat{S}_{22}| \tag{2.74}$$

$$|\hat{S}_{12}| = |\hat{S}_{21}| \tag{2.75}$$

$$\theta_{11} + \theta_{22} = \theta_{12} + \theta_{21} \mp \pi \quad \text{rad} \tag{2.76}$$

The above relations are true for any two-port network—reciprocal or non-reciprocal. Equations 2.74 and 2.75 along with (2.73) show that for a lossless network (reciprocal or nonreciprocal) one needs to know the amplitude of one scattering element to determine the amplitude of all other scattering elements. (2.75) also shows that even if the two-port network is composed of non-reciprocal media, $|\hat{S}_{12}| = |\hat{S}_{21}|$. In physical terms, this relation states that the ratio of the power coupled from port 1 to port 2 when the power is incident at port 1 is the same as that coupled from port 2 to port 1 when power is incident at port 2. This is an interesting result; one might then wonder about how a two-port isolator works. An isolator is supposed to have a very small attenuation between its ports when the power is incident at one port, and a large attenuation when the power is incident at the other port. The difficulty is resolved by noting that the unitary property is valid only for lossless networks. On the other hand, an isolator employs some kind of lossy elements to achieve different values of $|\hat{S}_{12}|$ and $|\hat{S}_{21}|$.

When the network is reciprocal, $\hat{S}_{12} = \hat{S}_{21}$, which leads to $\theta_{12} = \theta_{21}$, or from (2.76):

$$\theta_{12} = (\theta_{11} + \theta_{22} \pm \pi)/2 \quad \text{rad}$$

Using the above equation and (2.74) and (2.75), the scattering matrix of a lossless, reciprocal two-port network can be expressed in the form

$$[\hat{S}] = \begin{bmatrix} |\hat{S}_{11}|e^{j\theta_{11}} & \sqrt{1-|\hat{S}_{11}|^2}e^{j(\theta_{11}+\theta_{22}\pm\pi)/2} \\ \sqrt{1-|\hat{S}_{11}|^2}e^{j(\theta_{11}+\theta_{22}\pm\pi)/2} & |\hat{S}_{11}|e^{j\theta_{22}} \end{bmatrix} \quad (2.77)$$

Further, noting that $|\hat{S}_{11}| \leq 1$, (2.77) reduces to

$$[\hat{S}] = \begin{bmatrix} \cos\alpha\, e^{j\theta_{11}} & \sin\alpha\, e^{j(\theta_{11}+\theta_{22}\pm\pi)/2} \\ \sin\alpha\, e^{j(\theta_{11}+\theta_{22}\pm\pi)/2} & \cos\alpha\, e^{j\theta_{22}} \end{bmatrix} \quad (2.78)$$

where $\cos\alpha = |\hat{S}_{11}|$.

For a two-port network, any desired values of θ_{11} and θ_{22} can always be chosen by changing the location of the terminal planes. If θ_{11} and θ_{22} are chosen to be zero by appropriately locating the terminal planes, then the scattering matrix of the two-port network becomes

$$[\hat{S}] = \begin{bmatrix} \hat{S}_{11} & \hat{S}_{12} \\ \hat{S}_{21} & \hat{S}_{22} \end{bmatrix} = \begin{bmatrix} \cos\alpha & \pm j\sin\alpha \\ \pm j\sin\alpha & \cos\alpha \end{bmatrix} \quad (2.79)$$

2.4.2 Three-Port Reciprocal Networks

For networks having three ports, the unitary property of the scattering matrix leads to the result that it is impossible to match a passive, lossless reciprocal network at all its ports simultaneously. Therefore, a three-port network enclosing reciprocal media *cannot* have a scattering matrix of the form

$$[\hat{S}] = \begin{bmatrix} 0 & \hat{S}_{12} & \hat{S}_{13} \\ \hat{S}_{12} & 0 & \hat{S}_{23} \\ \hat{S}_{13} & \hat{S}_{23} & 0 \end{bmatrix} \quad (2.80)$$

The above condition holds only for a lossless reciprocal network, but by incorporating lossy elements in the network (such as in a Wilkinson's power divider), a three-port network can be matched at all its ports simultaneously.

For a reciprocal three-port network, the unitary property also leads to another important result. If two of the three ports of the network are completely matched, then the third port is completely isolated from the other two ports. The scattering matrix of this network is then given by

$$[\hat{S}] = \begin{bmatrix} 0 & 1 & 0 \\ 1 & 0 & 0 \\ 0 & 0 & 1 \end{bmatrix} \quad (2.81)$$

where it is assumed that the phase of nonzero scattering elements has been adjusted to zero.

2.4.3 Three-Port Nonreciprocal Networks

The unitary property leads to the result that a passive, lossless three-port nonreciprocal network (Figure 2.5) can be matched at all its ports simultaneously. With this, it can behave only as a circulator. The scattering matrix of a circulator is of the following form:

$$[\hat{S}] = \begin{bmatrix} 0 & 0 & e^{j\theta_{13}} \\ e^{j\theta_{21}} & 0 & 0 \\ 0 & e^{j\theta_{32}} & 0 \end{bmatrix} \quad (2.82)$$

or $|\hat{S}_{13}| = |\hat{S}_{21}| = |\hat{S}_{32}| = 1$. The lossless circulator has thus an important property that if power is incident at port 1 then all the power is transmitted to port 2. If the power is incident at port 2, all the power is transmitted to port 3, and if the power is incident at port 3, all the power is transmitted to port 1. This is shown schematically in Figure 2.5.

2.4.4 Four-Port Reciprocal Networks

Figure 2.6 shows a four-port network. For a passive, lossless reciprocal four-port network, the unitary property of the scattering matrix leads to the result that it is possible to match all the four ports of the network simultaneously. If all four ports are matched, the network behaves like a directional coupler. The scattering

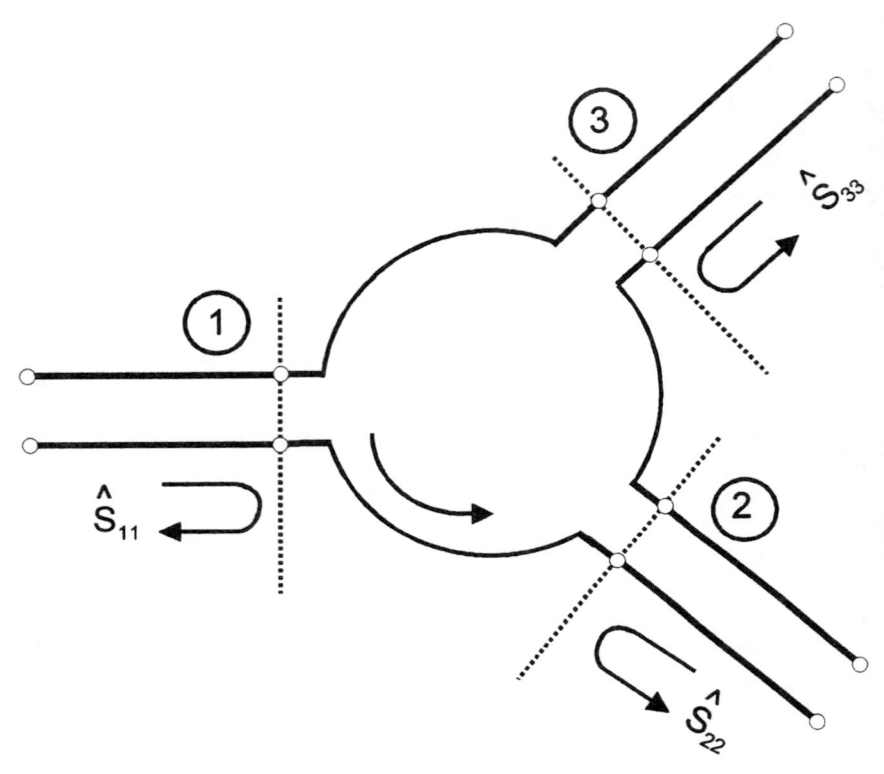

Figure 2.5 A passive, lossless, nonreciprocal three-port network. The network acts as a circulator if it is matched at all its ports.

matrix of a directional coupler is of the form

$$[\hat{S}] = \begin{bmatrix} 0 & 0 & \hat{S}_{13} & \hat{S}_{14} \\ 0 & 0 & \hat{S}_{23} & \hat{S}_{24} \\ \hat{S}_{13} & \hat{S}_{23} & 0 & 0 \\ \hat{S}_{14} & \hat{S}_{24} & 0 & 0 \end{bmatrix} \qquad (2.83)$$

The directional coupler can be considered to be composed of two pairs of ports, with ports of each pair matched and isolated from each other. As seen from (2.83), ports 1 and 2 of the network are matched and isolated from each other. Similarly, ports 3 and 4 are matched and isolated from each other.

Microwave Network Theory 45

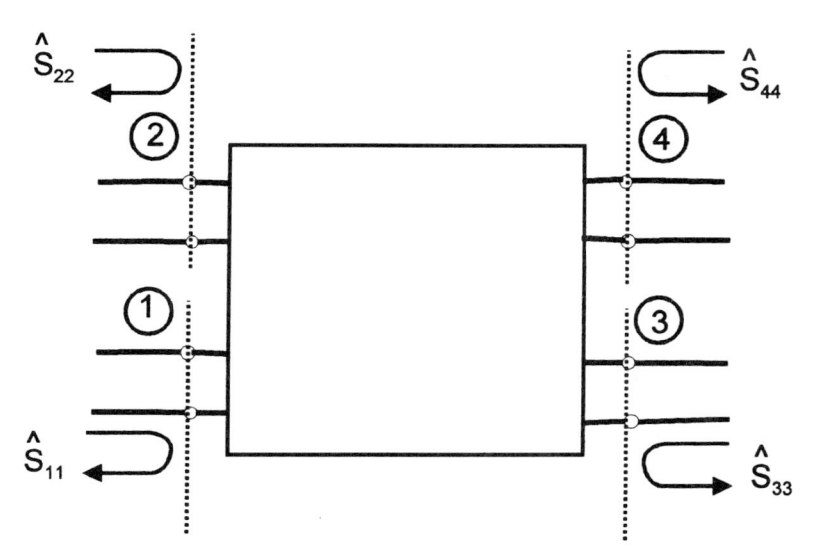

Figure 2.6 A passive, lossless, reciprocal four-port network. The network behaves like a directional coupler if it is matched at all ports.

The elements of the scattering matrix of a directional coupler as given by (2.83) also satisfy the following relationships:

$$|\hat{S}_{13}| = |\hat{S}_{24}|, \qquad |\hat{S}_{14}| = |\hat{S}_{23}|$$

Using the results of the above two equations, the elements of the scattering matrix can be expressed as

$$\hat{S}_{13} = C_1 e^{j\theta_{13}}, \quad \hat{S}_{23} = C_2 e^{j\theta_{23}}, \quad \hat{S}_{14} = C_2 e^{j\theta_{14}}, \quad \hat{S}_{24} = C_1 e^{j\theta_{24}}$$

where $C_1 = |\hat{S}_{13}| = |\hat{S}_{24}|$ and $C_2 = |\hat{S}_{14}| = |\hat{S}_{23}|$. Thus, (2.83) can be expressed as

$$[\hat{\mathbf{S}}] = \begin{bmatrix} 0 & 0 & C_1 e^{j\theta_{13}} & C_2 e^{j\theta_{14}} \\ 0 & 0 & C_2 e^{j\theta_{23}} & C_1 e^{j\theta_{24}} \\ C_1 e^{j\theta_{13}} & C_2 e^{j\theta_{23}} & 0 & 0 \\ C_2 e^{j\theta_{14}} & C_1 e^{j\theta_{24}} & 0 & 0 \end{bmatrix} \qquad (2.84)$$

We can easily show that all the phase factors of the various scattering elements θ_{13}, θ_{23}, θ_{14}, and θ_{24} cannot be chosen independently. Assume that desired values of θ_{13} and θ_{14} have been chosen by varying the positions of ports 3 and 4 respectively. The phase factor θ_{23} can be independently chosen by controlling the position of port 2. The remaining phase factor θ_{24} cannot be changed now because both ports 2 and 4 have already been adjusted. From the unitary property of the scattering matrix, one obtains the value of θ_{24} in terms of other phase factors as

$$\theta_{24} = \theta_{14} + \theta_{23} - \theta_{13} \pm \pi \quad \text{rad} \tag{2.85}$$

The amplitudes of scattering elements satisfy

$$C_1^2 + C_2^2 = 1 \tag{2.86}$$

The two forms of matrices to which the scattering matrix of a directional coupler can always be reduced by appropriately locating the position of terminal planes are now derived. Let us first choose $\theta_{13} = \theta_{14} = \theta_{23} = 0$. From (2.85) we obtain $\theta_{24} = \pm \pi$ rad, or the scattering matrix of a directional coupler takes the form

$$[\hat{S}] = \begin{bmatrix} 0 & 0 & C_1 & C_2 \\ 0 & 0 & C_2 & -C_1 \\ C_1 & C_2 & 0 & 0 \\ C_2 & -C_1 & 0 & 0 \end{bmatrix} \tag{2.87}$$

For deriving the second form, we choose $\theta_{13} = 0$ and $\theta_{23} = \theta_{14} = \pm \pi/2$ rad. From (2.85), we then obtain $\theta_{24} = 0$, or $\theta_{24} = \pm 2\pi$ rad. The scattering matrix thus reduces to

$$[\hat{S}] = \begin{bmatrix} 0 & 0 & C_1 & \pm jC_2 \\ 0 & 0 & \pm jC_2 & C_1 \\ C_1 & \pm jC_2 & 0 & 0 \\ \pm jC_2 & C_1 & 0 & 0 \end{bmatrix} \tag{2.88}$$

We see in later chapters that the scattering matrix of many useful four-port microwave networks can be represented either by (2.87) or (2.88). For example, a rat-race hybrid can be represented by (2.87), whereas quadrature hybrids, Lange couplers, and so on can be represented by (2.88).

2.5 Special Representation of Two-Port Networks

A typical microwave subsystem consists of a cascade of two-port networks such that the output of one network is connected to the input of the next and so on. The two-port networks can be represented by their impedance, admittance, or scattering parameters. It is often more useful, however, to represent two-port networks by ABCD parameters because knowing the ABCD parameters, the matrix of the overall cascaded network can be computed by multiplying the matrices of the individual networks.

2.5.1 ABCD Parameters

Figure 2.7 shows a two-port network. In the ABCD matrix representation, the voltage and current flowing into the network at the input of the network are related to the voltage and current flowing away from the network at the output as follows:

$$V_1 = AV_2 + BI_2$$
$$I_1 = CV_2 + DI_2 \tag{2.89}$$

or in matrix form

$$\begin{bmatrix} V_1 \\ I_1 \end{bmatrix} = \begin{bmatrix} A & B \\ C & D \end{bmatrix} \begin{bmatrix} V_2 \\ I_2 \end{bmatrix} \tag{2.90}$$

Note that in the ABCD matrix representation, the direction of positive current flow at the output as shown in Figure 2.7 is taken in a opposite sense than what is done in the impedance, or admittance matrix representation.

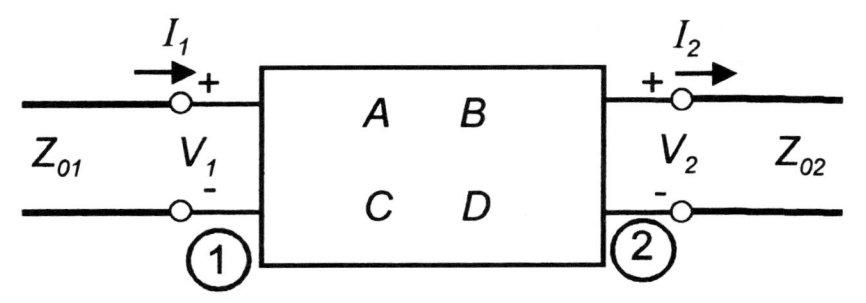

Figure 2.7 A two-port network. In the ABCD matrix representation, the direction of positive current flow at the output is opposite of that used in impedance and admittance matrix representation.

To explain the advantage of representation in terms of ABCD parameters, consider the two-port networks cascaded together in Figure 2.8. The ABCD parameters of the individual networks are also shown in the same figure. We are interested in finding the relationship between the input and output of the overall cascaded network. The voltages and currents at the input and output of the first network are related by the following matrix equation:

$$\begin{bmatrix} V_1 \\ I_1 \end{bmatrix} = \begin{bmatrix} A_1 & B_1 \\ C_1 & D_1 \end{bmatrix} \begin{bmatrix} V_2 \\ I_2 \end{bmatrix} \tag{2.91}$$

For the second network, V_2 represents the input voltage and I_2 represents the input current. So we can write

$$\begin{bmatrix} V_2 \\ I_2 \end{bmatrix} = \begin{bmatrix} A_2 & B_2 \\ C_2 & D_2 \end{bmatrix} \begin{bmatrix} V_3 \\ I_3 \end{bmatrix} \tag{2.92}$$

Substituting V_2 and I_2 into (2.91):

$$\begin{bmatrix} V_1 \\ I_1 \end{bmatrix} = \begin{bmatrix} A_1 & B_1 \\ C_1 & D_1 \end{bmatrix} \begin{bmatrix} A_2 & B_2 \\ C_2 & D_2 \end{bmatrix} \begin{bmatrix} V_3 \\ I_3 \end{bmatrix} \tag{2.93}$$

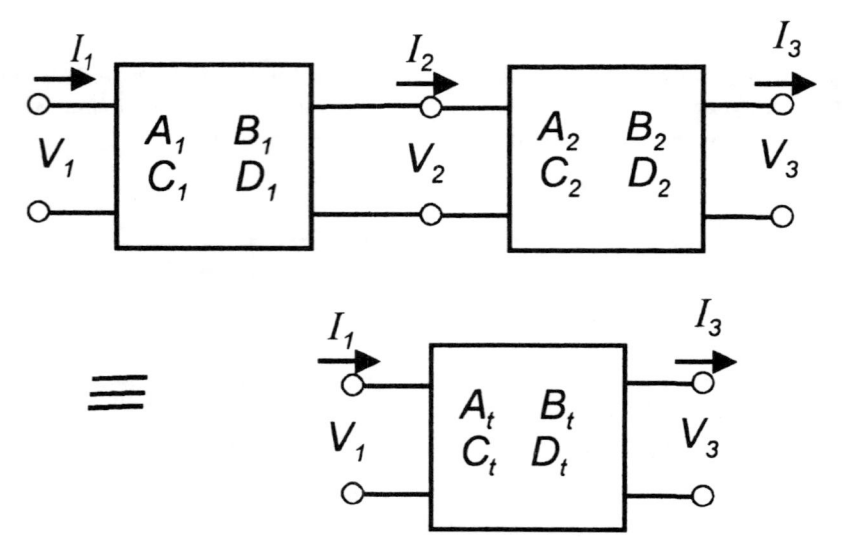

Figure 2.8 Cascade of two two-port networks and their equivalent representation.

or the ABCD matrix of the overall network between ports 1 and 3 can be expressed as

$$\begin{bmatrix} A_t & B_t \\ C_t & D_t \end{bmatrix} = \begin{bmatrix} A_1 & B_1 \\ C_1 & D_1 \end{bmatrix} \begin{bmatrix} A_2 & B_2 \\ C_2 & D_2 \end{bmatrix} \quad (2.94)$$

Therefore, the ABCD matrix of the overall network is the product of the ABCD matrices of the individual networks. The same is true for any number of two-port networks connected in cascade.

Properties of ABCD Parameters

Consider the two-port network as shown in Figure 2.7. Let Z_{01} and Z_{02} denote the characteristic impedances of ports 1 and 2, respectively. The normalized and unnormalized ABCD parameters are related as follows:

$$\hat{A} = A\sqrt{\frac{Z_{02}}{Z_{01}}}$$

$$\hat{B} = \frac{B}{\sqrt{Z_{01}Z_{02}}} \quad (2.95)$$

$$\hat{C} = C\sqrt{Z_{02}Z_{01}}$$

$$\hat{D} = D\sqrt{\frac{Z_{01}}{Z_{02}}}$$

For a passive, lossless, two-port reciprocal network:

$$AD - BC = 1 \quad (2.96)$$

The above relation is also satisfied by *normalized* ABCD parameters. In general, any relations satisfied by unnormalized parameters is also satisfied by normalized parameters.

For a lossless, two-port symmetrical network:

$$\hat{A} = \hat{D} \quad (2.97)$$

2.5.2 Reflection and Transmission Coefficients in Terms of ABCD Parameters

Consider the reciprocal two-port network shown in Figure 2.9. The characteristic impedances of transmission lines of the input and output ports are assumed to be Z_0 and both ports are assumed to be match-terminated. The reflection coefficient

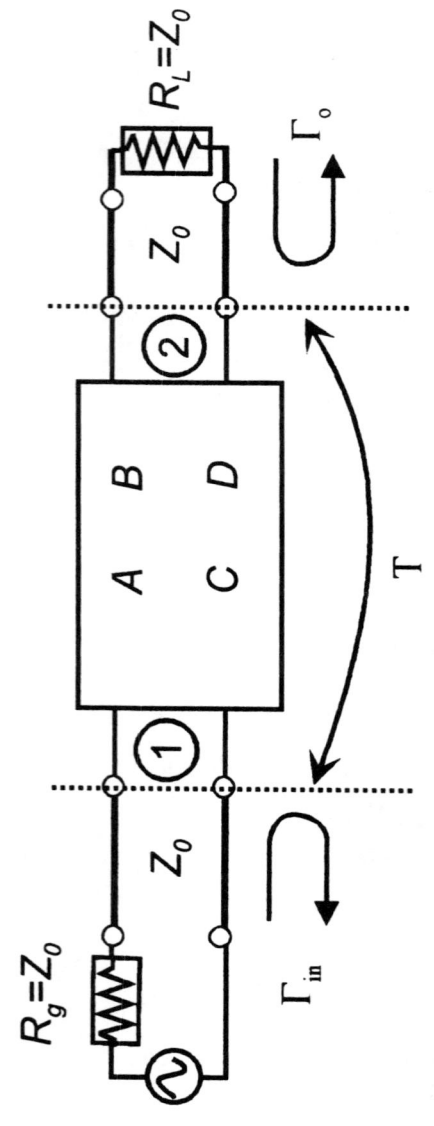

Figure 2.9 Reflection and transmission coefficients of a two-port network.

at the input port is given by

$$\Gamma_{in} = \frac{A + B/Z_0 - CZ_0 - D}{A + B/Z_0 + CZ_0 + D} \quad (2.98)$$

Further, the reflection coefficient at output port is given by

$$\Gamma_o = \frac{-A + B/Z_0 - CZ_0 + D}{A + B/Z_0 + CZ_0 + D} \quad (2.99)$$

The return loss (in decibels, which is a positive quantity) is given by (2.25). The transmission coefficient between input and output ports is given by,

$$T = \frac{\hat{V}_2^-}{\hat{V}_1^+} = \frac{V_2^-}{V_1^+} = \frac{2}{A + B/Z_0 + CZ_0 + D} \quad (2.100)$$

The insertion loss between input and output (in decibels, which is a positive quantity) is given by

$$\text{Insertion loss(dB)} = -20 \log |T| \quad (2.101)$$

For a lossless network,

$$|\Gamma_{in}|^2 + T^2 = |\Gamma_o|^2 + T^2 = 1 \quad (2.102)$$

In the above example, we have assumed that the characteristic impedances of the input and output ports are the same. In case they are not the same, the unnormalized scattering parameters of the network can be obtained from the unnormalized ABCD parameters using the conversion relation given in Table 2.1. The unnormalized scattering parameters can then be normalized using the equations given in Section 2.3.4. The elements $\hat{S}_{11}$ and $\hat{S}_{21}$ of the normalized scattering matrix directly give the reflection and transmission coefficients.

The unnormalized and normalized ABCD parameters of some elementary networks are given in Table 2.2. More complex networks can be obtained by cascading a number of elementary networks. The ABCD matrix of the overall cascaded network can then be determined by multiplying the ABCD matrices of elementary networks. It is quite simple to determine the ABCD parameters of elementary networks. Consider, for example, the circuit shown in Figure 2.10. Inspecting the circuit, the following equations are obtained:

$$V_1 = V_2 + ZI_2$$
$$I_1 = I_2 \quad (2.103)$$

Table 2.1
Conversion relationships between various representative matrices of two-port networks

$$\begin{bmatrix} Z_{11} & Z_{12} \\ Z_{21} & Z_{22} \end{bmatrix} = \frac{1}{Y_{11}Y_{22} - Y_{12}Y_{21}} \begin{bmatrix} Y_{22} & Y_{12} \\ -Y_{21} & Y_{11} \end{bmatrix}$$

$$\begin{bmatrix} Y_{11} & Y_{12} \\ Y_{21} & Y_{22} \end{bmatrix} = \frac{1}{Z_{11}Z_{22} - Z_{12}Z_{21}} \begin{bmatrix} Z_{22} & Z_{12} \\ -Z_{21} & Z_{11} \end{bmatrix}$$

$$\begin{bmatrix} Z_{11} & Z_{12} \\ Z_{21} & Z_{22} \end{bmatrix} = \frac{1}{C} \begin{bmatrix} A & AD - BC \\ 1 & D \end{bmatrix}$$

$$\begin{bmatrix} Y_{11} & Y_{12} \\ Y_{21} & Y_{22} \end{bmatrix} = \frac{1}{B} \begin{bmatrix} D & -(AD - BC) \\ -1 & A \end{bmatrix}$$

$$\begin{bmatrix} A & B \\ C & D \end{bmatrix} = \frac{1}{Z_{21}} \begin{bmatrix} Z_{11} & (Z_{11}Z_{22} - Z_{12}Z_{21}) \\ 1 & Z_{22} \end{bmatrix}$$

$$\begin{bmatrix} A & B \\ C & D \end{bmatrix} = \frac{1}{-Y_{21}} \begin{bmatrix} Y_{22} & 1 \\ Y_{11}Y_{22} - Y_{12}Y_{21} & Y_{11} \end{bmatrix}$$

$$\begin{bmatrix} S_{11} & S_{12} \\ S_{21} & S_{22} \end{bmatrix} = \frac{1}{\left(\frac{Z_{11}}{Z_{01}} + 1\right)\left(\frac{Z_{22}}{Z_{02}} + 1\right) - \frac{Z_{12}Z_{21}}{Z_{01}Z_{02}}}$$
$$\times \begin{bmatrix} \left(\frac{Z_{11}}{Z_{01}} - 1\right)\left(\frac{Z_{22}}{Z_{02}} + 1\right) - \frac{Z_{12}Z_{21}}{Z_{01}Z_{02}} & 2\frac{Z_{12}}{Z_{02}} \\ 2\frac{Z_{21}}{Z_{01}} & \left(\frac{Z_{11}}{Z_{01}} + 1\right)\left(\frac{Z_{22}}{Z_{02}} - 1\right) - \frac{Z_{12}Z_{21}}{Z_{01}Z_{02}} \end{bmatrix}$$

$$\begin{bmatrix} S_{11} & S_{12} \\ S_{21} & S_{22} \end{bmatrix} = \frac{1}{\left(1 + \frac{Y_{11}}{Y_{01}}\right)\left(1 + \frac{Y_{22}}{Y_{02}}\right) - \frac{Y_{12}Y_{21}}{Y_{01}Y_{02}}}$$
$$\times \begin{bmatrix} \left(1 - \frac{Y_{11}}{Y_{01}}\right)\left(1 + \frac{Y_{22}}{Y_{02}}\right) - \frac{Y_{12}Y_{21}}{Y_{01}Y_{02}} & -2\frac{Y_{12}}{Y_{01}} \\ -2\frac{Y_{21}}{Y_{02}} & \left(1 + \frac{Y_{11}}{Y_{01}}\right)\left(1 - \frac{Y_{22}}{Y_{02}}\right) + \frac{Y_{12}Y_{21}}{Y_{01}Y_{02}} \end{bmatrix}$$

$$\begin{bmatrix} Y_{11} & Y_{12} \\ Y_{21} & Y_{22} \end{bmatrix} = \frac{1}{(1 + S_{11})(1 + S_{22}) - (S_{12}S_{21})}$$
$$\times \begin{bmatrix} Y_{01}[(1 - S_{11})(1 + S_{22}) + S_{12}S_{21}] & -2Y_{01}S_{12} \\ -2Y_{02}S_{21} & Y_{02}[(1 + S_{11})(1 - S_{22}) + S_{12}S_{21}] \end{bmatrix}$$

(Continued)

Table 2.1 (Continued)

$$\begin{bmatrix} A & B \\ C & D \end{bmatrix} = \frac{1}{(2S_{21})} \begin{bmatrix} [(1+S_{11})(1-S_{22}) + S_{12}S_{21}] & Z_{02}[(1+S_{11})(1+S_{22}) - S_{12}S_{21}] \\ \frac{1}{Z_{01}}[(1-S_{11})(1-S_{22}) - S_{12}S_{21}] & \frac{Z_{01}}{Z_{02}}[(1-S_{11})(1+S_{22}) + S_{12}S_{21}] \end{bmatrix}$$

$$\begin{bmatrix} S_{11} & S_{12} \\ S_{21} & S_{22} \end{bmatrix} = \frac{1}{(B + CZ_{01}Z_{02}) + (AZ_{02} + DZ_{01})}$$

$$\times \begin{bmatrix} (B - CZ_{01}Z_{02}) + (AZ_{02} - DZ_{01}) & 2Z_{01}(AD - BC) \\ 2Z_{02} & (B - CZ_{01}Z_{02}) - (AZ_{02} - DZ_{01}) \end{bmatrix}$$

$$\begin{bmatrix} Z_{11} & Z_{12} \\ Z_{21} & Z_{22} \end{bmatrix} = \frac{1}{(1-S_{11})(1-S_{22}) - S_{12}S_{21}}$$

$$\times \begin{bmatrix} Z_{01}[(1+S_{11})(1-S_{22}) + S_{12}S_{21}] & 2Z_{02}S_{12} \\ 2Z_{01}S_{21} & Z_{02}[(1-S_{11})(1+S_{22}) + S_{12}S_{21}] \end{bmatrix}$$

By substituting $Z_{01} = Z_{02} = 1$, the above relations can be used for the conversion of normalized parameters.

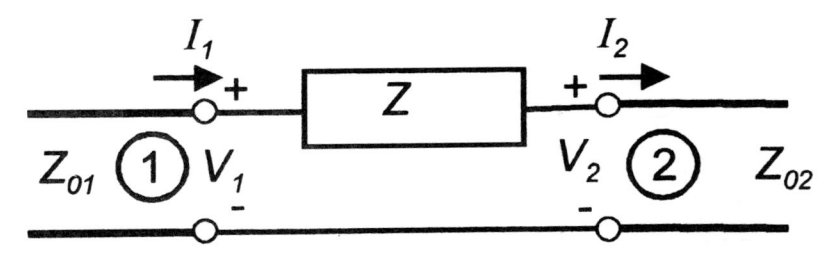

Figure 2.10 An impedance Z in series between two transmission lines.

By comparing the above set of equations with (2.89), we obtain

$$A = 1, \quad B = Z, \quad C = 0, \quad \text{and} \quad D = 1$$

Further, using (2.95), the normalized ABCD parameters can be found as follows:

$$\hat{A} = \sqrt{\frac{Z_{02}}{Z_{01}}}, \quad \hat{B} = \frac{Z}{\sqrt{Z_{01}Z_{02}}}, \quad \hat{C} = 0, \quad \text{and} \quad \hat{D} = \sqrt{\frac{Z_{01}}{Z_{02}}}$$

2.5.3 Equivalent T and Π Networks of Two-Port Circuits

If the impedance parameters of a two-port reciprocal network are known, the network can be represented as shown in Figure 2.11(a). Similarly, if the admittance parameters of a two-port reciprocal network are known, the circuit

Table 2.2
ABCD parameters of elementary two-port networks.

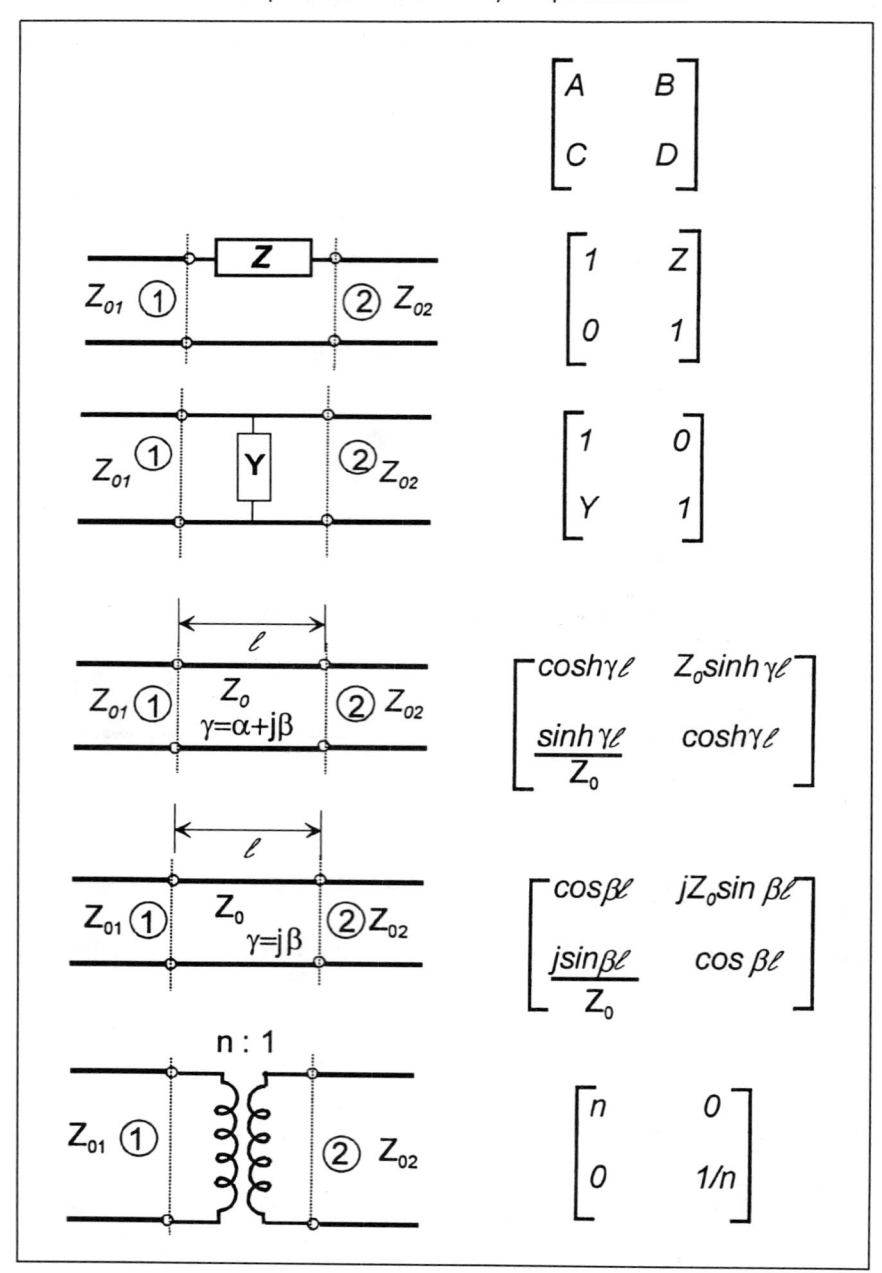

$$\begin{bmatrix} A & B \\ C & D \end{bmatrix}$$

$$\begin{bmatrix} 1 & Z \\ 0 & 1 \end{bmatrix}$$

$$\begin{bmatrix} 1 & 0 \\ Y & 1 \end{bmatrix}$$

$$\begin{bmatrix} \cosh\gamma\ell & Z_0\sinh\gamma\ell \\ \dfrac{\sinh\gamma\ell}{Z_0} & \cosh\gamma\ell \end{bmatrix}$$

$$\begin{bmatrix} \cos\beta\ell & jZ_0\sin\beta\ell \\ \dfrac{j\sin\beta\ell}{Z_0} & \cos\beta\ell \end{bmatrix}$$

$$\begin{bmatrix} n & 0 \\ 0 & 1/n \end{bmatrix}$$

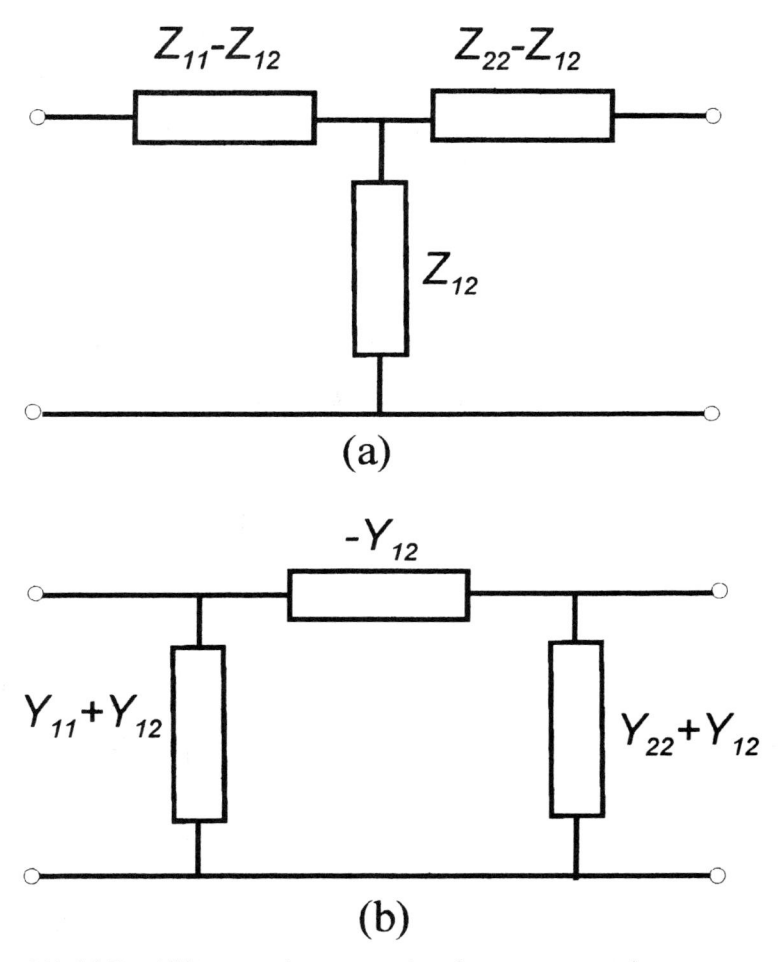

Figure 2.11 (a) T and (b) Π network representation of a two-port network.

can be represented as shown in Figure 2.11(b). T and Π forms are only two of many possible ways in which the equivalent circuit of a two-port network can be expressed. The other forms may contain a combination of a length of a transmission line, transformer, reactance and susceptance elements, and the like [1].

2.6 Conversion Relations

In the following, conversion relations among admittance, impedance, and scattering matrices are given. The conversion relations between a scattering matrix and impedance and admittance matrices are given by assuming that the respective matrices are normalized. To convert unnormalized matrices, the unnormalized parameters should be first normalized using the equations given earlier in various sections. The normalized matrix can then be converted from one type to the desired type, and later unnormalized if necessary.

$$[Z] = [Y]^{-1} \tag{2.104}$$

$$[Y] = [Z]^{-1} \tag{2.105}$$

The above equations are also valid for normalized parameters.

$$[\hat{S}] = ([\hat{Z}] - [U])([\hat{Z}] + [U])^{-1} \tag{2.106}$$

Another expression for $[\hat{S}]$ in terms of $[\hat{Z}]$ is,

$$[\hat{S}] = ([\hat{Z}] + [U])^{-1}([\hat{Z}] - [U]) \tag{2.107}$$

$$[\hat{Z}] = ([U] - [\hat{S}])^{-1}([U] + [\hat{S}]) \tag{2.108}$$

$$[\hat{Y}] = ([U] + [\hat{S}])^{-1}([U] - [\hat{S}]) \tag{2.109}$$

In some cases, the conversion formulas cannot be used. For example, if the determinant of matrix $([U] - [\hat{S}])$ is zero, the impedance matrix becomes indeterminant. Similarly, if the determinant of matrix $([U] + [\hat{S}])$ is zero, the admittance matrix becomes indeterminant.

Frequently, it is required to convert one form of matrix into another. The conversion relations between two-port matrices are given in Table 2.1 [3]. These relations are general and valid for nonreciprocal networks also. By substituting $Z_{01} = Z_{02} = 1$ in these equations, conversion between normalized parameters can be obtained.

2.7 Scattering Matrix of Interconnected Networks

A typical microwave system or subsystem results after interconnection of many intermediate networks. Consider two networks as shown in Figure 2.12. Networks I and II are assumed to have $M + P$ and $P + N$ ports, respectively. The P ports of each network are directly connected to each other. The overall network has therefore $M + N$ accessible ports, and its scattering matrix is of the order $(M + N) \times (M + N)$. Given the scattering matrices of networks I and II, the scattering matrix of the overall network can be determined. Let the ports of network I be so numbered that its M ports ($m = 1, \ldots, M$) represent

Microwave Network Theory

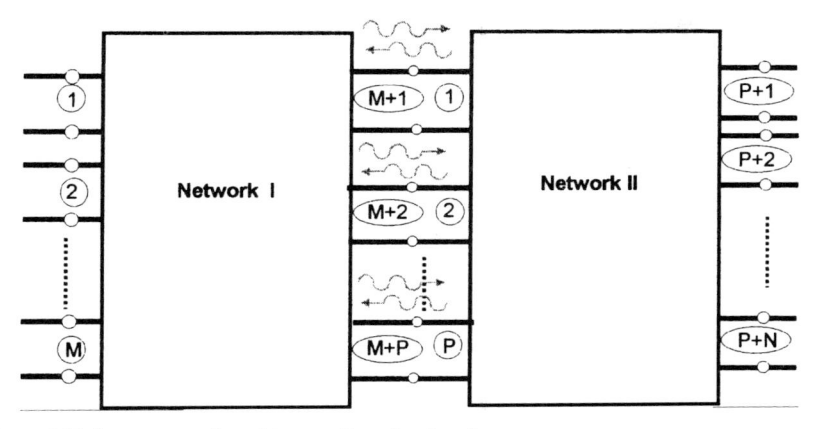

Figure 2.12 Interconnection of two multi-port networks.

the accessible ports, and the remaining P ports ($m = M + 1, \ldots, M + P$) represent those connected to network II. All the accessible ports are assumed to be terminated in matched loads. The scattering matrix of network I can be expressed as

$$\hat{S}_I = \begin{bmatrix} [\hat{S}_{AA}] & [\hat{S}_{AB}] \\ [\hat{S}_{BA}] & [\hat{S}_{BB}] \end{bmatrix} \quad (2.110)$$

where $[\hat{S}_{AA}]$ and $[\hat{S}_{BB}]$ are matrices of order $M \times M$, and $P \times P$ respectively. $[\hat{S}_{AB}]$ $[\hat{S}_{BA}]$ are matrices of order $M \times P$ and $P \times M$, respectively.

Further, let the ports of network II be numbered in a manner that its P ports ($m = 1, \ldots, P$) represent those connected to network I and the remaining N ports ($m = P + 1, \ldots, P + N$) represent the free ports. The scattering matrix of network II can be represented as

$$[\hat{S}_{II}] = \begin{bmatrix} [\hat{S}_{CC}] & [\hat{S}_{CD}] \\ [\hat{S}_{DC}] & [\hat{S}_{DD}] \end{bmatrix} \quad (2.111)$$

where $[\hat{S}_{CC}]$ and $[\hat{S}_{DD}]$ denote the matrices of order $P \times P$ and $N \times N$ respectively. $[\hat{S}_{CD}]$ and $[\hat{S}_{DC}]$ denote matrices of order $P \times N$ and $N \times P$, respectively.

The scattering matrix of the overall network of $(M + N)$ ports can be easily derived using matrix algebra [4, 5]. It follows that the scattering matrix of the overall network denoted as $\hat{S}_R$ can be expressed in concise notation as

$$[\hat{S}_R] = \begin{bmatrix} [\hat{S}_1] & [\hat{S}_2] \\ [\hat{S}_3] & [\hat{S}_4] \end{bmatrix} \quad (2.112)$$

where matrices $[\hat{S}_1]$, $[\hat{S}_2]$, $[\hat{S}_3]$, and $[\hat{S}_4]$ are given by

$$\begin{aligned}
[\hat{S}_1] &= [\hat{S}_{AA}] + [\hat{S}_{AB}](U - [\hat{S}_{CC}][\hat{S}_{BB}])^{-1}[\hat{S}_{CC}][\hat{S}_{BA}] \\
[\hat{S}_2] &= [\hat{S}_{AB}]([U] - [\hat{S}_{CC}][\hat{S}_{BB}])^{-1}[\hat{S}_{CD}] \\
[\hat{S}_3] &= [\hat{S}_{DC}]([U] - [\hat{S}_{BB}][\hat{S}_{CC}])^{-1}[\hat{S}_{BA}] \\
[\hat{S}_4] &= [\hat{S}_{DD}] + [\hat{S}_{DC}]([U] - [\hat{S}_{BB}][\hat{S}_{CC}])^{-1}[\hat{S}_{BB}][\hat{S}_{CD}]
\end{aligned} \quad (2.113)$$

In (2.113) $[U]$ denotes a unit matrix and $()^{-1}$ denotes the inverse of a matrix. A unit matrix is a square matrix. All the diagonal elements of a unit matrix are unity, while all its nondiagonal elements are zero. The scattering matrix of the overall network $[\hat{S}_R]$ is a square matrix of order $M + N$.

2.7.1 Scattering Parameters of Reduced Networks

The scattering parameters of a network are defined by assuming that all its ports are terminated in matched loads. The ports of a network are usually match-terminated when connected in a system. The elements of the scattering matrix thus directly give the reflection coefficient at each port and coupling between various ports. When one or more ports of a network are not match-terminated, however, reflections take place from these ports, and the scattering parameters of the remaining network are modified. For example, if the "direct" and "coupled" ports of a backward TEM directional coupler are open-circuited, the resulting two-port network behaves as a bandpass filter. Consider a network of $M + P$ ports as shown in Figure 2.13, whose scattering matrix is assumed to be given by (2.110). The ports $i = 1, \ldots, M$ of the network are assumed to be terminated in matched loads, while the ports $i = M + 1, \ldots, M + P$ are assumed to be connected in loads of reflection coefficient $\Gamma_{L1}, \Gamma_{L2}, \ldots \Gamma_{LP}$, respectively. It is required to find the scattering matrix of the reduced network. The scattering matrix of the reduced network will be of the order $M \times M$.

We can assume that the given network and the loads shown in Figure 2.13 represent networks I and II of Figure 2.12 respectively, where $N = 0$. The matrices $[\hat{S}_{CD}]$, $[\hat{S}_{CD}]$, and $[\hat{S}_{DD}]$ are therefore null matrices, and the matrix $[\hat{S}_{CC}]$ is a square diagonal matrix given by

$$[\hat{S}_{CC}] = \begin{bmatrix} \Gamma_{L1} & 0 & \cdots & 0 \\ 0 & \Gamma_{L2} & \cdots & 0 \\ \vdots & \vdots & \vdots & \vdots \\ 0 & 0 & \cdots & \Gamma_{LP} \end{bmatrix} \quad (2.114)$$

[Figure 2.13 diagram]

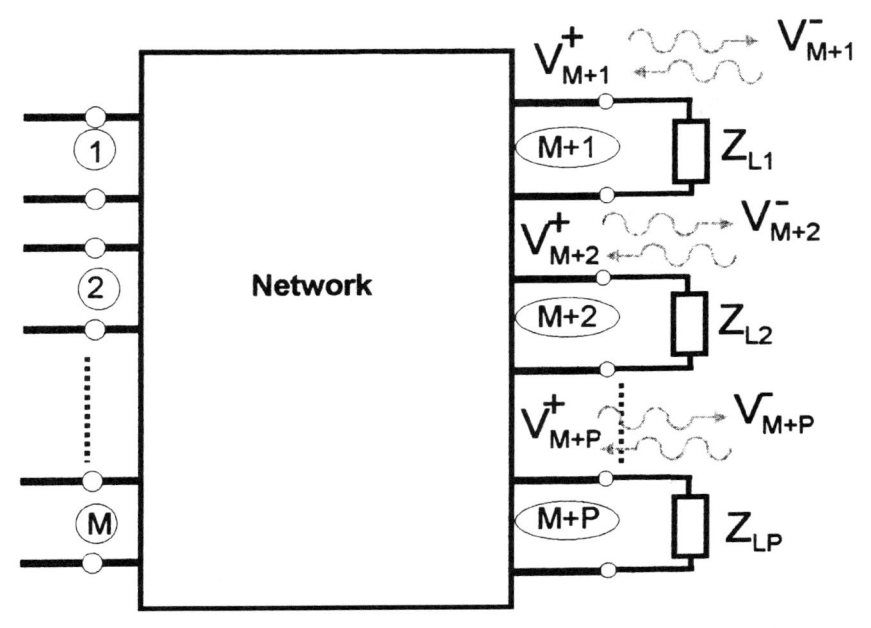

Figure 2.13 A $(M + P)$ port network whose P ports are terminated in arbitrary loads.

The scattering matrix of the reduced network can be found using (2.113). It is found that the scattering matrix of the reduced M-port network is given by

$$[\hat{S}_R] = [\hat{S}_1] \qquad (2.115)$$

where

$$[\hat{S}_1] = [\hat{S}_{AA}] + [\hat{S}_{AB}]([U] - [\hat{S}_{CC}][\hat{S}_{BB}])^{-1}[\hat{S}_{CC}][\hat{S}_{BA}] \qquad (2.116)$$

In the above equation $[\hat{S}_{AA}]$, $[\hat{S}_{AB}]$, $[\hat{S}_{BB}]$, and $[\hat{S}_{CC}]$ denote the partitioned matrices of the original $M + P$ port network as defined by (2.110). $[\hat{S}_{CC}]$ is given by (2.114).

2.7.2 Reduction of a Three-Port Network Into a Two-Port Network

The use of the above formulas is first demonstrated by finding the modified scattering matrix of a three-port network, one of whose ports is not match-terminated as shown in Figure 2.14.

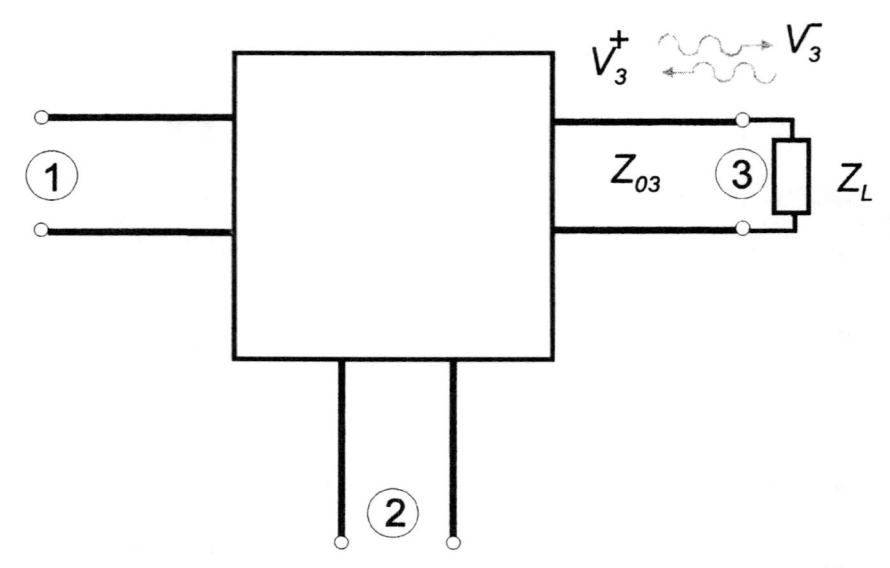

Figure 2.14 A three-port network with its port three terminated in arbitrary load.

Let the scattering matrix of the three-port network shown in Figure 2.14 be given by

$$[S] = \begin{bmatrix} \hat{S}_{11} & \hat{S}_{12} & \hat{S}_{13} \\ \hat{S}_{21} & \hat{S}_{22} & \hat{S}_{23} \\ \hat{S}_{31} & \hat{S}_{32} & \hat{S}_{33} \end{bmatrix} \qquad (2.117)$$

Referring to the notation used earlier in this section, we have

$$[\hat{S}_{AA}] = \begin{bmatrix} \hat{S}_{11} & \hat{S}_{12} \\ \hat{S}_{21} & \hat{S}_{22} \end{bmatrix} \qquad (2.118)$$

$$[\hat{S}_{AB}] = \begin{bmatrix} \hat{S}_{13} \\ \hat{S}_{23} \end{bmatrix} \qquad (2.119)$$

$$[\hat{S}_{BA}] = \begin{bmatrix} \hat{S}_{31} & \hat{S}_{32} \end{bmatrix} \qquad (2.120)$$

$$[\hat{S}_{BB}] = \begin{bmatrix} \hat{S}_{33} \end{bmatrix} \qquad (2.121)$$

and

$$[\hat{S}_{CC}] = \begin{bmatrix} \Gamma_{L3} \end{bmatrix} \qquad (2.122)$$

where

$$\Gamma_{L3} = \frac{\hat{V}_3^+}{\hat{V}_3^-} = \frac{Z_L - Z_{03}}{Z_L + Z_{03}} \quad (2.123)$$

The matrices $[\hat{S}_{CD}]$, $[\hat{S}_{DC}]$, and $[\hat{S}_{DD}]$ are null matrices. Substituting the values of the above matrices in (2.115) and (2.116), we find that the scattering matrix of the reduced two-port network is given by

$$[\hat{S}_R] = \begin{bmatrix} \hat{S}_{11} & \hat{S}_{12} \\ \hat{S}_{21} & \hat{S}_{22} \end{bmatrix} + \begin{bmatrix} \hat{S}_{13} \\ \hat{S}_{23} \end{bmatrix} (1 - \hat{S}_{33}\Gamma_{L3})^{-1} \Gamma_{L3} [\hat{S}_{31} \quad \hat{S}_{32}] \quad (2.124)$$

which leads to

$$[\hat{S}_R] = \begin{bmatrix} \hat{S}_{11} + \frac{\hat{S}_{13}\hat{S}_{31}\Gamma_{L3}}{1-\hat{S}_{33}\Gamma_{L3}} & \hat{S}_{12} + \frac{\hat{S}_{13}\hat{S}_{32}\Gamma_{L3}}{1-\hat{S}_{33}\Gamma_{L3}} \\ \hat{S}_{21} + \frac{\hat{S}_{23}\hat{S}_{31}\Gamma_{L3}}{1-\hat{S}_{33}\Gamma_{L3}} & \hat{S}_{22} + \frac{\hat{S}_{23}\hat{S}_{32}\Gamma_{L3}}{1-\hat{S}_{33}\Gamma_{L3}} \end{bmatrix} \quad (2.125)$$

Equation 2.125 thus describes the scattering matrix of the reduced two-port network. Note that the scattering matrix of the reduced two-port does not satisfy unitary conditions if the load connected to port 3 is lossy.

In the following, we give only the *final* equations describing how the normal scattering parameters of two- and four-port networks are modified when some of their ports are not match-terminated.

2.7.3 Reduction of a Two-Port Network Into a One-Port Network

Let the scattering matrix of a two-port network as shown in Figure 2.15 be given by

$$[S] = \begin{bmatrix} \hat{S}_{11} & \hat{S}_{12} \\ \hat{S}_{21} & \hat{S}_{22} \end{bmatrix} \quad (2.126)$$

If port 2 of the network is terminated in a load of reflection coefficient Γ_L, then the reflection coefficient at port 1 of the network is modified as

$$\hat{S}'_{11} = \hat{S}_{11} + \frac{\hat{S}_{12}\Gamma_L\hat{S}_{21}}{1 - \hat{S}_{22}\Gamma_L} \quad (2.127)$$

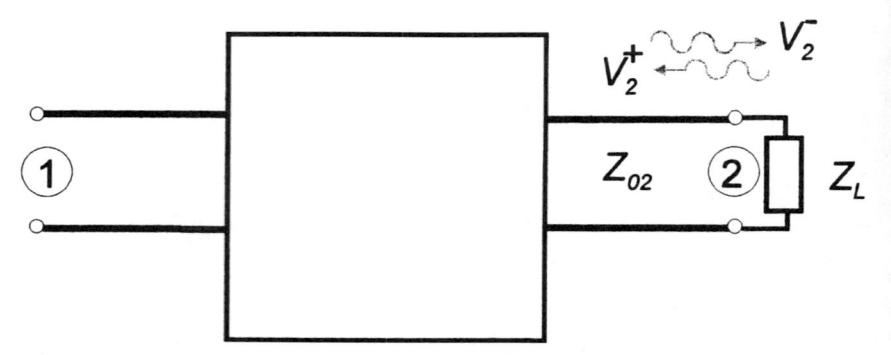

Figure 2.15 A two-port network with its port two terminated in arbitrary load.

where $\hat{S}'_{11}$ denotes the reflection coefficient at port 1 of the reduced one-port network and

$$\Gamma_L = \frac{Z_L - Z_{02}}{Z_L + Z_{02}} \qquad (2.128)$$

denotes the reflection coefficient at port 2.

2.7.4 Reduction of a Four-Port Network Into a Two-Port Network

Let the scattering elements of the four-port network as shown in Figure 2.16 be expressed as

$$[S] = \begin{bmatrix} \hat{S}_{11} & \hat{S}_{12} & \hat{S}_{13} & \hat{S}_{14} \\ \hat{S}_{21} & \hat{S}_{22} & \hat{S}_{23} & \hat{S}_{24} \\ \hat{S}_{31} & \hat{S}_{32} & \hat{S}_{33} & \hat{S}_{34} \\ \hat{S}_{41} & \hat{S}_{42} & \hat{S}_{43} & \hat{S}_{44} \end{bmatrix} \qquad (2.129)$$

Assume that ports 3 and 4 of the network are terminated in arbitrary loads. The reflection coefficients are given by

$$\Gamma_{L3} = \frac{\hat{V}_3^+}{\hat{V}_3^-} = \frac{Z_{L3} - Z_{03}}{Z_{L3} + Z_{03}} \qquad (2.130)$$

Microwave Network Theory 63

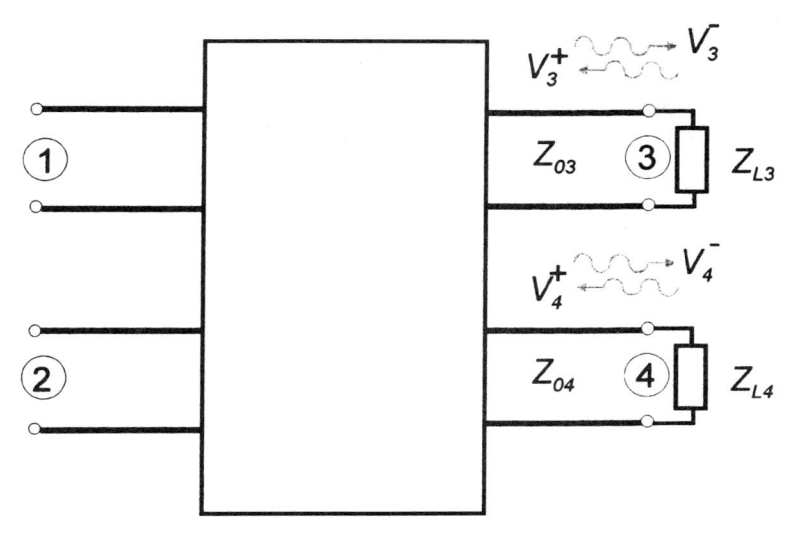

Figure 2.16 A four-port network with its two ports terminated in arbitrary load.

and

$$\Gamma_{L4} = \frac{\hat{V}_4^+}{\hat{V}_4^-} = \frac{Z_{L4} - Z_{04}}{Z_{L4} + Z_{04}} \qquad (2.131)$$

Using (2.115) and (2.116), the scattering parameters of the resulting two-port network are found as [6]

$$\hat{S}'_{ij} = \hat{S}_{ij} + \frac{\Gamma_{L3} \begin{vmatrix} \hat{S}_{i3}\hat{S}_{3j} & -\hat{S}_{i3}\hat{S}_{34}\Gamma_{L4} \\ \hat{S}_{4j} & 1 - \hat{S}_{44}\Gamma_{L4} \end{vmatrix} + \Gamma_{L4} \begin{vmatrix} \hat{S}_{3j} & 1 - \hat{S}_{33}\Gamma_{L3} \\ -\hat{S}_{i4}\hat{S}_{4j} & \hat{S}_{i4}\hat{S}_{43}\Gamma_{L3} \end{vmatrix}}{\begin{vmatrix} 1 - \hat{S}_{33}\Gamma_{L3} & -\hat{S}_{34}\Gamma_{L4} \\ -\hat{S}_{43}\Gamma_{L3} & 1 - \hat{S}_{44}\Gamma_{L4} \end{vmatrix}} \qquad (2.132)$$

where $i = 1, 2$ and $j = 1, 2$ and $\hat{S}'_{ij}$ denotes the element of the ith row and j^{th} column of the scattering matrix of the remaining network.

References

[1] Collin, R. E., *Foundations for Microwave Engineering*, McGraw Hill, New York 1966.

[2] Kurokawa, K., "Power-Waves and the Scattering Matrix," *IEEE Trans*, Vol. MTT-13, pp. 194–202, Mar. 1965.

[3] Beatty, R. W., and D. M., Kerns, "Relationships Between Different Kinds of Network Parameters, Not Assuming Reciprocity or Equality of the Waveguide or Transmission Line Characteristic Impedance," *Proc. IEEE*, Vol. 52, Jan. 1964, p. 84, Corrections: Apr. 1964, p. 420.

[4] Sazanov, D. M., A. N., Gridin and B. A., Mishustin *Microwave Circuits*, Mir Publishers, Moscow, CIS, 1982.

[5] von Abele Thomas-Alfred, "Uber die streumatrix allgemein zusammengeschalteter mehrpole," ("The Scattering Matrix of a General Interconnection of Multipoles"), Arch. Elek. Ubertragung, Vol. 14, pt. 6, 1960, pp. 262–268.

[6] Otoshi, T. Y., "On the Scattering Parameters of a Reduced Multiport," *IEEE Trans.*, MTT-17, Sept. 1969, pp. 722–724.

3

Characteristics of Planar Transmission Lines

Transmission lines used at microwave frequencies can be broadly divided into two categories: those that can support a TEM (or quasi-TEM) mode of propagation and those that cannot. For TEM (or quasi-TEM) modes, the determination of important electrical characteristics (such as characteristic impedance and phase velocity) of single and coupled lines reduces to finding the capacitances associated with the structure. Further, the conductor loss of TEM (or quasi-TEM) mode transmission lines can be determined in terms of variation of the characteristic impedance with respect to the geometrical parameters. This chapter discusses the general characteristics of single and coupled planar TEM and quasi-TEM mode transmission lines. Further, design equations of some single and coupled popular planar integrated transmission lines are given. The transmission lines considered are a stripline, microstrip line, coplanar waveguide, and their variants. Of all the planar transmission lines, microstrip is still the most popular for realizing microwave integrated circuits. We discuss characteristics of the microstrip line therefore in greater detail.

3.1 General Characteristics of TEM and Quasi-TEM Modes

It is an important property of any two-conductor lossless transmission line placed in a homogeneous dielectric medium that it supports a pure TEM mode of propagation. Common examples of these lines are a twin-wire line, coaxial line, and shielded stripline as shown in Figures 3.1(a) to (c). If a two-conductor

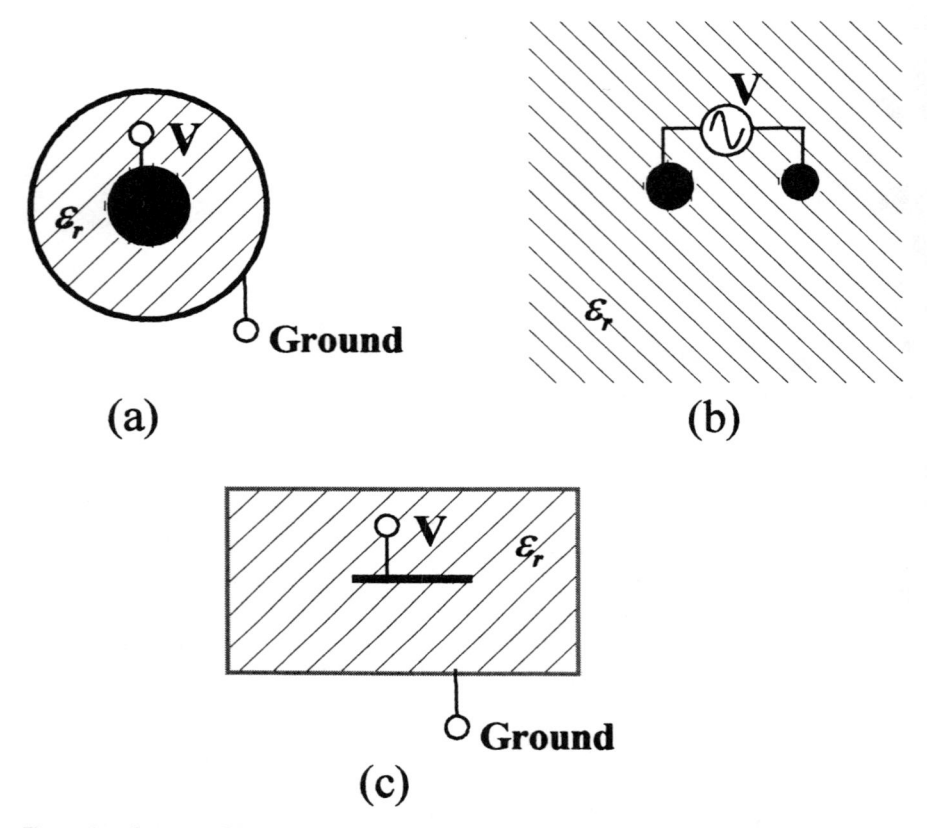

Figure 3.1 Common TEM mode transmission lines (a) coaxial line, (b) twin wire line, and (c) shielded stripline.

transmission line is enclosed in an inhomogeneous dielectric medium, the mode of propagation is *pure* TEM only in the limit of zero frequency. The most common example of such a transmission line is a microstrip line as shown in Figure 3.2(a). Some other examples of inhomogeneous transmission lines are a slotline and a coplanar waveguide (CPW) as shown in Figures 3.2(b) and (c), respectively. If the separation between the conductors of an inhomogeneous transmission line is very small compared to the wavelength, the mode of propagation on the line can be considered to be close to TEM. This mode is called a quasi-TEM mode.

The characteristic impedance and complex propagation constant of a TEM or a quasi-TEM mode transmission line can be described in terms of basic parameters of the line (i.e., its *per unit length* resistance R, inductance L, capacitance C, and conductance G). The equivalent circuit of a transmission line of length dz is shown in Figure 3.3. For a transmission line placed in an

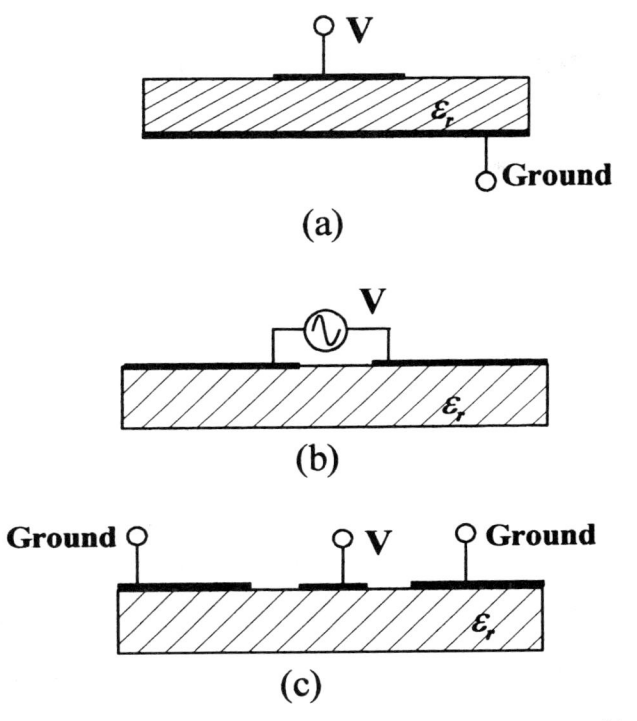

Figure 3.2 Common quasi-TEM mode transmission lines: (a) microstrip line, (b) slot line, and (c) coplanar waveguide.

inhomogeneous medium, the relations given are valid in the quasistatic limit, which means that the operating frequency is assumed to be low enough so that the distance between the conductors of the transmission line is very small compared to the wavelength ($\approx \lambda_g/20$ or smaller). For the present discussion, we assume that the conductors of the transmission line have a finite but very high conductivity. It is also assumed that the dielectric loss in the material surrounding the conductors of the transmission line is finite but small.

The parameters of interest for a transmission line are its characteristic impedance Z_0, phase constant β (or phase velocity v_p), and attenuation constant α. In terms of parameters R, G, L, and C expressed per unit length, the characteristic impedance of a transmission line is given by [1]

$$Z_0 = \sqrt{\frac{R + j\omega L}{G + j\omega C}} \qquad (3.1)$$

At microwave frequencies, low-loss conditions $\omega L \gg R$ and $\omega C \gg G$ are usually satisfied for transmission line conductors fabricated out of normal metals and

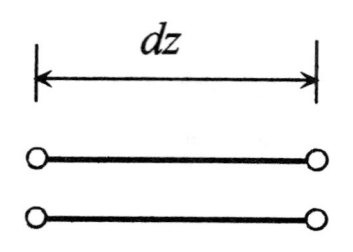

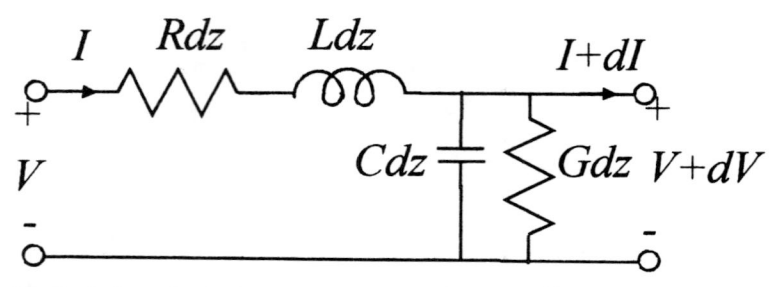

Figure 3.3 Equivalent circuit of a TEM or quasi-TEM transmission line of length dz.

enclosed in a low dielectric loss medium. Equation 3.1 then reduces to

$$Z_0 \approx \sqrt{\frac{L}{C}} \tag{3.2}$$

The incident and reflected voltage and current waves on a transmission line can be expressed as

$$V^{\pm}(z) = V_0^{\pm} e^{\mp \gamma z} \tag{3.3}$$

$$I^{\pm}(z) = I_0^{\pm} e^{\mp \gamma z} \tag{3.4}$$

where it is assumed that the axis of the transmission line lies along the z-direction. The complex propagation constant γ in (3.3) and (3.4) is given by

$$\gamma = \alpha + j\beta = \frac{1}{2}\left(\frac{R}{Z_0} + GZ_0\right) + j\omega\sqrt{LC} \tag{3.5}$$

where $\omega = 2\pi f$ denotes the angular frequency. From (3.5):

$$\beta = \frac{\omega}{v_p} = \omega\sqrt{LC} \quad \text{rad/unit length} \tag{3.6}$$

where β and v_p denote the phase constant and phase velocity, respectively, along the direction of propagation. The attenuation constant α is given by

$$\alpha = \frac{1}{2}\left(\frac{R}{Z_0} + GZ_0\right) \quad \text{Np/unit length} \qquad (3.7)$$

It is common to express the attenuation in decibels (dB) rather than in nepers (Np). The loss in dB is obtained by multiplying the loss in Np by 8.686. The attenuation of the transmission line can therefore also be expressed as

$$\alpha = 4.343\left(\frac{R}{Z_0} + GZ_0\right) \quad \text{dB/unit length} \qquad (3.8)$$

Relation Between Characteristic Impedance Z_0, Line Capacitance C, and Phase Velocity v_p

Eliminating L from (3.2) and (3.6) leads to the following very significant result:

$$Z_0 = \frac{1}{v_p C} \qquad (3.9)$$

Equation 3.9 shows that the characteristic impedance of a transmission line is related to the phase velocity along the transmission line and the capacitance (per unit length) between the conductors of the transmission line. It is also possible to express the phase velocity in terms of the ratio of the actual capacitance of the transmission line to the capacitance of the same transmission line obtained by assuming the dielectric constant of the medium in which it is placed to be unity. Therefore, the problem of determining the characteristic impedance and phase velocity of the structure reduces essentially to the problem of finding the capacitance of the structure.

Q-Factor

Equation 3.7 shows that the total line attenuation is due to two factors: nonzero series resistance R and nonzero shunt conductance G. The total attenuation can therefore be expressed as

$$\alpha = \alpha_c + \alpha_d \qquad (3.10)$$

where

$$\alpha_c = \frac{R}{2Z_0} \quad \text{Np/unit length} \qquad (3.11)$$

denotes the conductor loss, and

$$\alpha_d = \frac{GZ_0}{2} \quad \text{Np/unit length} \tag{3.12}$$

denotes the dielectric loss.

The attenuation of a transmission line can also be expressed in terms of the Q-factor. The Q-factor of a half-wavelength transmission line resonator is given by

$$Q = \frac{\beta}{2\alpha} = \frac{\beta}{2(\alpha_c + \alpha_d)} \tag{3.13}$$

where the attenuation α is expressed in Np/unit length. We can also define the Q-factors for conductor (Q_c) and dielectric loss (Q_d) separately as

$$Q_c = \frac{\beta}{2\alpha_c} \tag{3.14}$$

and

$$Q_d = \frac{\beta}{2\alpha_d} \tag{3.15}$$

Using (3.13) and (3.14), the overall Q-factor can be expressed as

$$\frac{1}{Q} = \frac{1}{Q_c} + \frac{1}{Q_d}$$

For dispersive lines, (3.13) and (3.14) are incorrect and require that the term β in (3.13) and (3.14) be replaced by ω/v_g, where v_g denotes the group velocity [1]. For example, (3.13) then becomes

$$Q = \frac{\omega}{2v_g\alpha} = \frac{\omega}{2v_g(\alpha_c + \alpha_d)} \tag{3.16}$$

Equation 3.16 is more general than (3.13) and is valid for non-TEM modes as well.

3.1.1 TEM Modes

We now specialize some of the above equations for the case when the transmission line is placed in a homogeneous dielectric medium. Some examples of these types of lines are shown in Figure 3.1(a) to (c). For these lines, the velocity of propagation v_p along the transmission line is independent of the type of the transmission line and frequency of operation and is given by

$$v_p = \frac{c}{\sqrt{\epsilon_r}} \qquad (3.17)$$

where c is the velocity of light in freespace and ϵ_r denotes the dielectric constant (relative permittivity) of the medium. The phase constant along the transmission line is therefore given by

$$\beta = \frac{\omega}{v_p} = \frac{\omega\sqrt{\epsilon_r}}{c} = k_0\sqrt{\epsilon_r} \quad \text{rad/unit length} \qquad (3.18)$$

where $k_0 = 2\pi/\lambda_0$ denotes the free-space propagation constant and λ_0 denotes the free-space wavelength.

Substituting the value of phase velocity from (3.18) in (3.9) leads to the following relation between the characteristic impedance and capacitance of the line:

$$Z_0 = \frac{\sqrt{\epsilon_r}}{cC} = \frac{\sqrt{\epsilon_r}}{c\epsilon_r C_0} = \frac{1}{c\sqrt{\epsilon_r}C_0} \qquad (3.19)$$

where C_0 denotes the capacitance between the conductors of the transmission line assuming that the transmission line is placed in a medium of a unity dielectric constant.

The determination of dielectric loss α_d is also straightforward in this case. It is given by

$$\alpha_d = \frac{\beta}{2}\tan\delta = \frac{k_0\sqrt{\epsilon_r}}{2}\tan\delta \quad \text{Np/unit length} \qquad (3.20)$$

or

$$\alpha_d = 4.343\beta\tan\delta = 4.343 k_0\sqrt{\epsilon_r}\tan\delta = 27.3\sqrt{\epsilon_r}\frac{\tan\delta}{\lambda_0} \quad \text{dB/unit length} \qquad (3.21)$$

where $\tan\delta$ denotes the loss tangent of the dielectric material. In general, the loss tangent $\tan\delta$ is also a function of frequency.

On the other hand, conductor loss factor α_c depends on the type of line, the conductivity of conductors of the transmission line, the frequency of operation, and geometrical parameters of the line and is discussed in detail in a later section.

3.1.2 Quasi-TEM Modes

Some examples of quasi-TEM mode transmission lines are shown in Figure 3.2 (a–c). For quasi-TEM modes, the effective dielectric constant ϵ_{re} is defined as follows:

$$\epsilon_{re} = \frac{c^2}{v_p^2} \qquad (3.22)$$

In qualitative terms, the effective dielectric constant ϵ_{re} takes into account the relative distribution of electric energy in the various regions of the inhomogeneous medium. The relation between phase constant β, effective dielectric constant ϵ_{re}, and phase velocity v_p is

$$\beta = \frac{\omega}{v_p} = \frac{\omega\sqrt{\epsilon_{re}}}{c} = \sqrt{\epsilon_{re}}k_0 \qquad (3.23)$$

ϵ_{re} is a function of frequency and strictly speaking should be evaluated using (3.22) where the phase velocity v_p is computed using some rigorous method based on Maxwell's equations. However, in the quasistatic limit, ϵ_{re} can be assumed to be

$$\epsilon_{re} = \frac{C}{C_0} \qquad (3.24)$$

where C denotes the capacitance between conductors of the transmission line in the presence of inhomogeneous dielectric medium and C_0 denotes the capacitance between the same conductors in a homogeneous dielectric medium of unity dielectric constant. Using (3.9), (3.23), and (3.24) the characteristic impedance of a quasi-TEM mode transmission line can be expressed as

$$Z_0 = \frac{\sqrt{\epsilon_{re}}}{cC} = \frac{1}{c\sqrt{CC_0}} = \frac{1}{c\sqrt{\epsilon_{re}}C_0} = \frac{Z_{0a}}{\sqrt{\epsilon_{re}}} \qquad (3.25)$$

where $Z_{0a} = 1/cC_0$ denotes the characteristic impedance of the same transmission line placed in a medium of unity dielectric constant.

The dielectric loss of a quasi-TEM mode transmission line is given by

$$\alpha_d = 27.3 \frac{\epsilon_r}{\sqrt{\epsilon_{re}}} \frac{(\epsilon_{re} - 1)}{(\epsilon_r - 1)} \frac{\tan\delta}{\lambda_0} \quad \text{dB/unit length} \qquad (3.26)$$

where $\tan\delta$ denotes the loss tangent of the substrate material and other symbols have their usual meaning.

The conductance G of a transmission line can be expressed in terms of loss tangent $\tan\delta$ as follows:

$$G = \frac{2\pi}{Z_0} \frac{\epsilon_r}{\sqrt{\epsilon_{re}}} \frac{(\epsilon_{re} - 1)}{(\epsilon_r - 1)} \frac{\tan\delta}{\lambda_0} \qquad (3.27)$$

Further, it is customary to define the effective filling fraction q of a quasi-TEM mode transmission line as follows:

$$q = \frac{\epsilon_{re} - 1}{\epsilon_r - 1} \tag{3.28}$$

3.1.3 Skin Depth and Surface Impedance of Imperfect Conductors

At high frequencies, the current flowing in a conductor tends to get confined near the outer surface of the conductor. The skin depth of a conductor is defined as the distance in the conductor (along the direction of the normal to the surface) in which the current density drops to 37% of its value at the surface (the current decays to a negligible value in a distance of about 4 to 5 skin depths) and is given by

$$\delta_s = \sqrt{\frac{2}{\omega\mu\sigma}} \tag{3.29}$$

where $\mu = \mu_r\mu_0$, $\mu_0 = 4\pi \times 10^{-7}$ H/m denotes the permeability of free space, and σ denotes the conductivity (Siemens/meter) of the conductor. Further, μ_r denotes the relative permeability of the material. Its value is almost equal to unity except for magnetic materials. (3.29) shows that the skin depth of a perfect conductor ($\sigma = \infty$) is zero. The conductivity of normal metals (which are used as conductors) is very high, although finite. For normal metals, the skin depth is therefore very small at microwave frequencies (e.g., the conductivity of copper is 5.8×10^7 S/m and the skin depth at a frequency of 10 GHz is 0.66 m). The tangential electric field at the surface of a conductor is not zero due to finite conductivity of the conductors of a transmission line.[1] The surface impedance (ohms/square) of a conductor (defined as the ratio of tangential electric and magnetic fields at the surface) is given by

$$Z_s = R_s + j\omega L_s = \frac{1+j}{\sigma\delta_s} \tag{3.30}$$

or

$$R_s = \omega L_s = \frac{1}{\sigma\delta_s} \tag{3.31}$$

(e.g., the surface impedance of copper at a frequency of 10 GHz is $0.026 + j0.026$ ohms/square).

---

1. The tangential electric field at the surface of a perfect conductor is zero.

3.1.4 Conductor Loss of TEM and Quasi-TEM Modes

An ingenious way to determine the conductor loss of TEM or quasi-TEM mode transmission lines was given by Wheeler more than half a century ago [2]. This method of determining the conductor loss is also known as the *incremental inductance rule*. The rule is valid only if the thickness of conductors of the transmission lines and the radius of curvature of conductor surfaces are at least five to six times the skin depth. These conditions can usually be satisfied at microwave frequencies except near very sharp edges. According to this rule, the conductor Q-factor of a transmission line is given by [3]

$$Q_c = \frac{Z'_{0a}}{(Z'_{0a} - Z_{0a})} \tag{3.32}$$

where Z_{0a} has been defined before and Z'_{0a} denotes the impedance of the same transmission line placed in a medium of unity dielectric constant, but assuming that the thickness of all conductors is reduced by $\delta_s/2$ from each surface where the fields are present. This is shown in Figure 3.4, where solid lines show the surfaces of the actual microstrip transmission line of a finite strip thickness and dashed lines show the surfaces of the fictitious microstrip transmission line obtained by removing a depth of $\delta_s/2$ from each surface of the conductor of the original transmission line. Note that the conductor Q-factor (but not the attenuation per unit length) of a TEM or quasi-TEM mode transmission line is independent of the dielectric constant of the medium in which it is placed. Using (3.14), (3.23), and (3.32), the attenuation factor α_c can be expressed as

$$\alpha_c = \sqrt{\epsilon_{re}} k_0 \frac{(Z'_{0a} - Z_{0a})}{2 Z'_{0a}} = \frac{\pi \sqrt{\epsilon_{re}} f (Z'_{0a} - Z_{0a})}{c} \frac{}{Z'_{0a}} \tag{3.33}$$

where f is the frequency of operation.

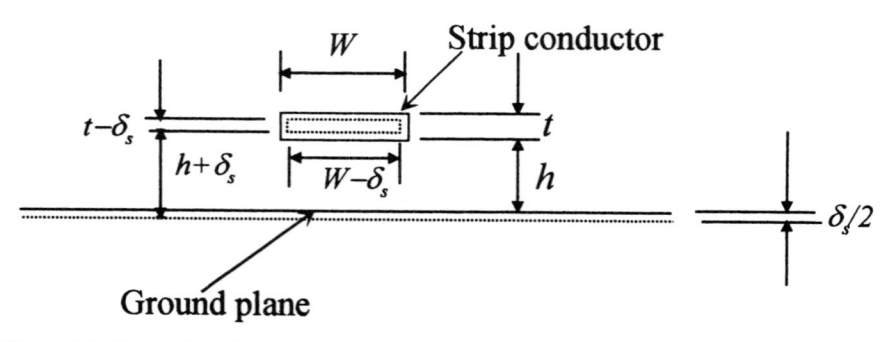

Figure 3.4 Illustration of incremental inductance rule showing the original and perturbed geometry of a microstrip line.

3.2 Representation of Capacitances of Coupled Lines

The coupling between lines can be expressed in terms of self- and mutual capacitances. It is therefore useful to discuss the representation of capacitances of coupled transmission lines.

Figure 3.5 shows the cross section of two coupled transmission lines having a common ground conductor with the capacitances associated with the coupled structure as shown. If Q_1 and Q_2 denote the charges and V_1 and V_2 denote the voltages of conductors 1 and 2, respectively, the charges Q_1 and Q_2 can be expressed in terms of voltages and capacitances as

$$Q_1 = C_a V_1 + C_m(V_1 - V_2) = (C_a + C_m) V_1 - C_m V_2 \quad (3.34)$$

$$Q_2 = C_m(V_2 - V_1) + C_b V_2 = -C_m V_1 + (C_b + C_m) V_2 \quad (3.35)$$

The capacitance matrix of two coupled transmission lines is represented as, [4]

$$[\mathbf{C}] = \begin{bmatrix} C_{11} & C_{12} \\ C_{21} & C_{22} \end{bmatrix} \quad (3.36)$$

where C_{11} and C_{22} are defined as self-capacitances of lines 1 and 2, respectively, in the presence of each other. The capacitance matrix denotes the relation between charges and voltages on the two transmission lines as follows:

$$Q_1 = C_{11} V_1 + C_{12} V_2 \quad (3.37)$$

$$Q_2 = C_{21} V_1 + C_{22} V_2 \quad (3.38)$$

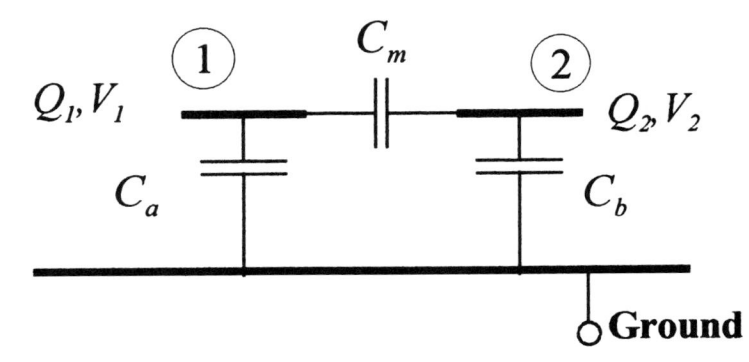

Figure 3.5 Representation of capacitances of coupled lines.

and hence the capacitance matrix of coupled lines can be expressed as

$$[\mathbf{C}] = \begin{bmatrix} C_{11} & C_{12} \\ C_{21} & C_{22} \end{bmatrix} = \begin{bmatrix} C_a + C_m & -C_m \\ -C_m & C_b + C_m \end{bmatrix} \qquad (3.39)$$

The inductance matrix of coupled line is given by

$$[\mathbf{L}] = \mu_0 \epsilon_0 [\mathbf{C}_0]^{-1} \qquad (3.40)$$

where $[\mathbf{C}_0]$ denotes the capacitance matrix of the transmission lines obtained by assuming that these are placed in a medium of unity dielectric constant.

3.2.1 Even- and Odd-Mode Capacitances of Symmetrical Coupled Lines

When the coupled lines are identical (also called *symmetrical coupled lines*), their capacitance matrix can be expressed in terms of even- and odd-mode capacitances.

Even-Mode Excitation

A cross section of uniformly coupled symmetrical lines is shown in Figure 3.6(a). In this case, the capacitance C_m shown in Figure 3.5 has been broken into two capacitances of values $2C_m$ each in series. With even-mode excitation, equal and in-phase voltages ($V_1 = V_2 = V_e$) are applied to both lines. Because the geometry under consideration is symmetrical, it is clear that if equal voltages of the same polarity are applied to both the lines, the charges on the two lines would also be the same (i.e., $Q_1 = Q_2 = Q_e$). Denoting the ratio Q_e/V_e by C_e, (3.34) and (3.35) then reduced to

$$C_a = C_b = \frac{Q_e}{V_e} = C_e \qquad (3.41)$$

Odd-Mode Excitation

In the odd-mode excitation, equal but out-of-phase voltages ($V_1 = -V_2 = V_o$) are applied to the two lines as shown in Figure 3.6(b). It follows from the symmetry of the structure that if equal voltages but of opposite polarity are applied to the symmetrical lines, equal charges but of opposite polarity will be induced on the two lines (i.e., $Q_1 = -Q_2 = Q_o$). With the ratio Q_o/V_o denoted by C_o, (3.34) and (3.35) reduced to

$$Q_o = (C_a + 2C_m) V_o \qquad (3.42)$$

Characteristics of Planar Transmission Lines

[Figure: Even-mode excitation circuit with PP'-Magnetic Wall (Open circuit)]

(a)

[Figure: Odd-mode excitation circuit with PP'-Electric Wall (Short circuit)]

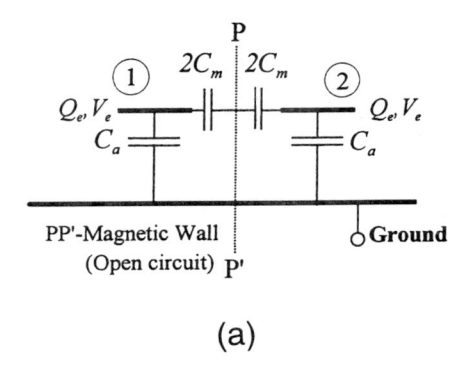

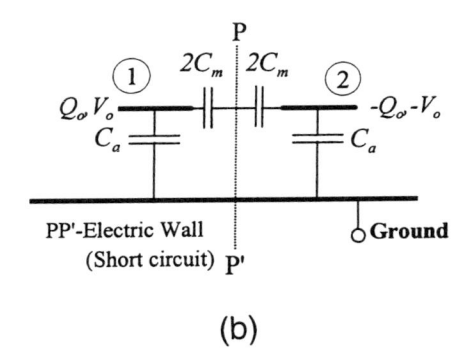

(b)

Figure 3.6 (a) Even- and (b) odd-mode excitation of symmetrical coupled lines.

or

$$C_a + 2C_m = \frac{Q_o}{V_o} = C_o \quad (3.43)$$

Substituting the value of C_a from (3.41) in (3.43), we obtain,

$$C_m = \frac{C_o - C_e}{2} \quad (3.44)$$

Therefore, once the even- and odd-mode capacitance parameters of coupled symmetrical lines are known, C_a, C_b, and C_m can be determined from (3.41) and (3.44). The capacitance matrix of coupled lines can then be determined using (3.39).

In the even-mode of excitation, the symmetry plane PP' as shown in Figure 3.6(a) acts as a magnetic wall (open circuit). The determination of the even-mode capacitance reduces to finding the capacitance of either line with the plane of symmetry PP' replaced by a magnetic wall such as shown in Figure 3.7(a). This results in a great simplification of the problem. Similarly, in the odd-mode of excitation, the symmetry plane behaves as an electric wall (short circuit). The determination of the odd-mode capacitance reduces to finding the capacitance of either line by replacing the plane of symmetry by an electric wall as shown in Figure 3.7(b).

The relationships between the even- and odd-mode capacitances and impedances are given by

$$Z_{0e} = \frac{1}{v_{pe} C_e} = \frac{\omega}{\beta_e C_e} \tag{3.45}$$

$C_e = C_a$ $2C_m$ PP'-Open circuit

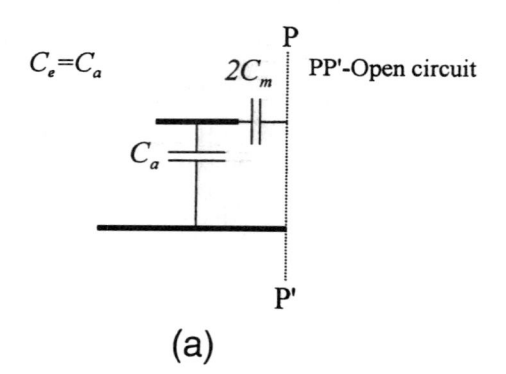

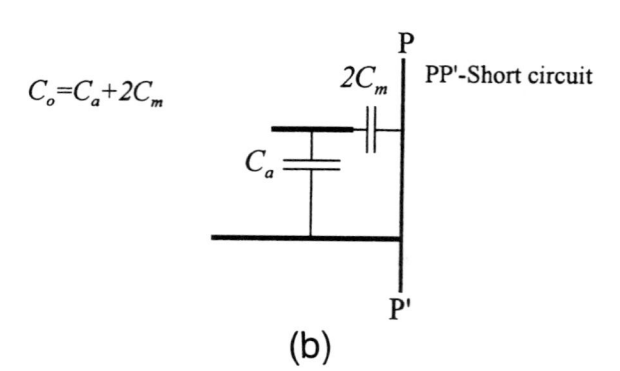

C_a

(a)

$C_o = C_a + 2C_m$ $2C_m$ PP'-Short circuit

C_a

(b)

Figure 3.7 Representation of capacitances of (a) even- and (b) odd-modes of symmetrical coupled lines.

and

$$Z_{0o} = \frac{1}{v_{po} C_o} = \frac{\omega}{\beta_o C_o} \qquad (3.46)$$

where Z_{0e}, v_{pe}, and β_e denote the characteristic impedance, phase velocity, and phase constant, respectively, of the even mode of the coupled lines; and Z_{0o}, v_{po}, and β_o denote the same quantities for the odd mode. If the lines are placed in a homogeneous medium of dielectric constant ϵ_r, the even- and odd-mode phase velocities are equal and are given by

$$v_{pe} = v_{po} = \frac{c}{\sqrt{\epsilon_r}} \qquad (3.47)$$

However, if the lines are placed in an inhomogeneous dielectric media (such as coupled microstrip lines), the even- and odd-mode phase velocities are, in general, different and are given by

$$v_{pe} = \frac{c}{\sqrt{\epsilon_{ree}}} \qquad (3.48)$$

and

$$v_{po} = \frac{c}{\sqrt{\epsilon_{reo}}} \qquad (3.49)$$

where ϵ_{ree} and ϵ_{reo} are defined as the even- and odd-mode effective dielectric constants, respectively. These can be determined using

$$\epsilon_{ree} = \frac{C_e}{C_{0e}} \qquad (3.50)$$

and

$$\epsilon_{reo} = \frac{C_o}{C_{0o}} \qquad (3.51)$$

where C_{0e} and C_{0o} denote, respectively, the even- and odd-mode capacitance of either line obtained by replacing the relative permittivity of the surrounding dielectric material by unity. C_e and C_o denote the corresponding capacitances in the presence of the inhomogeneous dielectric medium. Using (3.48) to (3.51), (3.45) and (3.46) reduce to

$$Z_{0e} = \frac{1}{c\sqrt{C_e C_{0e}}} \qquad (3.52)$$

and

$$Z_{0o} = \frac{1}{c\sqrt{C_o C_{0o}}} \qquad (3.53)$$

3.2.2 Parallel-Plate and Fringing Capacitances of Single and Coupled Planar Transmission Lines

So far we have discussed the capacitances of single and coupled transmission lines in general. In many cases, it is possible to visualize the various components of the total capacitance of the structure. This helps in obtaining a physical understanding of the problem and its analysis. For example, for planar transmission lines, the total capacitance can be broken into its various components such as parallel-plate and fringing capacitances. To explain the various components, we consider for simplicity single and coupled microstrip lines.

Single Line

The electric field distribution of a single microstrip line is shown in Figure 3.8(a). Because of the finite width of the microstrip line, fields not only exist directly below the strip conductor, but extend to the surrounding regions as well. The latter are known as *fringing fields*. The capacitance that results from the electric fields in the region directly below the strip is known as the *parallel-plate capacitance*, while that resulting from the fringing fields is known as the *fringing capacitance*. The total capacitance associated with a single microstrip line can be represented as shown in Figure 3.8(b) and is given by

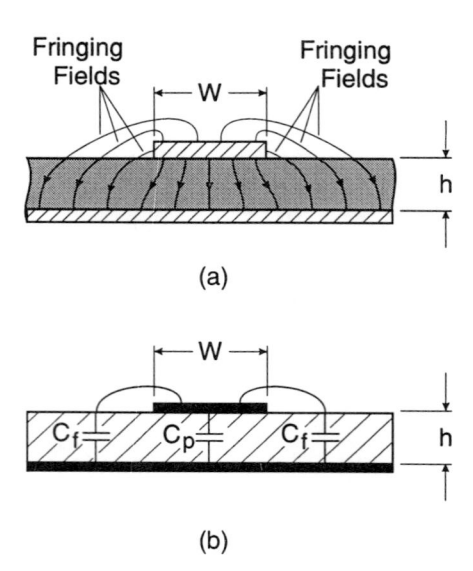

$$C = C_p + 2C_f \tag{3.54}$$

Figure 3.8 (a) Electric field distribution of single microstrip line, and (b) equivalent capacitance representation.

where

$$C_p = \frac{\epsilon_0 \epsilon_r W}{h} \tag{3.55}$$

denotes the parallel-plate capacitance and C_f denotes the fringing capacitance from either edge of the microstrip line. Once the value of C_f is known, the total capacitance of the line can be determined using (3.54) and (3.55). Conversely, if the characteristic impedance and effective dielectric constant of a microstrip line are known, the capacitance C can be found using (3.25), and using (3.54) and (3.55), the fringing capacitance C_f can be determined.

Symmetrical Coupled Lines

A cross-section of symmetrical coupled microstrip lines is shown in Figure 3.9(a) with the electric field distribution in Figure 3.9(b) for one-half of the structure for the case of even-mode excitation. In this case, the normal component of the electric field at the plane of symmetry PP' is zero, because the plane of symmetry behaves like a magnetic wall. The even-mode capacitance of either of the coupled lines, which can be represented as shown in Figure 3.9(c), is given by,

$$C_e = C_p + C_f + C_{fe} \tag{3.56}$$

where C_p is given by (3.55). If the two lines are not of very narrow width, it can be assumed that the value of C_f is the same as that of a single microstrip line having the same width as that of either of the coupled lines.

The electric field distribution of one-half of the coupled structure is shown in Figure 3.9(d) for the case of odd-mode excitation. The tangential electric field at the plane of symmetry PP' is zero, because in this case, the plane of symmetry behaves like an electric wall. The odd-mode capacitance of either of the coupled lines is given by

$$C_o = C_p + C_f + C_{fo} \tag{3.57}$$

where C_{fo} denotes the fringing capacitance from the inner edges of the coupled lines. When the spacing between the lines is small ($S/2$ is small compared to the height of the substrate h), nearly all the fringing fields that start from the inner edge of one of the lines terminate on the mid plane PP'. In that case, the capacitance of either of the coupled lines, which can be represented as shown in Figure 3.9(e), is given by,

$$C_o = C_p + C_f + C_{fo} = C_p + C_f + C_{ga} + C_{gd} \tag{3.58}$$

where the fringing capacitance C_{fo} is assumed to consist of two capacitances

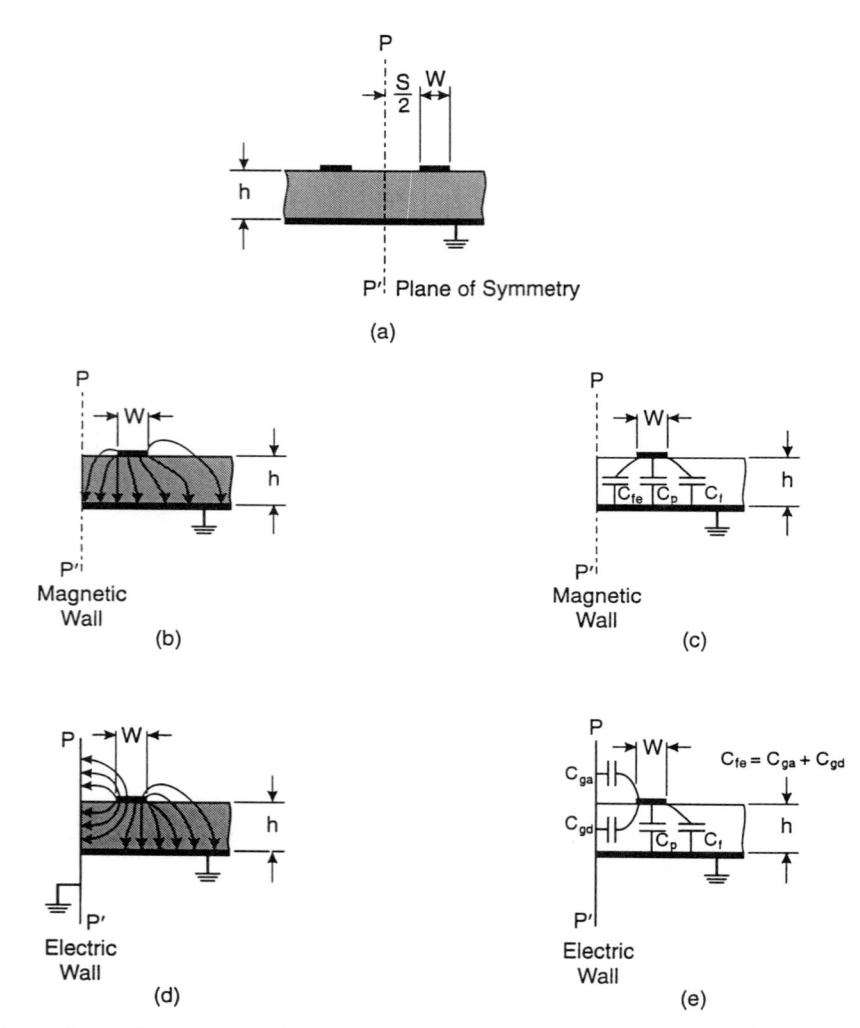

Figure 3.9 (a) Cross section of symmetrical coupled lines, (b)–(c) electric field distribution and capacitance representation of one-half of the structure for even-mode excitation, (d)–(e) electric field distribution and capacitance representation of one-half of the structure for odd-mode excitation.

C_{ga} and C_{gd} in parallel; that is,

$$C_{fo} = C_{ga} + C_{gd} \tag{3.59}$$

The characteristic impedances and phase velocities of the even and odd modes can be found by determining the parallel plate and fringing capacitances associated with the structure. Conversely, if the characteristic impedances and phase velocities of the even and odd modes are known, the corresponding fringing capacitances can be determined.

The breakup of the total capacitance into various components involves certain approximations. For example, (3.55) can only be termed as approximate because the fields under the strip are not exactly vertical, especially near the edges. The breakup of the total capacitance into various components, however, helps in a simple, although approximate design of various coupled-line components such as a Lange coupler.

3.3 Characteristics of Single and Coupled Striplines

A commonly used stripline is shown in Figure 3.10. The strip conductor is sandwiched between two flat dielectric substrates having the same dielectric constant. The outer surfaces of the dielectric substrates are metallized and serve as ground conductors. The signal is applied between the strip conductor and ground. In a commonly used fabrication technique, the strip conductor is etched on one of the dielectric substrates by the process of photolithography. The thickness of both the substrates is generally the same, although it is not essential. A lossless stripline can support a pure TEM mode of propagation at all frequencies. Two striplines can be coupled by placing strip conductors side by side as shown in Figure 3.11 in the edge-coupled configuration. In this configuration, both the strip conductors lie in the same plane, and although it is very convenient for realizing coupled line circuits, it has the disadvantage that tight coupling between the lines cannot be achieved. For tight coupling (about 8 dB or tighter), the width of the strip conductors and the spacing between them becomes quite small, making fabrication difficult. Further, the small dimensions lead to large current densities on the strip conductor leading to higher conductor loss. The

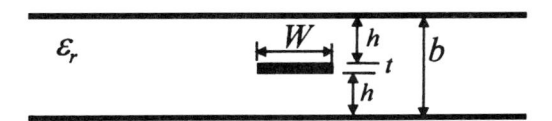

Figure 3.10 Cross section of a stripline.

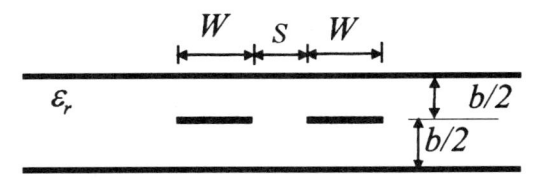

Figure 3.11 Cross section of edge-coupled striplines.

edge-coupled configuration is thus suitable for the design of loose couplers having coupling of 8 to 30 dB. Tighter coupling (e.g., 3 dB) can be obtained using broadside-coupled striplines, as discussed in Section 3.7.

3.3.1 Single Stripline

Based upon the Schwartz-Christoffel transformation, the exact expression for the characteristic impedance of a *lossless* stripline of zero thickness is given by [5, 6]

$$Z_0\sqrt{\epsilon_r} = 30\pi \frac{K(k)}{K(k')} \tag{3.60}$$

where

$$k = \text{sech}\left(\frac{\pi}{2}\frac{W}{b}\right); \quad k' = \sqrt{1-k^2}$$

In the above expression, K denotes the complete elliptic integral of the first kind and *sech* denotes the *hyperbolic secant*. An approximate expression for $K(k)/K(k')$ ($k' = \sqrt{1-k^2}$), which is accurate to within 8 ppm, is given by [7]

$$\frac{K(k)}{K(k')} = \frac{K(k)}{K'(k)} = \frac{1}{\pi} \ln\left(2\frac{1+\sqrt{k}}{1-\sqrt{k}}\right), \quad 0.5 \leq k^2 \leq 1$$

$$= \frac{\pi}{\ln\left(2\frac{1+\sqrt{k'}}{1-\sqrt{k'}}\right)}, \quad 0 \leq k^2 \leq 0.5 \tag{3.61}$$

Effect of Finite Strip Thickness on Characteristic Impedance

The capacitance between the strip conductor and ground is increased if the thickness of the strip is finite. This happens because of the additional capacitance resulting from the strip conductor edges of finite thickness. The characteristic impedance of a stripline of finite thickness is given by Wheeler [8] as,

$$Z_0\sqrt{\epsilon_r} = 30\ln\left(1+\frac{1}{2}(16h/\pi\ W')\left[(16h/\pi\ W')+\sqrt{(16h/\pi\ W')^2+6.27}\right]\right) \tag{3.62}$$

where W' denotes the effective width of the stripline. When the thickness of the strip conductor is zero, the effective width is the same as the physical width. For finite strip thickness, the effective width W' is given by

$$W' = W + \Delta W \tag{3.63}$$

where

$$\frac{\Delta W}{t} = \frac{1}{\pi} \ln \frac{2.718}{\sqrt{\left[\frac{1}{4h/t+1}\right]^2 + \left[\frac{1/4\pi}{W/t+1.1}\right]^m}}$$

$$m = \frac{6}{3+t/h}$$

The error in Wheeler's equation is expected to be less than 1%.

Closed-form expressions for the synthesis of a stripline have also been given by Wheeler [8]. These are as follows:

$$W = W' - \Delta W' \tag{3.64}$$

where

$$W'/h = \frac{16}{\pi} \frac{\sqrt{(e^{4\pi r} - 1) + 1.568}}{(e^{4\pi r} - 1)}$$

$$r = \sqrt{\epsilon_r} Z_0 / 376.7$$

$$\frac{\Delta W'}{t} = \frac{1}{\pi} \ln \frac{2.718}{\sqrt{\left[\frac{1}{4h/t+1}\right]^2 + \left[\frac{1/4\pi}{W'/t-0.26}\right]^m}}$$

$$m = \frac{6}{3+t/h}$$

The characteristic impedance of a stripline as a function of strip width for various values of t is shown in Figure 3.12 together with the characteristic impedance of striplines with strip conductor of square and circular cross section (Figure 3.13).

The propagation constant and dielectric loss of a stripline are given by (3.18) and (3.21), respectively.

The conductor loss depends on geometrical parameters of the line and can be computed in a rather simple manner using Wheeler's incremental inductance rule as discussed earlier [2]. For a stripline, the application of the rule leads to the following expression for the conductor loss [9]:

$$\alpha_c = \frac{R_s \sqrt{\epsilon_r}}{376.7 Z_0} \left[\frac{\partial Z_0}{\partial b} - \frac{\partial Z_0}{\partial W} - \frac{\partial Z_0}{\partial t}\right] \text{ Np/m} \tag{3.65}$$

where R_s is the surface resistance (ohms/square) given by (3.31) and ∂ denotes the partial derivative.

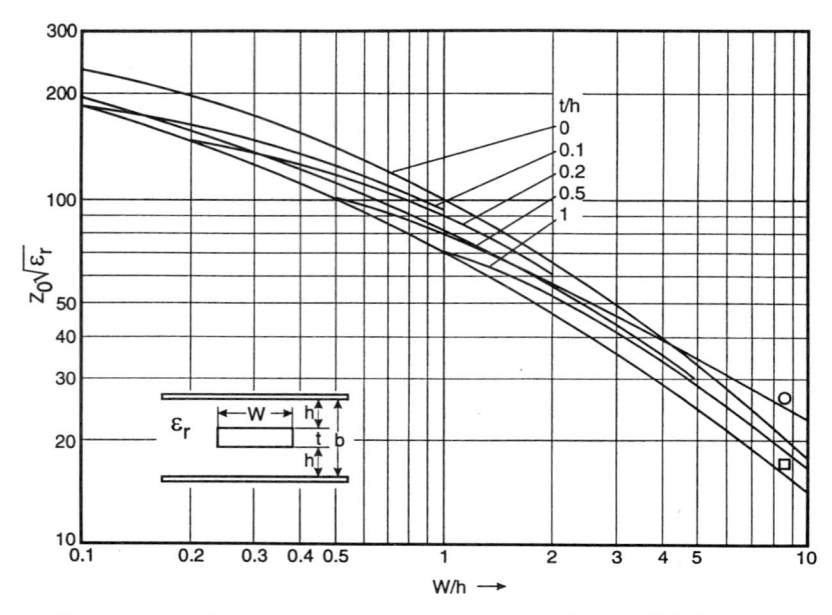

Figure 3.12 Characteristic impedance of finite thickness stripline vs W/h for various values of t/h. Curves marked □ and ○ are valid for stripline having strip-conductor of square or circular cross section respectively as shown in Figure 3.13. (From [8], ©1978 IEEE. Reprinted with permission.)

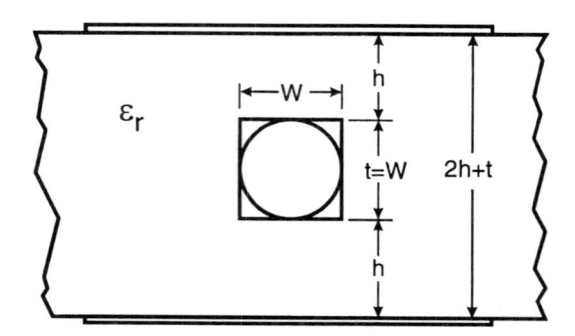

Figure 3.13 Stripline with strip-conductor of square or circular cross section.

The normalized conductor loss ($h/Q_c\delta_s$) of a stripline is shown in Figure 3.14 where Q_c and δ_s denote the conductor Q-factor and skin depth, respectively. The skin depth δ_s is given by (3.29). For given values of W and h, the conductor Q-factor Q_c can be determined using Figure 3.14. The conductor loss in Np/m can then be determined using (3.14). It may be noted that the normalized conductor loss ($h/Q_c\delta_s$) is independent of the dielectric constant of the substrate.

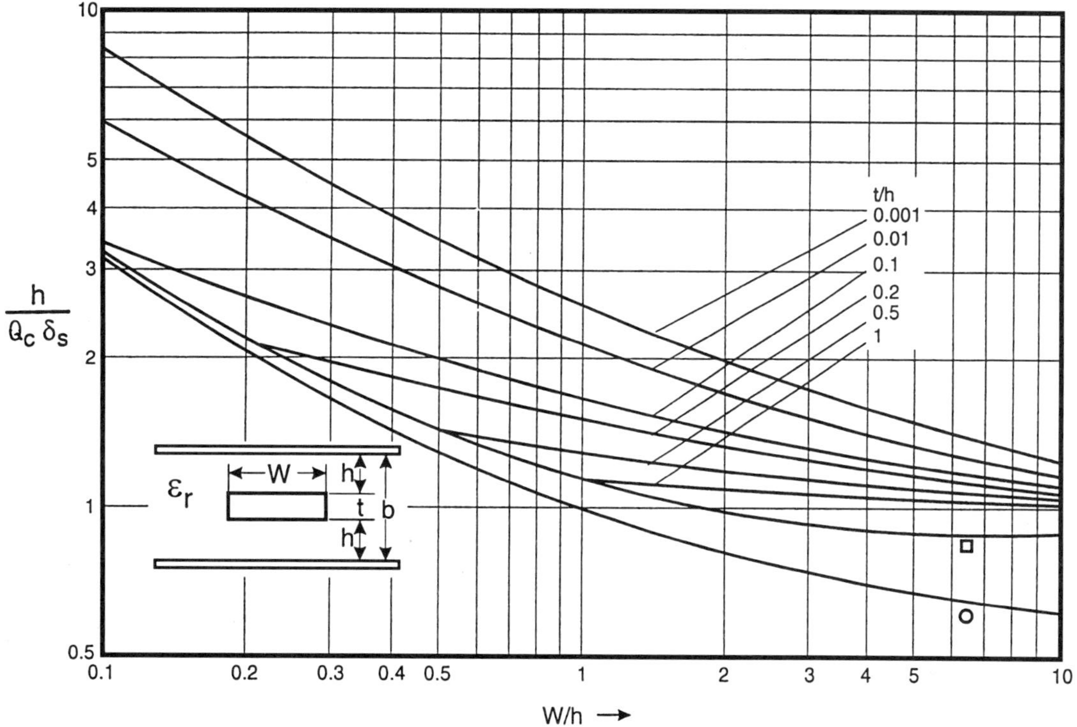

Figure 3.14 Normalized conductor Q-factor ($h/Q_c \delta_s$) of a stripline vs W/h for various values of t/h (valid for arbitrary value of ϵ_r). Curves marked □ and ○ are valid for stripline having strip-conductor of square or circular cross section respectively shown in Figure 3.13. (From [8], © 1978 IEEE. Reprinted with permission.)

For a wide stripline, the cutoff frequency of the first higher order mode is given by [10]

$$f_c = \frac{15}{b\sqrt{\epsilon_r}} \frac{1}{(W/b + \pi/4)} \text{ GHz} \qquad (3.66)$$

where W and b are in centimeters.

3.3.2 Edge-Coupled Striplines

The even- and odd-mode characteristic impedances of edge-coupled striplines of zero thickness (Figure 3.11) are given by the following expressions, which are exact [11]:

$$Z_{0e}\sqrt{\epsilon_r} = 30\pi \frac{K(k'_e)}{K(k_e)} \qquad (3.67)$$

where

$$k_e = \tanh\left(\frac{\pi}{2}\frac{W}{b}\right) \tanh\left[\frac{\pi}{2}\frac{(W+S)}{b}\right]; \quad k'_e = \sqrt{1 - k_e^2}$$

and

$$Z_{0o}\sqrt{\epsilon_r} = 30\pi \frac{K(k'_o)}{K(k_o)} \qquad (3.68)$$

where

$$k_o = \tanh\left(\frac{\pi}{2}\frac{W}{b}\right) \coth\left[\frac{\pi}{2}\frac{(W+S)}{b}\right]; \quad k'_o = \sqrt{1 - k_o^2}$$

The functions $K(k_e)/K(k'_e)$ and $K(k_o)/K(k'_o)$ can be evaluated using (3.61). The impedances (3.67) and (3.68) have been plotted in Figure 3.15. It is seen that when $S/b \to \infty$, both Z_{0e} and Z_{0o} approach the same value equal to the characteristic impedance of a single stripline of width W.

Synthesis equations for the design of zero-thickness coupled striplines are also given by Cohn [11]. These are as follows:

$$\frac{W}{b} = \frac{2}{\pi} \tanh^{-1} \sqrt{k_e k_o} \qquad (3.69)$$

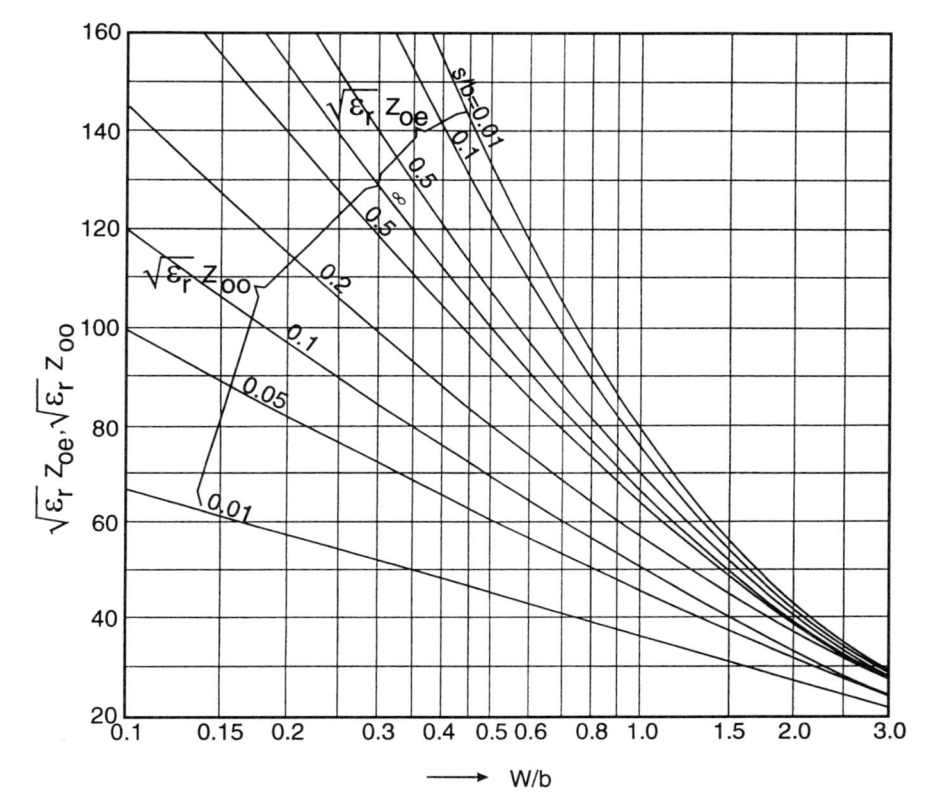

Figure 3.15 Characteristic impedances of even- and odd-modes of edge coupled striplines. (From [11], ©1995 IEEE. Reprinted with permission.)

and

$$\frac{S}{b} = \frac{2}{\pi} \tanh^{-1} \left(\frac{1-k_o}{1-k_e} \sqrt{\frac{k_e}{k_o}} \right) \quad (3.70)$$

For given values of Z_{0e} and Z_{0o}, the values of k_e and k_o can be determined by numerically solving (3.67) and (3.68), respectively, with the aid of (3.61). For example, if (3.67) is plotted as a function of k_e with the aid of (3.61), the value of k_e for a given value of Z_{0e} can be found. The values of W/b and S/b can then be determined by substituting the values of k_e and k_o in (3.69) and (3.70), respectively.

The propagation constant is the same for even and odd modes and is given by (3.18). Similarly, the dielectric loss is also the same for even and odd modes and is given by (3.21). The conductor losses, however, are different for the two modes. Expressions for these are given in [12].

The effect of finite thickness of strip conductors on the characteristic impedances of even and odd modes has also been reported in [11].

3.4 Characteristics of Single and Coupled Microstrip Lines

A microstrip transmission line is shown in Figure 3.16. Compared with a stripline, a microstrip line uses only a single dielectric substrate. The height of the substrate is chosen to be much smaller than the wavelength in the dielectric. The mode of propagation along a microstrip line is not pure TEM. Because of the simplicity of the structure and ease of fabrication, however, the microstrip line is the most popular transmission line for realizing microwave integrated circuits (MIC). Commonly used substrates are alumina (Al_2O_3), teflon-based materials such as RT duroid™ for hybrid MICs, and gallium arsenide (GaAs) for monolithic MICs. Numerous methods have been used to determine the characteristics of single and coupled microstrip lines. For design purposes, many simple closed-form empirical expressions have been reported in the literature [13–22]. A microstrip line is dispersive in nature and its characteristic impedance and effective dielectric constant vary with frequency. The accuracy of various dispersion formulas has also been experimentally verified [23, 24].

The characteristics of microstrip lines are generally described by two sets of equations. One gives the characteristics that are valid in the quasistatic limit, while the other set accounts for dispersion. Quasistatic relations are sufficiently accurate as long as the height of the dielectric substrate is very small compared with the wavelength.

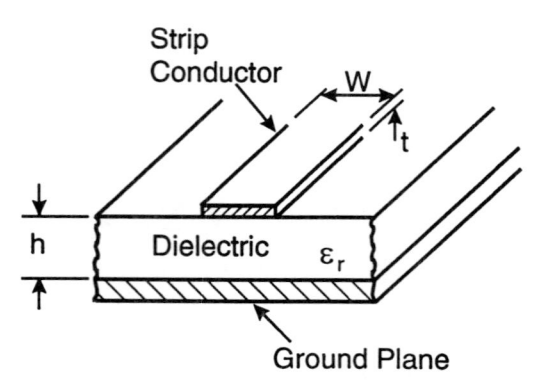

Figure 3.16 A microstrip line.

3.4.1 Single Microstrip

The quasistatic problem of a single microstrip line was solved analytically by Wheeler [14]. Over the years, this model has generally served as the basis for deriving simple semiempirical relations for the characteristics of a microstrip line. An accurate expression for effective dielectric constant defined according to (3.22) is given by [18]

$$\epsilon_{re}(0) = \frac{\epsilon_r + 1}{2} + \frac{\epsilon_r - 1}{2}\left(1 + \frac{10}{u}\right)^{-AB} \quad (3.71)$$

where $u = W/h$, and

$$A = 1 + \frac{1}{49}\ln\left[\frac{u^4 + (u/52)^2}{u^4 + 0.432}\right] + \frac{1}{18.7}\ln\left[1 + \left(\frac{u}{18.1}\right)^3\right]$$

$$B = 0.564\left(\frac{\epsilon_r - 0.9}{\epsilon_r + 3}\right)^{0.053}$$

The stated accuracy of (3.71) is better than 0.2% at least for $\epsilon_r \leq 128$ and $0.01 \leq u \leq 100$.

The quasistatic characteristic impedance of a microstrip line of zero thickness can be expressed as

$$Z_0(0) = \frac{60}{\sqrt{\epsilon_{re}}}\ln\left[\frac{f(u)}{u} + \sqrt{1 + \left(\frac{2}{u}\right)^2}\right] \quad (3.72)$$

where

$$f(u) = 6 + (2\pi - 6)\exp\left[-\left(\frac{30.666}{u}\right)^{0.7528}\right] \quad (3.73)$$

The symbol (0) in $\epsilon_{re}(0)$ and $Z_0(0)$ denotes that the formula is valid in the quasistatic limit. For $\epsilon_r = 1$, the stated accuracy of (3.72) is better than 0.01% for $u \leq 1$ and 0.03% for $u \leq 1{,}000$. For other values of ϵ_r, the accuracy of the expression is determined by the accuracy with which the effective dielectric constant is known.

Finite-Thickness Microstrip

Improved and more general equations for the analysis and synthesis of microstrip lines have also been given by Wheeler [3]. The characteristic impedance of a microstrip line of finite thickness is given by

$$Z_0 = \frac{42.4}{\sqrt{\epsilon_r + 1}} \ln \left(1 + \left(\frac{4h}{W'}\right) \left[\left(\frac{14 + 8/\epsilon_r}{11}\right) \left(\frac{4h}{W'}\right) \right.\right.$$

$$\left.\left.+ \sqrt{\left(\frac{14 + 8/\epsilon_r}{11}\right)^2 \left(\frac{4h}{W'}\right)^2 + \frac{1 + 1/\epsilon_r}{2}\pi^2}\right]\right) \quad (3.74)$$

where W' denotes the effective width of the microstrip. The effect of finite thickness of the strip is taken into account by assuming that the effective width of the microstrip W' is greater than its physical width W and is given by

$$W' = W + \Delta W \quad (3.75)$$

where

$$\frac{\Delta W}{t} = \left(\frac{1 + 1/\epsilon_r}{2\pi}\right) \ln \frac{10.872}{\sqrt{\left(\frac{t}{h}\right)^2 + \left(\frac{1/\pi}{W/t + 1.1}\right)^2}}$$

The above equations are supposed to be accurate for arbitrary values of ϵ_r and aspect ratio W/h (within the quasistatic limit). Note that if a general expression is available to determine the characteristic impedance of a microstrip line as a function of the dielectric constant of the substrate, it is not necessary to have a separate formula for the effective dielectric constant. The effective dielectric constant can then be determined using (3.25), that is:

$$\epsilon_{re} = \left(\frac{Z_{0a}}{Z_0}\right)^2$$

where Z_{0a} denotes the characteristic impedance of the microstrip line assuming that the dielectric constant of the substrate is unity.

The synthesis equations for a microstrip line are as follows:

$$W = W' - \Delta W' \quad (3.76)$$

where

$$\frac{W'}{h} = 8 \frac{\sqrt{\left[e^{\left(\frac{Z_0}{42.4}\sqrt{\epsilon_r + 1}\right)} - 1\right]\frac{7 + 4/\epsilon_r}{11} + \frac{1 + 1/\epsilon_r}{0.81}}}{\left[e^{\left(\frac{Z_0}{42.4}\sqrt{\epsilon_r + 1}\right)} - 1\right]}$$

and

$$\frac{\Delta W'}{t} = \left(\frac{1+1/\epsilon_r}{2\pi}\right) \ln \frac{10.872}{\sqrt{\left(\frac{t}{h}\right)^2 + \left(\frac{1/\pi}{W'/t-0.26}\right)^2}}$$

The characteristic impedance of a microstrip line is shown in Figure 3.17(a) as a function of effective width W' for various values of ϵ_r [3]. For a microstrip line of given thickness and width, the effective width can be found using (3.75). Figure 3.17(a) can therefore be used to find the characteristics of a microstrip line of finite thickness. The results shown in Figure 3.17(a) can also be used to find the effective dielectric constant of a microstrip line using (3.25). The characteristics of microstrip line printed on some commonly used dielectric substrates are shown in Figure 3.17(b).

Example 3.1

Determine the dimension ratio W/h and effective dielectric constant of a microstrip line ($t = 0$) of characteristic impedance of 50 Ω printed on a dielectric substrate of dielectric constant ($\epsilon_r = 2$).

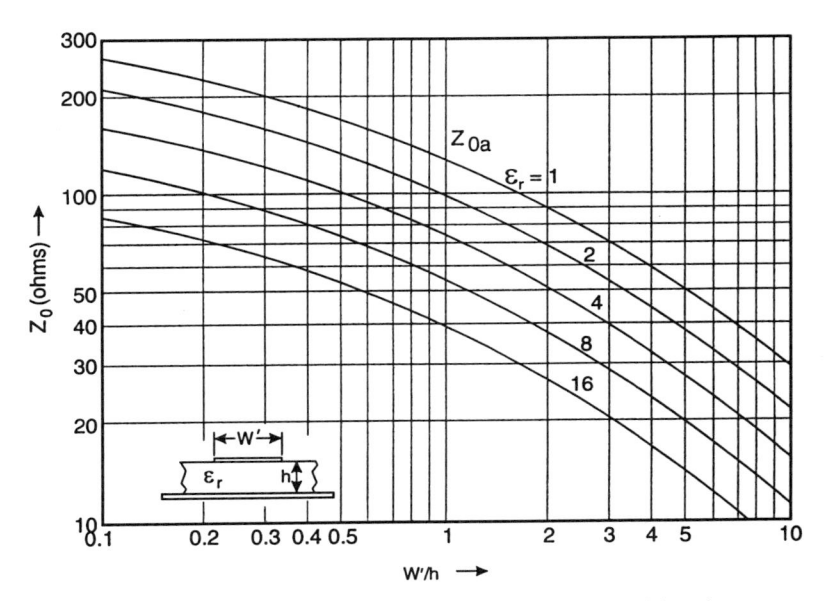

Figure 3.17 (a) Quasi-static characteristic impedance and effective dielectric constant of a microstrip line for various values of ϵ_r. Here $\epsilon_{re} = (\frac{Z_{0a}}{Z_0})^2$. (From [3], ©1977 IEEE. Reprinted with permission.)

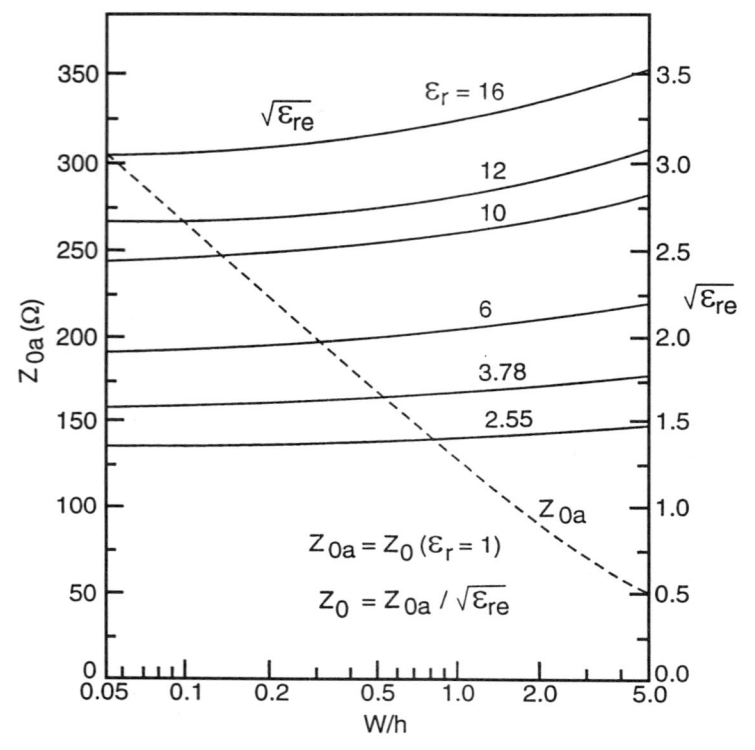

Figure 3.17 (b) Quasi-static characteristic impedance and effective dielectric constant of a microstrip line for some commonly used dielectric substrates. (From [12], © 1988 John Wiley and Sons. Reprinted with permission.)

From the curve labeled as $\epsilon_r = 2$ in Figure 3.17(a), we find that W/h should be chosen as equal to 3.3 to obtain a characteristic impedance (Z_0) of 50 Ω. Further, for the same value of W/h and $\epsilon_r = 1$, we find from the graph that $Z_{0a} = 66$ Ω. Using (3.25), we then find that

$$\epsilon_{re} = \left(\frac{66}{50}\right)^2 = 1.74$$

Conductor and Dielectric Loss

Once a general expression for the characteristic impedance of a microstrip is available in terms of the structural parameters, the conductor loss can be easily computed using (3.14) and (3.32). The conductor and other losses of a microstrip line on dielectric and ferrite substrates are discussed in [17]. The dielectric loss of a microstrip line is given by (3.26).

The normalized conductor loss ($h/Q_c\delta_s$) of a microstrip line (which is independent of the dielectric constant of the substrate) is shown in Figure 3.18(a) where Q_c and δ_s denote the conductor Q-factor and skin depth, respectively [3]. For given values of W and h, this figure can be used to determine the conductor Q-factor Q_c and hence the conductor loss in Np/m from (3.14). The conductor and dielectric loss of microstrip line on some commonly used dielectric substrates is shown in Figure 3.18(b).

Microstrip Dispersion

If the frequency of operation is high, so that the height of the substrate is not very small compared with the wavelength in the dielectric, the quasistatic expressions presented above are not accurate enough. In some other situations also, these expressions may be inadequate. For example, if a digital signal with short rise time propagates along a microstrip line, the signal can be assumed to contain a number of high-frequency harmonics. An accurate dispersion model for propagation along microstrip lines is therefore required. The frequency dependence of the effective dielectric constant of a microstrip line is well represented by the following relation [19]:

$$\epsilon_{re}(f) = \epsilon_r - \frac{\epsilon_r - \epsilon_{re}(0)}{1 + (f/f_{50})^m} \tag{3.77}$$

where

$$f_{50} = \frac{f_{K,TM_0}}{0.75 + \left(0.75 - \frac{0.332}{\epsilon_r^{1.73}}\right)\frac{W}{h}} \tag{3.78}$$

$$f_{K,TM_0} = \frac{c\tan^{-1}\left(\epsilon_r\sqrt{\frac{\epsilon_{re}(0)-1}{\epsilon_r-\epsilon_{re}(0)}}\right)}{2\pi h\sqrt{\epsilon_r - \epsilon_{re}(0)}} \tag{3.79}$$

$$m = m_0 m_c$$

$$m_0 = 1 + \frac{1}{1+\sqrt{W/h}} + 0.32\left(\frac{1}{1+\sqrt{W/h}}\right)^3$$

$$m_c = \begin{cases} 1 + \frac{1.4}{1+\sqrt{W/h}}\left\{0.15 - 0.235\exp\left(\frac{-0.45f}{f_{50}}\right)\right\} & \text{for } W/h \leq 0.7 \\ 1 & \text{for } w/h > 0.7 \end{cases}$$

Here c is the velocity of light. It is claimed that the above model has an accuracy of better than 0.6% in the range $0.1 \leq W/h \leq 10$, $1 \leq \epsilon_r \leq 128$ and for any value of h/λ_0.

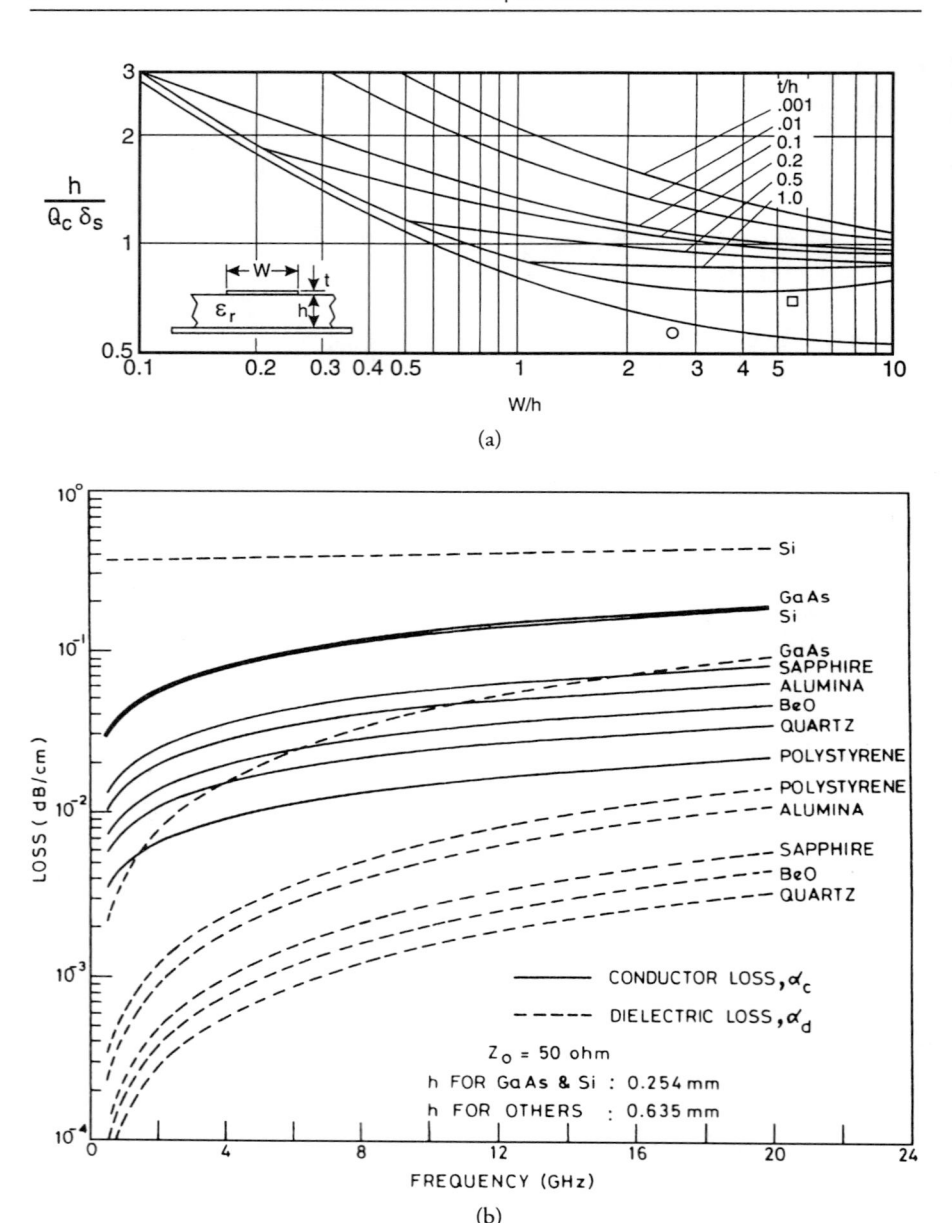

Figure 3.18 (a) Normalized conductor Q-factor of a microstrip line (valid for arbitrary value of ε_r). Curves marked □ and ○ are valid for microstrip line having strip-conductor of square or circular cross section respectively. (From [3], ©1977 IEEE, Reprinted with permission.) (b) Conductor and dielectric loss of microstrip line on some commonly used dielectric substrates. (From [13], © Artech House. Reprinted with permission.)

The dispersion of a microstrip line fabricated on a dielectric substrate of $\epsilon_r = 8$ is shown in Figure 3.19. The characteristic impedance of a microstrip line is also frequency dependent and shows a small positive increase with an increase in frequency. The following relation describes this dependence quite accurately [18]:

$$Z_0(f) = Z_0(0)\sqrt{\frac{\epsilon_{re}(0)}{\epsilon_{re}(f)} \frac{(\epsilon_{re}(f) - 1)}{(\epsilon_{re}(0) - 1)}} \qquad (3.80)$$

For many practical purposes, the variation of characteristic impedance with frequency can be neglected.

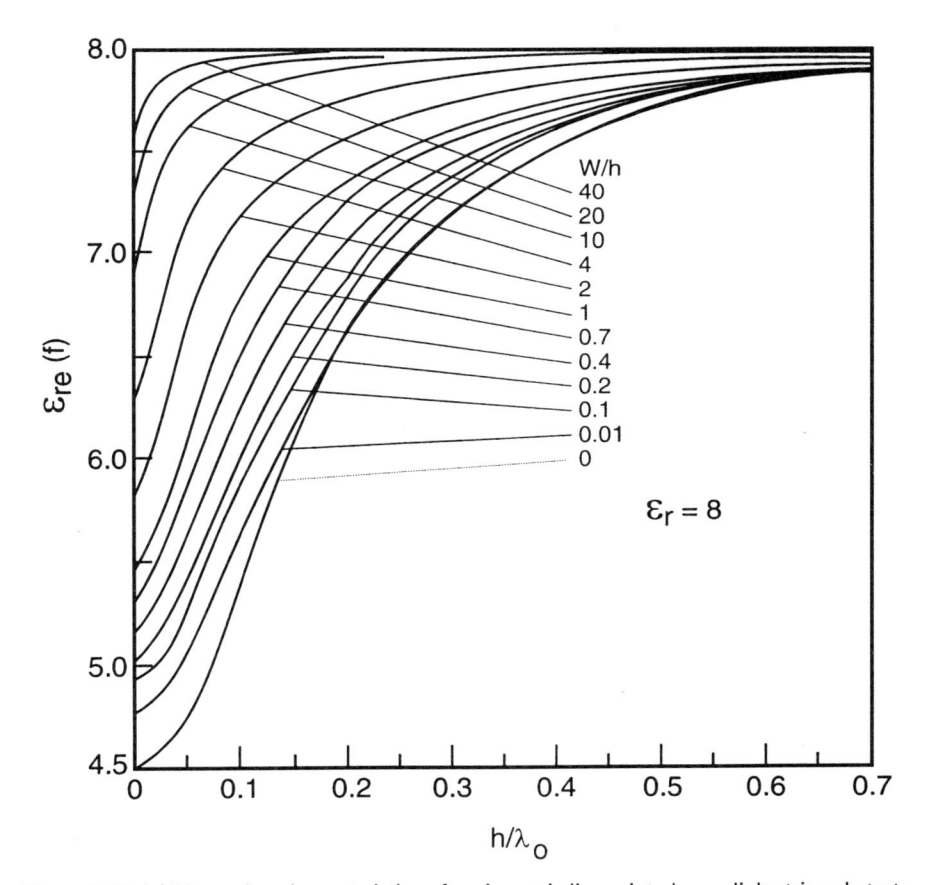

Figure 3.19 (a) Dispersion characteristics of a microstrip line printed on a dielectric substrate of dielectric constant $\epsilon_r = 8$. (From [19], © 1988 IEEE. Reprinted with permission.)

3.4.2 Coupled Microstrip Lines

Coupled microstrip lines are shown in Figure 3.20. The equations for the quasistatic characteristics of coupled microstrip lines have been given by many authors including Hammerstad and Jenson [18], Garg and Bahl [13, 21], and others. For the odd-mode case, the Hammerstad and Jenson equations were later modified by Kirschning and Jansen [22] to incorporate the effect of dispersion. Although the latter expressions are lengthy, these are simple to program and believed to be to be accurate to within 1% in the range of parameters $0.1 \leq u \leq 10$, $0.1 \leq g \leq 10$, and $1 \leq \epsilon_r \leq 18$. The quasistatic even-mode effective dielectric constant for coupled microstrip lines for zero conductor thickness is given by

$$\epsilon_{ree}(0) = 0.5(\epsilon_r + 1) + 0.5(\epsilon_r - 1).(1 + 10/v)^{-a_e(v) \cdot b_e(\epsilon_r)} \quad (3.81)$$

where

$$v = u(20 + g^2)/(10 + g^2) + g \cdot \exp(-g)$$

$$a_e(v) = 1 + \ln((v^4 + (v/52)^2)/(v^4 + 0.432))/49$$
$$+ \ln(1 + (v/18.1)^3)/18.7$$

$$b_e(\epsilon_r) = 0.564((\epsilon_r - 0.9)/(\epsilon_r + 3))^{0.053}$$

and $u = W/h$, $g = S/h$. The quasistatic odd-mode effective dielectric constant for zero conductor thickness is similarly given by

$$\epsilon_{reo}(0) = [0.5(\epsilon_r + 1) + a_o(u, \epsilon_r) - \epsilon_{re}(0)] \cdot \exp(-c_o g^{d_o}) + \epsilon_{re}(0) \quad (3.82)$$

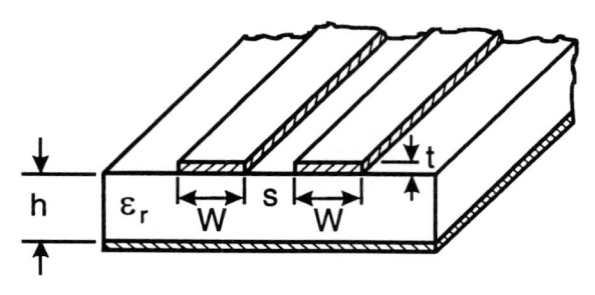

Figure 3.20 Edge-coupled microstrip lines.

where

$$a_o(u, \epsilon_r) = 0.7287(\epsilon_{re}(0) - 0.5(\epsilon_r + 1))\cdot(1 - \exp(-0.179u))$$
$$b_o(\epsilon_r) = 0.747\epsilon_r/(0.15 + \epsilon_r)$$
$$c_o = b_o(\epsilon_r) - (b_o(\epsilon_r) - 0.207)\cdot\exp(-0.414u)$$
$$d_o = 0.593 + 0.694\cdot\exp(-0.562u)$$

and $\epsilon_{re}(0)$ denotes the effective dielectric constant of a single microstrip of width W.

Quasistatic Even- and Odd-Mode Characteristic Impedances

The quasistatic even-mode characteristic impedance of coupled microstrip lines is given by [22]

$$Z_{0e}(0) = Z_0\sqrt{\frac{\epsilon_{re}(0)}{\epsilon_{ree}(0)}}\frac{1}{(1 - (Z_0(0)/377)(\epsilon_{re}(0))^{0.5}Q_4)} \quad (3.83)$$

where

$$Q_4 = (2Q_1/Q_2)$$
$$\cdot(\exp(-g)\cdot u^{Q_3} + (2 - \exp(-g))\cdot u^{-Q_3})^{-1}$$

with

$$Q_1 = 0.8695\cdot u^{0.194}$$
$$Q_2 = 1 + 0.7519g + 0.189\cdot g^{2.31}$$
$$Q_3 = 0.1975 + (16.6 + (8.4/g)^6)^{-0.387}$$
$$+ \ln(g^{10}/(1 + (g/3.4)^{10})/241$$

Similarly, the quasistatic odd-mode characteristic impedance of coupled microstrip lines is expressed by

$$Z_{0o}(0) = Z_0\sqrt{\frac{\epsilon_{re}(0)}{\epsilon_{reo}(0)}}\frac{1}{(1 - (Z_0(0)/377)(\epsilon_{re}(0))^{0.5}Q_{10})} \quad (3.84)$$

with

$$Q_5 = 1.794 + 1.14\cdot\ln(1 + 0.638/(g + 0.517g^{2.43}))$$
$$Q_6 = 0.2305 + \ln(g^{10}/(1 + (g/5.8)^{10}))/281.3$$
$$+ \ln(1 + 0.598g^{1.154})/5.1$$
$$Q_7 = (10 + 190g^2)/(1 + 82.3g^3)$$

$$Q_8 = \exp(-6.5 - 0.95\ln(g) - (g/0.15)^5)$$
$$Q_9 = \ln(Q_7) \cdot (Q_8 + 1/16.5)$$
$$Q_{10} = Q_2^{-1} \cdot (Q_2 Q_4 - Q_5 \cdot \exp(\ln(u) \cdot Q_6 \cdot u^{-Q_9}))$$

Equations for the frequency dependence of the even- and odd-mode effective dielectric constants and characteristic impedances were also given by Kirschning and Jansen [22]. These are quite involved, however, and are not repeated here.

The frequency-dependent characteristics of coupled microstrip lines printed on some typical dielectric substrates are shown in Figures 3.21 and 3.22 [22]. The results show that the variation of characteristic impedance with frequency is much smaller than the variation of an effective dielectric constant. Further, it is seen that the even-mode parameters show a greater variation with frequency than the odd-mode parameters.

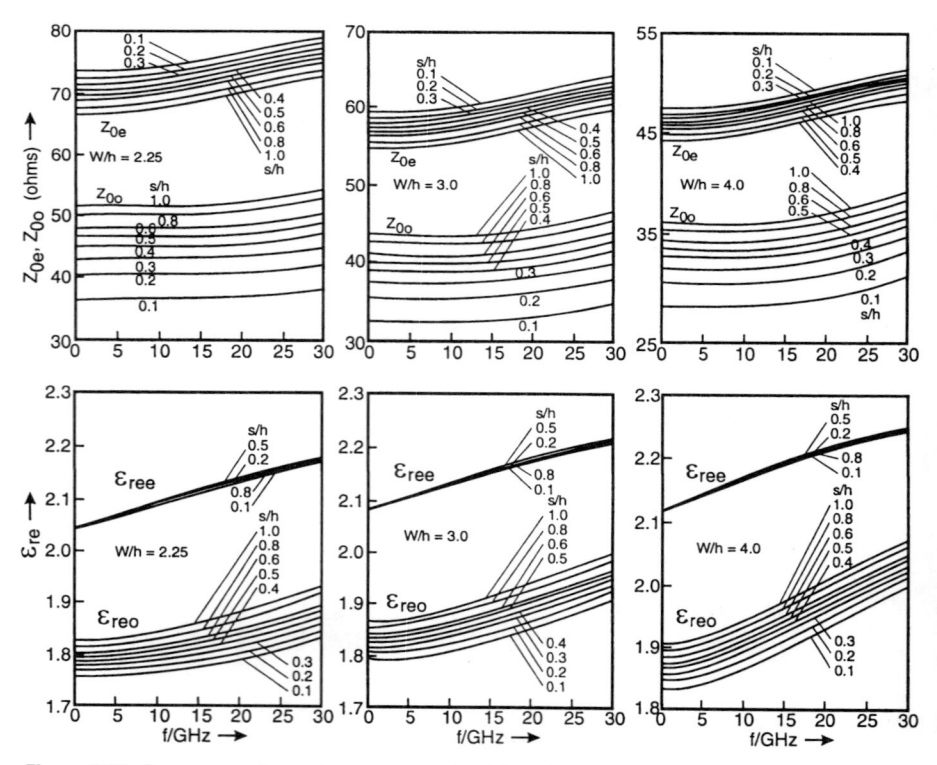

Figure 3.21 Frequency dependent even- and odd-mode characteristic impedances and effective dielectric constants of coupled microstrip lines printed on a substrate of dielectric constant $\epsilon_r = 2.35$, $h = 0.79$ mm. (From [22], © 1984 IEEE. Reprinted with permission.)

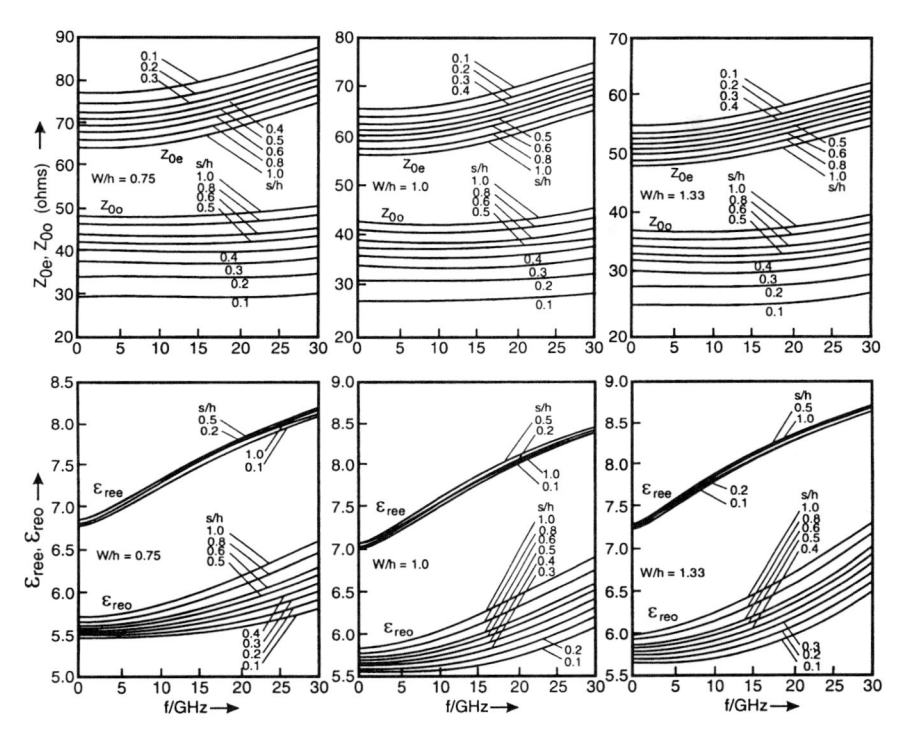

Figure 3.22 Frequency dependent even- and odd-mode characteristic impedances and effective dielectric constants of coupled microstrip lines printed on a substrate of dielectric constant $\epsilon_r = 9.7$, $h = 0.64$ mm. (From [22], © 1984 IEEE. Reprinted with permission.)

3.5 Single and Coupled Coplanar Wave Guides

A few coplanar waveguide (CPW) configurations are shown in Figures 3.23 to 3.25. In a coplanar waveguide, the signal is applied between the center conductor and two outer conductors that lie in the same plane [25, 26]. The outer conductors (which are of finite width in practice) are at the same (ground) potential. The coplanar waveguide has some inherent advantages, making it suitable for hybrid and monolithic microwave integrated circuits. Because both the center and ground conductors of a CPW lie in the same plane, active devices can be easily placed in a series or shunt across the transmission line without requiring a via hole geometry. The mode of propagation along a coplanar line is quasi-TEM. Both numerical and analytical methods have been used for the analysis in [27–34]. The dispersion in CPW is generally smaller than in a microstrip line [33].

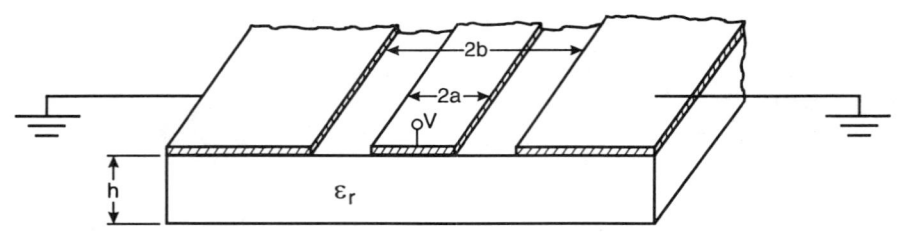

Figure 3.23 Coplanar waveguide on a finite thickness dielectric substrate.

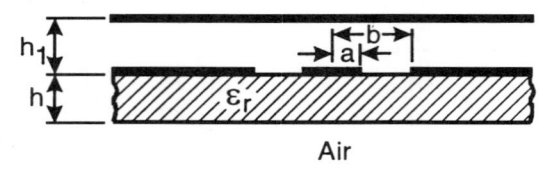

Figure 3.24 Cross section of a coplanar waveguide with upper shielding.

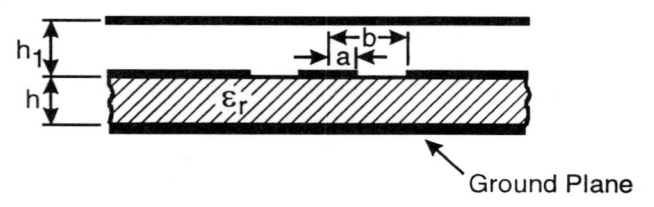

Figure 3.25 Cross section of a conductor-backed coplanar waveguide with upper shielding.

3.5.1 Coplanar Waveguide

Consider the CPW printed on a finite-thickness dielectric substrate (Figure 3.23). The width of the dielectric substrate is assumed to be infinite. The quasistatic parameters of this transmission line are given by [30]:

$$Z_0 = \frac{30\pi}{\sqrt{\epsilon_{re}}} \frac{K(k')}{K(k)} \quad (3.85)$$

$$\epsilon_{re} = 1 + \frac{\epsilon_r - 1}{2} \frac{K(k')K(k_1)}{K(k)K(k'_1)} \quad (3.86)$$

where

$$k = a/b; \quad k' = \sqrt{1-k^2}$$

$$k_1 = \sinh(\pi a/2h)/\sinh(\pi b/2h); \quad k'_1 = \sqrt{1-k_1^2}$$

Further, K is the complete integral of the first kind. The values of functions $K(k)/K(k')$ and $K(k_1)/K(k'_1)$ can be found using (3.61).

The effective dielectric constant and characteristic impedance of a CPW are shown in Figure 3.26 for a value of $\epsilon_r = 12.7$ (GaAs substrate).

Equations for the case of finite conductor thickness are presented by Wadell [35].

3.5.2 Coplanar Waveguide With Upper Shielding

The cross section of a CPW with upper shielding is shown in Figure 3.24. The quasistatic parameters of the transmission line are given by [31]

$$\epsilon_{re} = 1 + q(\epsilon_r - 1) \tag{3.87}$$

where q is the filling fraction given by

$$q = \frac{\frac{K(k_1)}{K(k'_1)}}{\frac{K(k_2)}{K(k'_2)} + \frac{K(k)}{K(k')}} \tag{3.88}$$

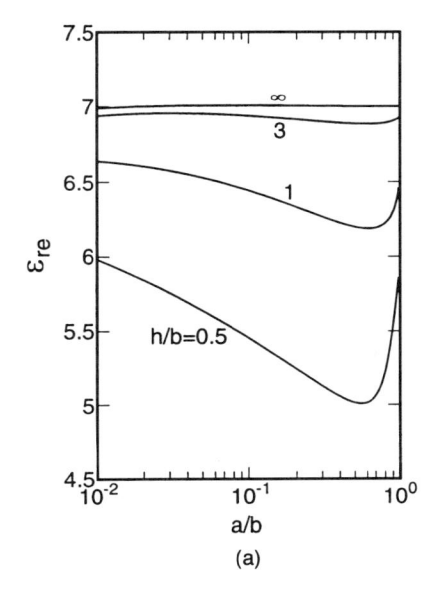

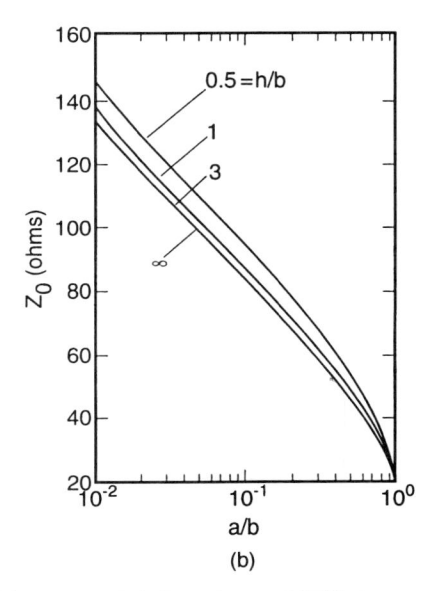

Figure 3.26 (a) Effective dielectric constant, and (b) characteristic impedance of CPW shown in Figure 3.23 as a function of aspect ration a/b for various values of h/b. $\epsilon_r = 13$. (From [13], ©1996 Artech House. Reprinted with permission.)

In (3.88), parameters k, k_1, and k_2 are given by

$$k = a/b; \quad k' = \sqrt{1-k^2}$$
$$k_1 = \sinh(\pi a/2h)/\sinh(\pi b/2h) \quad (3.89)$$
$$k_2 = \tanh(\pi a/2h_1)/\tanh(\pi b/2h_1)$$

and $k'_i = \sqrt{1-k_i^2}$. The values of functions $K(k)/K(k')$ and $K(k_i)/K(k'_i)$ can be found using (3.61). The characteristic impedance is then obtained from

$$Z_0 = \frac{60\pi}{\sqrt{\epsilon_{re}}} \frac{1}{\frac{K(k_2)}{K(k'_2)} + \frac{K(k)}{K(k')}} \quad (3.90)$$

The structure shown in Figure 3.24 reduces to that shown in Figure 3.23 when $h_1 \to \infty$.

3.5.3 Conductor-Backed Coplanar Waveguide With Upper Shielding

The cross section of a conductor-backed CPW with upper shielding is shown in Figure 3.25. This structure has a lower ground plane. The lower ground plane adds mechanical strength to the circuit and increases its power handling capability. The quasistatic effective dielectric constant of this transmission line can also be expressed in the form

$$\epsilon_{re} = 1 + q(\epsilon_r - 1)$$

where

$$q = \frac{\frac{K(k_3)}{K(k'_3)}}{\frac{K(k_3)}{K(k'_3)} + \frac{K(k_4)}{K(k'_4)}} \quad (3.91)$$

In (3.91), parameters k_3 and k_4 are given by [30]

$$k_3 = \tanh(\pi a/2h)/\tanh(\pi b/2h)$$
$$k_4 = \tanh(\pi a/2h_1)/\tanh(\pi b/2h_1) \quad (3.92)$$
$$k'_i = \sqrt{1-k_i^2}$$

The expression for the characteristic impedance is

$$Z_0 = \frac{60\pi}{\sqrt{\epsilon_{re}}} \frac{1}{\frac{K(k_3)}{K(k'_3)} + \frac{K(k_4)}{K(k'_4)}} \quad (3.93)$$

The characteristic impedance and effective dielectric constant of a conductor-backed CPW with upper shielding are shown in Figure 3.27 as a function of a/b for various values of a/b and $\epsilon_r = 10$.

3.5.4 Coupled Coplanar Waveguides

The cross section of coupled CPWs is shown in Figure 3.28. Unfortunately, very few simple formulas are available in the literature for the determination of parameters of coupled CPWs. For a value of $\epsilon_r = 12.9$, the even- and odd-mode characteristic impedances of coupled CPWs are shown in Figure 3.29 [32]. For the same parameters of the structure, the even- and odd-mode effective dielectric constants are shown in Figure 3.30.

3.6 Suspended and Inverted Microstrip Lines

The cross section of suspended and inverted microstrip lines is shown in Figures 3.31(a) and 3.31(b), respectively. These lines achieve a lower loss (or higher Q) than possible with microstrip lines. Further, these lines have a much lower effective dielectric constant (compared with that of a microstrip line), thus leading to their performance being less sensitive to dimensional tolerances at high frequencies. Further, the wide range of impedance values achievable using these lines makes them particularly suitable in realizing filters. The quasistatic equations for the design of suspended and inverted microstrip lines have been given by Pramanick and Bhartia [36] and Tomar and Bhartia [37]. The characteristic impedances of inverted and suspended microstrip lines ($t \ll h$) are given by the following expression [36]:

$$Z_0 = \frac{60}{\sqrt{\epsilon_{re}}} \ln\left[\frac{f(u)}{u} + \sqrt{1 + \left(\frac{2}{u}\right)^2}\right] \quad (3.94)$$

where

$$f(u) = 6 + (2\pi - 6)\exp\left[-\left(\frac{30.666}{u}\right)^{0.7528}\right]$$

The parameter is $u = W/(a+b)$ for suspended microstrips and $u = W/b$ for inverted microstrips.

The effective dielectric constant ϵ_{re} of a suspended microstrip is given by

$$\sqrt{\epsilon_{re}} = \left[1 + \frac{a}{b}\left(a_1 - b_1 \ln \frac{W}{b}\right)\left(\frac{1}{\sqrt{\epsilon_r - 1}}\right)\right]^{-1} \quad (3.95)$$

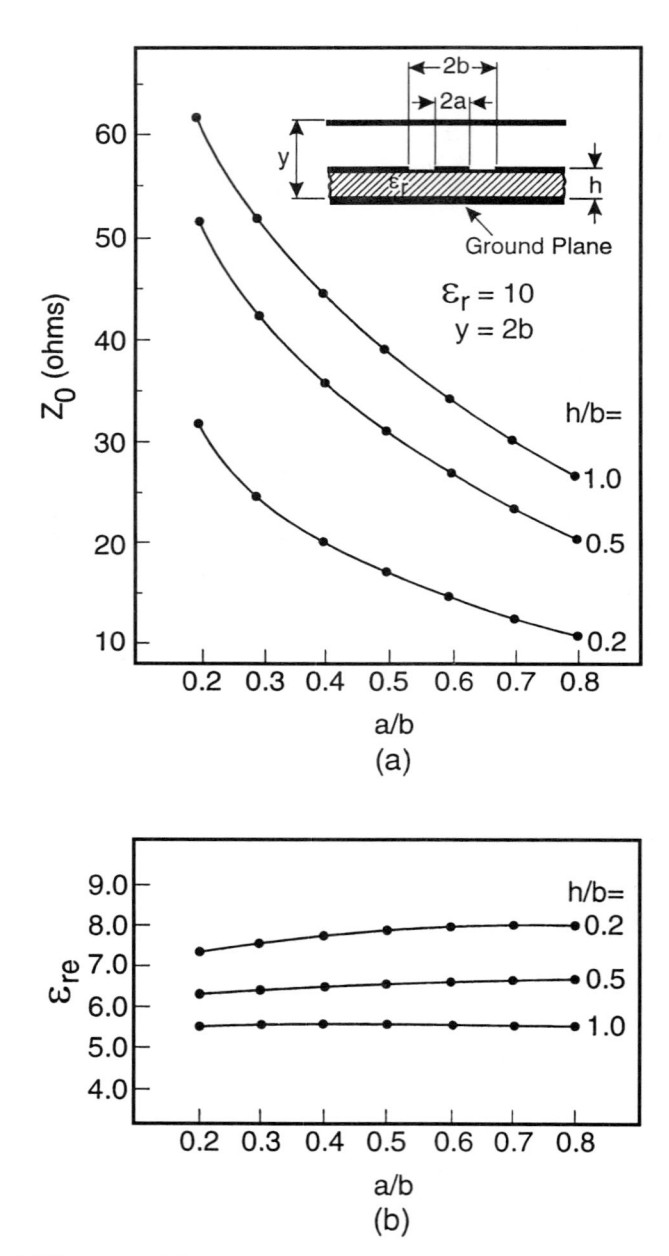

Figure 3.27 (a) Characteristic impedance, and (b) effective dielectric constant of a conductor-backed CPW with upper shielding shown in inset. (From [31], ©1987 IEEE. Reprinted with permission.)

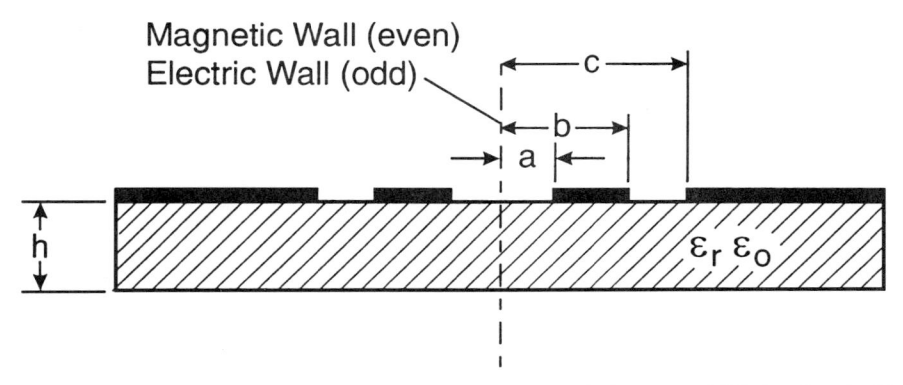

Figure 3.28 Cross section of coupled coplanar lines. (From [32], ©1996 IEEE. Reprinted with permission.)

where

$$a_1 = \left(0.8621 - 0.1251 \ln \frac{a}{b}\right)^4$$

$$b_1 = \left(0.4986 - 0.1397 \ln \frac{a}{b}\right)^4$$

The effective dielectric constant ϵ_{re} of an inverted microstrip is given by

$$\sqrt{\epsilon_{re}} = 1 + \frac{a}{b}\left(a_1 - b_1 \ln \frac{W}{b}\right)(\sqrt{\epsilon_r} - 1) \tag{3.96}$$

where

$$a_1 = \left(0.5173 - 0.1515 \ln \frac{a}{b}\right)^2$$

$$b_1 = \left(0.3092 - 0.1047 \ln \frac{a}{b}\right)^2$$

The above analysis equations are accurate to within 1% for $1 \leq W/b \leq 8$, $0.2 \leq a/b \leq 1$, and $\epsilon_r \leq 6$. For $\epsilon_r \approx 10$, the error is less than ±2%. The characteristic impedance and phase velocity of a suspended microstrip are shown in Figure 3.32(a) for a value of $\epsilon_r = 3.78$. The characteristic impedance and effective dielectric constant of an inverted microstrip line are shown in Figure 3.32(b) for various values of ϵ_r.

Closed-form formulas are generally not available for coupled suspended and inverted striplines. Electromagnetic simulators, however can be used to design circuits using such lines.

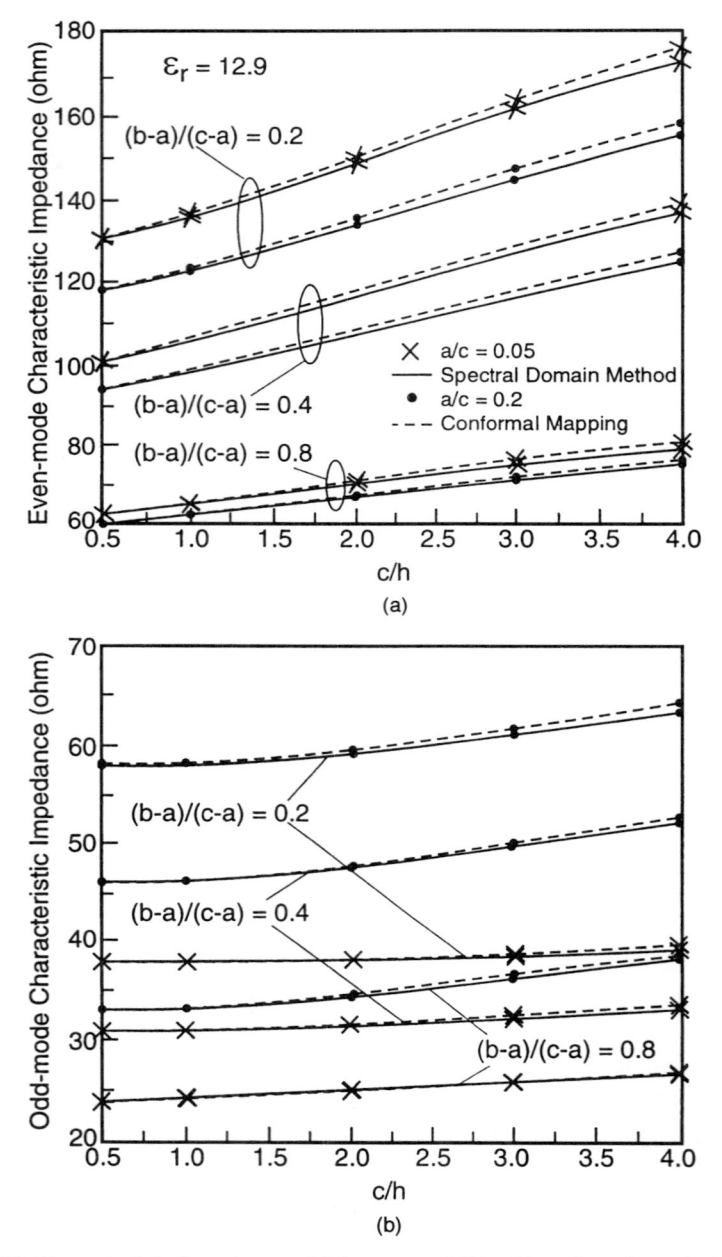

Figure 3.29 Characteristic impedance of (a) even- and (b) odd-modes of coupled coplanar lines shown in Figure 3.28. (From [32], ©1996 IEEE. Reprinted with permission.)

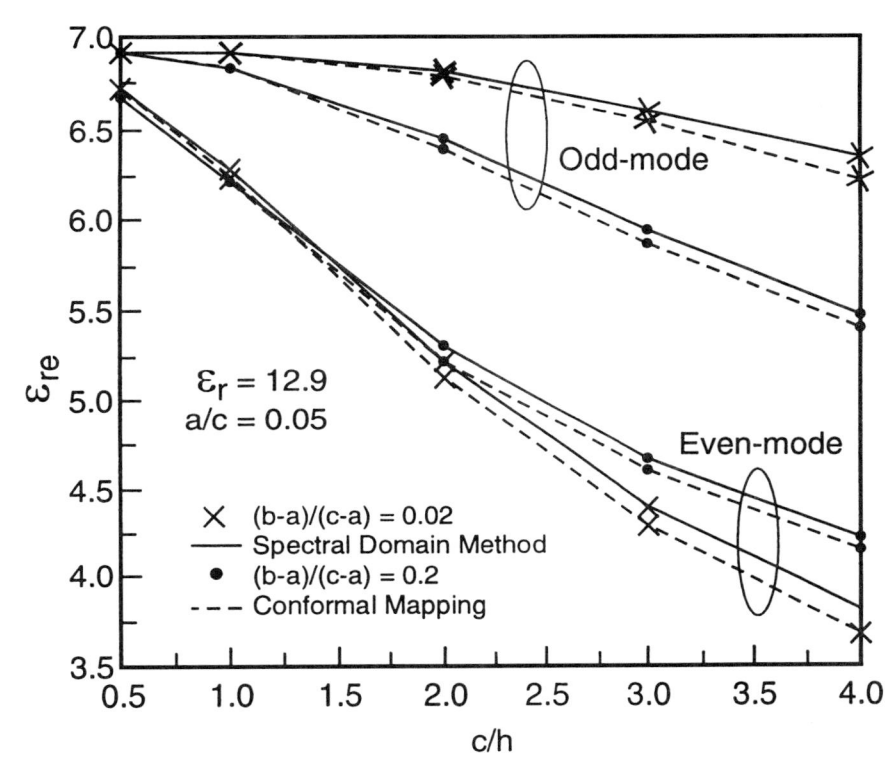

Figure 3.30 Effective dielectric constant of even- and odd-modes of coupled coplanar lines shown in Figure 3.28. (From [32], ©1996 IEEE. Reprinted with permission.)

3.7 Broadside-Coupled Lines

In earlier sections, we have considered coupling between lines that lie in the same plane (edge-coupled lines). Although edge-coupled lines are easier to fabricate, they are not suitable for tight coupling. Tight coupling can be obtained by placing coupled lines in a broadside manner. Further, in keeping with the trend toward miniaturization, multilayer microwave circuits that use broadside coupling between lines are becoming a reality and are described in Chapter 8.

3.7.1 Broadside-Coupled Striplines

General broadside-coupled microstrip lines are shown in Figure 3.33. For $\epsilon_{r1} = \epsilon_{r2} = \epsilon_r$, the structure reduces to broadside-coupled striplines. For coupled striplines, the even- and odd-mode effective dielectric constants are the same

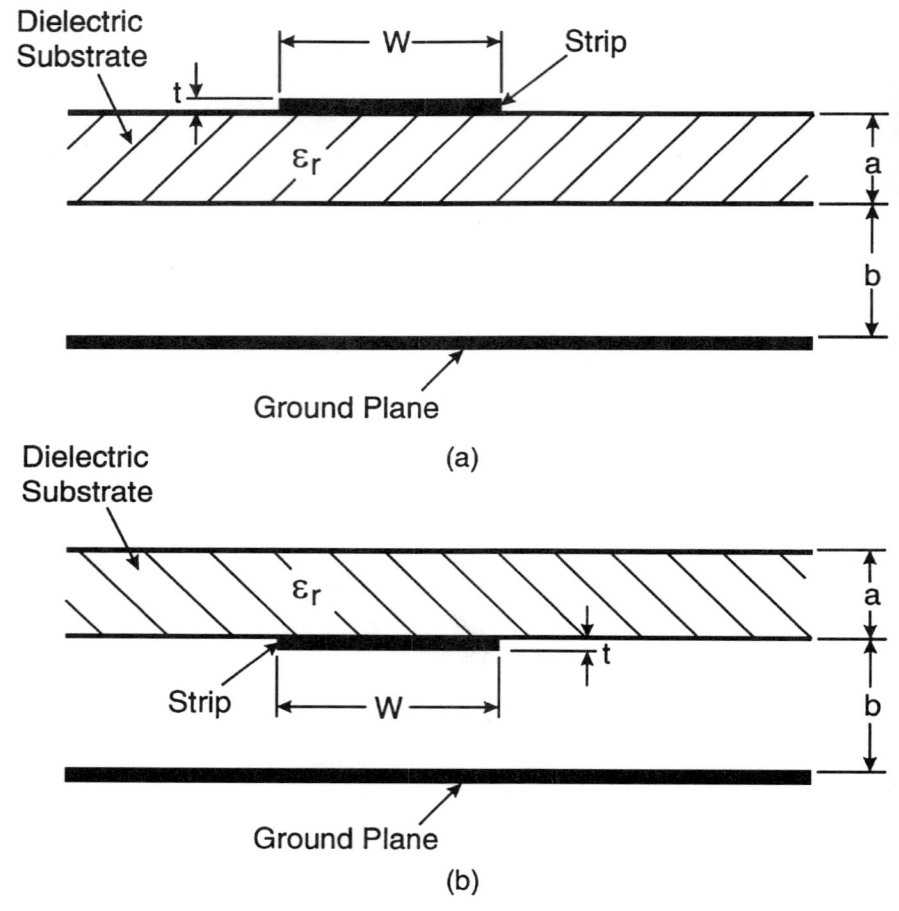

Figure 3.31 Cross section of (a) suspended, and (b) inverted microstrip lines.

and are given by

$$\epsilon_{ree} = \epsilon_{reo} = \epsilon_r$$

The characteristic impedance of the odd mode can be found using the following expression [38]:

$$Z_{0o}\sqrt{\epsilon_r} = Z_{0\infty}^a - \Delta Z_{0\infty}^a \qquad (3.97)$$

where

$$Z_{0\infty}^a = 60 \ln\left(\frac{3S}{W} + \sqrt{\left(\frac{S}{W}\right)^2 + 1}\right) \qquad (3.98)$$

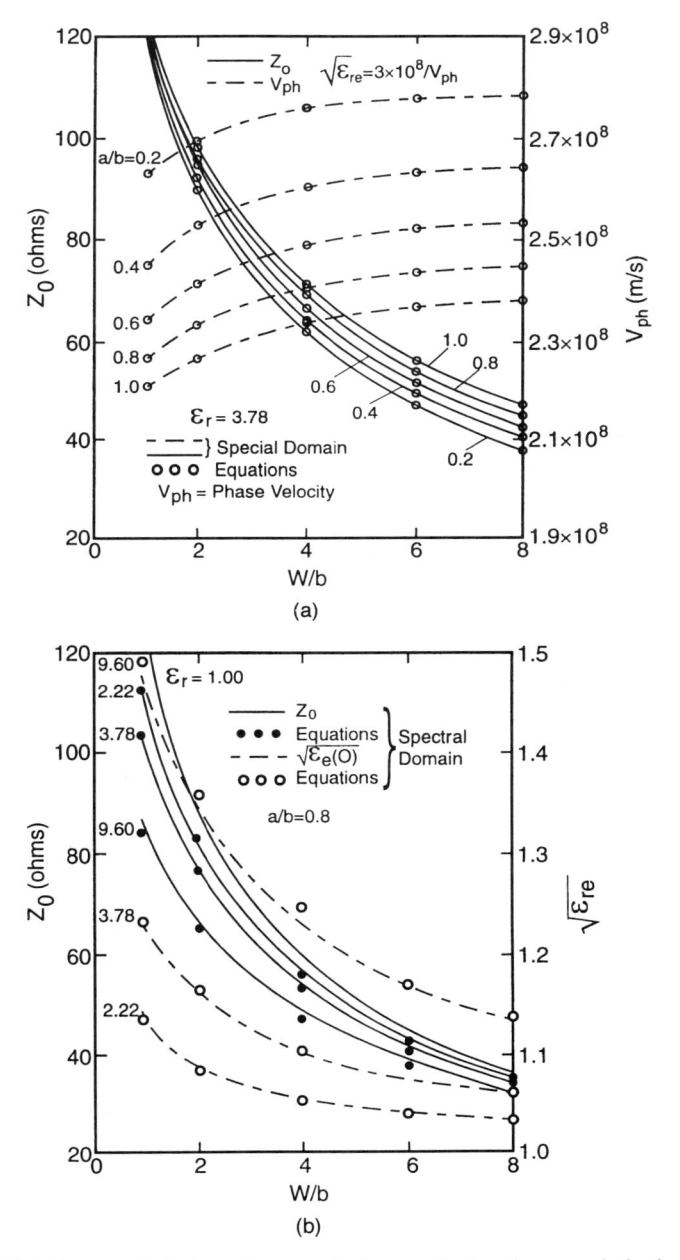

Figure 3.32 (a) Characteristic impedance and phase velocity of suspended microstrip line. $\epsilon_r = 3.78$. (b) Characteristic impedance and effective dielectric constant of inverted microstrip line. (From [36], ©1985 IEEE. Reprinted with permission.)

and

$$\Delta Z^a_{0\infty} = \begin{cases} P & \text{for } \dfrac{W}{S} \leq 1/2 \\ PQ & \text{for } \dfrac{W}{S} \geq 1/2 \end{cases}$$

$$P = 270\left[1 - \tanh\left(0.28 + 1.2\sqrt{\dfrac{b-S}{S}}\right)\right]$$

$$Q = 1 - \tanh^{-1}\left[\dfrac{0.48\sqrt{\dfrac{2W}{S}-1}}{\left(1+\dfrac{b-S}{S}\right)^2}\right]$$

Further, the even-mode characteristic impedance can be found using

$$Z_{0e} = \dfrac{60\pi}{\sqrt{\epsilon_r}} \dfrac{K(k')}{K(k)} \qquad (3.99)$$

where

$$k = \tanh\left(\dfrac{293.9 S/b}{Z_{0o}\sqrt{\epsilon_r}}\right) \qquad (3.100)$$

It has been reported that (3.97) and (3.99) offer an accuracy within 1% of spectral domain results [38].

3.7.2 Broadside-Coupled Suspended Microstrip Lines

The structure shown in Figure 3.33 reduces to broadside-coupled suspended microstrip lines for $\epsilon_{r1} = \epsilon_r \geq 1$, $\epsilon_{r2} = 1$. The even- and odd-mode characteristic impedances of broadside-coupled suspended microstrip lines are given by

$$Z_{0e} = \dfrac{Z^a_{0e}}{\sqrt{\epsilon_{ree}}} \qquad (3.101)$$

$$Z_{0o} = \dfrac{Z^a_{0o}}{\sqrt{\epsilon_{reo}}} \qquad (3.102)$$

where Z^a_{0e} and Z^a_{0o} are the even- and odd-mode characteristic impedances of the corresponding air-filled homogeneous broadside-coupled striplines. Their values can be found using (3.97) and (3.99), respectively. Further, the even- and

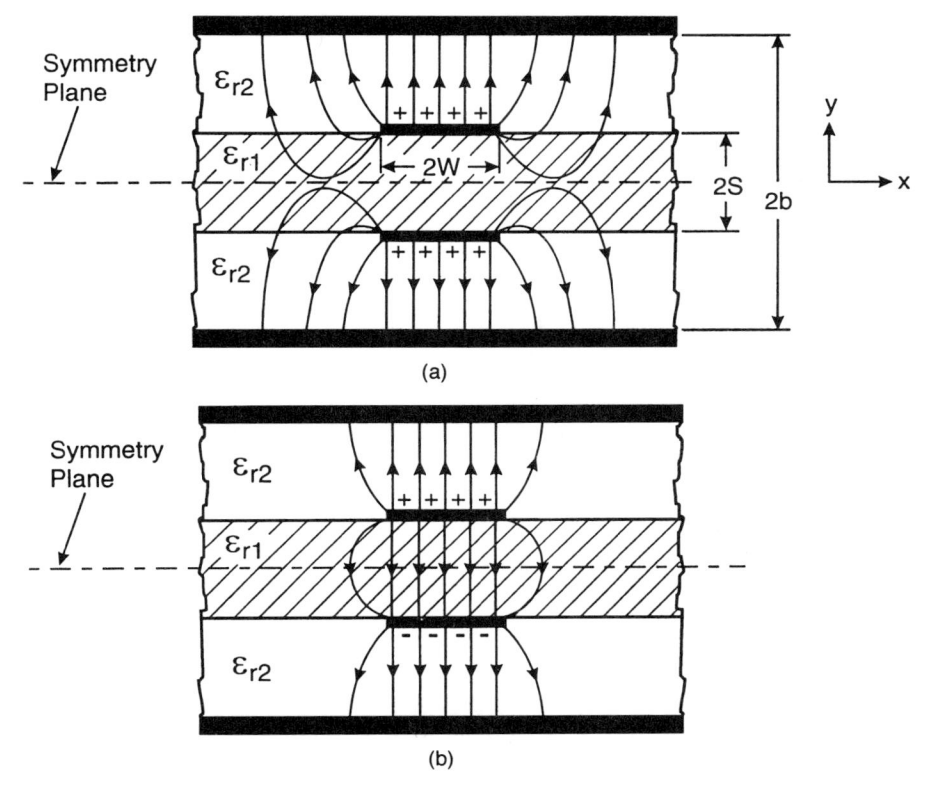

Characteristics of Planar Transmission Lines

Figure 3.33 (a) Even-mode, and (b) odd-mode field distribution of general broadside coupled microstrip lines. (From [38], ©1988 IEEE. Reprinted with permission.)

odd-mode effective dielectric constants are given by [38]

$$\epsilon_{reo} = \frac{1}{2}(\epsilon_r + 1) + q\frac{(\epsilon_r - 1)}{2} \qquad (3.103)$$

where

$$q = q_\infty q_c$$

$$q_\infty = \left(1 + \frac{5S}{W}\right)^{-a(U)b(\epsilon_r)}$$

$$a(U) = 1 + \frac{1}{49}\ln\left[\frac{U^4 + (U/52)^2}{U^4 + 0.432}\right] + \frac{1}{18.7}\ln\left[1 + \left(\frac{U}{18.1}\right)^3\right]$$

$$U = 2W/S$$

$$b(\epsilon_r) = 0.564\left(\frac{\epsilon_r - 0.9}{\epsilon_r + 3}\right)^{0.053}$$

and

$$q_c = \tanh\left[1.043 + 0.121\left(\frac{b-S}{S}\right) - 1.164\left(\frac{S}{b-S}\right)\right]$$

$$\epsilon_{ree} = \left[1 + \frac{S}{b}\left(a_1 - b_1 \ln\left(\frac{W}{b}\right)\right)(\sqrt{\epsilon_r} - 1)\right]^2 \quad (3.104)$$

where

$$a_1 = [0.8145 - 0.05824\ln(S/b)]^8$$
$$b_1 = [0.7581 - 0.07143\ln(S/b)]^8$$

The above equations offer an accuracy of about 1% for $\epsilon_r \leq 16$, $S/b \leq 0.4$, and $W/b \leq 1.2$. These conditions are usually met in practice. The even- and odd-mode characteristic impedances of coupled suspended microstrip lines are shown in Figure 3.34(a) for $\epsilon_r = 2.32$. For the same parameters, the effective dielectric constants are shown in Figure 3.34(b).

3.7.3 Broadside-Coupled Offset Striplines

Broadside-coupled offset striplines are shown in Figure 3.35. This structure is more general than the broadside-coupled striplines configuration discussed in section 3.7.1 or the edge-coupled stripline configuration shown in Figure 3.11. Shelton [39] has given closed-form expressions for the analysis and synthesis of broadside-coupled offset lines. Here, we present the synthesis equations only as they are more frequently used. Two sets of equations are given, one for tightly coupled lines and the other for loosely coupled lines. The conditions for tight and loose coupling are defined by

Tight coupling case: $\quad \dfrac{w'}{1-s'} \geq 0.35$

$$\frac{w'_c}{s'} \geq 0.7 \quad (3.105)$$

Loose coupling case: $\quad \dfrac{w'}{1-s'} \geq 0.35$

$$\frac{2w'_o}{1+s'} \geq 0.85 \quad (3.106)$$

In the above equations, $s' = S/b$, $w' = W/b$, $w'_c = W_c/b$, and $w'_o = W_o/b$ denote the normalized values. The coupling between TEM lines can be expressed

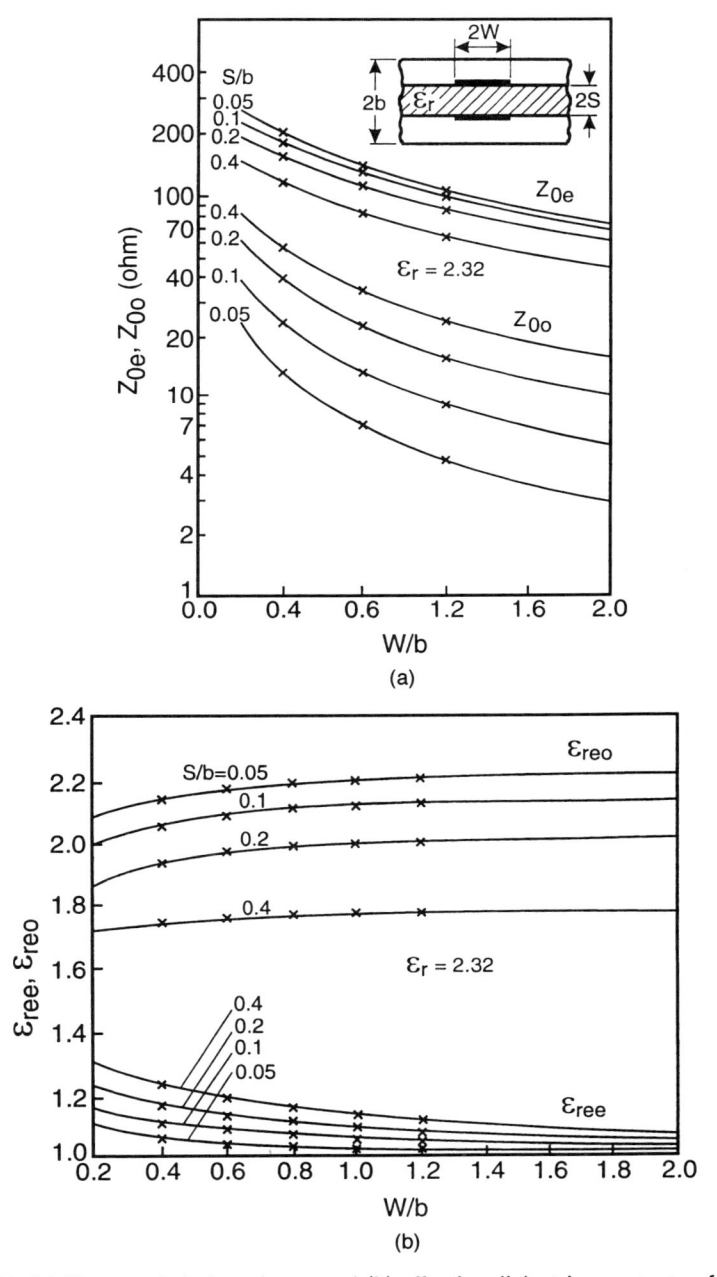

Figure 3.34 (a) Characteristic impedance and (b) effective dielectric constants of coupled broadside coupled suspended microstrip lines. $\epsilon_r = 2.32$. (From [38], ©1988 IEEE. Reprinted with permission.)

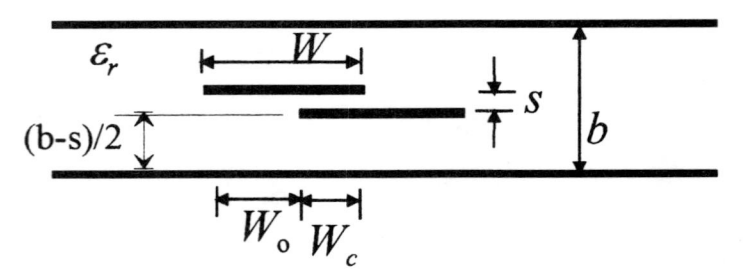

Figure 3.35 Broadside coupled off-set striplines.

in terms of even- and odd-mode characteristic impedances. For a TEM coupler that is matched at all its ports:

$$\frac{Z_{0e}}{Z_0} = \frac{Z_0}{Z_{0o}} \tag{3.107}$$

Defining ρ as

$$\sqrt{\rho} = \frac{Z_{0e}}{Z_0} = \frac{Z_0}{Z_{0o}} \tag{3.108}$$

we obtain the synthesis equations given below:

Tight coupling case: $A = \exp\left[\frac{60\pi^2}{\sqrt{\epsilon_r} Z_0}\left(\frac{1-\rho s'}{\sqrt{\rho}}\right)\right]$

$$B = \frac{A - 2 + \sqrt{A^2 - 4A}}{2}$$

$$p = \frac{(B-1)\left(\frac{1+s'}{2}\right) + \sqrt{(B-1)^2 \left(\frac{1+s'}{2}\right)^2 + 4s'B}}{2}$$

$$r = \frac{s'B}{p}$$

$$C_{fo} = \frac{1}{\pi}\left[-\frac{2}{1-s'}\ln s' + \frac{1}{s'}\ln\left(\frac{pr}{(p+s')(1+p)(r-s')(1-r)}\right)\right]$$

$$C_o = \frac{120\pi \sqrt{\rho}}{\sqrt{\epsilon_r} Z_0}$$

$$w' = \frac{s'(1-s')}{2}(C_o - C_{fo})$$

$$w'_o = \frac{1}{2\pi}\left[(1+s')\ln\frac{p}{r} + (1-s')\ln\left(\frac{(1+p)(r-s')}{(s'+p)(1-r)}\right)\right]$$

Loose coupling case:
$$C_o = \frac{120\pi \sqrt{\rho}}{\sqrt{\epsilon_r} Z_0} \tag{3.109}$$

$$\Delta C = \frac{120\pi}{\sqrt{\epsilon_r} Z_0} \frac{(\rho - 1)}{\sqrt{\rho}}$$

$$K = \frac{1}{\exp(\frac{\pi \Delta C}{2}) - 1}$$

$$a = \sqrt{\left(\frac{(s' - K)}{(s' + 1)}\right)^2 + K} - \frac{(s' - K)}{(s' + 1)}$$

$$q = \frac{K}{a}$$

$$C_{fo} = \frac{2}{\pi}\left[\frac{1}{1+s'}\ln\frac{1+a}{a(1-q)} - \frac{1}{1-s'}\ln q\right]$$

$$w'_c = \frac{1}{\pi}\left[(s'\ln\frac{q}{a} + (1-s')\ln\left(\frac{1-q}{1+a}\right)\right]$$

$$C_f(a = \infty) = -\frac{2}{\pi}\left[\frac{1}{1+s'}\ln\left(\frac{1-s'}{2}\right) + \frac{1}{1-s'}\ln\left(\frac{1+s'}{2}\right)\right]$$

$$w' = \frac{1-s'^2}{4}[C_o - C_{fo} - C_f(a=\infty)]$$

3.8 Slot-Coupled Microstrip Lines

Slot-coupled microstrip lines are shown in Figure 3.36. This configuration is useful for realizing coupling in multilayer MICs. Directional couplers realized using

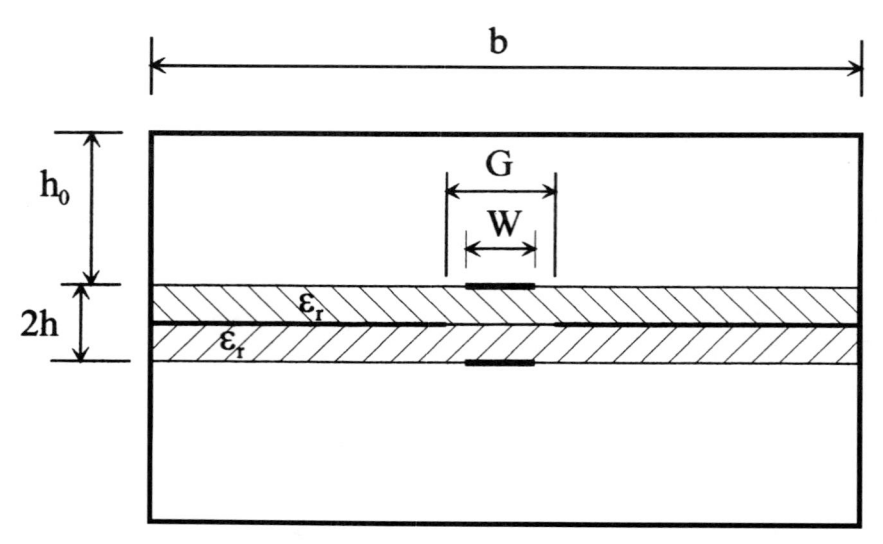

Figure 3.36 Slot-coupled microstrip lines.

this configuration can achieve both tight and loose coupling values. The quasistatic even-mode effective dielectric constant and the characteristic impedance of the structure are given by [40]

$$\epsilon_{ree} = \frac{\epsilon_r \frac{K'(k_1)}{K(k_1)} + \frac{K(k_2)}{K'(k_2)}}{\frac{K'(k_1)}{K(k_1)} + \frac{K(k_2)}{K'(k_2)}} \tag{3.110}$$

$$Z_{0e} = \frac{60\pi}{\sqrt{\epsilon_{ree}}} \frac{1}{\frac{K'(k_1)}{K(k_1)} + \frac{K(k_2)}{K'(k_2)}} \tag{3.111}$$

In (3.112), parameters k_1 and k_2 are given by

$$k_1 = \sqrt{\frac{\sinh^2(\pi G/4h)}{\sinh^2(\pi G/4h) + \cosh^2(\pi W/4h)}} \tag{3.112}$$

$$k_2 = \tanh(\pi W/4h_o) \tag{3.113}$$

Further, $K'(k_i) = K(k'_i)$, where $k'_i = \sqrt{1 - k_i^2}$.

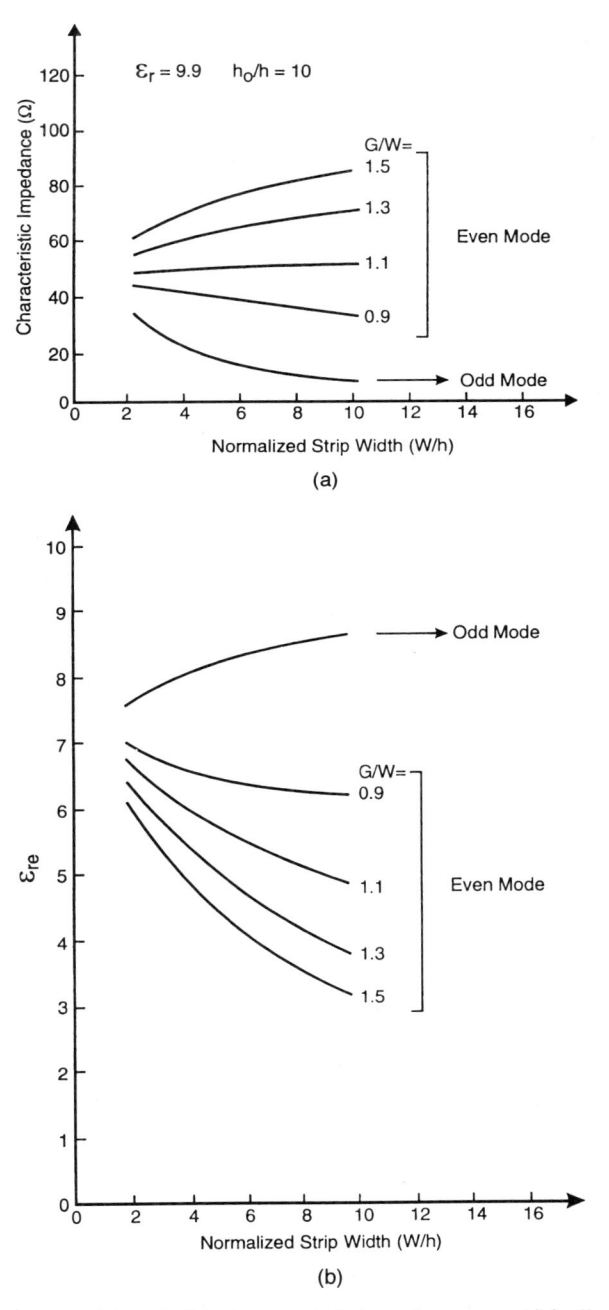

Figure 3.37 Even- and odd-mode (a) characteristic impedances, and (b) effective dielectric constants of slot-coupled microstrip lines shown in Figure 3.36. (From [40], ©1991 IEEE. Reprinted with permission.)

The quasistatic odd-mode effective dielectric constant and the characteristic impedance of the structure are given by

$$\epsilon_{reo} = \frac{\epsilon_r \frac{K(k_3)}{K'(k_3)} + \frac{K(k_4)}{K'(k_4)}}{\frac{K(k_3)}{K'(k_3)} + \frac{K(k_4)}{K'(k_4)}} \tag{3.114}$$

$$Z_{0e} = \frac{60\pi}{\sqrt{\epsilon_{reo}}} \frac{1}{\frac{K(k_3)}{K'(k_3)} + \frac{K(k_4)}{K'(k_4)}} \tag{3.115}$$

with parameters k_3 and k_4 given by

$$k_3 = \tanh(\pi W/4h) \tag{3.116}$$

$$k_4 = \tanh(\pi W/4h_o) \tag{3.117}$$

Figure 3.37(a) shows the variation of even- and odd-mode characteristic impedances of the structure as a function of strip width for a value of $\epsilon_r = 9.9$, while Figure 3.37(b) gives the variation of even- and odd-mode effective dielectric constants.

In this chapter, we have described some commonly used single and coupled strip transmission lines. Because it is not in the scope of this book to cover all transmission lines, readers are referred to the *Transmission Line Design Handbook* by Wadell [35], which provides a comprehensive treatment of printed transmission lines.

References

[1] Collin, R. E., *Field Theory of Guided Waves*, 2nd Ed., IEEE Press, New York, 1991.

[2] Wheeler, H. A., "Formulas for the Skin Effect," *Proc. IRE*, Vol. 30, 1942, pp. 412–424.

[3] Wheeler, H. A., "Transmission Line Properties of Strip on a Dielectric Sheet on a Plane," *IEEE Trans.*, Vol. MTT-25, Aug. 1977, pp. 631–647.

[4] Kammler, D. W., "Calculation of Characteristic Admittances and Coupling Coefficients for Strip Transmission Lines," *IEEE Trans.*, Vol. MTT-16, Nov. 1968, pp. 925–937.

[5] Howe, H., Jr., *Stripline Circuit Design*, Artech House, Dedham, MA, 1974.

[6] Cohn, S. B., "Characteristic Impedance of Shielded Strip Transmission Lines," *IRE Trans.*, Vol. MTT-2, July 1954, pp. 52–55.

[7] Hilberg, W., "From Approximations to Exact Relations for Characteristic Impedances," *IEEE Trans.*, Vol. MTT-17, May 1969, pp. 259–265.

[8] Wheeler, H. A., "Transmission Line Properties of a Stripline Between Parallel Planes," *IEEE Trans.*, Vol. MTT-26, Nov. 1978, pp. 866–876.

[9] Cohn, S. B., "Problems in Strip Transmission Line," *IRE Trans.*, Vol. MTT-3, Mar. 1955, pp. 119–126.

[10] Vendelin, G. D., "Limitations on Stripline Q," *Microwave J.*, Vol. 13, May 1970, pp. 63–69.

[11] Cohn, S. B., "Shielded Coupled Strip Transmission Line," *IRE Trans.*, Vol. MTT-3, Oct. 1955, pp. 29–38.

[12] Bahl, I. J., and P. Bhartia, *Microwave Solid-State Circuit Design*, John Wiley & Sons, NY, 1988.

[13] Gupta, K. C., et al., *Microstrip Lines and Slotlines*, 2nd Ed., Artech House, Norwood, MA, 1996.

[14] Wheeler, H. A., "Transmission Line Properties of Parallel Strips Separated by a Dielectric Sheet," *IEEE Trans.*, Vol. MTT-13, 1965, pp. 172–185.

[15] Schneider, M. V., "Microstrip Lines for Microwave Integrated Circuits," *Bell System Tech. J.*, Vol. 48, 1969, pp. 1,421–1,444.

[16] Pucel, R. A., et al., "Losses in Microstrip," *IEEE Trans.*, Vol. MTT-16, 1968, pp. 342–350. Corrections: *ibid*, MTT-16, 1968, p. 1,064.

[17] Denlinger, E. J., "Losses in Microstrip Lines," *IEEE Trans.*, Vol. MTT-28, June 1980, pp. 513–522.

[18] Hammerstad, E., and O. Jenson, "Accurate Models for Microstrip Computer-Aided Design," *IEEE MTT-S Int. Microwave Symp. Dig.*, 1980, pp. 407–409.

[19] Kobayashi, M., "A Dispersion Formula Satisfying Recent Requirements in Microstrip CAD," *IEEE Trans.*, Vol. MTT-36, Aug. 1988, pp. 1,246–1,250.

[20] Bianco, B., et al., "Frequency Dependence of Microstrip Parameters," *Alta Frequenza*, Vol. 43, 1974, pp. 413–416

[21] Garg, R., and I. J. Bahl, "Characteristics of Coupled Microstrip Lines," *IEEE Trans.*, Vol. MTT-27, July 1979, pp. 700–705.

[22] Kirschning, M., and R. H. Jansen, "Accurate Wide-Range Design Equations for the Frequency-Dependent Characteristics of Parallel Coupled Microstrip Lines," *IEEE Trans.*, Vol. MTT-32, Jan. 1984, pp. 83–90. Corrections: *ibid*, Mar. 1985, p. 288.

[23] Veghte, R. L., and C. A. Balanis, "Dispersion of Transient Signals in Microstrip Transmission Lines," *IEEE Trans.*, Vol. MTT-34, Dec. 1986, pp. 1,427–1,436.

[24] York, R. A., and R. C. Compton, "Experimental Evaluation of Existing CAD Models for Microstrip Dispersion," *IEEE Trans.*, Vol. MTT-38, Mar. 1990, pp. 327–328.

[25] Wen, C. P., "Coplanar Waveguide: A Surface Strip Transmission Line Suitable for Non-reciprocal Gyromagnetic Device Applications," *IEEE Trans.*, Vol. MTT-17, Dec. 1969, pp. 1,087–1,090.

[26] Wen, C. P., "Coplanar-Waveguide Directional Couplers," *IEEE Trans.*, Vol. MTT-18, June 1970, pp. 318–322.

[27] Ghione, G., and C. Naldi, "Parameters of Coplanar Waveguides With Lower Ground Planes," *Electronics Letters*, Vol. 19, Sept. 1983, pp. 734–735.

[28] Rowe, D. A., and B. Y. Lao, "Numerical Analysis of Shielded Coplanar Waveguides," *IEEE Trans.*, Vol. MTT-31, Nov. 1983, pp. 911–915.

[29] Leong, M. S., et al., "Effect of a Conducting Enclosure on the Characteristic Impedance of Coplanar Waveguides," *Microwave J.*, Vol. 29, Aug. 1986, pp. 105–108.

[30] Ghione, G., and C. Naldi, "Analytical Formulas for Coplanar Lines in Hybrid and Monolithic," *Electronics Letters*, Vol. 20, Feb. 1984, pp. 179–181.

[31] Ghione, G., and C. Naldi,"Coplanar Waveguides for MMIC Applications: Effect of Upper Shielding, Conductor Backing, Finite Extent Ground Planes and Line-to-Line Coupling," *IEEE Trans.*, Vol. MTT-35, Mar. 1987, pp. 260–267.

[32] Cheng, K. K. M., "Analysis and Synthesis of Coplanar Coupled Lines on Substrates of Finite Thicknesses, " *IEEE Trans.*, Vol. MTT-44, Apr. 1996, pp. 636–639.

[33] Shih, Y. C., and T. Itoh, "Analysis of Conductor-Backed Coplanar Waveguide," *Electronics Letters*, Vol. 18, June 1982, pp. 538–540.

[34] Kitazawa, T., and R. Mittra, "Quasi-Static Characteristics of Asymmetrical and Coupled Coplanar-Type Transmission Lines," *IEEE Trans.*, Vol. MTT-33, Sept. 1985, pp. 771–778.

[35] Wadell, B. C., *Transmission Line Design Handbook*, Artech House, Norwood, MA, 1991.

[36] Pramanick, P., and P. Bhartia, "CAD Models for Millimeter-Wave Finlines and Suspended-Substrate Microstrip Lines," *IEEE Trans.*, Vol. MTT-33, Dec. 1985, pp. 1,429–1,435.

[37] Tomar, R. S., and P. Bhartia, "New Quasi-Static Models for the Computer-Aided Design of Suspended and Inverted Microstrip Lines," *IEEE Trans.*, Vol. MTT-35, Apr. 1987, pp. 453–457. Corrections: *ibid*, Nov. 1987, p. 1,076.

[38] Bhartia, P., and P. Pramanick, "Computer-Aided Design Models for Broadside-Coupled Striplines and Millimeter-Wave Suspended Substrate Microstrip Lines," *IEEE Trans.*, Vol. MTT-36, Nov. 1988, pp. 1,476–1,481. Corrections: *ibid*, Oct. 1989, p. 1,658.

[39] Shelton, J. P., "Impedances of Offset Parallel-Coupled Strip Transmission Lines," *IEEE Trans.*, Vol. MTT-14, Jan. 1966, pp. 7–15. Corrections: *ibid*, 1996, p. 249.

[40] Wong, M. F., et al., "Analysis and Design of Slot-Coupled Directional Couplers Between Double-Sided Substrate Microstrip Lines," *IEEE Trans.*, Vol. MTT-39, Dec. 1991, pp. 2,123–2,128.

4

Analysis of Uniformly Coupled Lines

Traditionally, two approaches have been used to study coupling between transmission lines (i.e., normal-mode and coupled-mode). The normal-modes of symmetrical coupled lines are the even and odd modes. The coupling between symmetrical lines can be determined in terms of phase velocities and characteristic impedances of the even and odd modes of coupled lines [1–4]. When the coupled lines are asymmetrical, the even and odd modes are no longer the normal modes of the structure and are designated as the c and π modes. Knowing the c- and π-mode parameters, the coupling between asymmetrical lines can be determined.

The normal mode provides an exact method of analysis for coupled lines. In some cases (e.g., when two transmission lines in nonreciprocal media are coupled), however, the task of determining normal-mode parameters is very complicated. In this case, another approach known as the *coupled-mode theory* may prove to be easier and more intutive. This theory is discussed in the next chapter.

In this chapter, we discuss the normal-mode analysis of uniform symmetrical and asymmetrical coupled lines. We first show how the analysis of a symmetrical four-port network is reduced to analyzing two two-port networks using the even- and odd-mode analysis. We also examine conditions under which a four-port network composed of symmetrical coupled lines behaves as a *forward-wave* or *backward-wave* directional coupler as well as the unique properties of backward-wave and forward-wave directional couplers. Section 4.3 describes the normal-mode analysis of asymmetrical coupled lines. Also given are the relations between distributed line parameters (e.g., L, C, L_m, and C_m), phase velocities, and characteristic impedances of normal modes along with the

Z (impedance) parameters of a four-port network consisting of asymmetrical coupled lines. Once we know the Z-parameters of a linear network, we can determine the response of the network to any arbitrary excitation and termination.

Determining normal-mode parameters of asymmetrical coupled lines is generally quite complicated. Further, design data are available in the literature for only a few cases. In Section 4.3, we describe an approximate method from which the parameters of asymmetrical coupled lines are determined from the data of symmetrical coupled lines. Finally, we discuss the design of asymmetrical backward-wave and forward-wave directional couplers.

4.1 Even- and Odd-Mode Analysis of Symmetrical Networks

A four-port network (shown in Figure 4.1) is assumed to be symmetrical about the plane PP'. It is also assumed that the impedances terminating various ports are the same. When one or more ports of the network are connected to a source(s), waves propagating in either direction are generally set up on both the lines. These waves are referred to as *incident* or *reflected*.

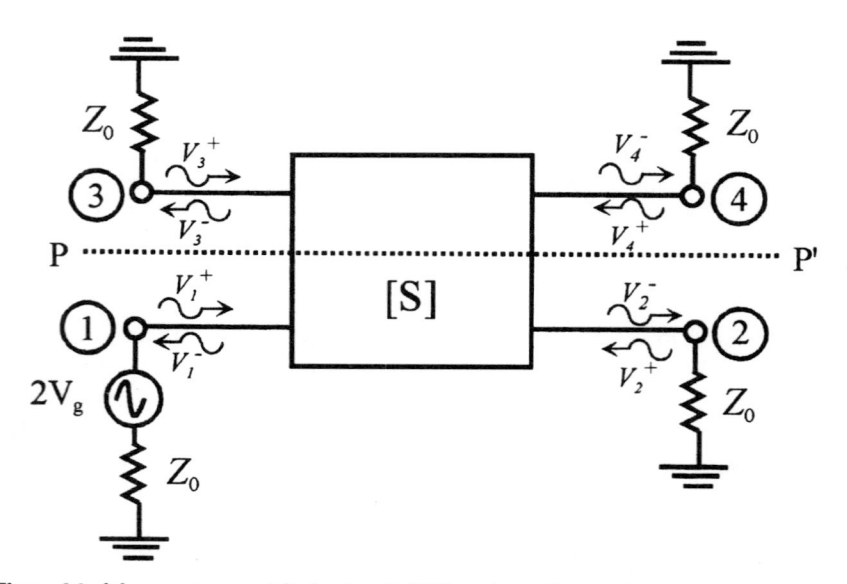

Figure 4.1 A four-port symmetrical network. PP' is a plane of symmetry.

The relationship between incident and reflected voltage waves at different ports of the network as shown in Figure 4.1 can be expressed as

$$\begin{bmatrix} V_1^- \\ V_2^- \\ V_3^- \\ V_4^- \end{bmatrix} = [\mathbf{S}] \begin{bmatrix} V_1^+ \\ V_2^+ \\ V_3^+ \\ V_4^+ \end{bmatrix} \qquad (4.1)$$

where

$$[\mathbf{S}] = \begin{bmatrix} S_{11} & S_{12} & S_{13} & S_{14} \\ S_{21} & S_{22} & S_{23} & S_{24} \\ S_{31} & S_{32} & S_{33} & S_{34} \\ S_{41} & S_{42} & S_{43} & S_{44} \end{bmatrix} \qquad (4.2)$$

denotes the scattering matrix of the network. Note that because all the ports of the network are assumed to be terminated in identical loads (denoted by impedance Z_0), the *unnormalized* scattering matrix is identical to the *normalized* scattering matrix as discussed in Section 2.3.4.

Not all the elements of the scattering matrix of the network shown in Figure 4.1 are independent. For example, because of the assumed symmetry and the reciprocal nature of the structure:

$$S_{21} = S_{12}, \ S_{31} = S_{13}, \ S_{41} = S_{14}, \ S_{32} = S_{23}$$

$$S_{42} = S_{24}, \ S_{43} = S_{34}$$

$$S_{33} = S_{11}, \ S_{44} = S_{22}, \ S_{34} = S_{12}, \ S_{23} = S_{14}$$

The scattering matrix of (4.2) can therefore be expressed as

$$[\mathbf{S}] = \begin{bmatrix} S_{11} & S_{21} & S_{31} & S_{41} \\ S_{21} & S_{22} & S_{41} & S_{42} \\ S_{31} & S_{41} & S_{11} & S_{21} \\ S_{41} & S_{42} & S_{21} & S_{22} \end{bmatrix} \qquad (4.3)$$

In a more compact notation, the matrix (4.3) can be expressed as

$$[\mathbf{S}] = \begin{bmatrix} [\mathbf{S}_A] & [\mathbf{S}_B] \\ [\mathbf{S}_B] & [\mathbf{S}_A] \end{bmatrix} \qquad (4.4)$$

where

$$[S_A] = \begin{bmatrix} S_{11} & S_{21} \\ S_{21} & S_{22} \end{bmatrix} \quad (4.5)$$

and

$$[S_B] = \begin{bmatrix} S_{31} & S_{41} \\ S_{41} & S_{42} \end{bmatrix} \quad (4.6)$$

(4.1) then becomes

$$\begin{bmatrix} V_1^- \\ V_2^- \\ V_3^- \\ V_4^- \end{bmatrix} = \begin{bmatrix} [S_A] & [S_B] \\ [S_B] & [S_A] \end{bmatrix} \begin{bmatrix} V_1^+ \\ V_2^+ \\ V_3^+ \\ V_4^+ \end{bmatrix} \quad (4.7)$$

It may be noted that symmetrical structures show some special electrical behavior. If the symmetrical ports 1 and 3 are connected to *equal magnitude* and *in-phase* sources, the voltages at ports 1 and 3 will be equal in magnitude and in phase. Similarly, the voltages will be equal in magnitude and in-phase at ports 2 and 4. This scheme is shown in Figure 4.2(a) and is called an *even-mode excitation*. Further, if the symmetrical ports 1 and 3 are connected to *equal magnitude* but *out-of-phase* sources, the voltages at ports 1 and 3 will be equal in magnitude but out of phase. The voltages at ports 2 and 4 will also be equal in magnitude but out of phase. This excitation scheme is shown in Figure 4.2(b) and is called an *odd mode* excitation. The algebraic sum of these two excitations is equivalent to the excitation scheme shown in Figure 4.1.

4.1.1 Even-Mode Excitation

Figure 4.2(a) shows a symmetrical structure excited by equal magnitude and in-phase sources at ports 1 and 3. The incident and reflected voltages set up at the different ports are also shown in the same figure. Let $V_3^\pm = V_1^\pm = V_{1e}^\pm$ and $V_4^\pm = V_2^\pm = V_{2e}^\pm$, where the suffix e has been used to denote the even mode. Further, by the substitution of $V_1^\pm$, $V_2^\pm$, $V_3^\pm$, and $V_4^\pm$ in (4.7), the reflected voltages can be expressed in terms of incident voltages as follows:

$$\begin{bmatrix} V_{1e}^- \\ V_{2e}^- \\ V_{1e}^- \\ V_{2e}^- \end{bmatrix} = \begin{bmatrix} [S_A] & [S_B] \\ [S_B] & [S_A] \end{bmatrix} \begin{bmatrix} V_{1e}^+ \\ V_{2e}^+ \\ V_{1e}^+ \\ V_{2e}^+ \end{bmatrix} \quad (4.8)$$

Analysis of Uniformly Coupled Lines 127

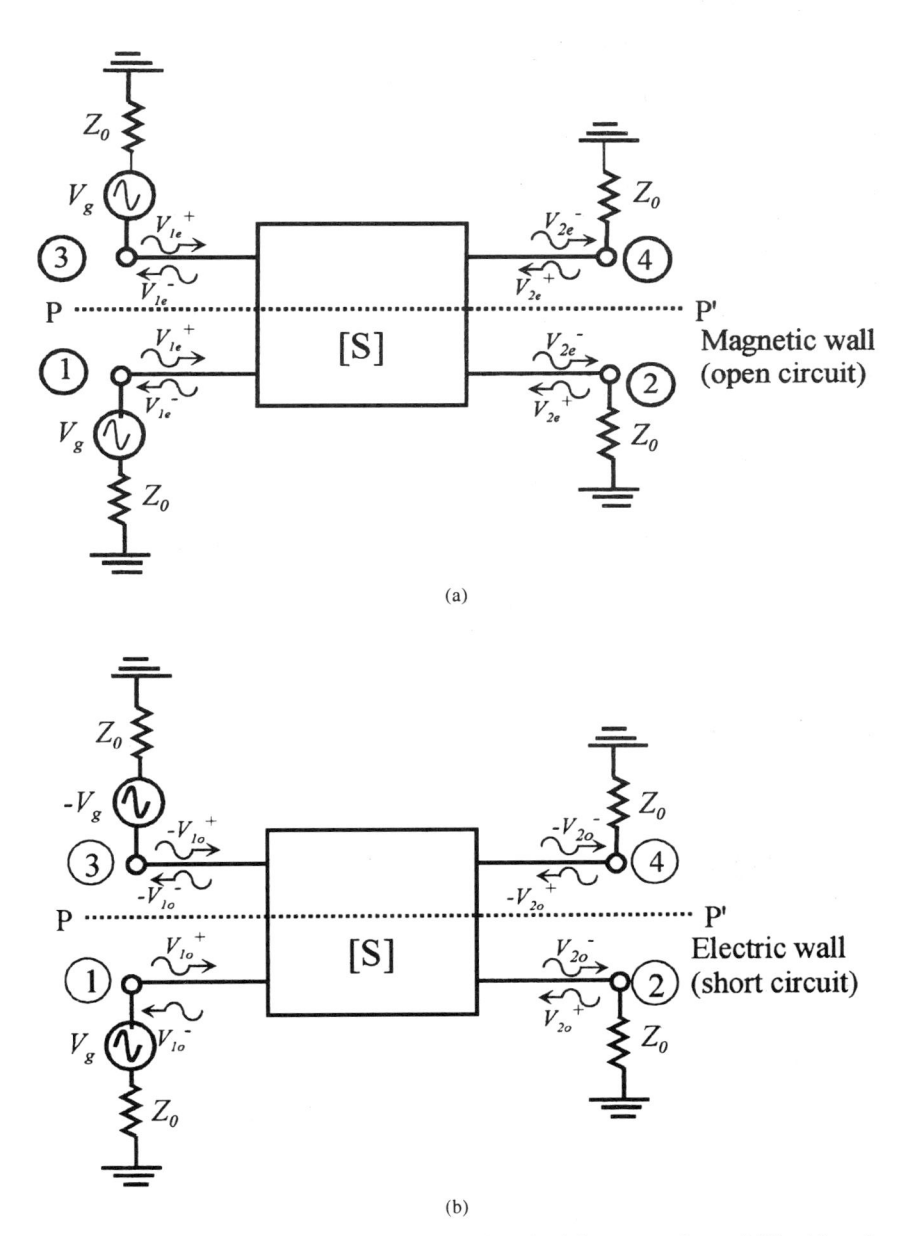

Figure 4.2 A four-port symmetrical network excited by (a) even-mode, and (b) odd-mode sources.

From (4.8), we obtain

$$\begin{bmatrix} V_{1e}^- \\ V_{2e}^- \end{bmatrix} = ([S_A] + [S_B]) \begin{bmatrix} V_{1e}^+ \\ V_{2e}^+ \end{bmatrix} \tag{4.9}$$

4.1.2 Odd-Mode Excitation

Figure 4.2(b) shows the symmetrical structure excited by equal magnitude but out-of-phase sources at ports 1 and 3. The incident and reflected voltages set up at various ports by the odd-mode sources are also shown in the same figure. Let $V_1^\pm = -V_3^\pm = V_{1o}^\pm$ and $V_2^\pm = -V_4^\pm = V_{2o}^\pm$, where the suffix o has been used to denote that the quantities correspond to the odd mode. Further, by substitution of $V_1^\pm$, $V_2^\pm$, $V_3^\pm$, and $V_4^\pm$ in (4.7), the reflected voltages at various ports can be expressed in terms of incident voltages as follows:

$$\begin{bmatrix} V_{1o}^- \\ V_{2o}^- \\ -V_{1o}^- \\ -V_{2o}^- \end{bmatrix} = \begin{bmatrix} [S_A] & [S_B] \\ [S_B] & [S_A] \end{bmatrix} \begin{bmatrix} V_{1o}^+ \\ V_{2o}^+ \\ -V_{1o}^+ \\ -V_{2o}^+ \end{bmatrix} \tag{4.10}$$

which gives

$$\begin{bmatrix} V_{1o}^- \\ V_{2o}^- \end{bmatrix} = ([S_A] - [S_B]) \begin{bmatrix} V_{1o}^+ \\ V_{2o}^+ \end{bmatrix} \tag{4.11}$$

Scattering Matrix in Terms of Even- and Odd-Mode Parameters

Equation 4.9 can be written as

$$\begin{bmatrix} V_{1e}^- \\ V_{2e}^- \end{bmatrix} = [S_e] \begin{bmatrix} V_{1e}^+ \\ V_{2e}^+ \end{bmatrix} \tag{4.12}$$

where

$$[S_e] = [S_A] + [S_B] \tag{4.13}$$

Similarly, (4.11) can be written as

$$\begin{bmatrix} V_{1o}^- \\ V_{2o}^- \end{bmatrix} = [S_o] \begin{bmatrix} V_{1o}^+ \\ V_{2o}^+ \end{bmatrix} \tag{4.14}$$

where

$$[S_o] = [S_A] - [S_B] \tag{4.15}$$

From (4.13) and (4.15) we obtain

$$[S_A] = \frac{[S_e] + [S_o]}{2} \tag{4.16}$$

and
$$[S_B] = \frac{[S_e] - [S_o]}{2} \tag{4.17}$$

Therefore, if the scattering matrices $[S_e]$ and $[S_o]$, which are matrices of order 2×2, are known, then the scattering matrices $[S_A]$ and $[S_B]$ can be determined using (4.16) and (4.17), respectively. Further, if the scattering matrices $[S_A]$ and $[S_B]$ are known, the complete scattering matrix of the four-port network can be determined using (4.4).

Let the elements of the scattering matrices $[S_e]$ and $[S_o]$ be given by

$$[S_e] = \begin{bmatrix} S_{11e} & S_{21e} \\ S_{21e} & S_{22e} \end{bmatrix} \tag{4.18}$$

and

$$[S_o] = \begin{bmatrix} S_{11o} & S_{21o} \\ S_{21o} & S_{22o} \end{bmatrix} \tag{4.19}$$

Using (4.16) and (4.17), the 2×2 scattering matrices $[S_A]$ and $[S_B]$ are then found to be

$$[S_A] = \begin{bmatrix} \frac{S_{11e}+S_{11o}}{2} & \frac{S_{21e}+S_{21o}}{2} \\ \frac{S_{21e}+S_{21o}}{2} & \frac{S_{22e}+S_{22o}}{2} \end{bmatrix} \tag{4.20}$$

and

$$[S_B] = \begin{bmatrix} \frac{S_{11e}-S_{11o}}{2} & \frac{S_{21e}-S_{21o}}{2} \\ \frac{S_{21e}-S_{21o}}{2} & \frac{S_{22e}-S_{22o}}{2} \end{bmatrix} \tag{4.21}$$

Further, using (4.4), the various elements of the scattering matrix of the four-port network are given as follows:

$$S_{11} = \frac{S_{11e} + S_{11o}}{2}, \quad S_{12} = S_{21}, \quad S_{13} = S_{31}, \quad S_{14} = S_{41}$$

$$S_{21} = \frac{S_{21e} + S_{21o}}{2}, \quad S_{22} = \frac{S_{22e} + S_{22o}}{2}, \quad S_{23} = S_{41},$$

$$S_{24} = \frac{S_{22e} - S_{22o}}{2}, \quad S_{31} = \frac{S_{11e} - S_{11o}}{2}, \quad S_{32} = S_{41}, \tag{4.22}$$

$$S_{33} = S_{11}, \quad S_{34} = S_{21}$$

$$S_{41} = \frac{S_{21e} - S_{21o}}{2}, \quad S_{42} = S_{24}, \quad S_{43} = S_{21}, \quad S_{44} = S_{22}$$

Determination of Even- and Odd-Mode Scattering Matrices [S_e] and [S_o]

The matrix [S_e] denotes the scattering matrix of one-half of the network (between ports 1 and 2) when the structure is excited by equal magnitude and in-phase sources as shown in Figure 4.2(a). It so happens that in this case (even-mode excitation), the plane of symmetry PP' behaves as a magnetic wall (open-circuit). The scattering matrix of one-half of the original network as shown in Figure 4.3 where the symmetrical plane PP' is replaced by a magnetic wall (open-circuit) denotes the scattering matrix [S_e].

Similarly, the matrix [S_o] denotes the scattering matrix of one-half of the network when equal magnitude but out-of-phase sources are applied at the symmetrical ports as shown in Figure 4.2(b). In this case (odd mode excitation), the symmetrical plane PP' behaves as an electric wall (short-circuit). The scattering matrix of one-half of the network as shown in Figure 4.3 where the symmetrical plane PP' is replaced by an electric wall (short-circuit) thus denotes the scattering matrix [S_o].

4.2 Directional Couplers Using Uniform Coupled Lines

Having described the basic theory behind even- and odd-mode analysis, we now discuss the conditions under which a symmetrical four-port network composed of uniformly coupled lines as shown in Figure 4.4 can act as a directional coupler. The scattering parameters of an ideal directional coupler were discussed in Chapter 2, which showed that if all ports of a four-port network are matched, then the network behaves like a directional coupler. Because the network as shown in Figure 4.4 is assumed to be symmetrical about the mid-plane PP', matching of ports 1 and 2 automatically ensures that the ports 3 and 4 are

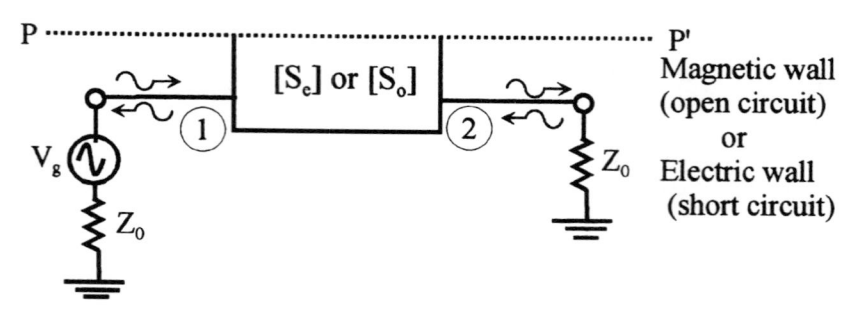

Figure 4.3 Reduced circuit for determining scattering matrices of the even- and odd-modes of the structure shown in Figure 4.1.

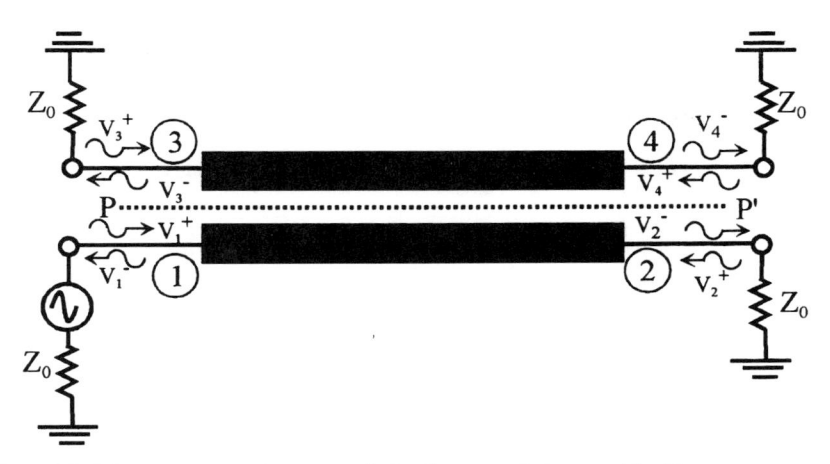

Figure 4.4 A four-port network composed of uniform coupled symmetrical lines.

also matched. Therefore, the condition

$$S_{11} = S_{22} = 0$$

leads to the result that the network is a directional coupler.

In terms of even- and odd-mode reflection coefficients, the scattering parameters S_{11} and S_{22} are given by (4.22) as

$$S_{11} = \frac{S_{11e} + S_{11o}}{2} \quad (4.23)$$

$$S_{22} = \frac{S_{22e} + S_{22o}}{2} \quad (4.24)$$

The equivalent circuit for determining the even-mode scattering parameters S_{11e} and S_{22e} is shown in Figure 4.5(a), where Z_{0e} and β_e denote the characteristic impedance and propagation constant of the even-mode of symmetrical coupled lines. Similarly, the equivalent circuit for determining the odd-mode scattering parameters S_{11o} and S_{22o} is shown in Figure 4.5(b), where Z_{0o} and β_o denote the characteristic impedance and propagation constant of the odd mode of symmetrical coupled lines.

To obtain $S_{11} = S_{22} = 0$, which are the conditions for realizing a directional coupler, the following possibilites exist:

Case I

$$S_{11e} = S_{11o} = S_{22e} = S_{22o} = 0 \quad (4.25)$$

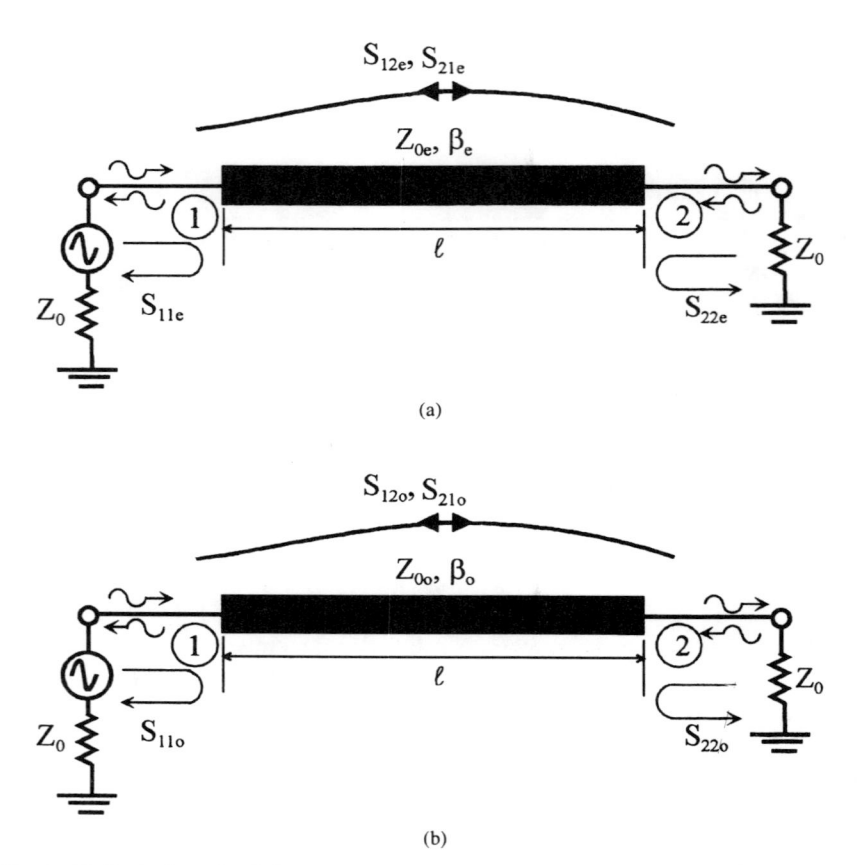

Figure 4.5 Equivalent circuit for determining scattering matrix of the (a) even-mode, and (b) odd-mode of the structure shown in Figure 4.4.

When the above values are substituted in (4.23) and (4.24), we obtain $S_{11} = S_{22} = 0$. Further, using (4.22) we obtain

$$S_{13} = S_{31} = S_{42} = S_{24} = 0 \tag{4.26}$$

Therefore, in this case (when (4.25) is satisfied), no power is coupled to the backward port. For example, if power is incident at port 1, then no power is coupled to port 3 on the coupled line. Similarly, no power is coupled between ports 2 and 4. However, power can be coupled between ports 1 and 4 (or between ports 2 and 3). These types of couplers are called *forward-wave* or *codirectional couplers* and are discussed further in this chapter.

Case II

From (4.23) and (4.24), we see that $S_{11} = 0$ and $S_{22} = 0$ can also be obtained if the following conditions are satisfied:

$$S_{11e} = -S_{11o} \tag{4.27}$$

and

$$S_{22e} = -S_{22o} \tag{4.28}$$

where $S_{11e}, S_{11o}, S_{22e}$, and S_{22o} are not equal to zero. Using (4.22), we then find that

$$S_{31} \neq 0, \quad S_{42} \neq 0$$

In this case, power is thus coupled to the backward port. From the properties of a directional coupler, we know that if $S_{31} \neq 0$, then either

$$S_{41} = 0 \text{ or } S_{21} = 0$$

To ensure that no power is coupled in the forward direction on the coupled line, it is required that $S_{41} = 0$. Using (4.22), we find that this condition is satisfied when

$$S_{21e} = S_{21o} \tag{4.29}$$

Therefore, when conditions given by (4.27) through (4.29) hold, the structure behaves as a backward-wave directional coupler.

4.2.1 Forward-Wave (or Codirectional) Directional Couplers

As discussed earlier, the four-port network as shown in Figure 4.4 behaves like a forward-wave directional coupler when (4.25) is satisfied. Referring to the equivalent circuits shown in Figures 4.5(a) and (b) for the even and odd modes, respectively, the above condition is satisfied for any arbitrary length l of the coupling section if

$$Z_{0e} = Z_{0o} = Z_0$$

The above condition can be *nearly* met in practice by keeping a relatively large spacing between the lines. Substituting

$$S_{11e} = S_{22e} = 0$$

in the following equation (this equation follows from the unitary property of the scattering matrix):

$$|S_{11e}|^2 + |S_{21e}|^2 = |S_{22e}|^2 + |S_{21e}|^2 = 1$$

we obtain

$$|S_{21e}| = 1$$

or
$$S_{21e} = e^{-j\psi_e} \tag{4.30}$$

where ψ_e denotes the phase difference between ports 1 and 2 for the even-mode signal. Similarly, by substituting

$$S_{11o} = S_{22o} = 0$$

in the equation

$$|S_{11o}|^2 + |S_{21o}|^2 = |S_{22o}|^2 + |S_{21o}|^2 = 1$$

we obtain

$$S_{21o} = e^{-j\psi_o} \tag{4.31}$$

where ψ_o denotes the phase difference between ports 1 and 2 for the odd-mode signal.

Because the coupled structure is assumed to be uniform, we can further write

$$\psi_e = \beta_e l \tag{4.32}$$

and

$$\psi_o = \beta_o l \tag{4.33}$$

where β_e and β_o denote the propagation constants of the even and odd-mode signals, respectively, and l is the length of the coupling section.

Further, using (4.22), the scattering parameters of an ideal forward-wave directional coupler are given by

$$S_{11} = S_{22} = S_{33} = S_{44} = 0 \tag{4.34}$$

$$S_{12} = S_{21} = S_{34} = S_{43} = \frac{S_{21e} + S_{21o}}{2} = \frac{e^{-j\beta_e l} + e^{-j\beta_o l}}{2}$$

$$= e^{-j(\beta_e + \beta_o)l/2} \cos\left[\frac{(\beta_e - \beta_o)l}{2}\right] \tag{4.35}$$

$$S_{14} = S_{41} = S_{23} = S_{32} = \frac{S_{21e} - S_{21o}}{2} = \frac{e^{-j\beta_e l} - e^{-j\beta_o l}}{2}$$

$$= -j e^{-j(\beta_e + \beta_o)l/2} \sin\left[\frac{(\beta_e - \beta_o)l}{2}\right] \tag{4.36}$$

$$S_{13} = S_{31} = S_{24} = S_{42} = 0 \tag{4.37}$$

The fractional power coupled from port 1 to port 4 is thus given by

$$\frac{P_4}{P_1} = |S_{41}|^2 = \sin^2\left[\frac{(\beta_e - \beta_o)l}{2}\right] \quad (4.38)$$

while the fractional power coupled from port 1 to port 2 is given by

$$\frac{P_2}{P_1} = |S_{21}|^2 = \cos^2\left[\frac{(\beta_e - \beta_o)l}{2}\right] \quad (4.39)$$

Notice that $|S_{21}|^2 + |S_{41}|^2 = 1$, accounting for all the incident power. It should be apparent that forward-wave directional couplers cannot be obtained using TEM mode lines such as coaxial lines because for the TEM mode, the propagation constants of the even and odd modes are equal, and therefore as shown by (4.36), there is no coupling between ports 1 and 4 or between ports 2 and 3. Forward-wave coupling exists only in non-TEM lines such as metallic waveguides, fin lines, and dielectric waveguides and can also exist in quasi-TEM-mode transmission lines such as microstrip lines at high frequencies. In these transmission line structures, in general, the phase velocities of the even and odd modes are not equal.

Remarks on Forward-Wave or Codirectional Couplers

1. From (4.38), we see that complete power can be transferred between lines if the length l of the directional coupler is chosen as

$$l = \frac{\pi}{|\beta_e - \beta_o|} = \frac{\lambda_0}{2|(\sqrt{\epsilon_{ree}} - \sqrt{\epsilon_{reo}})|} \quad (4.40)$$

   The above result is significant in the sense that even for arbitrarily small values of difference in the propagation constants of even and odd modes, complete power can be transferred between the lines if the length of the coupler is chosen according to (4.40). We show later that it is not possible to completely transfer power from one line to another in the case of backward-wave directional couplers.

2. By comparing (4.35) and (4.36), we find that the phase difference between S_{41} and S_{21} is 90 degree. The wave on the "coupled" line is thus 90 deg out of phase with the "direct" wave.

3. In deriving equations for the forward-wave coupling, we assumed that the condition given by (4.25) is satisfied, which leads to zero coupling between ports 1 and 3 or between ports 2 and 4. The directivity and isolation of the coupler are thus infinite. In general, however, the above condition cannot be completely satisfied. Therefore, some finite amount of backward-wave coupling, however small, always exists between coupled lines. The exact amount of backward-wave coupling

(S_{31} and S_{42}) can be determined using (4.22), if the values of S_{11e}, S_{11o}, S_{22e}, and S_{22o} are known.

4.2.2 Backward-Wave Directional Couplers

As discussed earlier, a symmetrical four-port network as shown in Figure 4.4 behaves like a backward-wave directional coupler if the following conditions are satisfied:

$$S_{11o} = -S_{11e}$$
$$S_{22o} = -S_{22e}$$

and

$$S_{21o} = S_{21e}$$

where S_{11o}, S_{11e}, S_{22o}, and S_{22e} are not equal to zero.

The above conditions are easily satisfied if the coupled lines are of the TEM type similar to striplines and the even- and odd-mode characteristic impedances of the lines are properly chosen. The equivalent circuits for the even and odd modes are shown in Figures 4.5(a) and (b) respectively. Using these equivalent ciruits, the ABCD matrices of the coupler can be determined for the even and odd modes. For example, using Table 2.2, the ABCD matrices for the even and odd modes are given, respectively, by

$$\begin{bmatrix} A_e & B_e \\ C_e & D_e \end{bmatrix} = \begin{bmatrix} \cos(\beta l) & jZ_{0e}\sin(\beta l) \\ \frac{j\sin(\beta l)}{Z_{0e}} & \cos(\beta l) \end{bmatrix} \quad (4.41)$$

and

$$\begin{bmatrix} A_o & B_o \\ C_o & D_o \end{bmatrix} = \begin{bmatrix} \cos(\beta l) & jZ_{0o}\sin(\beta l) \\ \frac{j\sin(\beta l)}{Z_{0o}} & \cos(\beta l) \end{bmatrix} \quad (4.42)$$

where we assume that the propagation constants are the same for the even and odd modes and are denoted by β. Because the coupled lines are terminated in an impedance of Z_0, the even- and odd-mode reflection coefficients can be shown from (2.98) and (2.99) to be,

$$S_{11e} = S_{22e} = \frac{j\left(\frac{Z_{0e}}{Z_0} - \frac{Z_0}{Z_{0e}}\right)\sin\beta l}{2\cos\beta l + j\left(\frac{Z_{0e}}{Z_0} + \frac{Z_0}{Z_{0e}}\right)\sin\beta l} \quad (4.43)$$

and

$$S_{11o} = S_{22o} = \frac{j\left(\frac{Z_{0o}}{Z_0} - \frac{Z_0}{Z_{0o}}\right)\sin\beta l}{2\cos\beta l + j\left(\frac{Z_{0o}}{Z_0} + \frac{Z_0}{Z_{0o}}\right)\sin\beta l} \quad (4.44)$$

Comparing (4.43) and (4.44), we find that the conditions $S_{11e} = -S_{11o}$ and $S_{22e} = -S_{22o}$ are satisfied for any *arbitrary* value of length l when

$$\frac{Z_{0e}}{Z_0} = \frac{Z_0}{Z_{0o}}$$

or

$$Z_{0e} Z_{0o} = Z_0^2 \tag{4.45}$$

The scattering parameters S_{21o} and S_{21e} can be computed from (2.100) as follows:

$$S_{21e} = \frac{2}{2\cos\beta l + j\left(\frac{Z_{0e}}{Z_0} + \frac{Z_0}{Z_{0e}}\right)\sin\beta l} \tag{4.46}$$

and

$$S_{21o} = \frac{2}{2\cos\beta l + j\left(\frac{Z_{0o}}{Z_0} + \frac{Z_0}{Z_{0o}}\right)\sin\beta l} \tag{4.47}$$

We see that when (4.45) holds, $S_{21e} = S_{21o}$ as required by (4.29). Therefore, (4.45) gives the necessary condition for TEM backward-wave directional couplers. Once the values of S_{11o}, S_{11e}, S_{21o}, and S_{21e} are known, the scattering parameters of an ideal backward-wave directional coupler can be easily determined using (4.22) as follows:

$$S_{11} = S_{22} = S_{33} = S_{44} = 0 \tag{4.48}$$

$$S_{14} = S_{41} = S_{23} = S_{32} = 0 \tag{4.49}$$

$$S_{12} = S_{21} = S_{34} = S_{43} = \frac{S_{21e} + S_{21o}}{2} = S_{21e} = S_{21o}$$

$$= \frac{2}{2\cos\beta l + j\left(\frac{Z_{0e}}{Z_0} + \frac{Z_{0o}}{Z_0}\right)\sin\beta l} \tag{4.50}$$

$$S_{13} = S_{31} = S_{24} = S_{42} = \frac{S_{11e} - S_{11o}}{2} = S_{11e} = -S_{11o}$$

$$= \frac{j\left(\frac{Z_{0e}}{Z_0} - \frac{Z_{0o}}{Z_0}\right)\sin\beta l}{2\cos\beta l + j\left(\frac{Z_{0e}}{Z_0} + \frac{Z_{0o}}{Z_0}\right)\sin\beta l} \tag{4.51}$$

From (4.45) and (4.50), we obtain

$$S_{21} = \frac{\sqrt{1-k^2}}{\sqrt{1-k^2}\cos\theta + j\sin\theta} \tag{4.52}$$

Further, from (4.45) and (4.51)

$$S_{31} = \frac{jk \sin \theta}{\sqrt{1 - k^2} \cos \theta + j \sin \theta} \qquad (4.53)$$

where $\theta = \beta l$, and

$$k = \frac{Z_{0e} - Z_{0o}}{Z_{0e} + Z_{0o}} \qquad (4.54)$$

The maximal amount of coupling between ports 1 and 3 (or between ports 2 and 4) occurs when

$$\theta = \beta l = \frac{\pi}{2} \quad \text{rads} \qquad (4.55)$$

or

$$l = \frac{\pi}{2\beta} = \frac{\lambda_g}{4}$$

where λ_g denotes the guide wavelength in the medium of the transmission line. The maximum value of coupling is found by substituting $\theta = \frac{\pi}{2}$ in (4.53), which gives

$$|S_{13}| = |S_{31}| = |S_{24}| = |S_{42}| = k \qquad (4.56)$$

Further, when $\theta = \frac{\pi}{2}$,

$$|S_{12}| = |S_{21}| = |S_{34}| = |S_{43}| = \sqrt{1 - k^2} \qquad (4.57)$$

Thus, at the frequency where $\theta = \beta l = \pi/2$, the scattering matrix of a backward-wave coupler can be represented as follows:

$$[\mathbf{S}] = \begin{bmatrix} 0 & -j\sqrt{1-k^2} & k & 0 \\ -j\sqrt{1-k^2} & 0 & 0 & k \\ k & 0 & 0 & -j\sqrt{1-k^2} \\ 0 & k & -j\sqrt{1-k^2} & 0 \end{bmatrix} \qquad (4.58)$$

The above scattering matrix is valid at the frequency where the length of the coupler is a quarter-wave long. We can calculate, however, the frequency response of the backward-wave coupler at any other frequency using (4.48), (4.49), (4.52), and (4.53). The frequency response of ideal backward-wave couplers of various coupling values is given in Chapter 6.

From (4.45) and (4.54), the relationships between even- and odd-mode characteristic impedances and the voltage coupling coefficient k are given by

$$Z_{0e} = Z_0 \sqrt{\frac{1+k}{1-k}}$$

$$Z_{0o} = Z_0 \sqrt{\frac{1-k}{1+k}}$$

(4.59)

Example 4.1

Compute even- and odd-mode characteristic impedances of a 20-dB quarter-wave, 50-ohm backward-wave coupler.

For a 20-dB coupler, $k = 10^{-20/20} = 0.1$. Given that the terminal impedance is 50 ohms, then from (4.59), $Z_{0e} = 55.3$ ohms and $Z_{0o} = 45.2$ ohms.

Coupling k in Terms of Capacitance Parameters

Substituting values of Z_{0e} and Z_{0o} from (3.45) and (3.46), respectively, along with (3.41) and (3.43) in (4.54), we obtain

$$k = \frac{C_m}{C_a + C_m}$$

(4.60)

where C_a and C_m denote the capacitances of coupled lines as shown in Figure 3.5 ($C_b = C_a$).

Remarks on Backward-Wave Directional Couplers

1. Equation 4.53 shows that there exists a maximum value of backward-wave coupling that can be achieved between two coupled lines. The maximum value of coupling which is given by (4.56) occurs when the length of the coupler is a quarter-wave long (or odd multiples thereof). This is unlike the symmetrical forward-wave directional couplers case where arbitrary coupling can be achieved between the lines by properly choosing the length of the coupling section.

2. By comparing (4.52) and (4.53), we find that the wave coupled to the "backward" port (S_{31}) is 90 deg out of phase with the wave coupled to the "direct" port (S_{21}). This relationship is independent of the electrical length of the coupling section. These types of couplers are thus capable of being used as quadrature phase-shifters over a wide frequency range.

4.3 Uniformly Coupled Asymmetrical Lines

Symmetrical coupled lines represent a very useful but *restricted* class of coupled lines. In many practical cases, it might be more useful or even necessary to design components using asymmetrical coupled lines. For example, the bandwidth of a forward-wave directional coupler using asymmetrical coupled lines is greater than one formed using symmetrical coupled lines. Also, in some situations, the terminal impedances of one of the coupled lines may be different from those of the other. It may then be more useful to choose two coupled lines with different characteristic impedances.

In this section, the analysis and design of asymmetrical coupled quasi-TEM mode lines is described. The normal-mode parameters of asymmetrical coupled lines are first defined. It is shown that six independent parameters are required to characterize asymmetrical coupled lines. The relation between normal-mode parameters (i.e., characteristic impedances, phase velocities) and line parameters (i.e., per unit length inductance, capacitance) are derived. Because symmetrical lines are a special case of asymmetrical coupled lines, various expressions given in the following sections can also be used for symmetrical coupled lines. A concise but excellent analysis of asymmetrical coupled lines is found in [5], which forms the principal basis for the analysis below.

4.3.1 Parameters of Asymmetrical Coupled Lines

A set of two coupled lines can support two fundamental independent modes of propagation (called normal modes). For asymmetrical coupled lines, the two normal modes of propagation are known as the c and π modes. Strictly speaking, a structure composed of two coupled lines can support four independent modes of propagation: two traveling in the backward direction and two traveling in the forward direction. The characteristics (phase velocity and characteristic impedance) of a backward-traveling mode, however, are the same as those of the corresponding forward-wave traveling mode.

c Mode

Figure 4.6 shows two uniformly coupled quasi-TEM mode asymmetrical transmission lines. The assumption of quasi-TEM mode is made because it is possible to define unique voltages and currents in this case as compared with non-TEM mode transmission lines. Let the voltage and current waves on asymmetrical coupled lines for the c mode be denoted as shown. The forward-traveling voltage waves on lines 1 and 2 are denoted by $V_{1c}^{+} e^{-\gamma_c z}$ and $V_{2c}^{+} e^{-\gamma_c z}$, respectively; and the corresponding current waves by $I_{1c}^{+} e^{-\gamma_c z}$ and $I_{2c}^{+} e^{-\gamma_c z}$, respectively.

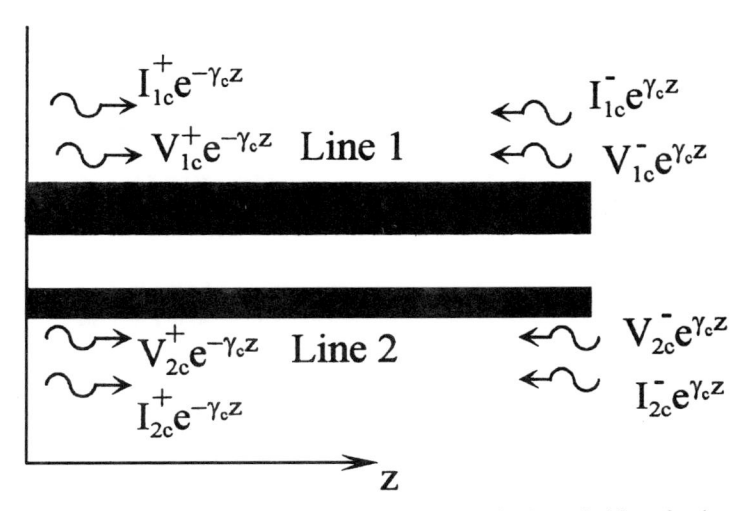

Figure 4.6 Voltage and current waves on uniform asymmetrical coupled lines for the c mode.

Similarly, $V_{1c}^- e^{\gamma_c z}$ and $V_{2c}^- e^{\gamma_c z}$, $I_{1c}^- e^{\gamma_c z}$ and $I_{2c}^- e^{\gamma_c z}$ denote the corresponding quantities for the backward-traveling mode.

The characteristic impedance of line 1 (in the presence of line 2) can be defined as

$$Z_{c1} = \frac{V_{1c}^+}{I_{1c}^+} = \frac{V_{1c}^-}{I_{1c}^-} \tag{4.61}$$

Similarly, for the same mode the characteristic impedance of line 2 (in the presence of line 1) can be defined as

$$Z_{c2} = \frac{V_{2c}^+}{I_{2c}^+} = \frac{V_{2c}^-}{I_{2c}^-} \tag{4.62}$$

Further, let the ratio of voltages on the two lines be defined by a parameter R_c as follows:

$$R_c = \frac{V_{2c}^+}{V_{1c}^+} = \frac{V_{2c}^-}{V_{1c}^-}$$

A c mode is therefore characterized by four parameters: γ_c, the propagation constant of the mode; Z_{c1} and Z_{c2}, which are, respectively, the characteristic impedances of lines 1 and 2; and R_c, the ratio of the voltages on the two lines.

π Mode

Similar to the c mode, a π mode is also characterized by four parameters: γ_π, the propagation constant of the mode; $Z_{\pi 1}$ and $Z_{\pi 2}$, which are, respectively, the characteristic impedances of lines 1 and 2; and R_π, the ratio of the voltages on the two lines.

Of the eight quantities discussed above (four each for c and π modes), only six are independent. The currents and voltages of the two modes satisfy the following relationships:

$$\frac{V_{2c}^+}{V_{1c}^+} = \frac{V_{2c}^-}{V_{1c}^-} = -\frac{I_{1\pi}^+}{I_{2\pi}^+} = -\frac{I_{1\pi}^-}{I_{2\pi}^-}$$

and

$$\frac{V_{2\pi}^+}{V_{1\pi}^+} = \frac{V_{2\pi}^-}{V_{1\pi}^-} = -\frac{I_{1c}^+}{I_{2c}^+} = -\frac{I_{1c}^-}{I_{2c}^-}$$

Therefore

$$R_c = \frac{V_{2c}^+}{V_{1c}^+} = \frac{V_{2c}^-}{V_{1c}^-} = -\frac{I_{1\pi}^+}{I_{2\pi}^+} = -\frac{I_{1\pi}^-}{I_{2\pi}^-} \qquad (4.63)$$

and

$$R_\pi = \frac{V_{2\pi}^+}{V_{1\pi}^+} = \frac{V_{2\pi}^-}{V_{1\pi}^-} = -\frac{I_{1c}^+}{I_{2c}^+} = -\frac{I_{1c}^-}{I_{2c}^-} \qquad (4.64)$$

Using (4.63) and (4.64), the relations between characteristic impedances Z_{c1}, Z_{c2}, $Z_{\pi 1}$, and $Z_{\pi 2}$ and ratio parameters R_c and R_π are found to be

$$\frac{Z_{c2}}{Z_{c1}} = \frac{Z_{\pi 2}}{Z_{\pi 1}} = -R_c R_\pi \qquad (4.65)$$

Therefore, a total number of six quantities (i.e., γ_c, γ_π, Z_{c1} [or Z_{c2}], $Z_{\pi 1}$ [or $Z_{\pi 2}$], R_c, and R_π) is required to characterize asymmetrical coupled lines. It is not necessary to specify both Z_{c1} and Z_{c2}, as they are related by (4.65). The same holds for $Z_{\pi 1}$ and $Z_{\pi 2}$. On the other hand, symmetrical coupled lines are completely characterized by four parameters: the even- and odd-mode characteristic impedances of any line (as both lines are identical) and the even- and odd-mode phase constants. For symmetrical coupled lines, $R_c = 1$ and $R_\pi = -1$.

Z- and Y-Parameters of a Four-Port Network in Terms of Normal-Mode Parameters

Figure 4.7 shows a four-port network composed of asymmetrical coupled lines. When one or more ports of the structure are excited, voltage and current waves are set up on both the lines. The voltage and current waves can be expressed

Analysis of Uniformly Coupled Lines   143

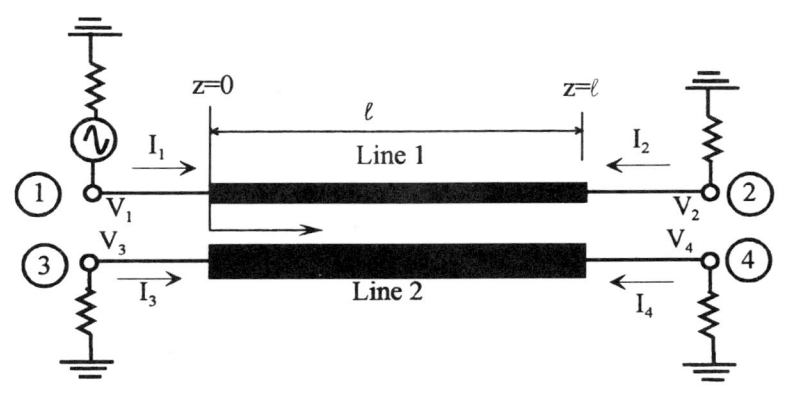

Figure 4.7 A four-port network composed of uniform asymmetrical coupled lines.

as a linear sum of forward- and backward-traveling c and π mode waves. For example, the voltage and current waves on the two lines can be represented as

$$V_1(z) = A_1 e^{-\gamma_c z} + A_2 e^{\gamma_c z} + A_3 e^{-\gamma_\pi z} + A_4 e^{\gamma_\pi z} \tag{4.66}$$

$$V_2(z) = A_1 R_c e^{-\gamma_c z} + A_2 R_c e^{\gamma_c z} + A_3 R_\pi e^{-\gamma_\pi z} + A_4 R_\pi e^{\gamma_\pi z} \tag{4.67}$$

$$I_1(z) = A_1 Y_{c1} e^{-\gamma_c z} - A_2 Y_{c1} e^{\gamma_c z} + A_3 Y_{\pi 1} e^{-\gamma_\pi z} - A_4 Y_{\pi 1} e^{\gamma_\pi z} \tag{4.68}$$

$$I_2(z) = A_1 R_c Y_{c2} e^{-\gamma_c z} - A_2 R_c Y_{c2} e^{\gamma_c z} + A_3 R_\pi Y_{\pi 2} e^{-\gamma_\pi z}$$
$$- A_4 R_\pi Y_{\pi 2} e^{\gamma_\pi z} \tag{4.69}$$

where $Y_{ci} = 1/Z_{ci}$; $Y_{\pi i} = 1/Z_{\pi i}$ ($i = 1, 2$) and A_1, A_2, A_3, and A_4 are constants depending on the sources and terminations. By substituting $z = 0$ and $z = l$ in (4.66) to (4.69), the voltages and currents at all the ports can be found. For example, the voltage at port 2(V_2) can be found by substituting $z = l$ in (4.66). The voltages and currents at various ports are found to be

$$V_1 = A_1 + A_2 + A_3 + A_4 \tag{4.70a}$$

$$V_2 = A_1 e^{-\gamma_c l} + A_2 e^{\gamma_c l} + A_3 e^{-\gamma_\pi l} + A_4 e^{\gamma_\pi l} \tag{4.70b}$$

$$V_3 = A_1 R_c + A_2 R_c + A_3 R_\pi + A_4 R_\pi \tag{4.70c}$$

$$V_4 = A_1 R_c e^{-\gamma_c l} + A_2 R_c e^{\gamma_c l} + A_3 R_\pi e^{-\gamma_\pi l} + A_4 R_\pi e^{\gamma_\pi l} \tag{4.70d}$$

$$I_1 = A_1 Y_{c1} - A_2 Y_{c1} + A_3 Y_{\pi 1} - A_4 Y_{\pi 1} \tag{4.71a}$$

$$-I_2 = A_1 Y_{c1} e^{-\gamma_c l} - A_2 Y_{c1} e^{\gamma_c l} + A_3 Y_{\pi 1} e^{-\gamma_\pi l} - A_4 Y_{\pi 1} e^{\gamma_\pi l} \tag{4.71b}$$

$$I_3 = A_1 R_c Y_{c2} - A_2 R_c Y_{c2} + A_3 R_\pi Y_{\pi 2} - A_4 R_\pi Y_{\pi 2} \tag{4.71c}$$

$$-I_4 = A_1 R_c Y_{c2} e^{-\gamma_c l} - A_2 R_c Y_{c2} e^{\gamma_c l} + A_3 R_\pi Y_{\pi 2} e^{-\gamma_\pi l} - A_4 R_\pi Y_{\pi 2} e^{\gamma_\pi l} \tag{4.71d}$$

The set of equations given by (4.71a) to (4.71d) can be solved to obtain the coefficients A_i in terms of I_1, I_2, I_3, and I_4. Further, substituting these in (4.70a) to (4.70d) the Z-parameters can be determined from the resulting equations by inspection. The Z-parameters of the four-port network are found in terms of normal-mode parameters to be:

$$Z_{11} = Z_{22} = \frac{Z_{c1} \coth \gamma_c l}{(1 - R_c/R_\pi)} + \frac{Z_{\pi 1} \coth \gamma_\pi l}{(1 - R_\pi/R_c)}$$

$$Z_{13} = Z_{31} = Z_{24} = Z_{42} = \frac{Z_{c1} R_c \coth \gamma_c l}{(1 - R_c/R_\pi)} + \frac{Z_{\pi 1} R_\pi \coth \gamma_\pi l}{(1 - R_\pi/R_c)}$$

$$= -\frac{Z_{c2} \coth \gamma_c l}{R_\pi (1 - R_c/R_\pi)} - \frac{Z_{\pi 2} \coth \gamma_\pi l}{R_c (1 - R_\pi/R_c)}$$

$$Z_{14} = Z_{41} = Z_{32} = Z_{23} = \frac{R_c Z_{c1}}{(1 - R_c/R_\pi)\sinh \gamma_c l} + \frac{R_\pi Z_{\pi 1}}{(1 - R_\pi/R_c)\sinh \gamma_\pi l}$$

$$Z_{12} = Z_{21} = \frac{Z_{c1}}{(1 - R_c/R_\pi)\sinh \gamma_c l} + \frac{Z_{\pi 1}}{(1 - R_\pi/R_c)\sinh \gamma_\pi l}$$

$$Z_{33} = Z_{44} = -\frac{R_c Z_{c2} \coth \gamma_c l}{R_\pi (1 - R_c/R_\pi)} - \frac{R_\pi Z_{\pi 2} \coth \gamma_\pi l}{R_c (1 - R_\pi/R_c)}$$

$$= \frac{R_c^2 Z_{c1} \coth \gamma_c l}{(1 - R_c/R_\pi)} + \frac{R_\pi^2 Z_{\pi 1} \coth \gamma_\pi l}{(1 - R_\pi/R_c)}$$

$$Z_{34} = Z_{43} = \frac{R_c^2 Z_{c1}}{(1 - R_c/R_\pi) \sinh \gamma_c l} + \frac{R_\pi^2 Z_{\pi 1}}{(1 - R_\pi/R_c) \sinh \gamma_\pi l}.$$

$$\tag{4.72}$$

Further, the Y-parameters of the four-port network shown in Figure 4.7 are given by,

$$Y_{11} = Y_{22} = \frac{Y_{c1} \coth\gamma_c l}{(1 - R_c/R_\pi)} + \frac{Y_{\pi 1} \coth\gamma_\pi l}{(1 - R_\pi/R_c)}$$

$$Y_{13} = Y_{31} = Y_{24} = Y_{42} = -\frac{Y_{c1} \coth\gamma_c l}{R_\pi(1 - R_c/R_\pi)} - \frac{Y_{\pi 1} \coth\gamma_\pi l}{R_c(1 - R_\pi/R_c)}$$

$$Y_{14} = Y_{41} = Y_{23} = Y_{32} = \frac{Y_{c1}}{(R_\pi - R_c) \sinh\gamma_c l} + \frac{Y_{\pi 1}}{(R_c - R_\pi) \sinh\gamma_\pi l}$$

$$Y_{12} = Y_{21} = -\frac{Y_{c1}}{(1 - R_c/R_\pi) \sinh\gamma_c l} - \frac{Y_{\pi 1}}{(1 - R_\pi/R_c) \sinh\gamma_\pi l}$$

$$Y_{33} = Y_{44} = -\frac{R_c Y_{c2} \coth\gamma_c l}{R_\pi(1 - R_c/R_\pi)} - \frac{R_\pi Y_{\pi 2} \coth\gamma_\pi l}{R_c(1 - R_\pi/R_c)}$$

$$Y_{34} = Y_{43} = \frac{R_c Y_{c2}}{R_\pi(1 - R_c/R_\pi) \sinh\gamma_c l} + \frac{R_\pi Y_{\pi 2}}{R_c(1 - R_\pi/R_c) \sinh\gamma_\pi l}.$$

(4.73)

Z-Parameters of Interdigital Two-Port Network

If the two ports (ports 2 and 3) of the four-port network shown in Figure 4.7 are terminated in an open circuit, a two-port network such as the one shown in Figure 4.8 results with the Z-parameters given by

$$\begin{bmatrix} Z_{11} & Z_{21} \\ Z_{21} & Z_{22} \end{bmatrix} = -j\frac{Z_{c1}}{(1 - R_c/R_\pi)} \begin{bmatrix} \cot\theta_c & R_c \csc\theta_c \\ R_c \csc\theta_c & R_c^2 \cot\theta_c \end{bmatrix}$$

$$- j\frac{Z_{\pi 1}}{(1 - R_\pi/R_c)} \begin{bmatrix} \cot\theta_\pi & R_\pi \csc\theta_\pi \\ R_\pi \csc\theta_\pi & R_\pi^2 \cot\theta_\pi \end{bmatrix} \quad (4.74)$$

where $\theta_c = \beta_c l$ and $\theta_\pi = \beta_\pi l$. This subcircuit finds extensive application in planar microwave circuits such as bandpass filters.

4.3.2 Distributed Equivalent Circuit of Coupled Lines

The distributed equivalent circuit of two uniformly coupled lossless asymmetrical transmission lines is shown in Figure 4.9. The voltages and currents on the

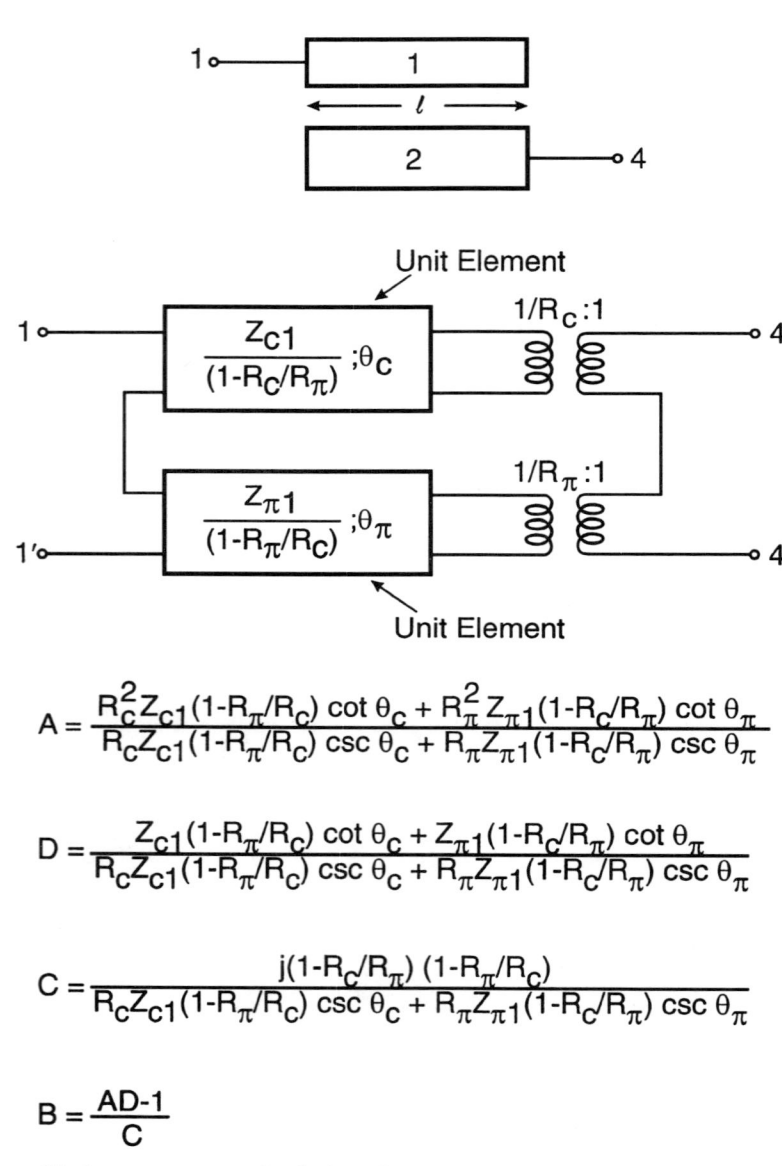

$$A = \frac{R_c^2 Z_{c1}(1-R_\pi/R_c)\cot\theta_c + R_\pi^2 Z_{\pi 1}(1-R_c/R_\pi)\cot\theta_\pi}{R_c Z_{c1}(1-R_\pi/R_c)\csc\theta_c + R_\pi Z_{\pi 1}(1-R_c/R_\pi)\csc\theta_\pi}$$

$$D = \frac{Z_{c1}(1-R_\pi/R_c)\cot\theta_c + Z_{\pi 1}(1-R_c/R_\pi)\cot\theta_\pi}{R_c Z_{c1}(1-R_\pi/R_c)\csc\theta_c + R_\pi Z_{\pi 1}(1-R_c/R_\pi)\csc\theta_\pi}$$

$$C = \frac{j(1-R_c/R_\pi)(1-R_\pi/R_c)}{R_c Z_{c1}(1-R_\pi/R_c)\csc\theta_c + R_\pi Z_{\pi 1}(1-R_c/R_\pi)\csc\theta_\pi}$$

$$B = \frac{AD-1}{C}$$

Figure 4.8 A prototype open circuited section composed of uniform asymmetrical coupled lines, its equivalent circuit and ABCD parameters. (From [5], © 1975 IEEE. Reprinted with permission.)

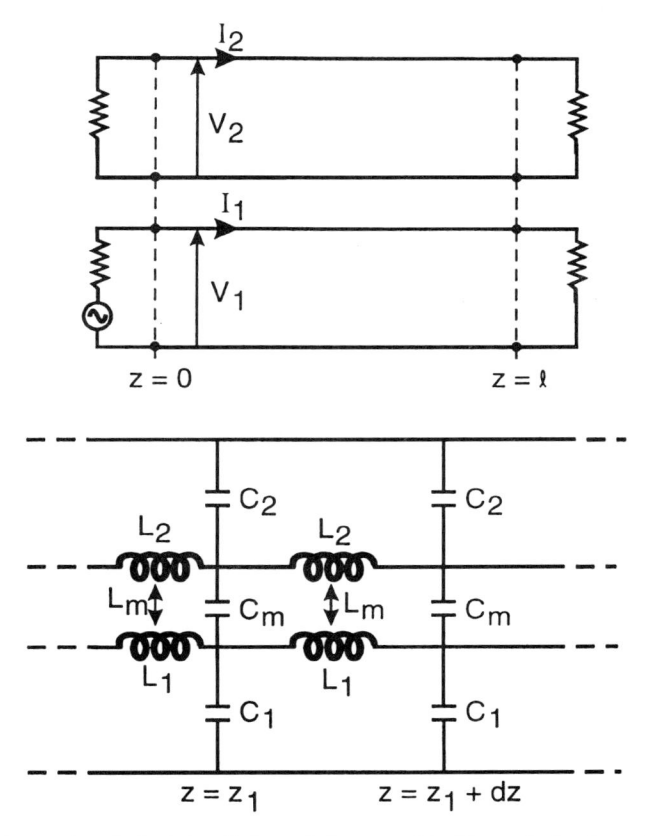

Figure 4.9 Lumped equivalent circuit of coupled asymmetrical transmission lines of differential length *dz*.

coupled lines are governed by the following differential equations, [6, 7]:

$$\frac{\partial V_1(z,t)}{\partial z} + L_1 \frac{\partial I_1(z,t)}{\partial t} + L_m \frac{\partial I_2(z,t)}{\partial t} = 0 \qquad (4.75)$$

$$\frac{\partial I_1(z,t)}{\partial z} + C_1 \frac{\partial V_1(z,t)}{\partial t} - C_m \frac{\partial V_2(z,t)}{\partial t} = 0 \qquad (4.76)$$

$$\frac{\partial V_2(z,t)}{\partial z} + L_2 \frac{\partial I_2(z,t)}{\partial t} + L_m \frac{\partial I_1(z,t)}{\partial t} = 0 \qquad (4.77)$$

$$\frac{\partial I_2(z,t)}{\partial z} + C_2 \frac{\partial V_2(z,t)}{\partial t} - C_m \frac{\partial V_1(z,t)}{\partial t} = 0 \qquad (4.78)$$

where $V_i(z, t)$ and $I_i(z, t)$ denote the voltage and current, respectively, on line i ($i = 1, 2$) as a function of distance z along the transmission line and time t. L_1 and C_1 denote the (per unit) self-inductance and -capacitance of line 1 in the presence of line 2. Similarly, L_2 and C_2 denote the self-inductance and -capacitance of line 2 in the presence of line 1. Further, L_m and C_m denote the mutual inductance and mutual capacitance between the lines, respectively. More specifically, self and mutual inductance and capacitance parameters are the elements of inductance and capacitance matrices [L] and [C], where

$$[\mathbf{L}] = \begin{bmatrix} L_{11} & L_{12} \\ L_{21} & L_{22} \end{bmatrix} = \begin{bmatrix} L_1 & L_m \\ L_m & L_2 \end{bmatrix} \tag{4.79}$$

$$[\mathbf{C}] = \begin{bmatrix} C_{11} & C_{12} \\ C_{21} & C_{22} \end{bmatrix} = \begin{bmatrix} C_1 & -C_m \\ -C_m & C_2 \end{bmatrix} \tag{4.80}$$

For a lossless case, the inductance matrix [L] is given by [8]

$$[\mathbf{L}] = \epsilon_0 \mu_0 [\mathbf{C}_0]^{-1} \tag{4.81}$$

where $[\mathbf{C}_0]$ denotes the free-space capacitance matrix of the coupled lines. (4.81) is general and is valid for any number of coupled lines.

In the frequency domain, (4.75) to (4.78) reduce to

$$\frac{\partial V_1(z)}{\partial z} + j\omega L_1 I_1(z) + j\omega L_m I_2(z) = 0 \tag{4.82}$$

$$\frac{\partial I_1(z)}{\partial z} + j\omega C_1 V_1(z) - j\omega C_m V_2(z) = 0 \tag{4.83}$$

$$\frac{\partial V_2(z)}{\partial z} + j\omega L_2 I_2(z) + j\omega L_m I_1(z) = 0 \tag{4.84}$$

$$\frac{\partial I_2(z)}{\partial z} + j\omega C_2 V_2(z) - j\omega C_m V_1(z) = 0 \tag{4.85}$$

By solving the set of coupled equations given by (4.82) to (4.85), the propagation constants and other parameters of asymmetrical coupled lines defined earlier in this section can be determined.

The complex propagation constants of the c and π modes are found to be

$$\gamma_c^2 = \frac{a_1 + a_2}{2} + \frac{1}{2}[(a_1 - a_2)^2 + 4b_1 b_2]^{1/2} \tag{4.86}$$

and

$$\gamma_\pi^2 = \frac{a_1 + a_2}{2} - \frac{1}{2}[(a_1 - a_2)^2 + 4b_1 b_2]^{1/2} \tag{4.87}$$

In the above equations

$$a_1 = y_1 z_1 + y_m z_m$$
$$a_2 = y_2 z_2 + y_m z_m$$
$$b_1 = z_1 y_m + y_2 z_m \quad (4.88)$$
$$b_2 = z_2 y_m + y_1 z_m$$

where

$$z_1 = j\omega L_1$$
$$z_2 = j\omega L_2 \quad (4.89)$$
$$z_m = j\omega L_m$$

and

$$y_1 = j\omega C_1$$
$$y_2 = j\omega C_2 \quad (4.90)$$
$$y_m = -j\omega C_m$$

By substituting values of a_1, a_2, b_1, and b_2 from (4.88) in (4.86) and (4.87), we obtain the phase velocities of the c and π modes:

$$v_c = \left[\frac{L_1 C_1 + L_2 C_2 - 2 L_m C_m + \sqrt{(L_1 C_1 - L_2 C_2)^2 + 4(L_m C_1 - L_2 C_m)(L_m C_2 - L_1 C_m)}}{2}\right]^{-1/2}$$

(4.91)

and

$$v_\pi = \left[\frac{L_1 C_1 + L_2 C_2 - 2 L_m C_m - \sqrt{(L_1 C_1 - L_2 C_2)^2 + 4(L_m C_1 - L_2 C_m)(L_m C_2 - L_1 C_m)}}{2}\right]^{-1/2}$$

(4.92)

where $v_{c,\pi} = \omega/\beta_{c,\pi}$ and $\beta_{c,\pi} = -j\gamma_{c,\pi}$. Further, the characteristic impedances and admittances of the lines for the c and π modes and R_c and R_π parameters are given by

$$Z_{c1} = \left(\frac{1}{\gamma_c}\right)\frac{z_1 z_2 - z_m^2}{z_2 - z_m R_c} = \frac{1}{Y_{c1}} \quad (4.93)$$

$$Z_{c2} = \left(\frac{R_c}{\gamma_c}\right)\frac{z_1 z_2 - z_m^2}{z_1 R_c - z_m} = \frac{1}{Y_{c2}} \quad (4.94)$$

$$Z_{\pi 1} = \left(\frac{1}{\gamma_\pi}\right) \frac{z_1 z_2 - z_m^2}{z_2 - z_m R_\pi} = \frac{1}{Y_{\pi 1}} \quad (4.95)$$

$$Z_{\pi 2} = \left(\frac{R_\pi}{\gamma_\pi}\right) \frac{z_1 z_2 - z_m^2}{z_1 R_\pi - z_m} = \frac{1}{Y_{\pi 2}} \quad (4.96)$$

$$R_c = \frac{(a_2 - a_1) + [(a_2 - a_1)^2 + 4b_1 b_2]^{1/2}}{2b_1} \quad (4.97)$$

and

$$R_\pi = \frac{(a_2 - a_1) - [(a_2 - a_1)^2 + 4b_1 b_2]^{1/2}}{2b_1} \quad (4.98)$$

Inductive and Capacitive Coupling Coefficients

The inductive coupling coefficient between the lines is defined by [6]

$$k_L = \frac{L_m}{\sqrt{L_1 L_2}} \quad (4.99)$$

whereas the capacitive coupling coefficient between the lines is given by

$$k_C = \frac{C_m}{\sqrt{C_1 C_2}} \quad (4.100)$$

4.3.3 Relation Between Normal Mode (c and π) and Distributed Line Parameters

Symmetrical Lines

The relations between c and π parameters and distributed line parameters (self- and mutual inductances and capacitances) reduce to simple forms for some special cases. For example, for symmetrical coupled lines, $L_1 = L_2$ and $C_1 = C_2$. In this case, the c and π modes reduce to even and odd modes, respectively. From (4.86) and (4.87),

$$\gamma_c = j\beta_e = j\omega\sqrt{(L_1 + L_m)(C_1 - C_m)} \quad (4.101)$$

and

$$\gamma_\pi = j\beta_o = j\omega\sqrt{(L_1 - L_m)(C_1 + C_m)} \quad (4.102)$$

Further, for symmetrical coupled lines $R_c = 1$ and $R_\pi = -1$ from (4.97) and (4.98). Substituting the values of $R_c = 1$, $R_\pi = -1$ and propagation constants γ_c and γ_π from (4.101) and (4.102) in (4.93) to (4.96), we obtain

$$Z_{c1} = Z_{c2} = Z_{0e} = \sqrt{\frac{(L_1 + L_m)}{(C_1 - C_m)}} \tag{4.103}$$

$$Z_{\pi 1} = Z_{\pi 2} = Z_{0o} = \sqrt{\frac{(L_1 - L_m)}{(C_1 + C_m)}} \tag{4.104}$$

Using (4.101) to (4.104), the distributed line parameters are found in terms of characteristic impedances and propagation constants of the even and odd modes to be

$$L_1 = L_2 = \frac{1}{2\omega}(\beta_e Z_{0e} + \beta_o Z_{0o}) \tag{4.105}$$

$$C_1 = C_2 = \frac{1}{2\omega}\left(\frac{\beta_o}{Z_{0o}} + \frac{\beta_e}{Z_{0e}}\right) \tag{4.106}$$

$$L_m = \frac{1}{2\omega}(\beta_e Z_{0e} - \beta_o Z_{0o}) \tag{4.107}$$

$$C_m = \frac{1}{2\omega}\left(\frac{\beta_o}{Z_{0o}} - \frac{\beta_e}{Z_{0e}}\right) \tag{4.108}$$

Note that for lines supporting pure TEM mode of propagation, the even- and odd-mode phase velocities are the same. (4.101) and (4.102) then lead to

$$\frac{L_m}{L_1} = \frac{C_m}{C_1} \tag{4.109}$$

or from (4.99) and (4.100):

$$k_C = k_L \tag{4.110}$$

The inductive and capacitive coupling coefficients are therefore equal. Equation 4.110 is true for TEM asymmetrical coupled lines as well.

Asymmetrical Coupled Lines

For *lossless* TEM-mode coupled lines, the propagation constants of both the c and π modes are the same, and are given by

$$\gamma_c = \gamma_\pi = j\beta = jk_0\sqrt{\epsilon_r} \tag{4.111}$$

The following relations are satisfied by line parameters in this case:

$$L_1 C_1 = L_2 C_2 \tag{4.112}$$

$$\frac{C_m}{\sqrt{C_1 C_2}} = \frac{L_m}{\sqrt{L_1 L_2}} \tag{4.113}$$

Further, R_c and R_π are given by

$$R_c = -R_\pi = \left(\frac{Z_2}{Z_1}\right)^{1/2} \tag{4.114}$$

where $Z_1 = (L_1/C_1)^{1/2}$ and $Z_2 = (L_2/C_2)^{1/2}$.

In this case, the Z-parameters of a four-port network as shown in Figure 4.7 are given by

$$\begin{aligned}
Z_{11} = Z_{22} &= -\frac{j}{2}\left(\frac{Z_1}{Z_2}\right)^{1/2}(Z_c + Z_\pi)\cot\theta \\
Z_{13} = Z_{31} = Z_{42} = Z_{24} &= -\frac{j}{2}(Z_c - Z_\pi)\cot\theta \\
Z_{14} = Z_{41} = Z_{32} = Z_{23} &= -\frac{j}{2}(Z_c - Z_\pi)\csc\theta \\
Z_{12} = Z_{21} &= -\frac{j}{2}\left(\frac{Z_1}{Z_2}\right)^{1/2}(Z_c + Z_\pi)\csc\theta \\
Z_{33} = Z_{44} &= -\frac{j}{2}\left(\frac{Z_2}{Z_1}\right)^{1/2}(Z_c + Z_\pi)\cot\theta \\
Z_{34} = Z_{43} &= -\frac{j}{2}\left(\frac{Z_2}{Z_1}\right)^{1/2}(Z_c + Z_\pi)\csc\theta
\end{aligned} \tag{4.115}$$

where

$$Z_c = (Z_1 Z_2)^{1/2}\left[\frac{1 + y_m/(y_1 y_2)^{1/2}}{1 - y_m/(y_1 y_2)^{1/2}}\right]^{1/2}$$

and

$$Z_\pi = (Z_1 Z_2)^{1/2}\left[\frac{1 - y_m/(y_1 y_2)^{1/2}}{1 + y_m/(y_1 y_2)^{1/2}}\right]^{1/2}$$

4.3.4 Approximate Distributed Line or Normal-Mode Parameters of Asymmetrical Coupled Lines

The performance of a network consisting of coupled asymmetrical lines can be determined if the distributed line parameters (i.e., L_1, L_2, C_1, C_2, L_m, and C_m) or the normal-mode parameters (i.e., Z_{c1} [or Z_{c2}], $Z_{\pi 1}$ [or $Z_{\pi 2}$], γ_c, γ_π, R_c, and R_π) are known. The computation of distributed line or normal-mode parameters, however, is quite complicated and can only be carried out using field theoretical methods [8]. Commercially available programs based on this method are also available [9, 10]. Ikalainen and Matthaei [11] have given an *approximate* technique from which the inductance and capacitance parameters of asymmetrical coupled lines can be determined from the characteristic impedances and effective dielectric constants of the even and odd modes of symmetrical coupled lines. This approach is useful in practice because the even- and odd-mode parameters of symmetrical coupled lines are generally more readily available.

Consider two coupled lines of width W_1 and W_2, each with separation S between them as shown in Figure 4.10(a). It is assumed that the mutual inductance and capacitance between the lines is the same as that between symmetrical lines of width $(W_1 + W_2)/2$, each with separation S as shown in Figure 4.10(b). Using the even- and odd-mode data of coupled symmmetrical lines as shown in Figure 4.10(b), the values of L_m and C_m can be computed using (4.107) and (4.108). Further, it is assumed that the self-inductance and capacitance of line 1 in the presence of line 2 is the same as if line 2 has the same width as line 1. Therefore, by using the even and odd-mode data of coupled symmetrical lines of width W_1 each and separated by a distance S as shown in Figure 4.10(c), the self-capacitance and -inductance of line 1 can be computed using (4.105) and (4.106). Similarly, by using even- and odd-mode data of coupled symmetrical lines of width W_2 each and separated by a distance S (Figure 4.10(d)), the self-capacitance and -inductance of line 2 can be computed. Once the distributed line parameters have been found, the normal-mode parameters can be found using using (4.86), (4.87), and (4.93) through (4.98).

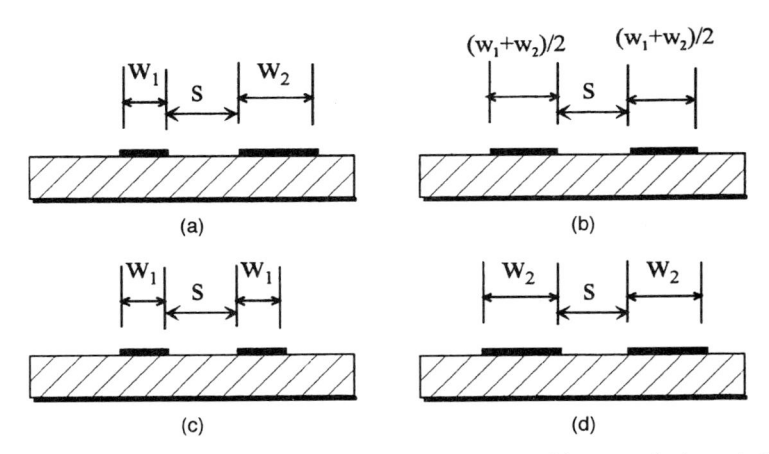

Figure 4.10 (a) Asymmetrical coupled lines of width w_1 and w_2, (b) symmetrical coupled lines of width $(w_1 + w_2)/2$, (c) symmetrical coupled lines of width w_1, (d) symmetrical coupled lines of width w_2.

4.4 Directional Couplers Using Asymmetrical Coupled Lines

4.4.1 Forward-Wave Directional Couplers

It is known that if the phase velocities of the two normal modes of asymmetrical coupled lines are different, energy is coupled from one line to another in the forward direction. Because a microstrip line is essentially a quasi-TEM line, the even- and odd-mode phase velocities of coupled microstrip lines are not equal. Therefore, coupling occurs both in the forward and backward directions. Usually, backward-wave couplers are realized in microstrip configuration by properly choosing the characteristic impedances of the even and odd modes. The directivity of microstrip backward-wave couplers is generally poor, however, because of the forward-wave coupling that takes place because of unequal even- and odd-mode phase velocities. The backward-wave coupling can be reduced to negligibly small values by choosing a relatively large separation between the lines. On the other hand, appreciable power can be made to couple in the forward direction, if the length of the coupling section is properly chosen. The bandwidth of an asymmetrical forward-wave coupler is larger than that of a symmetrical forward-wave coupler. This makes them useful in practice [11, 12]. Figure 4.11 shows an asymmetrical microstrip coupler. It is assumed that the backward-wave coupling between the lines is negligble and each line is terminated in a matched load. With unit power incident at port 1, the forward-traveling voltage waves on the two lines can be expressed as a linear combination of c- and π-mode voltage waves

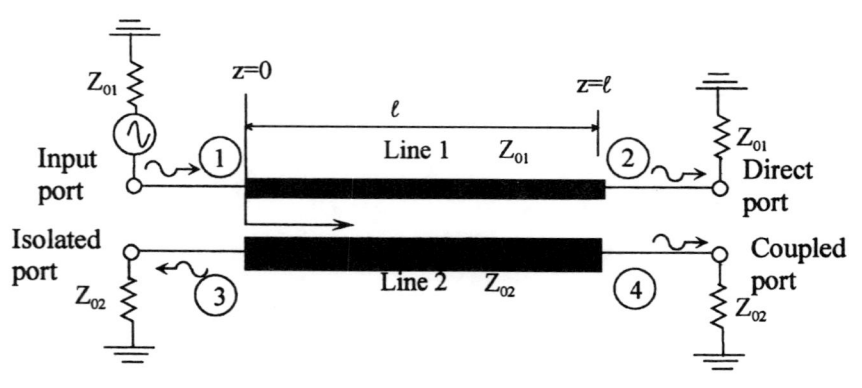

Figure 4.11 A forward (co-directional) directional coupler using uniform asymmetrical coupled lines.

Analysis of Uniformly Coupled Lines

as follows:

$$V_1^+(z) = A_1 e^{-\gamma_c z} + A_2 e^{-\gamma_\pi z} \tag{4.116}$$

$$V_2^+(z) = A_1 R_c e^{-\gamma_c z} + A_2 R_\pi e^{-\gamma_\pi z} \tag{4.117}$$

The voltages in the above equations denote their actual values. As discussed in Chapter 2, the concept of actual voltages and currents is restrictive and is applicable to TEM and quasi-TEM mode transmission lines only. On the other hand, the concept of normalized voltages is more general and can be applied to non-TEM mode transmission lines as well. Using (4.116), (4.117), and the conversion relations between normalized and actual voltages given by (2.39a) to (2.39d), the normalized voltage waves on the two lines can be expressed as

$$\hat{V}_1^+(z) = \frac{A_1}{\sqrt{Z_{c1}}} e^{-\gamma_c z} + \frac{A_2}{\sqrt{Z_{\pi 1}}} e^{-\gamma_\pi z} \tag{4.118}$$

$$\hat{V}_2^+(z) = \frac{A_1 R_c}{\sqrt{Z_{c2}}} e^{-\gamma_c z} + \frac{A_2 R_\pi}{\sqrt{Z_{\pi 2}}} e^{-\gamma_\pi z} \tag{4.119}$$

Because unit power is assumed to be incident at the input port, the initial conditions are $\hat{V}_1^+ = 1$ and $\hat{V}_2^+ = 0$ at $z = 0$. Substituting these conditions in (4.118) and (4.119), we obtain

$$A_1 = \frac{\sqrt{Z_{c1}}}{1 - \frac{R_c}{R_\pi}} \tag{4.120}$$

$$A_2 = -\frac{\sqrt{Z_{\pi 1}}}{1 - \frac{R_c}{R_\pi}} \frac{R_c}{R_\pi} \tag{4.121}$$

Further, substituting the values of A_1 and A_2 in (4.118) and (4.119), the normalized voltage wave on line 1 is

$$\hat{V}_1^+(z) = \frac{1}{\left(1 - \frac{R_c}{R_\pi}\right)} e^{-\gamma_c z} - \frac{1}{\left(1 - \frac{R_c}{R_\pi}\right)} \frac{R_c}{R_\pi} e^{-\gamma_\pi z}$$

After some straightforward algebraic manipulations, the above equation reduces to

$$\hat{V}_1(z) = \left[\cos \frac{(\beta_c - \beta_\pi)z}{2} - j \frac{1-p}{1+p} \sin \frac{(\beta_c - \beta_\pi)z}{2}\right] e^{-j(\beta_c + \beta_\pi)z/2} \tag{4.122}$$

where

$$p = -\frac{R_c}{R_\pi} \tag{4.123}$$

Similarly, by substituting the values of A_1 and A_2 from (4.120) and (4.121) in (4.119), the normalized voltage wave on line 2 is given by

$$\hat{V}_2(z) = -2j\frac{\sqrt{p}}{1+p}\sin\frac{(\beta_c - \beta_\pi)z}{2}e^{-j(\beta_c+\beta_\pi)z/2} \qquad (4.124)$$

Using (4.122) and (4.124), the scattering parameters between different ports of the coupler shown in Figure 4.11 can be determined. In deriving (4.122) and (4.124), we assumed that unit power was incident at port 1. The scattering parameters $\hat{S}_{21}$ and $\hat{S}_{41}$ are therefore given by

$$\hat{S}_{21} = \hat{V}_1^+(z)\Big|_{z=l} \qquad (4.125)$$

and

$$\hat{S}_{41} = \hat{V}_2^+(z)\Big|_{z=l} \qquad (4.126)$$

or

$$\frac{P_2}{P_1} = |\hat{S}_{21}|^2 \qquad (4.127)$$

and

$$\frac{P_4}{P_1} = |\hat{S}_{41}|^2 \qquad (4.128)$$

Note that (4.122) and (4.124) are quite general and are valid for both quasi-TEM and non-TEM mode asymmetrical coupled transmission lines, or for that matter any two coupled waves. For non-TEM modes or waves, however, it is not possible to determine p using (4.123). This is because for the non-TEM modes, R_c and R_π (which are defined on the basis of actual voltages (4.63) and (4.64), respectively) cannot be determined. In this case, the parameter p should be determined as described in [11].

Example 4.2

The design of a 3-dB ($|\hat{S}_{21}| = |\hat{S}_{41}| = \sqrt{\frac{1}{2}}$) directional coupler in microstrip form is now discussed. Let the length of the coupler be chosen as

$$l_g = \frac{\pi}{\beta_c - \beta_\pi} \qquad (4.129)$$

From (4.124) and (4.126), we obtain

$$\frac{2\sqrt{p}}{1+p} = \sqrt{\frac{1}{2}}$$

or

$$-\frac{R_c}{R_\pi} = p = 3 \pm \sqrt{8} = 5.828, \text{ or } 0.1715 \quad (4.130)$$

The width of the coupled lines and separation between them should be chosen such that the values of R_c and R_π satisfy (4.130).

4.4.2 Backward-Wave Directional Couplers

It may be of interest in certain applications to design backward-wave couplers using asymmetrical coupled lines. For example, if the terminating impedances are different for the two lines, it may be advantageous to choose different characteristic impedances for the two lines. Note, however, that unlike asymmetrical forward-wave couplers, backward-wave asymmetrical couplers do not offer any advantages over symmetrical couplers in terms of bandwidth [13]. Their main advantage is that one does not require an additional impedance transformer to match the impedance of a low- or high-impedance device to that of the coupler. Cristal [13] has given equations for the design of backward-wave couplers using asymmetrical coupled lines. Figure 4.12(a) shows such a coupler. Assume that lines 1 and 2 are terminated in conductances G_a and G_b, respectively. Further, assume that the different capacitances of the lines are as shown in Figure 4.12(b). The capacitance matrix of the coupled lines can be expressed as (see section 3.2)

$$[C] = \begin{bmatrix} C_1 & C_{12} \\ C_{21} & C_2 \end{bmatrix} = \begin{bmatrix} C_a + C_m & -C_m \\ -C_m & C_b + C_m \end{bmatrix} \quad (4.131)$$

If k^2 denotes the power coupling coefficient between the lines, then the values of C_a, C_b, and C_m should be chosen according to

$$\frac{C_a}{\epsilon} = \frac{376.7(G_a - k\sqrt{G_a G_b})}{\sqrt{\epsilon_r}\sqrt{1-k^2}}$$

$$\frac{C_b}{\epsilon} = \frac{376.7(G_b - k\sqrt{G_a G_b})}{\sqrt{\epsilon_r}\sqrt{1-k^2}} \quad (4.132)$$

$$\frac{C_m}{\epsilon} = \frac{376.7 k\sqrt{G_a G_b}}{\sqrt{\epsilon_r}\sqrt{1-k^2}}$$

where $\epsilon = \epsilon_0 \epsilon_r$, and ϵ_0 denotes the permittivity of the free-space.

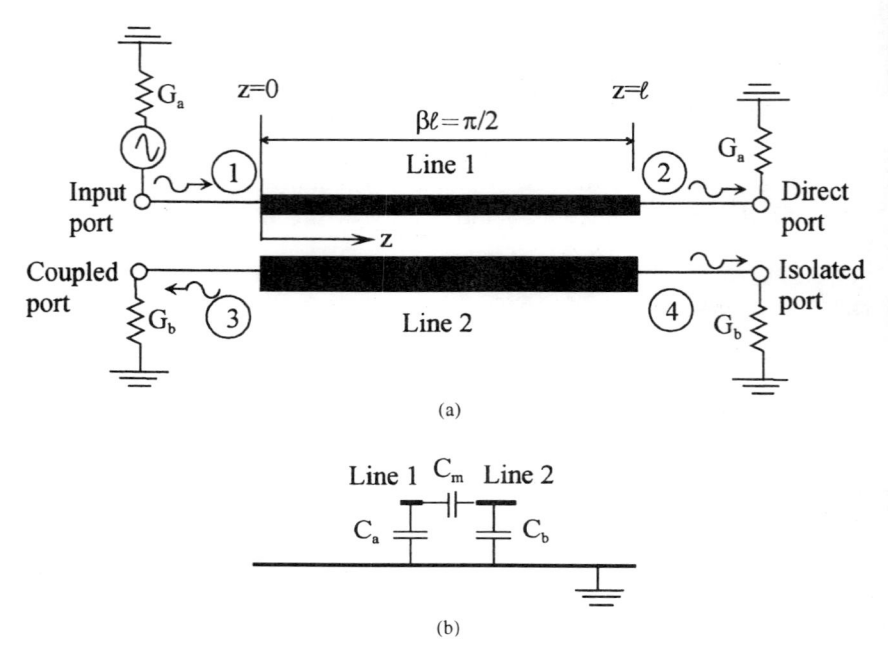

Figure 4.12 (a) A backward directional coupler using uniform asymmetrical coupled lines, (b) capacitance representation of coupled lines.

Example 4.3

Compute the per-unit length capacitances of a 10-dB asymmetrical coupler whose lines are terminated in loads of 50 ohms and 75 ohms, respectively.
    The given quantities are

$$k = 10^{(-10/20)} = 0.316$$

$$G_a = 1/50 = 0.02 \quad \text{ohms}$$

$$G_b = 1/75 = 0.0133 \quad \text{ohms}$$

From (4.132), we then obtain

$$\frac{C_a}{\epsilon} = 5.895$$

$$\frac{C_b}{\epsilon} = 3.244$$

$$\frac{C_m}{\epsilon} = 2.046$$

Once the capacitance parameters are known, the required physical dimensions of the coupler can be determined using data relevant to the transmission

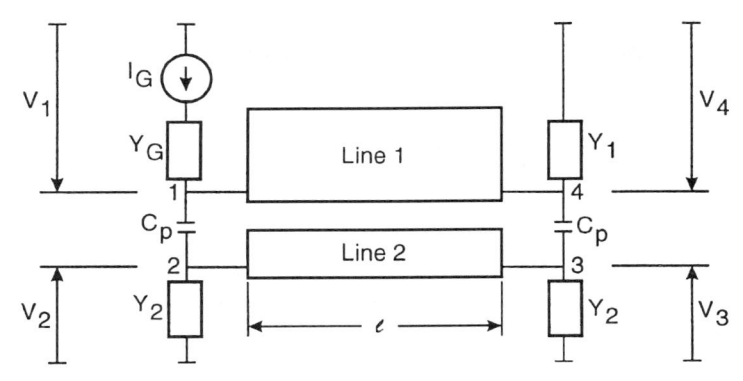

Figure 4.13 A general four-port uniform coupler with shunting capacitors and arbitrary terminations. (From [14], ©1990 IEEE. Reprinted with permission.)

line media in which the coupler is to be realized. Commercially available general programs can be used for this purpose [9, 10].

A general uniform four-port coupler with arbitrary terminations and shunting capacitors is shown in Figure 4.13. Formulas useful for the synthesis of a backward-wave directional coupler in the configuration shown in Figure 4.13 are reported in [14].

References

[1] Lippmann, B. A., "Theory of Directional Couplers," M.I.T. Rad. Lab. Rep., No. 860, Dec. 28, 1945.

[2] Reed, J., and G. J. Wheeler, "A Method of Analysis of Symmetrical Four-Port Networks," *IRE Trans.*, Vol. MTT-4, Oct. 1956, pp. 246–253.

[3] Sazanov, D. M., et al., *Microwave Circuits*, Mir Publishers, Moscow, CIS, 1982.

[4] Oliver, B. M., "Directional Electromagnetic Couplers," *Proc. IRE*, Nov. 1954, pp. 1,686–1,692.

[5] Tripathi V. K., "Asymmetric Coupled Transmission Lines in an Inhomogeneous Medium," *IEEE Trans.*, Vol. MTT-23, Sept. 1975, pp. 734–739.

[6] Krage, M. K., and G. I. Haddad, "Characteristics of Coupled Microstrip Transmission Lines-I: Coupled-Mode Formulation of Inhomogeneous Lines," *IEEE Trans.*, Vol. MTT-18, Apr. 1970, pp. 217–222.

[7] Krage, M. K., and G. I. Haddad, "Characteristics of Coupled Microstrip Transmission Lines-I: Evaluation of Coupled-Line Parameters," *IEEE Trans.*, Vol. MTT-18, Apr. 1970, pp. 222–228.

[8] Wei, C., et al., "Multiconductor Transmission Lines in Multilayered Dielectric Media," *IEEE Trans.*, Vol. MTT-32, Apr. 1984, pp. 439–450.

[9] Djordjevic, A. R., et al., *MULTLIN for Windows: Circuit Analysis Models for Multiconductor Transmission Lines, Software and User's Manual*, Artech House, Norwood, MA, 1996.

[10] Djordjevic, A. R., et al., *LINPAR for Windows: Matrix Parameters for Multiconductor Transmission Lines, Software and User's Manual*, Artech House, Norwood, MA, 1995.

[11] Ikalainen, P. K., and G. L. Matthaei, "Wideband, Forward-Coupling Microstrip Hybrids With High Directivity," *IEEE Trans.*, Vol. MTT-35, Aug. 1987, pp. 719–725.

[12] Ikalainen, P. K., and G. L. Matthaei, "Design of Broadband Dielectric Guide 3-dB Couplers," *IEEE Trans.*, Vol. MTT-35, July 1987, pp. 621–628.

[13] Cristal, E. G., "Coupled-Transmission-Line Directional Couplers With Coupled Lines of Unequal Characteristic Impedances," *IEEE Trans.*, Vol. MTT-14, July 1966, pp. 337–346.

[14] Sellberg, F., "Formulas Useful for the Synthesis and Optimization of General, Uniform Contradirectional Couplers," *IEEE Trans.*, Vol. MTT-38, Aug. 1990, pp. 1,000–1,010.

5

Broadband Forward-Wave Directional Couplers

As discussed in the previous chapter, forward-wave coupling exists between uniform symmetrical coupled lines if the even- and odd-mode phase velocities of the coupled lines are unequal. Further, the backward-wave coupling between these lines can be reduced to a very small value by keeping a relatively large separation between the lines (such that the even- and odd-mode characteristic impedances of the coupled lines are nearly equal). These types of couplers are known as *forward-wave directional couplers* and can be realized using non-TEM mode transmission lines such as metallic waveguides, dielectric guides, and the like. The bandwidth of forward-wave directional couplers realized using symmetrical coupled lines is generally small and can be increased by using asymmetrical coupled lines. In this chapter, the design and performance of forward-wave couplers realized using uniform asymmetrical lines is first discussed.

In the previous chapter, normal-mode analysis of symmetrical and asymmetrical coupled lines was discussed. Although the normal-mode analysis is rigorous, its application may prove to be very tedious in certain cases. Another approach that can be used to study coupled structures is known as the *coupled-mode theory*. In its simpler form (which is also its most useful form), the theory is valid for *weakly* coupled structures [1–3]. The theory has been used extensively in the past in numerous applications for the analysis of both passive and active coupled circuits. In its early development, the theory was used mainly for the analysis of microwave circuits such as mode conversion in multi-moded waveguides, parametric amplifiers, beamwave interaction in TWTs, and so forth. In recent years, the theory has also been used for the design and analysis of fiber optics and optoelectronics circuits and components. A good review of the coupled-mode theory and its applications has been given by Haus and Huang [3]. The theory is quite general and can be used to study coupling phenomenon between any two waves. The two waves may represent two modes of two different transmission lines or of the same transmission line. For example, if a transmission line is bent along its length, the coupled-mode theory can be used to study the conversion of power from one mode to another.

The coupled-mode theory leads to explicit expressions showing how individual waves are modified in the presence of coupling. The theory also leads to an important result that complete power can be transferred between two lossless lines (or two waves) only if both the lines (waves) have the same phase constants. The equivalence between normal- and coupled-mode theories is also discussed.

5.1 Forward-Wave Directional Couplers

The forward-wave coupling between ports 1 and 4 of symmetrical coupled lines shown in Figure 5.1 is given by

$$|S_{41}| = \left| \sin\left(\frac{\pi \Delta n_{eff} L}{c} f \right) \right| \tag{5.1}$$

where L is the length of the coupler, c is the velocity of light in free space, and f is the operating frequency. Δn_{eff} is the difference between the square roots of the effective dielectric constants of the even and odd modes:

$$\Delta n_{eff} = \sqrt{\epsilon_{ree}} - \sqrt{\epsilon_{reo}} \tag{5.2}$$

Figure 5.1 Forward-wave coupler using symmetrical coupled lines.

The direct coupling between ports 1 and 2 can be expressed as

$$|S_{21}| = \left|\cos\left(\frac{\pi \Delta n_{eff} L}{c} f\right)\right| \tag{5.3}$$

Equation 5.1 shows that the coupling is a function of frequency. Assuming that Δn_{eff} is independent of frequency, the forward-wave coupling becomes a sinusoidal function of frequency. In general, however, Δn_{eff} is also a function of frequency. Its variation with frequency depends on the type of transmission line and its parameters. It is also seen from (5.1) that the maximum coupling that can be obtained using symmetrical coupled lines by appropriately choosing the length L is 0 dB (complete power transfer). Equation 5.1 is plotted in Figure 5.2 where it is assumed that Δn_{eff} is independent of frequency. It is seen from Figure 5.2 that the coupling versus frequency curve is flat (zero first derivative) at the frequency where the coupling is 0 dB. For any other coupling value, however, the coupling versus frequency curve is not flat at the frequency where the desired coupling is obtained. For example, the coupling versus frequency response is not flat at

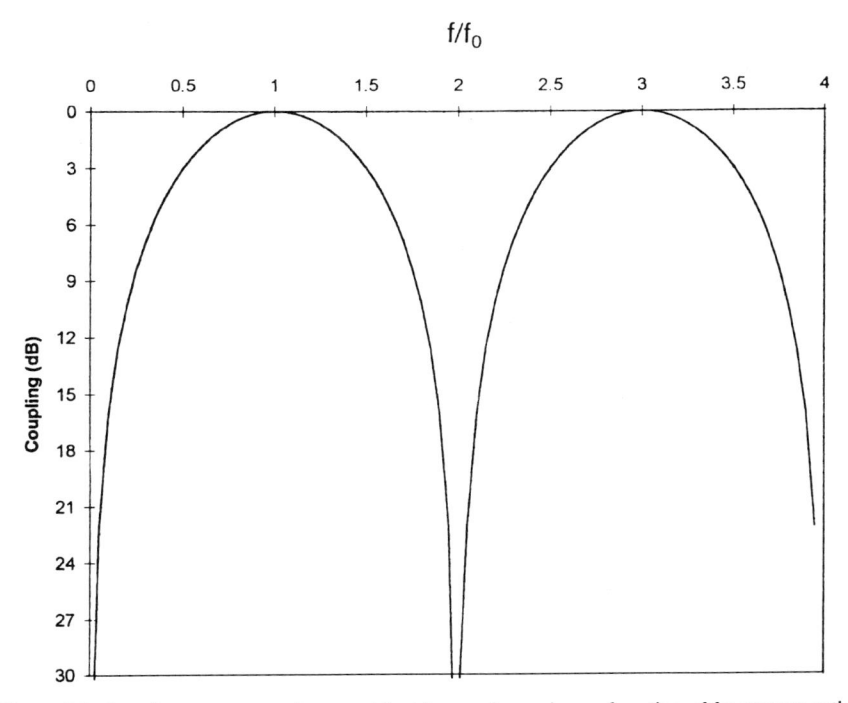

Figure 5.2 Coupling response of symmetrical forward coupler as function of frequency ratio f/f_0. f_0 denotes the frequency where maximum coupling is achieved.

the frequency where 3-dB coupling is obtained. It is therefore expected that a 0-dB symmetrical coupler has a wider bandwidth than does a coupler designed for any other coupling value.

It is known that it is possible to achieve complete power transfer only between symmetrical coupled lines. If the lines are asymmetrical,[1] the maximum coupling that can be achieved between lines is less than 0 dB. The amount of maximum coupling depends on the difference between the phase constants of the asymmetrical coupled lines. This is described in more detail later in this chapter while discussing the coupled-mode theory. The asymmetrical couplers will have a flat coupling versus frequency response at the frequency where maximum coupling is obtained. Therefore, if an asymmetrical coupler is designed such that the maximal coupling which can be obtained is equal to the desired coupling value, then such a coupler will have a wider bandwidth than a symmetrical coupler. This principle has been used to demonstrate wideband 3-dB forward-wave couplers [4]. A more comprehensive explanation on the broadband properties of asymmetrical couplers can be found in [5]. Most of the material in this section is based on the work reported in this reference.

5.1.1 3-dB Coupler Using Symmetrical Microstrip Lines

Usually, the microstrip configuration is used to realize quarter-wave backward-wave couplers by choosing suitable values of the even- and odd-mode impedances. It becomes, however, difficult because of fabrication tolerances to achieve very tight backward-wave coupling (e.g., 3-dB) in parallel-coupled microstrip lines. Further, the directivity of backward-wave couplers realized using microstrip lines tends to be quite poor. This is because the even- and odd-mode phase velocities of coupled microstrip lines are unequal, resulting in forward-wave coupling. In general, the directivity becomes poorer as the frequency is increased. By choosing a relatively large separation between coupled microstrip lines (such that the even- and odd-mode impedances are nearly equal), the backward-wave coupling can be reduced to a very small value. Further, a desired level of forward-wave coupling can be obtained by appropriately choosing the length of the coupler, which can be determined using (5.1).

The strip pattern of a 3-dB coupler using symmetrical coupled microstrip lines is shown in Figure 5.3. The spacing between the lines is tapered toward the ends of the coupler. This is done to avoid any abrupt physical discontinuities in the structure, which will lead to reflections causing power to couple to port 3, which is designed to be the isolated port.

---

1. By asymmetrical lines, we mean lines that have different phase constants when uncoupled.

Figure 5.3 Strip pattern of symmetrical 3-dB forward-wave coupler. (From [5], © 1987 IEEE. Reprinted with permission.)

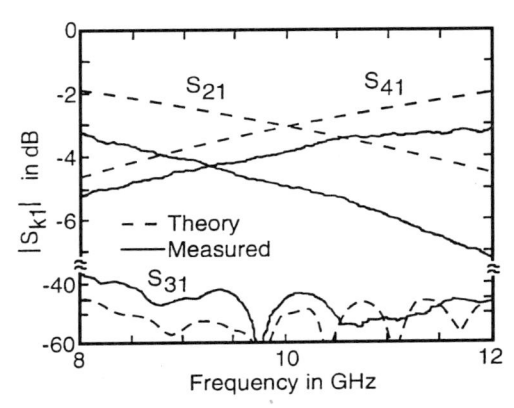

Figure 5.4 Theoretical and experimental response of a symmetrical 3-dB forward-wave coupler. (From [5], © 1987 IEEE. Reprinted with permission.)

The theoretical and experimental results of a 3-dB symmetrical forward-wave coupler are shown in Figure 5.4. The coupler was designed to operate at 10 GHz. The substrate material has a dielectric constant of 2.20 and a thickness of 0.762 mm. The width of the microstrip line corresponds to an impedance of 50 Ω and is a constant throughout. The length of the straight middle part is 113 mm, which is equivalent to 5.2 guide wavelengths at 10 GHz. The spacing between the lines in the middle section is twice the substrate thickness. The curved sections have a radius of curvature of 102 mm. The coupling of the straight section was determined using (5.1). To determine the coupling between curved sections, they were considered to consist of 20 small, straight segments. The coupling of each small segment was determined using (5.1), with the overall coupling determined by summing coupling from various sections. The coupled microstrip lines were analyzed using formulas given in [6].

Figure 5.4 shows that the shapes of the measured coupling curves match well with the theoretically predicted values, except that the measured center frequency is somewhat lower. The measured directivity of the coupler is about 40 dB. It is seen that in the frequency band considered, the coupling increases monotonically with frequency.

5.1.2 Design and Performance of 3-dB Asymmetrical Couplers

The design equations of a 3-dB directional coupler using asymmetrical coupled microstrip lines were given in Section 4.4.1 of the last chapter. The length of the coupler is given by

$$l_g = \frac{\pi}{\beta_c - \beta_\pi} \quad (5.4)$$

where β_c and β_π denote the phase constants of the c and π modes, respectively.

Further, the ratio of R_c and R_π should be chosen as

$$-\frac{R_c}{R_\pi} = 3 \pm \sqrt{8} = 5.828, \text{ or } 0.1715 \quad (5.5)$$

where R_c and R_π denote the voltage ratios on the two lines for the c and π modes, respectively.

In the design of the coupler, the width of asymmetrical coupled microstrip lines was chosen to correspond to (uncoupled) impedances of 50 and 100Ω. The design is completed by choosing the separation between coupled lines such that the ratio of R_c and R_π satisfies (5.5). For given width of the lines and assumed separation between them, the self and mutual inductance and capacitance parameters can be found using the coupled microstrip data [6] and the technique described in Section 4.3.4. Further, if the self and mutual inductance and capacitance parameters of the coupled lines are known, the values of R_c and R_π can be found using (4.97) and (4.98).

The coupler was fabricated on a substrate having the same parameters as that used for the symmetrical coupler discussed in the last section. The top view of the strip pattern of the coupled section and the input and output lines are shown in Figure 5.5. It was found that a 1.81-mm separation between the lines is needed to obtain the required value of R_c/R_π. In practice, a 1.65-mm separation was used, which gave more than 3-dB coupling at the center frequency but gave a wider bandwidth for 1-dB amplitude balance. The theoretical and experimental results of the coupling and isolation for this coupler are shown in Figure 5.6. The length of the coupler was found to be 220 mm using (5.4), where β_c and β_π were obtained from (4.91) and (4.92), respectively. Because the feed lines also contribute to some coupling in the fabricated coupler, the length

Figure 5.5 Strip pattern of asymmetrical forward-wave coupler. (From [5], © 1987 IEEE. Reprinted with permission.)

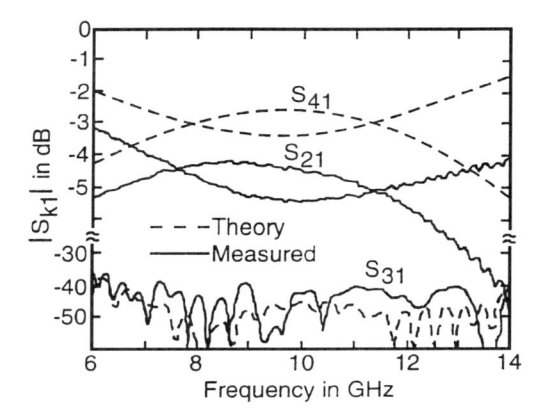

Figure 5.6 Theoretical and experimental response of an asymmetrical 3-dB forward-wave coupler. (From [5], © 1987 IEEE. Reprinted with permission.)

of the uniform coupled section was chosen to be 190 mm and the curved feed lines had a radius of curvature of 102 mm. It is seen from Figure 5.6 that the agreement between theory and experiment is quite good, considering that the conductor and dielectric losses were not accounted for in the theory. The coupler has a bandwidth of about 60% for 1-dB amplitude balance. The isolation is better than about 40 dB in the complete frequency range. It is easily verified by comparing Figures 5.4 and 5.6 that a uniform forward-wave asymmetrical coupler offers more bandwidth than a uniform symmetrical coupler.

Note, however, that unlike a symmetrical coupler, the phase difference between the output ports of an asymmetrical coupler is not 90 degrees. The theoretically computed and measured phase differences between the coupled and direct port of the asymmetrical coupler are shown in Figure 5.7(a). It is interesting to note that at the frequency where the phase difference between the output ports (ports 2 and 4) is 0 degrees when port 1 is the driven port, the phase difference between output ports is 180 degrees when the input is at port 3. In general, an asymmetrical coupler that has end-to-end symmetry satisfies the following phase-difference relationship:

$$(\angle S_{41} - \angle S_{21}) + (\angle S_{23} - \angle S_{43}) = 180 \text{ degrees} \quad (5.6)$$

It is possible to achieve approximately the desired phase difference between output ports over a reasonably wide bandwidth in an asymmetrical coupler using an extra length of line as a phase compensating element. For example, if one adds a length of line having a phase shift of 106 degrees at 9.6 GHz to ports 3 and 4, the outputs (S_{21} and S_{41}) will be in phase quadrature at that frequency. Further, the outputs are held in phase quadrature within about 12 degrees over a frequency range of 7.0 to 12.2 GHz. By adding another quarter-wave-long line

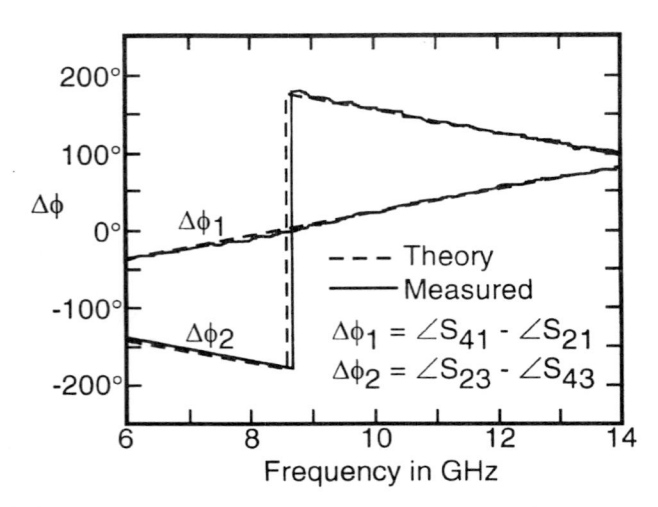

Figure 5.7(a) Computed phase difference between output ports of an asymmetrical 3-dB forward-wave coupler. (From [5], © 1987 IEEE. Reprinted with permission.)

section at ports 3 and 4, a phase difference of 180 degrees between the outputs can be obtained. This is shown in Figure 5.7(b), where the two cases are labeled as "quadrature-type" and "magic-T-type," respectively.

5.1.3 Ultra-Broadband Forward-Wave Directional Couplers

The bandwidth of forward-wave couplers realized using asymmetrical coupled lines is greater than those realized using symmetrical coupled lines. It is still not possible, however, to achieve a very broadband coupling (multioctave) using uniform asymmetrical coupled lines. Very broadband coupling can be achieved by continuously varying the phase constants of coupled lines (β_1 and β_2) and the coupling coefficient between them along the length of the structure. This results in a nonuniform structure. This principle of broadband coupling is known as the normal-mode warping [7] to [9]. The cross section of such a structure varies continuously along the length. The essential features of the normal-mode warping can be summarized as follows [8]:

- Adjust the geometrical parameters at the input of the structure such that the input excitation is identical with one of the normal modes of the coupled structure.
- Gradually warp the normal mode (the mode in which the power is launched) by continuously varying the structure along the longitudinal

Broadband Forward-Wave Directional Couplers 169

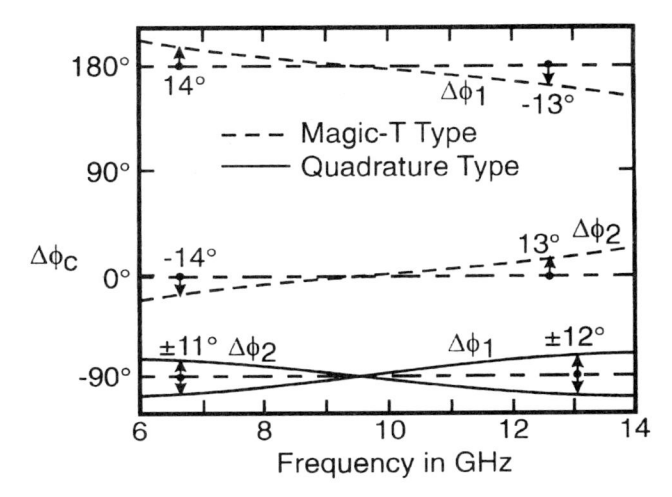

Figure 5.7(b) Theoretically computed phase difference between the coupled and through ports of the asymmetrical coupler with reference planes chosen to approximate quadrature or magic-T performance. $\Delta\phi_1$ and $\Delta\phi_2$ are defined in Figure 5.7(a). (From [5], © 1987 IEEE. Reprinted with permission.)

direction until the distribution of the normal mode is identical with the desired output. The normal mode at the cross section of the output now contains power in both the coupled modes in the desired ratio.

Note that nonuniform couplers tend to be very long (tens to hundreds of wavelengths). Therefore, they are mainly useful at high mm-wave and optical frequencies where their physical lengths can be kept reasonably small.

5.2 Coupled-Mode Theory

Consider two lines that are uniformly coupled over a certain length as shown in Figure 5.8. As already discussed, these lines may represent two actual transmission lines, or in a more general case, any two waves. The lines are assumed to be weakly coupled. By the term "weakly" coupled, we mean that the impedances of individual lines (or waves) are affected by a very small amount in the presence of coupling. There is, therefore, negligible coupling in the backward direction, and the predominant coupling takes place in the forward direction only. For example, if power is incident at port 1 as shown in Figure 5.8, the power coupled between

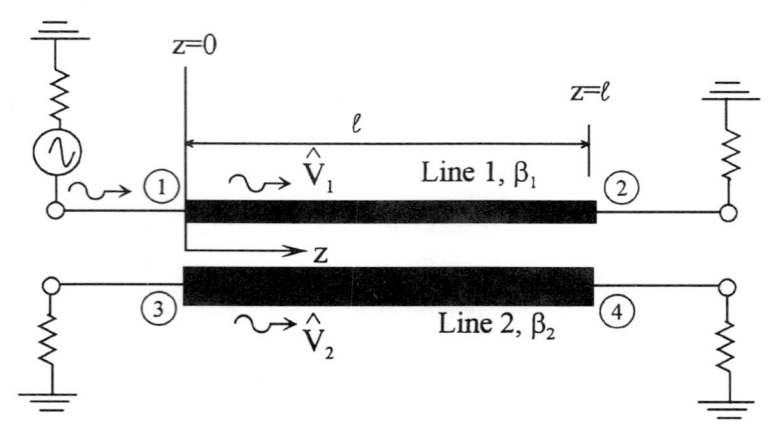

Figure 5.8 Uniformly coupled asymmetrical coupled lines.

the lines appears at port 4 only. The forward-wave coupling between the lines, *per-unit wavelength* of the coupling section, is also assumed to be small.

The forward-traveling waves on the two lines (in the absence of any coupling between them) can then be expressed respectively as

$$\hat{V}_1 = Ae^{-j\beta_1 z} \qquad (5.7)$$

and

$$\hat{V}_2 = Be^{-j\beta_2 z} \qquad (5.8)$$

where $\hat{V}_1$ and $\hat{V}_2$ denote the normalized voltage waves on lines 1 and 2, respectively, and are complex quantities. The power carried by lines 1 and 2 are thus given by $|\hat{V}_1|^2$ and $|\hat{V}_2|^2$, respectively. β_1 and β_2 denote the respective phase constants of lines 1 and 2. The coupled-mode theory as given by Miller [1] on which the present discussion is based is valid for complex values of β_1 and β_2, but for the sake of simplicity, we assume that these are real quantities.

By differentiating (5.7) and (5.8) with respect to z, we obtain, respectively

$$\frac{d\hat{V}_1}{dz} = -j\beta_1 \hat{V}_1 \qquad (5.9)$$

and

$$\frac{d\hat{V}_2}{dz} = -j\beta_2 \hat{V}_2 \qquad (5.10)$$

In the presence of coupling between the lines as shown in Figure 5.8, the existing voltage waves on both the lines are perturbed. According to coupled-mode theory, (5.9) and (5.10) representing voltage waves on the two lines are modified as follows in the presence of coupling:

$$\frac{d\hat{V}_1}{dz} = -j(\beta_1 + K_{11})\hat{V}_1 - jK_{12}\hat{V}_2 \quad (5.11)$$

and

$$\frac{d\hat{V}_2}{dz} = -jK_{21}\hat{V}_1 - j(\beta_2 + K_{22})\hat{V}_2 \quad (5.12)$$

where K_{11} and K_{22} are the self-coupling coefficients, and K_{12} and K_{21} are the mutual-coupling coefficients. Their dimensional unit is per-unit length. When the two lines are uncoupled, the propagation constants of the two lines are given by β_1 and β_2, respectively. When the two lines are brought closer, the propagation constant of each line changes because of the presence of the other line. The modified propagation constant of line 1 due to the presence of line 2 is denoted by $(\beta_1 + K_{11})$. Similarly, $(\beta_2 + K_{22})$ denotes the modified propagation constant of line 2 in the presence of line 1. For weak coupling between lines:

$$\begin{aligned} |K_{11}| &\ll \beta_1 \\ |K_{22}| &\ll \beta_2 \\ |K_{12}|, |K_{21}| &\ll \beta_1, \beta_2 \end{aligned} \quad (5.13)$$

With the above assumptions, (5.11) and (5.12) reduce to

$$\frac{d\hat{V}_1}{dz} = -j\beta_1 \hat{V}_1 - jK_{12}\hat{V}_2 \quad (5.14)$$

and

$$\frac{d\hat{V}_2}{dz} = -jK_{21}\hat{V}_1 - j\beta_2\hat{V}_2 \quad (5.15)$$

Note that while deriving (5.14) and (5.15), the terms containing K_{12} and K_{21} in (5.11) and (5.12) have been retained, while those containing K_{11} and K_{22} have been neglected. The reason for this is that although $|K_{12}|$ and $|K_{21}|$ are much smaller than β_1 and β_2, it is not necessary that $|K_{12}\hat{V}_2| \ll |\beta_1\hat{V}_1|$ or $|K_{21}\hat{V}_1| \ll |\beta_2\hat{V}_2|$ for all values of z.

Nature of Coupling Coefficient K_{12} and K_{21}

The total power on the two lines at any cross section is given by

$$W = |\hat{V}_1|^2 + |\hat{V}_2|^2 \quad (5.16)$$

The principle of conservation of power requires that if the lines are lossless, the total power remains the same at all cross sections. In mathematical terms:

$$\frac{d}{dz}(|\hat{V}_1|^2 + |\hat{V}_2|^2) = 0 \tag{5.17}$$

or

$$\frac{d}{dz}(\hat{V}_1\hat{V}_1^* + \hat{V}_2\hat{V}_2^*) = 0 \tag{5.18}$$

or

$$\hat{V}_1\frac{d\hat{V}_1^*}{dz} + \hat{V}_1^*\frac{d\hat{V}_1}{dz} + \hat{V}_2\frac{d\hat{V}_2^*}{dz} + \hat{V}_2^*\frac{d\hat{V}_2}{dz} = 0 \tag{5.19}$$

Substituting values of first derivatives in (5.19) from (5.14) and (5.15), we obtain

$$K_{12} = K_{21} = K \tag{5.20}$$

where K is purely real and denotes the *coupling coefficient* between the lines.

Waves on Lines 1 and 2 in the Presence of Coupling

Let us assume that initially there is a wave carrying unit power on line 1 only; that is:

$$\hat{V}_1 = 1, \quad \hat{V}_2 = 0 \text{ at } z = 0 \tag{5.21}$$

Solution of coupled equations (5.14) and (5.15) with the initial condition of (5.21) gives

$$\hat{V}_1 = \left[\frac{1}{2} + \frac{(\beta_1 - \beta_2)}{2\sqrt{(\beta_1 - \beta_2)^2 + 4K^2}}\right]e^{-j\beta_s z}$$

$$+ \left[\frac{1}{2} - \frac{(\beta_1 - \beta_2)}{2\sqrt{(\beta_1 - \beta_2)^2 + 4K^2}}\right]e^{-j\beta_f z} \tag{5.22}$$

and

$$\hat{V}_2 = \frac{K}{\sqrt{(\beta_1 - \beta_2)^2 + 4K^2}}e^{-j\beta_s z} - \frac{K}{\sqrt{(\beta_1 - \beta_2)^2 + 4K^2}}e^{-j\beta_f z} \tag{5.23}$$

where

$$\beta_s = \frac{(\beta_1 + \beta_2)}{2} + \frac{\sqrt{(\beta_1 - \beta_2)^2 + 4K^2}}{2} \tag{5.24}$$

and

$$\beta_f = \frac{(\beta_1 + \beta_2)}{2} - \frac{\sqrt{(\beta_1 - \beta_2)^2 + 4K^2}}{2} \qquad (5.25)$$

The above equations show that in the presence of coupling, the waves on the two lines can be represented as interference between two waves having different phase constants from those of the uncoupled waves. One of these waves (which can be termed as a *slow* wave) has a phase constant equal to β_s; while the other (which can be termed as a *fast* wave) has a phase constant equal to β_f. The phase constants β_s and β_f depend on the phase constants of individual lines (when the lines are uncoupled) and the coupling coefficient K. These waves with phase constants β_s and β_f may be considered to represent two normal modes of the coupled structure. (5.22) and (5.23) express the wave on each line in terms of interference between *normal* modes.

The coupling coefficient K depends on the specifics of the structure, and its determination requires the use of field theoretic methods [10]. Irrespective of the value of the coupling coefficient K, however, (5.22) and (5.23) can be used to draw some significant conclusions on uniform coupling between two lines (or waves). For example, these equations lead to a significant conclusion that if the phase constants of two lines (waves) are different ($\beta_1 \neq \beta_2$), it is not possible to completely transfer power between the two lines(waves).

Coupling Between Symmetrical Lines

Let the two coupled lines be symmetrical ($\beta_1 = \beta_2$). Using (5.24) and (5.25), we obtain

$$\beta_s = \beta_0 + K \qquad (5.26)$$

and

$$\beta_f = \beta_0 - K \qquad (5.27)$$

where

$$\beta_0 = \beta_1 = \beta_2$$

Substituting the values of β_s and β_f from (5.26) and (5.27) in (5.22) and (5.23), we obtain

$$\hat{V}_1 = \frac{(e^{-jKz} + e^{jKz})}{2} e^{-j\beta_0 z}$$

$$= \cos(Kz) e^{-j\beta_0 z} \qquad (5.28)$$

$$\hat{V}_2 = \frac{(e^{-jKz} - e^{jKz})}{2} e^{-\beta_0 z}$$
$$= -j\sin(Kz)e^{-j\beta_0 z} \qquad (5.29)$$

The fraction of power coupled from line 1 to line 2 over a length z of the coupling section is then given by

$$r = \frac{|\hat{V}_2(z)|^2}{|\hat{V}_1(z=0)|^2} = \sin^2(Kz) \qquad (5.30)$$

where it is assumed that all the power is in line 1 at $z = 0$. The power coupled to line 2 appears at port 4.

Coupled-Mode Theory and Even- and Odd-Mode Analysis

Using (5.26) and (5.27):

$$K = \frac{\beta_s - \beta_f}{2} \qquad (5.31)$$

and

$$\beta_0 = \frac{\beta_s + \beta_f}{2} \qquad (5.32)$$

Substituting the values of K and β_0 from (5.31) and (5.32) in (5.28) and (5.29), we obtain

$$\hat{V}_1 = \cos\left[\frac{(\beta_s - \beta_f)z}{2}\right] e^{-j\frac{(\beta_s + \beta_f)}{2}z} \qquad (5.33)$$

and

$$\hat{V}_2 = -j\sin\left[\frac{(\beta_s - \beta_f)z}{2}\right] e^{-j\frac{(\beta_s + \beta_f)}{2}z} \qquad (5.34)$$

It is interesting to find that (5.33) and (5.34) are identical to (4.35) and (4.36), respectively, if β_s and β_f are assumed to denote the phase constants of even and odd modes, respectively. (4.35) and (4.36) were derived using the even- and odd-mode approach, while (5.33) and (5.34) were derived using the coupled-mode theory.

Coupling Between Asymmetrical Lines

Let the phase constants of two asymmetrical coupled lines be β_1 and β_2, respectively. It is assumed that initially a unit amount of power is incident in line 1 (i.e., $\hat{V}_1 = 1$ at $z = 0$). Using (5.22) and (5.23), the voltage wave on line 1 is given by

$$\hat{V}_1 = e^{-j\frac{(\beta_1+\beta_2)}{2}z} V_1' \tag{5.35}$$

where

$$V_1' = \cos\left[\left(\sqrt{\frac{(\beta_1-\beta_2)^2}{4K^2}+1}\right)Kz\right]$$

$$-j\frac{(\beta_1-\beta_2)}{2K}\frac{1}{\left[\sqrt{\frac{(\beta_1-\beta_2)^2}{4K^2}+1}\right]}\sin\left[\left(\sqrt{\frac{(\beta_1-\beta_2)^2}{4K^2}+1}\right)Kz\right]$$

The voltage wave on line 2 is expressed as

$$\hat{V}_2 = e^{-j\frac{(\beta_1+\beta_2)}{2}z} V_2' \tag{5.36}$$

where

$$V_2' = -\frac{j}{\sqrt{\frac{(\beta_1-\beta_2)^2}{4K^2}+1}}\sin\left[\left(\sqrt{\frac{(\beta_1-\beta_2)^2}{4K^2}+1}\right)Kz\right]$$

The maximum power transfer from line 1 to line 2 takes place when

$$\left(\sqrt{\frac{(\beta_1-\beta_2)^2}{4K^2}+1}\right)Kz = \frac{\pi}{2} \tag{5.37}$$

The maximum fractional power coupled between lines 1 and 2 is given by

$$r_{max} = |\hat{V}_2|^2_{max} = \frac{1}{\frac{(\beta_1-\beta_2)^2}{4K^2}+1} \tag{5.38}$$

It is seen from (5.38) that r_{max} (which represents the maximum power that can be coupled between lines) is less than unity if the values of β_1 and β_2 are different from each other. The value of r_{max} is plotted as a function of normalized difference in phase velocities of the two lines in Figure 5.9. It is seen that if the difference in phase velocities of the two lines is much greater than

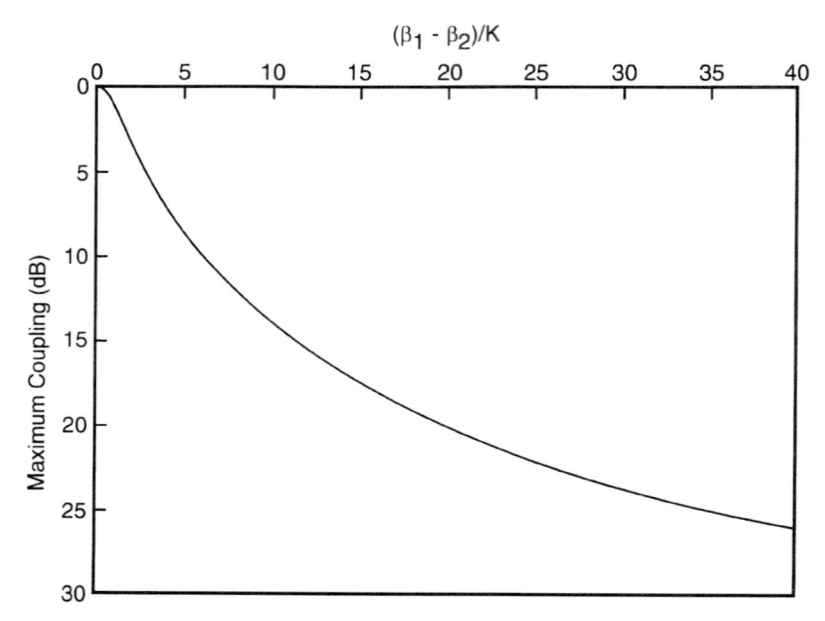

Figure 5.9 Maximum power coupled between asymmetrical coupled lines.

the coupling coefficient $((\beta_1 - \beta_2) \gg K)$, only a small amount of power can be coupled between the lines. For example, for a value of $(\beta_1 - \beta_2)/K = 10$, the maximal coupling that can be achieved between the lines is only 14.2 dB.

After having discussed the normal (Chapter 4) and coupled modes, it is useful to summarize the essential difference between the two. The normal modes of a uniform structure are those that can propagate independent of each other along the structure. Each normal mode is characterized by a unique phase velocity and field distribution. For example, the various TE_{mn} and TM_{mn} modes are the normal modes of a straight, uniform rectangular metal waveguide. Similarly, the even and odd modes are the two normal modes of symmetrical coupled lines. An important property of normal modes is that there is no conversion of energy from one normal mode to another. For example, if an even mode is launched along a symmetrical coupled structure, the energy remains in the even mode all along the length of the structure. On the other hand, energy is continuously exchanged between coupled modes. The individual waves on two coupled lines are an example of two coupled modes. There is a continuous exchange of energy between the waves on the two lines. As another example of coupled modes, consider a rectangular waveguide that is uniformly bent along the direction of propagation. The various TE_{mn} and TM_{mn} modes that are the normal modes of a straight waveguide are no longer the normal modes of the bent waveguide. The various TE_{mn} and TM_{mn} modes are now the coupled modes because energy will continuously be exchanged between these modes along the bend of the waveguide. Usually, it is sufficient to consider coupling between two or three modes only to determine the state of a weakly coupled system.

5.3 Coupled-Mode Theory for Weakly Coupled Resonators

The theory of weakly coupled resonators can be developed in a similar fashion as is for the weakly coupled lines. Consider two isolated resonators having resonant frequencies as ω_1 and ω_2, respectively. The time-varying amplitudes of two *uncoupled* resonators are given by [3]

$$\frac{d\hat{a}_1}{dt} = j\omega_1 \hat{a}_1 \tag{5.39}$$

and

$$\frac{d\hat{a}_2}{dt} = j\omega_2 \hat{a}_2 \tag{5.40}$$

where $\hat{a}_1$ and $\hat{a}_2$ denote the *instantaneous* normalized amplitudes of resonators 1 and 2, respectively. When two resonators are weakly coupled, the governing equations for the resonator amplitudes are modified as

$$\frac{d\hat{a}_1}{dt} = j\omega_1 \hat{a}_1 + jK\hat{a}_2 \tag{5.41}$$

and

$$\frac{d\hat{a}_2}{dt} = jK\hat{a}_1 + j\omega_2 \hat{a}_2 \tag{5.42}$$

where K denotes the *coupling coefficient* between the resonators. The dimensional unit of K in this case is *per-unit time*. It may be noted than in case of coupled lines, the dimensional unit of K is *per-unit length*. (5.41) and (5.42) are similar to (5.14) and (5.15), respectively. The solution of the coupled equations (5.41) and (5.42) is therefore similar to the solution of coupled equations (5.14) and (5.15). The coupling between resonators causes the resonant frequencies of the normal modes of the coupled system to be different from ω_1 and ω_2. More specifically, the resonant frequencies of normal modes (denoted by ω_a and ω_b) are given by

$$\omega_a = \frac{(\omega_1 + \omega_2)}{2} + \frac{\sqrt{(\omega_1 - \omega_2)^2 + 4K^2}}{2} \tag{5.43}$$

and

$$\omega_b = \frac{(\omega_1 + \omega_2)}{2} - \frac{\sqrt{(\omega_1 - \omega_2)^2 + 4K^2}}{2} \tag{5.44}$$

When two resonators having the same resonant frequency ($\omega_0 = \omega_1 = \omega_2$) are coupled, the resonant frequencies of the coupled resonators become

$$\omega_a = \omega_{even} = \omega_0 + K \tag{5.45}$$

$$\omega_b = \omega_{odd} = \omega_0 - K \tag{5.46}$$

It may be noted that in microwave circuits, it is quite common to denote coupling between resonators using a dimensionless quantity k known as the *voltage coupling coefficient* [11, 12]. In terms of the voltage coupling coefficient k, the even- and odd-mode resonant frequencies of two identical coupled resonators are given by [12]

$$\omega_{even}^2 = \frac{\omega_0^2}{1+k} \tag{5.47}$$

and

$$\omega_{odd}^2 = \frac{\omega_0^2}{1-k} \tag{5.48}$$

By comparing (5.45) and (5.46) with (5.47) and (5.48), respectively, we can determine the relation between coupling coefficients K and k.

References

[1] Miller, S. E., "Coupled Wave Theory and Waveguide Applications," *Bell Syst. Tech. J.*, Vol. 33, 1954, pp. 661–719.

[2] Louisell, W. H., *Coupled-Mode and Parametric Electronics*, Wiley, New York, 1960.

[3] Haus, H. A., and W. Huang, "Coupled-Mode Theory," *Proc. IEEE*, Vol. 79, Oct. 1991, pp. 1,505–1,518.

[4] Ikalainen, P. K., and G. L. Matthaei, "Design of Broadband Dielectric Waveguide 3-dB Couplers," *IEEE Trans.*, Vol. MTT-35, July 1987, pp. 621–628.

[5] Ikalainen, P. K., and G. L. Matthaei, "Wideband, Forward-Coupling Microstrip Hybrids With High Directivity," *IEEE Trans.*, Vol. MTT-35, Aug. 1987, pp. 719–725.

[6] Kirsching, M., and R. H. Jansen, "Accurate Wide-Range Design Equations for the Frequency-Dependent Characteristics of Parallel-Coupled Microstrip Lines," *IEEE Trans.*, Vol. MTT-32, Jan. 1984, pp. 83–90. Corrections, Vol. MTT-33, Mar. 1985, p. 288.

[7] Cook, J. S., "Tapered Velocity Couplers," *Bell Syst. Tech. J.*, Vol. 34, July 1955, pp. 807–822.

[8] Fox, A. G., "Wave Coupling by Warped Normal Modes," *Bell Syst. Tech. J.*, Vol. 34, July 1955, pp. 823–852. Also see Fox, A. G., "Wave Coupling by Warped Normal Modes," *IRE Trans.*, Vol. 3, Dec. 1955, pp. 2–6.

[9] Louisell, W. H., "Analysis of Single Tapered-Mode Coupler," *Bell Syst. Tech. J.*, Vol. 34, July 1955, pp. 853–870.

[10] Yariv, A., "Coupled-Mode Theory for Guided Wave Optics," *IEEE J.*, Vol. QE-9, Sept. 1973, pp. 919–933.

[11] Cohn, S. B., "Bandpass Filters Containing High Q Dielectric Resonators," *IEEE Trans.*, Vol. MTT-16, Apr. 1968, pp. 218–227.

[12] Van Bladel, J., "Weakly Coupled Dielectric Resonators," *IEEE Trans.*, Vol. MTT-30, 1982, pp. 1907–1914.

6

Parallel-Coupled TEM Directional Couplers

Introduction

In Chapter 4 it was demonstrated that if two identical TEM lines are parallel-coupled as shown in Figure 6.1, then by properly choosing the even- and odd-mode impedances of the coupled lines, a four-port directional coupler can be obtained. The coupler shown in Figure 6.1 is also called a *backward-wave directional coupler* because the coupling takes place in the backward direction on the coupled line. For example, if power is incident at port 1, power is coupled to port 3 on the coupled line. The maximum coupling between ports 1 and 3 (or between ports 2 and 4) takes place at a frequency where the coupler is a quarter-wave long (or odd multiples thereof). Because the electrical length of a coupler varies with frequency, the coupling also varies with frequency. The variation of coupling with frequency can be reduced by employing multisection TEM couplers. In a multisection coupler, a number of coupled sections are cascaded. Each coupled section is a quarter-wave long at the center frequency and has different even- and odd-mode impedances compared with those of the adjacent sections. Multisection couplers that have end-to-end symmetry are known as *symmetrical couplers*, while those that do not have end-to-end symmetry are known as *asymmetrical couplers*.

This chapter discusses the theory and design of single- and multisection parallel-coupled TEM directional couplers. Simple analytical expressions for the design of single-section couplers exist. Unfortunately, no simple analytical formulas are possible for the design of optimal multisection couplers. Some design tables are available in the literature for equal-ripple symmetrical and asymmetrical couplers [1–3].

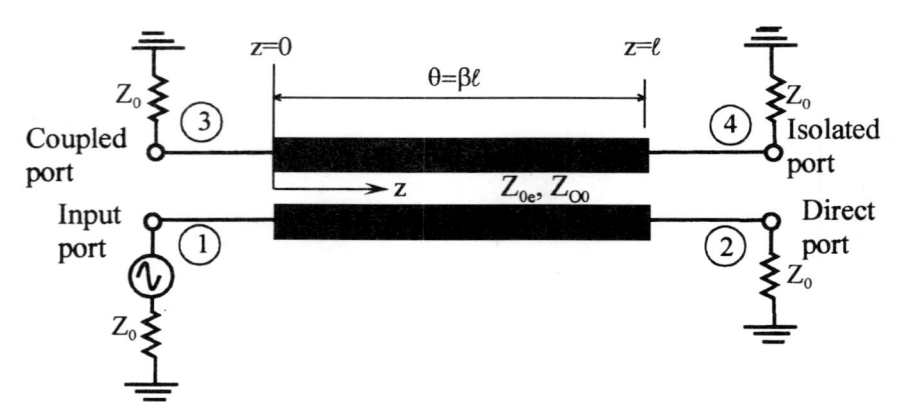

Figure 6.1 A single section TEM coupler.

The microstrip line is the most popular transmission line for realizing microwave integrated circuit (MIC) components. When microstrip is used in backward-wave couplers, however, the directivity is generally poor because the even- and odd-mode phase velocities of coupled microstrip lines are unequal. In this chapter, various techniques that can be used to equalize the even- and odd-mode phase velocities of coupled microstrip lines are also discussed.

6.1 Coupler Parameters

The schematic of a four-port directional coupler is shown in Figure 6.2. The four ports are labeled as "input," "direct" (through), "coupled," and "isolated." Two important factors that characterize a directional coupler are its coupling and directivity, defined below:

$$\text{Coupling (dB)} = 10 \log \frac{P_1}{P_3} \qquad (6.1)$$

$$\text{Directivity (dB)} = 10 \log \frac{P_3}{P_4} \qquad (6.2)$$

where P_1 is the power input at port 1 and P_3 and P_4 are the power outputs at ports 3 and 4, respectively. Note that all the ports are assumed to be match-terminated. There is no power at port 4 in an ideal case; in practice, a small amount of power is always coupled to this port.

If the coupling and directivity are known, the isolation of the coupler can be determined. The *isolation* is defined as

$$\text{Isolation (dB)} = 10 \log \frac{P_1}{P_4} \qquad (6.3)$$

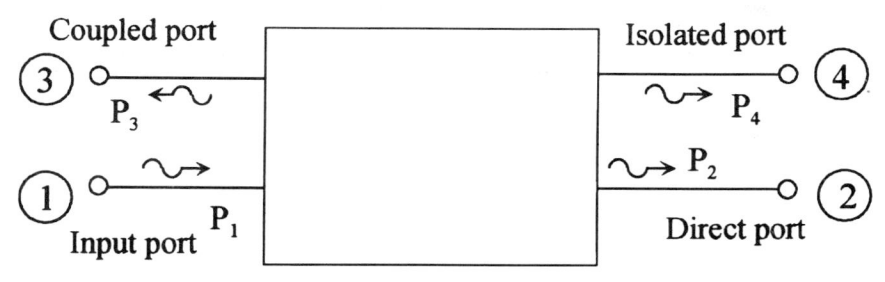

Figure 6.2 Schematic of a four-port directional coupler.

or

$$\text{Isolation (dB)} = \text{Coupling (dB)} + \text{Directivity (dB)}$$

All the parameters described above are normally expressed in decibels and are defined here as positive quantities.

6.2 Single-Section Directional Coupler

6.2.1 Frequency Response

In Section 4.2.2, it was shown that if two identical parallel TEM lines are coupled over a length l as shown in Figure 6.1, then under the condition

$$Z_{0e} Z_{0o} = Z_0^2 \tag{6.4}$$

the scattering parameters of the network are given by

$$S_{11} = S_{22} = S_{33} = S_{44} = 0 \tag{6.5}$$

$$S_{14} = S_{41} = S_{23} = S_{32} = 0 \tag{6.6}$$

$$S_{12} = S_{21} = S_{34} = S_{43} = S_{21e}$$

$$= \frac{\sqrt{1-k^2}}{\sqrt{1-k^2}\cos\theta + j\sin\theta} \tag{6.7}$$

$$S_{13} = S_{31} = S_{24} = S_{42} = S_{11e}$$

$$= \frac{jk\sin\theta}{\sqrt{1-k^2}\cos\theta + j\sin\theta} \tag{6.8}$$

where $\theta = \beta l$ denotes the electrical length of the coupler and k is given by

$$k = \frac{Z_{0e} - Z_{0o}}{Z_{0e} + Z_{0o}} \tag{6.9}$$

Further, S_{11e} and S_{21e} denote the reflection and transmission coefficients, respectively, of the coupled lines for the case of even-mode excitation. The maximum amount of coupling between ports 1 and 3 (or between ports 2 and 4) occurs when

$$\theta = \beta l = \frac{\pi}{2} \text{ rads} \qquad (6.10)$$

or

$$l = \frac{\pi}{2\beta} = \frac{\lambda_g}{4}$$

The properties of an ideal parallel-coupled TEM directional coupler were described in Chapter 4.

Frequency Bandwidth Ratio

The frequency bandwidth ratio B of a directional coupler (single- or multisection) is defined as

$$B = \frac{f_2}{f_1} \qquad (6.11)$$

where f_2 and f_1 are the upper and lower frequencies in between which the coupling is within the tolerance amount δ compared with its midband value as shown in Figure 6.3. The tolerance amount δ can be arbitrarily specified.

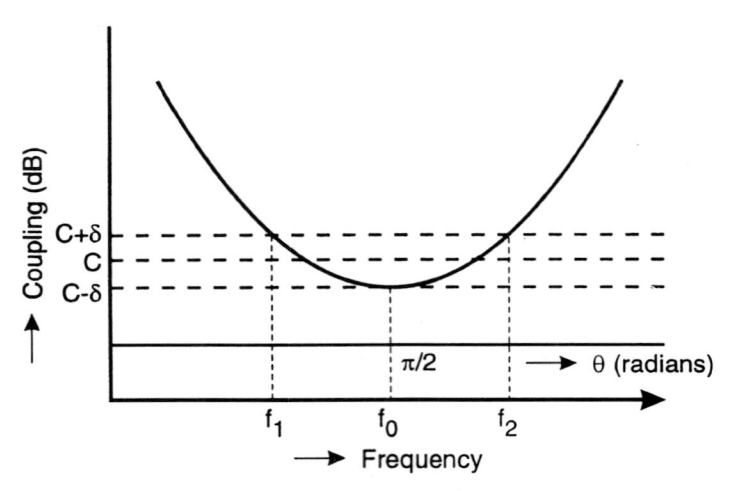

Figure 6.3 Typical variation of coupling in a single section TEM coupler.

Fractional Bandwidth

The fractional bandwidth w of a directional coupler is defined as

$$w = \frac{f_2 - f_1}{f_0} \tag{6.12}$$

where

$$f_0 = \frac{f_1 + f_2}{2} \tag{6.13}$$

is the center frequency of the coupler.

The frequency bandwidth ratio B and the fractional bandwidth w are related by

$$w = 2\frac{B-1}{B+1} \tag{6.14}$$

and

$$B = \frac{1 + w/2}{1 - w/2} \tag{6.15}$$

Useful Operating Bandwidth

The variation of coupling to the "direct" and "coupled" ports of an ideal 3 ± 0.3-dB coupler is shown in Figure 6.4(a) as a function of frequency ratio f/f_0, where f_0 refers to the frequency where the coupler is a quarter-wave long. Because we have assumed a tolerance amount of ± 0.3 dB, this coupler is designed to have a coupling (to the "coupled" port) of 2.7 dB at the midband. The performance of 6 ± 0.3 and 10 ± 0.5-dB single-section couplers is shown in Figures 6.4(b) and (c). The variation of coupling to the "coupled" port of a 20 ± 0.5-dB coupler is shown in Figure 6.4(d). In this case, the power coupled to the direct port is nearly 0 dB. The useful frequency operating range of single-section couplers can be determined by referring to these plots. For example, it is seen from Figure 6.4(a) that for a tolerance in coupling of ± 0.3 dB, a 3-dB coupler can be operated over a frequency bandwidth ratio of about 2 ($f_2 : f_1 \approx 2 : 1$). Single-section couplers are generally useful for operation over a frequency bandwidth ratio (B) of approximately 2.

6.2.2 Design

From (6.9), we can write Z_{0e}/Z_{0o} in terms of a voltage coupling coefficient k as

$$\frac{Z_{0e}}{Z_{0o}} = \frac{1 + k}{1 - k} \tag{6.16}$$

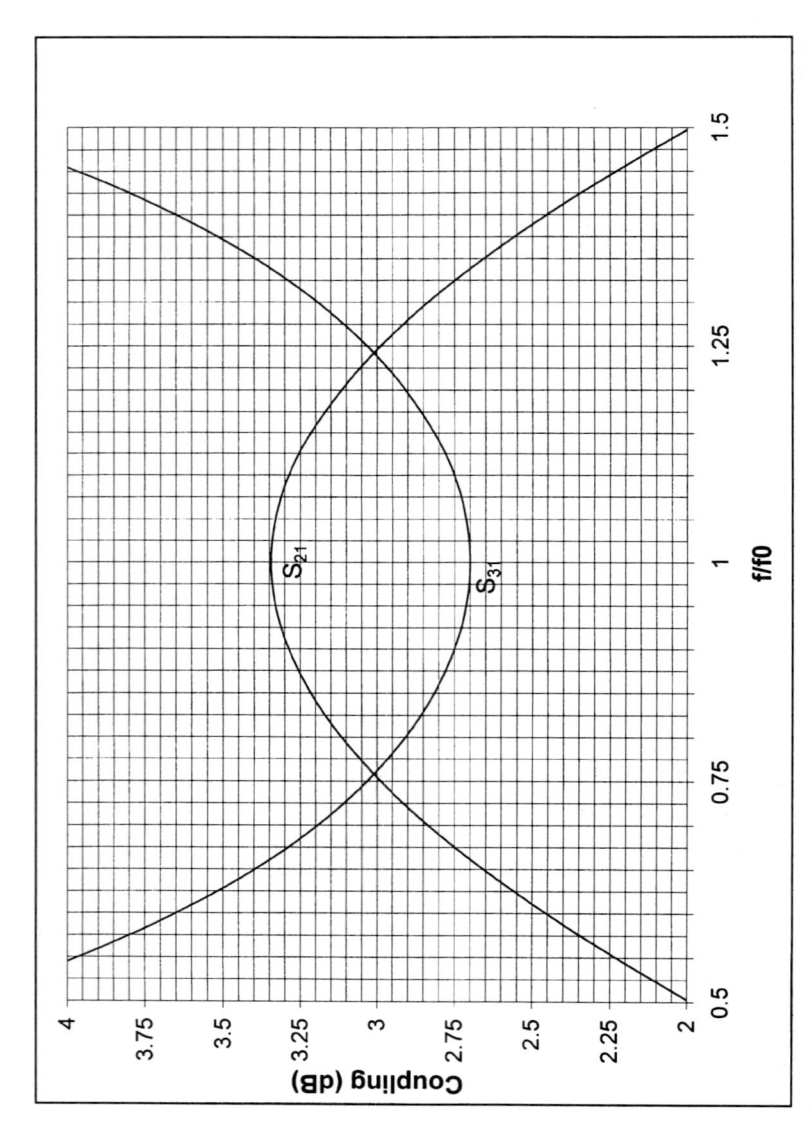

Figure 6.4 (a) Variation of coupling to the direct and coupled ports of a TEM coupler designed for nominal coupling of 3 dB.

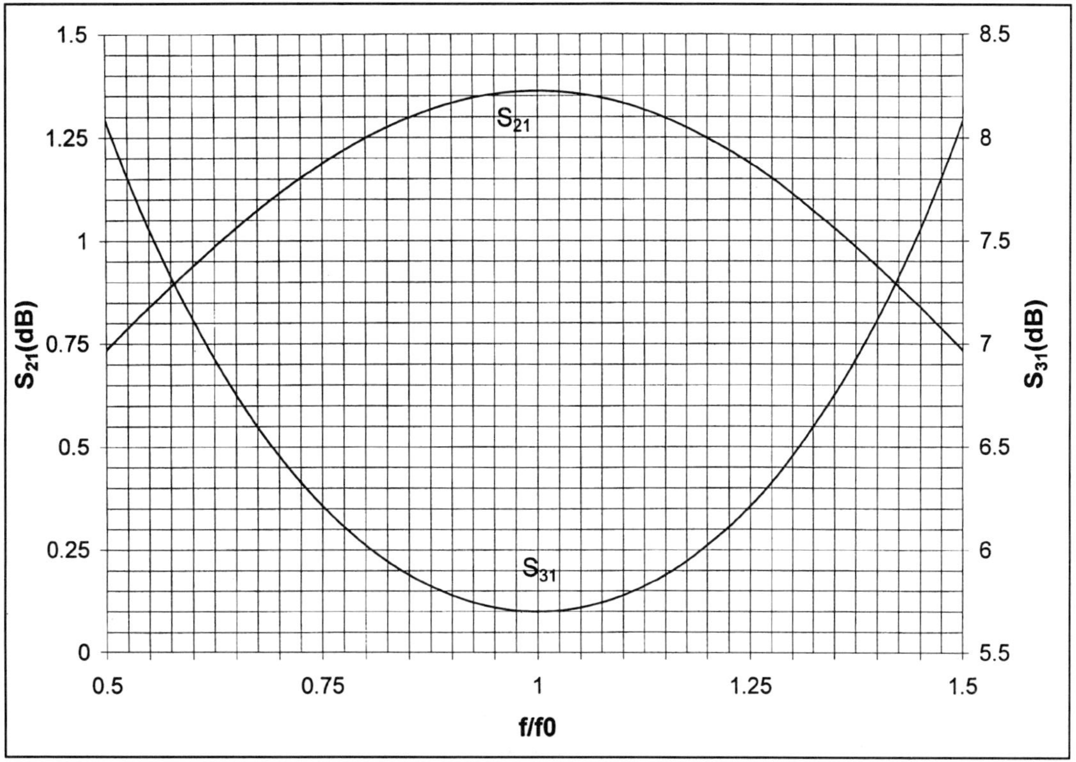

Figure 6.4 *(continued)* (b) Variation of coupling to the direct and coupled ports of a TEM coupler designed for 6-dB nominal coupling.

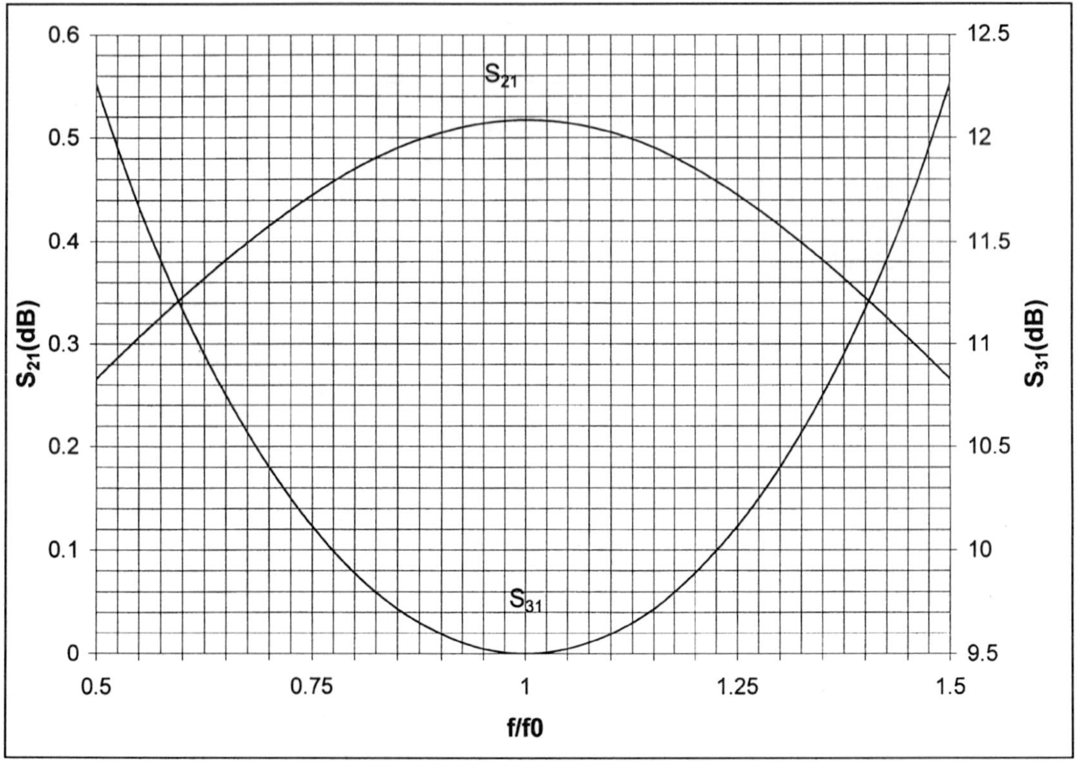

Figure 6.4 *(continued)* (c) Variation of coupling to the direct and coupled ports of a TEM coupler designed for nominal coupling of 10 dB.

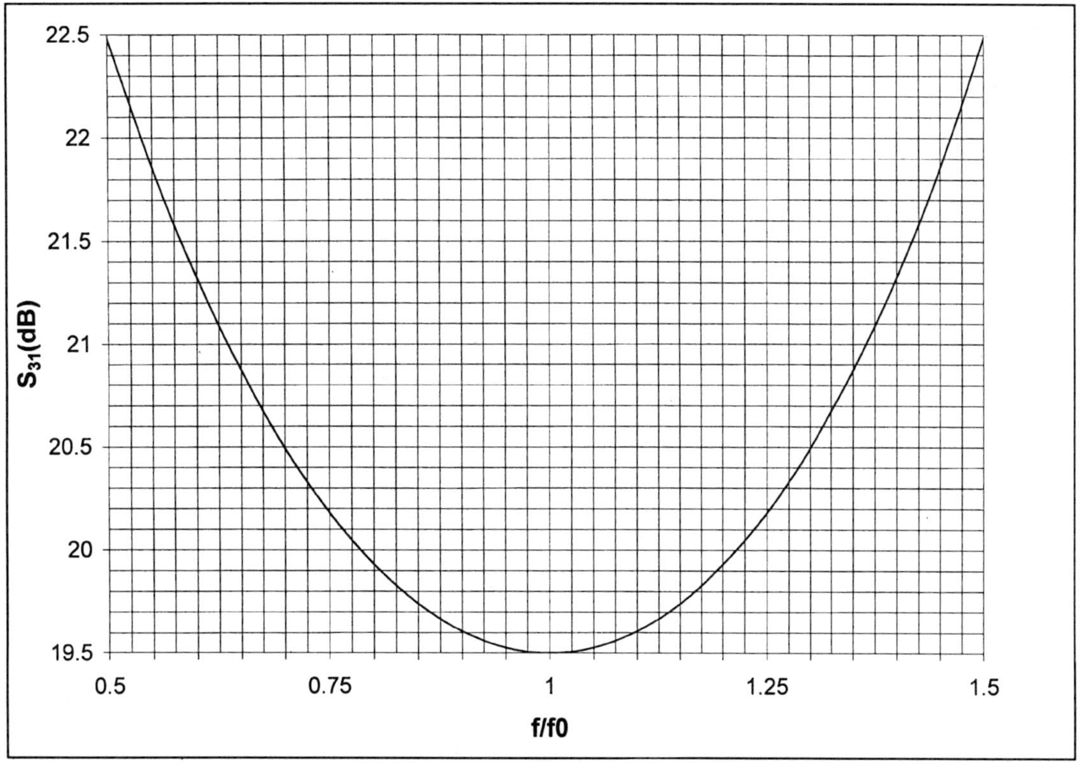

Figure 6.4 *(continued)* (d) Variation of coupling to the coupled port of a TEM coupler designed for a nominal coupling of 20 dB.

Further, simultaneous solution of (6.4) and (6.16) gives

$$Z_{0e} = Z_0 \sqrt{\frac{1+k}{1-k}} \qquad (6.17)$$

and

$$Z_{0o} = Z_0 \sqrt{\frac{1-k}{1+k}} \qquad (6.18)$$

In a coupler design, for a given voltage coupling coefficient k and characteristic impedance Z_0, we first determine the even- and odd-mode impedances using (6.17) and (6.18), respectively. The dimensions of the coupler are then calculated using the physical data of coupled transmission lines such as those discussed in Chapter 3 for various transmission lines. The physical length l of the coupler is chosen as

$$l = \frac{\lambda_g}{4} \qquad (6.19)$$

where λ_g is the guide wavelength of the TEM wave in the transmission line medium at the design frequency f_0.

If the coupling from port 1 to port 3 is given as C dB (where C is a positive quantity), then k is related to C as

$$k = 10^{-C/20} \qquad (6.20)$$

Example 6.1

A directional coupler of 10 ± 0.5-dB coupling is desired in the configuration as shown in Figure 6.1 at a frequency of 10 GHz. Determine the physical dimensions of the coupler assuming that ports are terminated in an impedance of 50 Ω and the coupler is realized in a stripline medium of $\epsilon_r = 2.25$.

The coupler is designed to have a coupling of 9.5 dB at the midband because a tolerance of ± 0.5 dB in the coupling value has been specified. Using (6.20), the voltage coupling coefficient k is found to be

$$k = 10^{-9.5/20} = 0.335$$

The even- and odd-mode impedances of the coupled lines are obtained as,

$$Z_{0e} = Z_0 \sqrt{\frac{1+k}{1-k}} = 50 \sqrt{\frac{1+0.335}{1-0.335}} = 70.84 \, \Omega$$

and

$$Z_{0o} = Z_0\sqrt{\frac{1-k}{1+k}} = 50\sqrt{\frac{1-0.335}{1+0.335}} = 35.29\,\Omega$$

Therefore

$$\sqrt{\epsilon_r}Z_{0e} = 106.26\,\Omega$$
$$\sqrt{\epsilon_r}Z_{0o} = 52.94\,\Omega$$

From Figure 3.15, $S/b \approx 0.03$ and $W/b \approx 0.65$ if we assume that the thickness of the strip conductors is negligble. Thus, if the separation between the ground planes is 1 mm, the gap between conductors, $S \approx 0.03$ mm, and conductor width, $W \approx 0.65$ mm.

A stripline supports a pure TEM mode of propagation. In a medium of $\epsilon_r = 2.25$, the guide wavelength in the medium at a frequency of 10 GHz is given by

$$\lambda_g = \frac{\lambda_0}{\sqrt{\epsilon_r}} = \frac{30}{\sqrt{2.25}} = 20\,\text{mm}$$

The physical length l of the coupler is therefore given by

$$l = \frac{\lambda_g}{4} = 0.005\,\text{m} = 5\,\text{mm}$$

The useful bandwidth of the ideal coupler is found to be 62.5% from Figure 6.4(c).

Example 6.2

Design a 20 ± 0.5-dB directional coupler in the microstrip configuration at 5 GHz. Determine physical dimensions of the coupler realized on 0.635 mm-thick alumina substrate having $\epsilon_r = 9.7$.

The coupler is designed to have a 19.5-dB midband coupling. From (6.20), the voltage coupling coefficient k is given by

$$k = 10^{-19.5/20} = 0.106$$

Further, from (6.17) and (6.18), the even- and odd-mode impedances of the coupled lines are obtained as

$$Z_{0e} = 55.6\,\Omega$$

and

$$Z_{0o} = 45.0\,\Omega$$

From Figure 3.22, $W/h \approx 0.95$ and $S/h \approx 1.3$ (approximately extrapolated). For a 0.635-mm-thick substrate, $W \approx 0.6$ mm and $S \approx 0.83$ mm. The physical length of the coupler is calculated using

$$\theta = \frac{\theta_e + \theta_o}{2} = \frac{2\pi}{\lambda_0} \frac{(\sqrt{\epsilon_{ree}} + \sqrt{\epsilon_{reo}})}{2} l = 90 \text{ deg}$$

or

$$\frac{360}{60} \frac{(\sqrt{7.2} + \sqrt{6})}{2} l = 90 \text{ deg}$$

which gives $l \approx 5.85$ mm. The useful bandwidth of the ideal coupler is found to be 60% from Figure 6.4(d).

6.2.3 Compact Couplers

When size and cost requirements are stringent, compact directional couplers are mandatory. The coupled line approaches for such couplers are reported in literature [4–6]. For such couplers, we can use either MIC or MMIC technology. Basically, there are two techniques to design such MIC couplers: one is to use high-dielectric constant ($\epsilon_r \approx 30 - 100$) substrate to reduce the size, and second to fold the coupler length in different shapes such as spiral and meander. MMIC couplers that generally use GaAs substrates ($\epsilon_r = 12.9$) employ the latter technique to make compact couplers. One of the important applications of these couplers is in wireless communications. The design, fabrication, and test results of these couplers are discussed in Chapter 8.

6.2.4 Equivalent Circuit of a Quarter-Wave Coupler

The coupling to the direct and backward ports of a parallel-coupled TEM coupler is given in terms of even-mode parameters by (6.7) and (6.8), as follows:

$$S_{21} = S_{21e}$$
$$S_{31} = S_{11e}$$

where S_{11e} and S_{21e} denote the reflection and transmission coefficients, respectively, of the coupled lines for the case of even-mode excitation.

The equivalent circuit of an ideal parallel-coupled TEM directional coupler is therefore as shown in Figure 6.5, where Z_{0e} denotes the even-mode characteristic impedance of the coupled lines [7]. S_{11e} and S_{21e}, which are, respectively, the reflection and transmission coefficients of the two-port circuit shown in Figure 6.5 give, respectively, the coupling to the backward (S_{31}) and direct (S_{21})

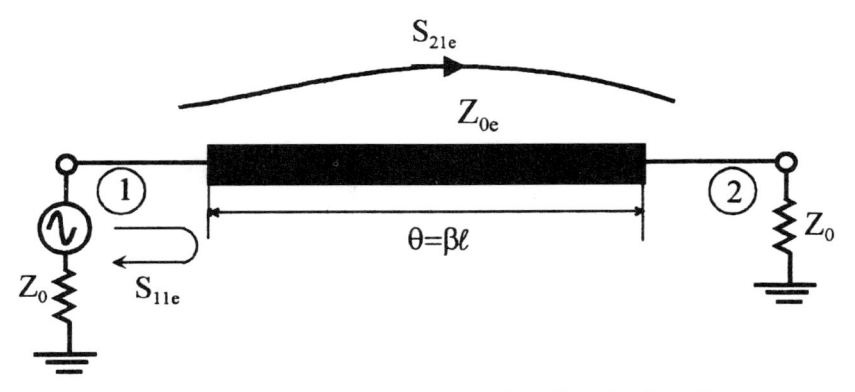

Figure 6.5 Equivalent circuit of an ideal single-section TEM directional coupler.

ports of the four-port coupler shown in Figure 6.1. We thus see that the analysis of a parallel-coupled directional coupler reduces to analyzing a simple circuit consisting of a length of transmission line of characteristic impedance Z_{0e} terminated by an impedance of Z_0 at either of its ends. This analogy is extremely useful in the design and synthesis of parallel-coupled directional couplers, as it is relatively much simpler to analyze single transmission line circuits.

6.3 Multisection Directional Couplers

6.3.1 Theory and Synthesis

To obtain a near-constant coupling over a wider frequency bandwidth than is possible using a single-section coupler, a number of coupled sections must be cascaded, as shown in Figure 6.6(a). Each section is a quarter-wave long at the center frequency. By properly choosing the even- and odd-mode impedances of the various sections, the bandwidth of the coupler can be increased. By analogy with the equivalent circuit of a single-section coupler, the equivalent cascaded transmission line circuit for finding the coupling to the backward and direct ports of the multisection coupler of Figure 6.6(a) is shown in Figure 6.6(b). The reflection and transmission coefficients of the circuit shown in Figure 6.6(b) give, respectively, the coupling to ports 3 and 2 of the coupler shown in Figure 6.6(a). The even- and odd-mode impedances of the ith section of the multisection coupler are related by

$$Z_{0oi} = \frac{Z_0^2}{Z_{0ei}} \qquad (6.21)$$

where Z_0 denotes the impedance terminating the ports of the directional coupler.

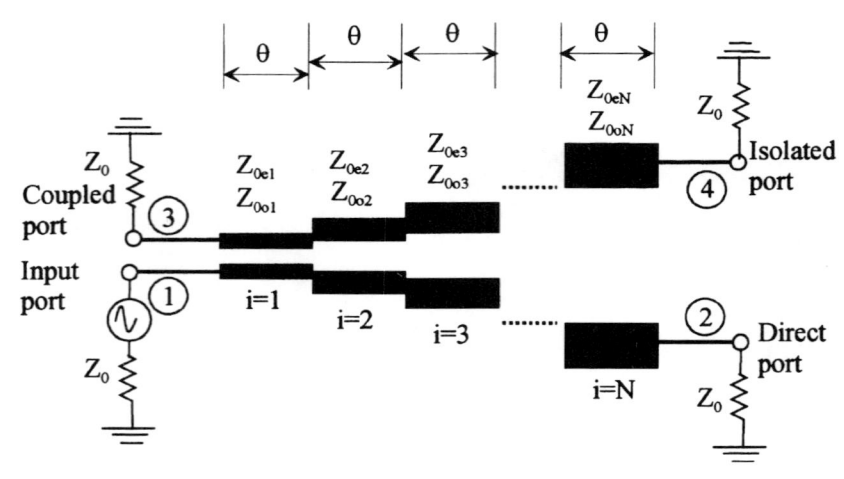

Figure 6.6 (a) An N-section asymmetrical parallel-coupled multisection directional coupler.

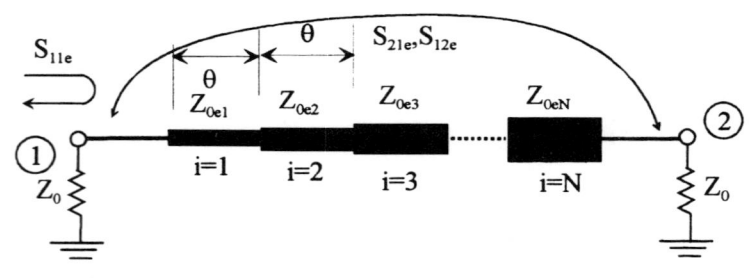

Figure 6.6 (b) Equivalent circuit of directional coupler shown in Figure 6.6(a).

A multisection coupler can be either symmetrical or asymmetrical. In multisection couplers, the term *symmetrical* is used to denote a coupler that has end-to-end symmetry. A symmetrical coupler employs an odd number of sections. In a symmetrical coupler, the ith section will be identical to the $N+1-i^{th}$ section as shown in Figure 6.7(a). The equivalent circuit for analyzing the symmetrical coupler is shown in Figure 6.7(b). If the coupler does not have end-to-end symmetry (Figure 6.6(a)), it is referred to as an *asymmetrical* coupler. An asymmetrical coupler can employ an even or odd number of sections. A significant property of symmetrical couplers is that in their case, the signal coupled to the direct port is 90 degree out of phase with the signal coupled to the backward port ($\angle S_{31} = \angle S_{21} + 90$ degree). This phase relationship is independent of the frequency. Because of this property, 3-dB symmetrical directional couplers find extensive use in diplexers, multiplexers, directional filters, balanced mixers, and in other devices where the 90-degree phase difference property is required.

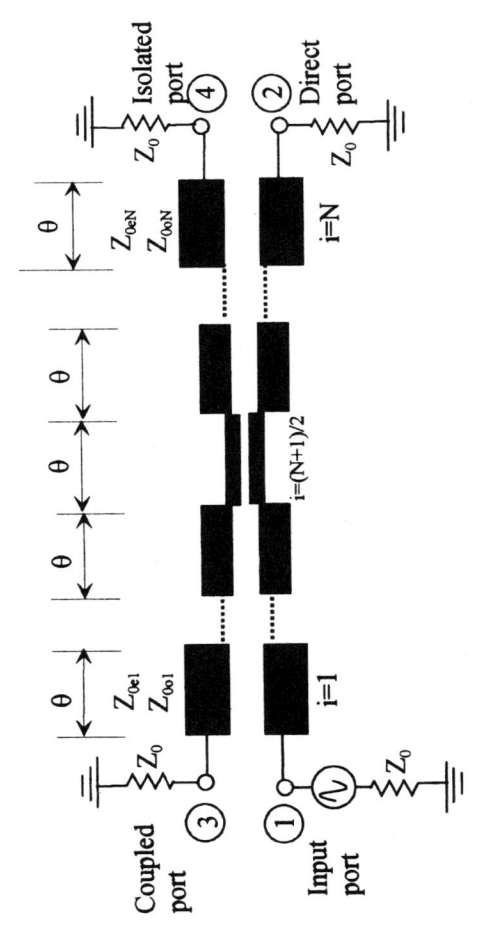

Figure 6.7 (a) An N-section symmetrical parallel-coupled multisection directional coupler.

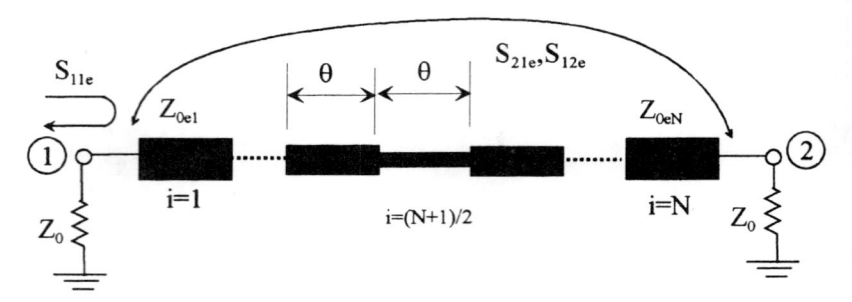

Figure 6.7 (b) Equivalent circuit of directional coupler shown in Figure 6.7(a).

Asymmetrical couplers do not exhibit the phase property of symmetrical couplers and are generally used where couplers are designed to obtain broadband power division only.

The response of an optimal five-section symmetrical coupler is shown in Figure 6.8.

Before discussing the synthesis of multisection TEM couplers we need to define the power loss ratio of a directional coupler.

Power Loss Ratio

The power loss ratio of a directional coupler is defined as

$$L = \frac{1}{|S_{21}|^2} \tag{6.22}$$

where S_{21} is the transmission coefficient between the input and direct ports. The power loss ratio is a positive quantity and is always greater than or equal to unity. For example, for the directional couplers shown in Figures 6.6 and 6.7, the power loss ratio L can be expressed as

$$L = \frac{1}{|S_{21}|^2} = \frac{1}{|S_{21e}|^2} \tag{6.23}$$

Note that S_{21} denotes the scattering parameter between ports 1 and 2 of the four-port directional coupler, whereas S_{21e} denotes the scattering parameter between ports 1 and 2 of the equivalent two-port network.

In decibels:

$$10 \log L = -20 \log |S_{21}| = -20 \log |S_{21e}| \tag{6.24}$$

The quantities on the right-hand side in the above equation denote the insertion loss in decibels between the input and direct ports. The function L is, therefore, also called the *insertion loss function*. The relationship between the scattering parameters of an ideal directional coupler ($S_{11} = S_{41} = 0$) is given by

$$|S_{21}|^2 + |S_{31}|^2 = 1 \tag{6.25}$$

Parallel-Coupled TEM Directional Couplers

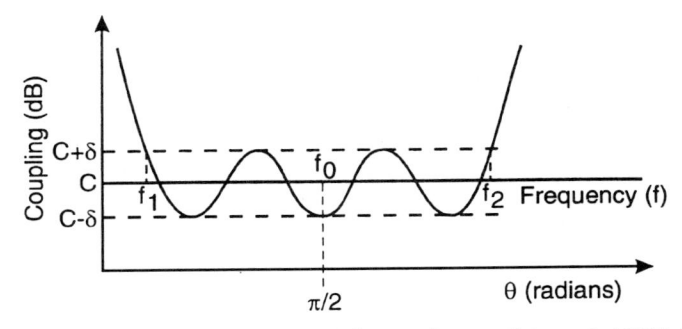

Figure 6.8 Typical response of a symmetrical five-section parallel-coupled TEM directional coupler.

or using (6.22) and (6.25), we obtain

$$|S_{31}|^2 = 1 - \frac{1}{L} \tag{6.26}$$

In terms of ABCD parameters, the power loss ratio L of a network is given by [1, 3]

$$L = 1 + \frac{1}{4}\left[(A-D)^2 - \left(\frac{B}{Z_0} - CZ_0\right)^2\right] \tag{6.27}$$

The equivalent circuit of a single-section coupler is shown in Figure 6.5. Its ABCD parameters (for a lossless case) are given by,

$$A = \cos\theta, \quad B = jZ_{0e}\sin\theta, \quad C = \frac{j\sin\theta}{Z_{0e}}, \quad D = \cos\theta$$

The power loss ratio of a single-section coupler is therefore given by

$$L = 1 + \frac{1}{4}\left(\frac{Z_{0e}}{Z_0} - \frac{Z_0}{Z_{0e}}\right)^2 \sin^2\theta \tag{6.28}$$

Similarly, the power loss ratio of a symmetrical three section coupler is given by

$$L = 1 + \frac{1}{4}\left[\left\{2\left(\frac{Z_{0e1}}{Z_0} - \frac{Z_0}{Z_{0e1}}\right) + \left(\frac{Z_{0e2}}{Z_0} - \frac{Z_0}{Z_{0e2}}\right)\right\}\sin\theta\cos^2\theta \\ - \left(\frac{Z_{0e1}^2}{Z_0 Z_{0e2}} - \frac{Z_0 Z_{0e2}}{Z_{0e1}^2}\right)\sin^3\theta\right]^2 \tag{6.29}$$

where Z_{0e1} denotes the even-mode impedance of the first and third sections and Z_{0e2} denotes the even-mode impedance of the middle section.

Power Loss Ratio of an Ideal Directional Coupler

The characteristics expected of an ideal directional coupler are that over a given frequency band, the values of $|S_{21}|$ and $|S_{31}|$ be constant. This requires the function L to be also constant over this frequency band. For example, if it is required to have a 3-dB flat coupling over 5:1 frequency bandwidth ratio, the form of the corresponding function L in the frequency band of interest is as shown in Figure 6.9. Outside this frequency range, the function L can take any form.

It is impossible to realize any arbitrary power loss ratio function using physical networks. For example, the power loss ratio function L of the form shown in Figure 6.9 cannot be realized. Siedel and Rosen [8] have stated the necessary and sufficient conditions for the form of power loss ratio function L that can be realized using homogeneous stepped impedance networks as shown in Figures 6.6 and 6.7. For example, the power loss ratio function L of the form

$$L = P_N(\sin^2 \theta) \tag{6.30}$$

can be realized using an asymmetrical network as shown in Figure 6.6(b), where P_N is a polynomial of degree N whose value is greater than or equal to unity for all real values of θ. If the physical network is to be symmetrical as shown in Figure 6.7(b), an extra condition is imposed on the power loss function that can be realized. The necessary and sufficient condition that a power loss ratio function L can be realized using a symmetrical homogeneous stepped impedance network

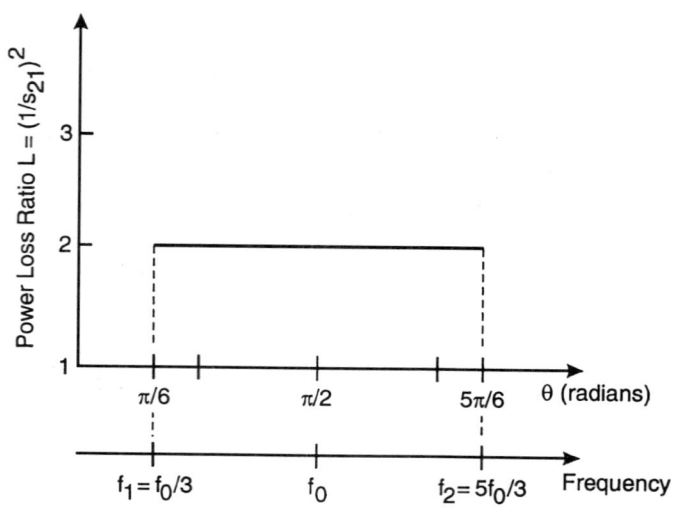

Figure 6.9 Power loss ratio L of an ideal directional coupler having a frequency bandwidth ratio of $B = 5$.

of N equal-length sections as shown in Figure 6.7(b) is that it be of the form

$$L = 1 + [P_N(\sin\theta)]^2 \tag{6.31}$$

where P_N is an odd polynomial in $\sin\theta$ of degree N.

It may be verified that the power loss ratio of a three-section symmetrical coupler as given by (6.29) satisfies the condition of (6.31).

Based on the discussion so far, the synthesis of a multisection coupler can be summarized as follows:

1. Find an optimal polynomial L as a function of electrical length θ that satisfies the conditions imposed by (6.30) if the coupler is asymmetrical or (6.31) if the coupler is to be symmetrical. The polynomial L is to be optimal in the sense that for a given number of sections, response type (equal-ripple or maximally flat), a given mean coupling, and a given coupling tolerance (ripple level), the polynomial should exhibit maximal bandwidth.
2. Compute the impedance of each section of the network using network synthesis techniques once the optimal polynomial L has been found.

The above approach has been used by Levy [1] to design optimal asymmetrical couplers and by Cristal and Young [3], and Toulios and Todd [9] to design symmetrical couplers. Unfortunately, analytical design expressions tend to be very cumbersome even for couplers with a small number of sections [1, 9]. For aiding designers, Levy has generated design tables for equal-ripple asymmetrical couplers for various values of coupling and bandwidth for up to six sections [2]. For equal-ripple and maximally flat symmetrical couplers, Cristal and Young have generated similar tables for up to nine sections [3]. For equal-ripple response symmetrical couplers, these are reproduced in Table 6.1.

Example 6.3

Design a symmetrical multisection TEM coupler with the following specifications:

Mean coupling = 6 dB
Maximal ripple level = ±0.1 dB
Frequency bandwidth ratio B = 8

The schematic of a symmetrical coupler is shown in Figure 6.7(a). To design a symmetrical coupler, we use the design Tables 6.1(q) [3]. We find that a 6-dB, nine-section coupler designed for a ripple level of ±0.1 dB exhibits a

Table 6.1
Tables of parameters for symmetrical TEM-mode coupled-transmission-line directional couplers

δ	Z_1	Z_2	w	B
(a) Normalized even-mode impedances for equal-ripple symmetrical 3.01-db couplers of three sections ($Z_{4-i} = Z_i$)					
0.05	1.14888	3.16095	0.86101	2.51187
0.10	1.17135	3.25984	1.00760	3.03063
0.15	1.9039	3.34049	1.10168	3.45275
0.20	1.20776	3.41242	1.17199	3.83085
0.25	1.22415	3.47932	1.22844	4.18429
0.30	1.23992	3.54311	1.27572	4.52271
0.35	1.25528	3.60495	1.31645	4.85178
0.40	1.27036	3.66560	1.35225	5.17521
0.45	1.28527	3.72563	1.38420	5.49559
0.50	1.30008	3.78546	1.41305	5.81489
0.60	1.32964	3.90585	1.46353	6.45616
0.70	1.35942	4.02894	1.50670	7.10860
0.80	1.38970	4.15648	1.54440	7.77966
0.90	1.42073	4.29005	1.57788	8.47591
1.00	1.45274	4.43120	1.60798	9.20361
(b) Normalized even-mode impedances for equal-ripple symmetrical 6-db couplers of three sections ($Z_{4-i} = Z_i$)					
0.10	1.10298	2.09445	0.91996	2.70356
0.20	1.12090	2.14693	1.07404	3.31984
0.30	1.13625	2.18999	1.17223	3.83226
0.40	1.15038	2.22865	1.24518	4.29931
0.50	1.16381	2.26488	1.30345	4.74258
0.60	1.17680	2.29968	1.35201	5.17291
0.70	1.18952	2.33366	1.39364	5.59673
0.80	1.20208	2.36724	1.43006	6.01830
0.90	1.21454	2.40072	1.46241	6.44068
1.00	1.22698	2.43431	1.49150	6.86621
(c) Normalized even-mode impedances for equal-ripple symmetrical 8.34-db couplers of three sections ($Z_{4-i} = Z_i$)					
0.05	1.06661	1.69824	0.76021	2.22636
0.10	1.07434	1.71858	0.89286	2.61290

(*Continued*)

Table 6.1 (*Continued*)

δ	Z_1	Z_2	w	B
0.15	1.08073	1.73468	0.97882	2.91703
0.20	1.08644	1.74864	1.04355	3.18211
0.25	1.09171	1.76127	1.09583	3.42397
0.30	1.09670	1.77299	1.13986	3.65041
0.35	1.10146	1.78405	1.17796	3.86595
0.40	1.10606	1.79461	1.21159	4.07347
0.45	1.11054	1.80478	1.24170	4.27495
0.50	1.11490	1.81463	1.26898	4.47178
0.55	1.11918	1.82424	1.29391	4.66502
0.60	1.12339	1.83365	1.31688	4.85550
0.65	1.12754	1.84289	1.33817	5.04386
0.70	1.13164	1.85200	1.35801	5.23063
0.75	1.13570	1.86101	1.37658	5.41623
0.80	1.13973	1.86993	1.39403	5.60103
0.85	1.14373	1.87878	1.41049	5.78534
0.90	1.14770	1.88759	1.42607	5.96943
0.95	1.15166	1.89636	1.44084	6.15354
1.00	1.15560	1.90510	1.45488	6.33787
(d) Normalized even-mode impedances for equal-ripple symmetrical 10-db couplers of three sections ($Z_{4-i} = Z_i$)			
0.20	1.06945	1.57423	1.03140	3.12968
0.40	1.08475	1.60708	1.19816	3.98852
0.60	1.09817	1.63470	1.30282	4.73738
0.80	1.11075	1.66014	1.37959	5.44739
1.00	1.12290	1.68458	1.44020	6.14545
(e) Normalized even-mode impedances for equal-ripple symmetrical 20-db couplers of three sections ($Z_{4-i} = Z_i$)			
0.20	1.02070	1.14914	1.00980	3.03958
0.40	1.02497	1.15617	1.17423	3.84396
0.60	1.02866	1.16197	1.27772	4.53804
0.80	1.03208	1.16720	1.35381	5.19011
1.00	1.03534	1.17213	1.41398	5.82570

$Z_{oe1} = Z_1 Z_0$, $Z_{oe2} = Z_2 Z_0$, $Z_{oe3} = Z_3 Z_0$

(*Continued*)

Table 6.1 (*Continued*)

δ	Z_1	Z_2	Z_3	w	B
(f) Normalized even-mode impedances for equal-ripple symmetrical 3.01-db couplers of five sections ($Z_{6-i} = Z_i$)					
0.05	1.05972	1.32624	3.81243	1.20488	4.03071
0.10	1.07851	1.37268	3.97615	1.32559	4.93114
0.15	1.09451	1.40890	4.10191	1.39889	5.65437
0.20	1.10921	1.44029	4.21023	1.45184	6.29714
0.25	1.12314	1.46883	4.30864	1.49333	6.89474
0.30	1.13659	1.49551	4.40089	1.52744	7.46462
0.35	1.14973	1.52091	4.48917	1.55639	8.01698
0.40	1.16266	1.54541	4.57491	1.58152	8.55845
0.45	1.17547	1.56926	4.65912	1.60371	9.09367
0.50	1.18822	1.59265	4.74253	1.62357	9.62609
0.60	1.21370	1.63864	4.90924	1.65791	10.69292
0.70	1.23941	1.68425	5.07867	1.68691	11.77568
0.80	1.26555	1.73013	5.25363	1.71196	12.88720
0.90	1.29235	1.77678	5.43655	1.73402	14.03860
1.00	1.31998	1.82466	5.62978	1.75370	15.24047
(g) Normalized even-mode impedances for equal-ripple symmetrical 6-db couplers of five sections ($Z_{6-i} = Z_i$)					
0.10	1.04501	1.21972	2.38181	1.25446	4.34522
0.20	1.06052	1.25302	2.46010	1.37766	5.45738
0.30	1.07392	1.27919	2.52068	1.45202	6.29953
0.40	1.08633	1.30203	2.57332	1.50548	7.08866
0.50	1.09818	1.32294	2.62159	1.54720	7.83386
0.60	1.10969	1.34262	2.66727	1.58135	8.55462
0.70	1.12099	1.36148	2.71142	1.61023	9.26242
0.80	1.13217	1.37978	2.75470	1.63520	9.96482
0.90	1.14328	1.39772	2.79760	1.65716	10.66721
1.00	1.15438	1.41542	2.84048	1.67673	11.37370
(h) Normalized even-mode impedances for equal-ripple symmetrical 8.34-db couplers of five sections ($Z_{6-i} = Z_i$)					
0.05	1.02538	1.14102	1.85802	1.11764	3.53328
0.10	1.03211	1.15690	1.89019	1.23184	4.20727
0.15	1.03770	1.16899	1.91418	1.30256	4.73524

(*Continued*)

Table 6.1 (*Continued*)

δ	Z_1	Z_2	Z_3	w	B
0.20	1.04271	1.17918	1.93414	1.35395	5.19150
0.25	1.04737	1.18822	1.95170	1.39442	5.60527
0.30	1.05179	1.19648	1.96764	1.42783	5.99090
0.35	1.05602	1.20417	1.98243	1.45627	6.35662
0.40	1.06012	1.21142	1.99635	1.48104	6.70767
0.45	1.06412	1.21833	2.00960	1.50296	7.04761
0.50	1.06803	1.22497	2.02232	1.52262	7.37898
0.55	1.07187	1.23138	2.03462	1.54043	7.70370
0.60	1.07565	1.23760	2.04658	1.55670	8.02323
0.65	1.07939	1.24367	2.05826	1.57168	8.33872
0.70	1.08309	1.24960	2.06971	1.58554	8.65112
0.75	1.08675	1.25542	2.08098	1.59844	8.96119
0.80	1.09039	1.26114	2.09210	1.61050	9.26959
0.85	1.09401	1.26678	2.10310	1.62182	9.57685
0.90	1.09761	1.27235	2.11401	1.63247	9.88347
0.95	1.10119	1.27785	2.12484	1.64253	10.18985

(i) Normalized even-mode impedances for equal-ripple symmetrical 10-db couplers of five sections ($Z_{6-i} = Z_i$)

1.00	1.10476	1.28331	2.13562	1.65206	10.49635
0.20	1.03418	1.14316	1.70922	1.34442	5.10148
0.40	1.4784	1.16808	1.75305	1.47118	6.56407
0.60	1.05996	1.18815	1.78805	1.54675	7.82513
0.80	1.07140	1.20606	1.81943	1.60053	9.01322
1.00	1.08249	1.22280	1.84912	1.64210	10.17639

(j) Normalized even-mode impedances for equal-ripple symmetrical 20-db couplers of five sections ($Z_{6-i} = Z_i$)

0.20	1.01016	1.04183	1.17873	1.32734	4.94656
0.40	1.01406	1.04855	1.18767	1.45350	6.31936
0.60	1.01747	1.05386	1.19463	1.52888	7.49038
0.80	1.02066	1.05851	1.20073	1.58261	8.58338
1.00	1.02371	1.06280	1.20638	1.62420	9.64410

$Z_{oe1} = Z_1 Z_0$, $Z_{oe2} = Z_2 Z_0$, $Z_{oe3} = Z_3 Z_0$
$Z_{oe4} = Z_4 Z_0$, $Z_{oe5} = Z_5 Z_0$

(*Continued*)

Table 6.1 (*Continued*)

δ	Z_1	Z_2	Z_3	Z_4	w	B
(k) Normalized even-mode impedances for equal-ripple symmetrical 3.01-db couplers of seven sections ($Z_{8-i} = Z_i$)							
0.05	1.03635	1.14905	1.50280	4.39954	1.40024	5.6693
0.10	1.05240	1.18406	1.56753	4.61180	1.49705	6.9531
0.15	1.06643	1.21166	1.61640	4.77112	1.55447	7.9780
0.20	1.07950	1.23581	1.65795	4.90662	1.59539	8.8860
0.25	1.09201	1.25786	1.69523	5.02872	1.62715	9.7283
0.30	1.10419	1.27860	1.72975	5.14254	1.65308	10.5302
0.35	1.11615	1.29840	1.76238	5.25103	1.67497	11.3064
0.40	1.12798	1.31754	1.79367	5.35611	1.69388	12.0666
0.45	1.13975	1.33622	1.82400	5.45909	1.71051	12.8174
0.50	1.15149	1.35457	1.85365	5.56097	1.72534	13.5637
0.60	1.17505	1.39069	1.91172	5.76434	1.75090	15.0578
0.70	1.19890	1.42654	1.96915	5.97094	1.77238	16.5728
0.80	1.22323	1.46258	2.02682	6.18437	1.79087	18.1270
0.90	1.24820	1.49918	2.08545	6.40775	1.80710	19.7360
1.00	1.27399	1.53668	2.14566	6.64407	1.82155	21.4147
(l) Normalized even-mode impedances for equal-ripple symmetrical 6-db couplers of seven sections ($Z_{8-i} = Z_i$)							
0.10	1.02686	1.10756	1.32930	2.62516	1.44052	6.1494
0.20	1.04246	1.13419	1.37278	2.72038	1.53802	7.6583
0.30	1.05449	1.15540	1.40584	2.79246	1.59558	8.8908
0.40	1.06580	1.17408	1.43416	2.85438	1.63645	10.0026
0.50	1.07670	1.19128	1.45977	2.91078	1.66806	11.0505
0.60	1.08735	1.20755	1.48367	2.96391	1.69378	12.0626
0.70	1.09787	1.22318	1.50642	3.01511	1.71541	13.0553
0.80	1.10831	1.23839	1.52841	3.06523	1.73404	14.0396
0.90	1.11872	1.25331	1.54989	3.11486	1.75036	15.0232
1.00	1.12915	1.26805	1.57104	3.16446	1.76487	16.0119
(m) Normalized even-mode impedances for equal-ripple symmetrical 8.34-db couplers of seven sections ($Z_{8-i} = Z_i$)							
0.05	1.01460	1.06403	1.21141	1.99183	1.32568	4.9319
0.10	1.02033	1.07694	1.23301	2.03194	1.42127	5.9117

(*Continued*)

Table 6.1 (*Continued*)

δ	Z_1	Z_2	Z_3	Z_4	w	B
0.15	1.02519	1.08680	1.24872	2.06076	1.47818	6.6655
0.20	1.02963	1.09518	1.26167	2.08436	1.51889	7.3140
0.25	1.03379	1.10266	1.27297	2.10489	1.55059	7.9005
0.30	1.03778	1.10953	1.28316	2.12339	1.57653	8.4458
0.35	1.04163	1.11595	1.29256	2.14044	1.59848	8.9622
0.40	1.04538	1.12204	1.30136	2.15641	1.61749	9.4572
0.45	1.04905	1.12786	1.30969	2.17156	1.63423	9.9360
0.50	1.05265	1.13346	1.31764	2.18606	1.64919	10.4022
0.55	1.05621	1.13889	1.32528	2.20004	1.66270	10.8588
0.60	1.05972	1.14417	1.33267	2.21361	1.67500	11.3077
0.65	1.06320	1.14933	1.33984	2.22683	1.68629	11.7507
0.70	1.06666	1.15438	1.34684	2.23979	1.69672	12.1891
0.75	1.07009	1.15934	1.35368	2.25251	1.70640	12.6240
0.80	1.07350	1.16423	1.36040	2.26506	1.71543	13.0563
0.85	1.07689	1.16904	1.36700	2.27746	1.72389	13.4870
0.90	1.08028	1.17381	1.37351	2.28975	1.73184	13.9165
0.95	1.08365	1.17852	1.37993	2.30194	1.73934	14.3456
1.00	1.08702	1.18319	1.38629	2.31408	1.74643	14.7747

(n) Normalized even-mode impedances for equal-ripple symmetrical 10-db couplers of seven sections ($Z_{8-i} = Z_i$)

0.20	1.02360	1.07622	1.20802	1.81699	1.51198	7.1965
0.40	1.03597	1.09725	1.23839	1.86715	1.61028	9.2638
0.60	1.04718	1.11444	1.26213	1.90649	1.66773	11.0383
0.80	1.05786	1.12991	1.28298	1.94149	1.70815	12.7059
1.00	1.06834	1.14444	1.30229	1.97446	1.73917	14.3359

(o) Normalized even-mode impedances for equal-ripple symmetrical 20-db couplers of seven sections ($Z_{8-i} = Z_i$)

0.20	1.00697	1.02256	1.05976	1.20128	1.49853	6.9766
0.40	1.01052	1.02846	1.06767	1.21112	1.59672	8.9188
0.60	1.01369	1.03320	1.07372	1.21863	1.65421	10.5678
0.80	1.01669	1.03740	1.07894	1.22515	1.69472	12.1029
1.00	1.01958	1.04129	1.08368	1.23116	1.72584	13.5903

$Z_{oei} = Z_i Z_0$

(*Continued*)

Table 6.1 (*Continued*)

δ	Z_1	Z_2	Z_3	Z_4	Z_5	w	B
(p) Normalized even-mode impedances for equal-ripple symmetrical 3.01-db couplers of nine sections ($Z_{10-i} = Z_i$)							
0.05	1.02680	1.09163	1.24706	1.66958	4.93133	1.5218	7.365
0.10	1.04112	1.12024	1.29488	1.74863	5.18240	1.6012	9.030
0.15	1.05391	1.14328	1.33137	1.80742	5.36886	1.6478	10.356
0.20	1.06598	1.16366	1.36260	1.85696	5.52654	1.6807	11.528
0.25	1.07763	1.18248	1.39075	1.90116	5.66814	1.7062	12.615
0.30	1.08904	1.20027	1.41691	1.94192	5.79985	1.7269	13.649
0.35	1.10030	1.21737	1.44168	1.98035	5.92523	1.7444	14.648
0.40	1.11149	1.23397	1.46548	2.01711	6.04655	1.7594	15.627
0.45	1.12264	1.25023	1.48856	2.05271	6.16540	1.7726	16.594
0.50	1.13379	1.26625	1.51114	2.08747	6.28296	1.7844	17.554
0.60	1.15624	1.29789	1.55536	2.15551	6.51769	1.8046	19.475
0.70	1.17904	1.32941	1.59902	2.22278	6.75634	1.8216	21.423
0.80	1.20234	1.36117	1.64277	2.29038	7.00316	1.8362	23.421
0.90	1.22630	1.39348	1.68712	2.35918	7.26188	1.8490	25.488
1.00	1.25107	1.42660	1.73250	2.42995	7.53602	1.8604	27.644
(q) Normalized even-mode impedances for equal-ripple symmetrical 6-db couplers of nine sections ($Z_{10-i} = Z_i$)							
0.10	1.02201	1.06888	1.17282	1.42807	2.83542	1.5550	7.989
0.20	1.03437	1.09137	1.20736	1.47877	2.94305	1.6345	9.943
0.30	1.04554	1.10967	1.23393	1.51676	3.02373	1.6809	11.535
0.40	1.05615	1.12599	1.25686	1.54902	3.09269	1.7136	12.969
0.50	1.06645	1.14117	1.27768	1.57805	3.15533	1.7389	14.319
0.60	1.07658	1.15561	1.29716	1.60504	3.21427	1.7594	15.622
0.70	1.08662	1.16957	1.31574	1.63070	3.27103	1.7765	16.900
0.80	1.09661	1.18320	1.33370	1.65546	3.32658	1.7913	18.166
0.90	1.10661	1.19661	1.35125	1.67962	3.38161	1.8042	19.431
1.00	1.11663	1.20989	1.36852	1.70342	3.43663	1.8157	20.702
(r) Normalized even-mode impedances for equal-ripple symmetrical 8.34-db couplers of nine sections ($Z_{10-i} = Z_i$)							
0.05	1.01032	1.03838	1.10598	1.27508	2.10668	1.4599	6.406
0.10	1.01536	1.04904	1.12341	1.30048	2.15200	1.5392	7.681

(*Continued*)

Table 6.1 (*Continued*)

δ	Z_1	Z_2	Z_3	Z_4	Z_5	w	B
0.15	1.01974	1.05735	1.13622	1.31862	2.18413	1.5858	8.658
0.20	1.02379	1.06452	1.14687	1.33341	2.21025	1.6190	9.498
0.25	1.02764	1.07099	1.15622	1.34622	2.23285	1.6446	10.256
0.30	1.03134	1.07697	1.16469	1.35771	2.25315	1.6656	10.960
0.35	1.03494	1.08261	1.17253	1.36825	2.27180	1.6832	11.627
0.40	1.03846	1.08798	1.17989	1.37809	2.28925	1.6985	12.265
0.45	1.04193	1.09314	1.18687	1.38738	2.30577	1.7119	12.883
0.50	1.04534	1.09813	1.19356	1.39622	2.32156	1.7238	13.484
0.55	1.04872	1.10298	1.19999	1.40471	2.33677	1.7346	14.072
0.60	1.05206	1.10771	1.20622	1.41290	2.35152	1.7444	14.650
0.65	1.05538	1.11234	1.21228	1.42085	2.36589	1.7534	15.221
0.70	1.05868	1.11689	1.21819	1.42858	2.37996	1.7617	15.785
0.75	1.06196	1.12137	1.22398	1.43615	2.39378	1.7694	16.345
0.80	1.06523	1.12579	1.22966	1.44356	2.40740	1.7766	16.901
0.85	1.06849	1.13016	1.23525	1.45084	2.42087	1.7833	17.455
0.90	1.07174	1.13448	1.24076	1.45802	2.43421	1.7896	18.008
0.95	1.07498	1.13877	1.24620	1.46511	2.44745	1.7955	18.560
1.00	1.07823	1.14302	1.25158	1.47211	2.46063	1.8011	19.111

(s) Normalized even-mode impedances for equal-ripple symmetrical 10-db couplers of nine sections ($Z_{10-i} = Z_i$)

0.20	1.01889	1.05161	1.11743	1.26387	1.90628	1.6133	9.345
0.40	1.03041	1.07004	1.14313	1.29777	1.96074	1.6927	12.016
0.60	1.04103	1.08543	1.16344	1.32390	2.00313	1.7386	14.303
0.80	1.05127	1.09945	1.18139	1.34672	2.04073	1.7708	16.450
1.00	1.06133	1.11271	1.19805	1.36779	2.07614	1.7954	18.547

(t) Normalized even-mode impedances for equal-ripple symmetrical 20-db couplers of nine sections ($Z_{10-i} = Z_i$)

0.20	1.00555	1.01529	1.03447	1.07471	1.21931	1.6024	9.061
0.40	1.00886	1.02054	1.04153	1.08328	1.22965	1.6818	11.571
0.60	1.01187	1.02485	1.04700	1.08974	1.23748	1.7278	13.697
0.80	1.01474	1.02871	1.05175	1.09527	1.24426	1.7601	15.674
1.00	1.01753	1.03232	1.05608	1.10028	1.25049	1.7848	17.588

$Z_{oe1} = Z_i Z_0$

frequency bandwidth ratio $B = 7.99$. Because this figure of B is very close to the specified value of $B = 8$, it is sufficient to employ a nine-section coupler to achieve the desired specifications. From the same table, we determine even-mode impedances of the various sections. Once the even-mode impedances are known, we can find the odd-mode impedances using (6.21). The even- and odd-mode impedances, the voltage coupling coefficient k, and the coupling in decibels of various sections of the coupler are given in Table 6.2. We see that the coupling of the center section is 2.17 dB, which is much tighter than the overall coupling for which the coupler has been designed (6 dB). The coupling of other sections is smaller than the overall coupling.

6.3.2 Limitations of Multisection Couplers

One of the major limitations of a multisection coupler is that the coupling of at least one of the sections is much tighter than the overall coupling as shown in the above example. This can create fabrication problems in microstrip technology where it is difficult to achieve tight coupling. Further, because the even- and odd-mode impedances of each section of a multisection coupler are different from those of the adjacent ones, the dimensions of the coupler abruptly change at the start and end of each section. Because of practical considerations, it may be necessary to join adjacent sections using small lengths of tapered transmission lines as shown in Figure 6.10. If the operating frequency is high, the extra reactances produced by these abrupt discontinuities or extra lengths of joining transmission lines can reduce the input match and directivity of the coupler. In this case, a better solution may be to use nonuniform couplers, which are discussed in the next chapter.

Table 6.2
Parameters of a nine-section symmetrical coupler

Section	Z_{0e}/Z_0	Z_{0o}/Z_0	k	C (dB)
1	1.02201	0.97846	0.02177	33.24
2	1.06888	0.93556	0.06651	23.54
3	1.17282	0.85265	0.15807	16.02
4	1.42807	0.70025	0.34197	9.32
5	2.83542	0.35268	0.77875	2.17
6	1.42807	0.70025	0.34197	9.32
7	1.17282	0.85265	0.15807	16.02
8	1.06888	0.93556	0.06651	23.54
9	1.02201	0.97846	0.02177	33.24

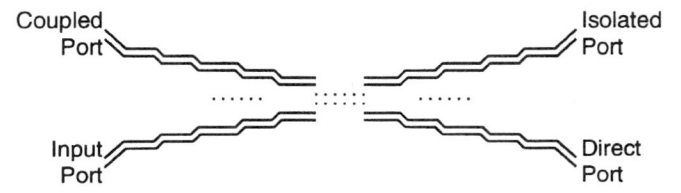

Figure 6.10 Typical physical layout of a symmetrical multisection coupler. Adjacent quarter-wave sections are joined by small length of transmission line sections.

6.4 Techniques to Improve Directivity of Microstrip Couplers

There are mainly three techniques for improving the directivity of microstrip couplers [10]:

1. By adding lumped capacitances at the ends of the coupled lines.
2. By using a dielectric overlay on top of the coupled lines.
3. By using wiggly lines.

6.4.1 Lumped Compensation

In this technique, lumped capacitances are added at the ends of a coupler as shown in Figure 6.11 [10–13]. With the addition of lumped capacitances, the electrical length of the coupler can be made to be equal for the even and odd modes at the design frequency. The addition of lumped capacitance does not affect the even-mode signal, but it affects the odd-mode signal. This can be easily explained.

In the case of even-mode excitation, the midplane PP' as shown in Figure 6.11(b) behaves like an open circuit. This was discussed in Section 4.1. The equivalent circuit of one-half of the network for the even mode is shown in Figure 6.11(c). We see that the lumped capacitance has no effect on the overall capacitance between the strip and the ground because one of the ends of the capacitor is open circuited. On the other hand, in the case of odd-mode excitation, the midplane PP' behaves like a short circuit. The equivalent circuit of one-half of the network for the case of odd-mode excitation is shown in Figure 6.11(d). In this case, the lumped capacitance is parallel with the capacitance between the strip and ground conductors. The overall capacitance between the strip and ground is therefore increased. Because the phase velocity along a line is related to the capacitance as given by (3.6), the phase velocity of the odd mode is reduced because of the additional lumped capacitance. We can show that the electrical length of the coupler can be made equal for the even- and

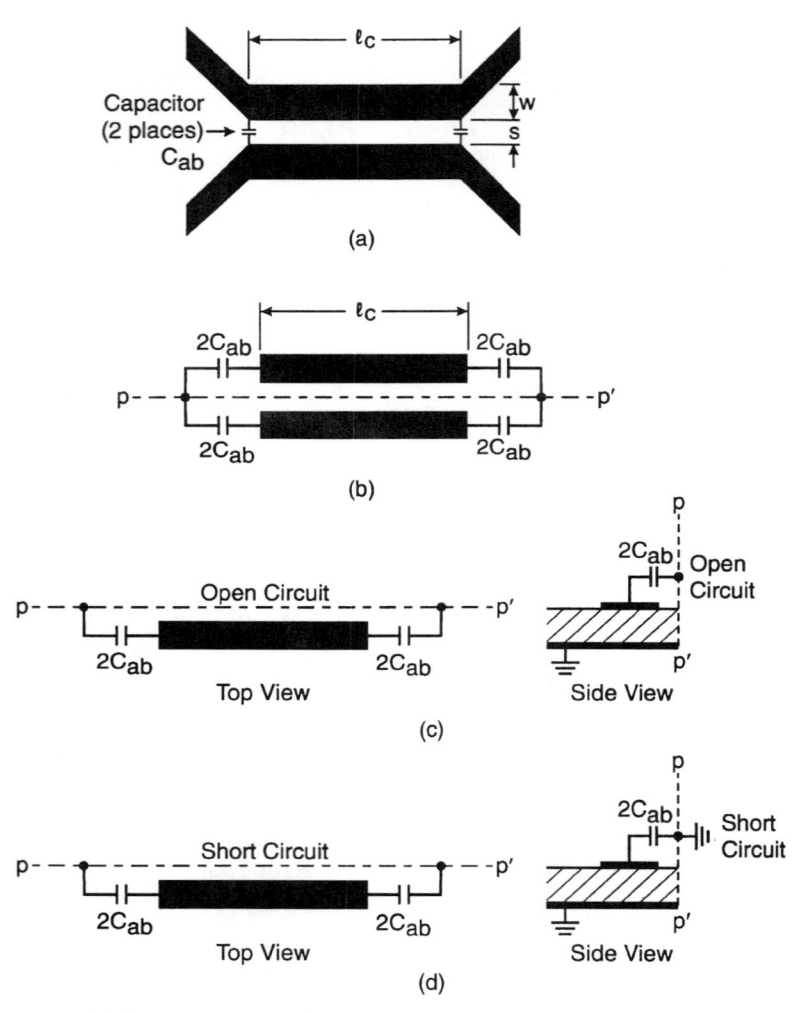

Figure 6.11 (a) Top view of lumped capacitor compensated microstrip coupler, (b) top view of coupled section showing plane of symmetry pp' and lumped capacitors, (c) equivalent circuit for even-mode excitation, and (d) equivalent circuit for odd-mode excitation.

odd-mode signals by choosing the value of the lumped capacitance C_{ab} as [12]

$$C_{ab} = \frac{1}{4\pi f_0 Z_{0o} \tan \theta_0} \tag{6.32}$$

where

$$\theta_0 = \frac{\pi}{2} \sqrt{\frac{\epsilon_{reo}}{\epsilon_{ree}}} \text{ radians}$$

In the above equation, ϵ_{ree} and ϵ_{reo} denote the even- and odd-mode effective dielectric constants of the coupled microstrip lines, respectively, Z_{0o} denotes the odd-mode impedance of the coupled structure, and f_0 is the design center frequency.

The physical length of a capacitor compensated quarter-wave coupler is given by [10]

$$l_c = \frac{\frac{\pi}{2} - \tan^{-1}(\pi f_0 C_{ab} Z_{0e})}{k_0 \sqrt{\epsilon_{ree}}} \tag{6.33}$$

where k_0 is the free space propagation constant.

Experiments have shown that (6.32) and (6.33) lead to a fairly accurate design if the coupling required between the lines is tight (10 dB or tighter) [10]. For weaker coupling, (6.32) and (6.33) may not yield a very accurate design. In that case, it is more useful to optimize the value of the capacitance C_{ab} and the length l_c of the coupler using computer simulation programs such as LIBRA™ or SUPER COMPACT™. The values given by (6.32) and (6.33) can be used as starting values for optimization purposes.

The typical improvement in the directivity of a lumped capacitor compensated 15.7-dB coupler on an alumina substrate is shown in Figure 6.12 [10].

Example 6.4

Given the parameters of coupled microstrip lines as $\epsilon_{ree} = 6.7713$, $\epsilon_{reo} = 5.5194$, $Z_{0e} = 88.83\ \Omega$, $Z_{0o} = 28.14\ \Omega$, and $f_0 = 3$ GHz, compute the value of compensating capacitance C_{ab} and length l_c of a quarter-wave coupler.

Using (6.32) and (6.33), we obtain $C_{ab} = 0.145$ pF and $l_c = 8.86$ mm. On the other hand, the physical length of a uncompensated coupler is 10.09 mm.

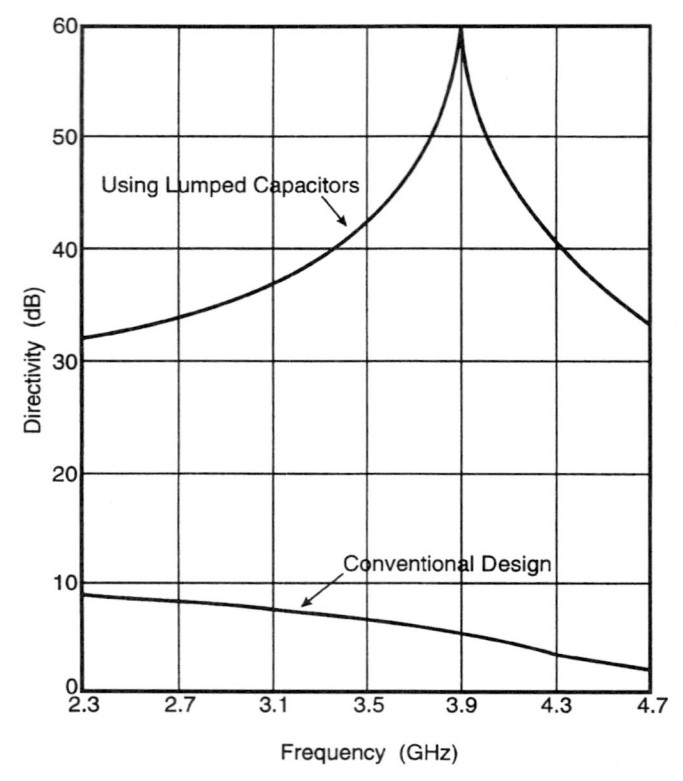

Figure 6.12 Directivity improvement of a lumped capacitor compensated 15.7-dB microstrip coupler on alumina substrate. (From [10], ©1982 IEEE. Reprinted with permission.)

6.4.2 Use of Dielectric Overlays

If an additional layer of dielectric is deposited over coupled microstrip lines as shown in Figure 6.13, then by properly choosing the thickness and dielectric constant of the layer, near equalization of even- and odd-mode phase velocities can be achieved over a reasonably wide frequency band [14–17]. The dielectric overlay can completely cover the bottom dielectric layer or may only cover the region containing the strips. The dielectric overlay also affects quite significantly the backward-wave coupling between the lines. Therefore, the effect of overlay should be considered while computing the dimensions of the strips, the spacing between them, and the length of the coupler. A successful design of the overlay coupler depends on the availability of accurate data on the phase velocities and the characterisitic impedances of the even and odd modes of the overlay structure.

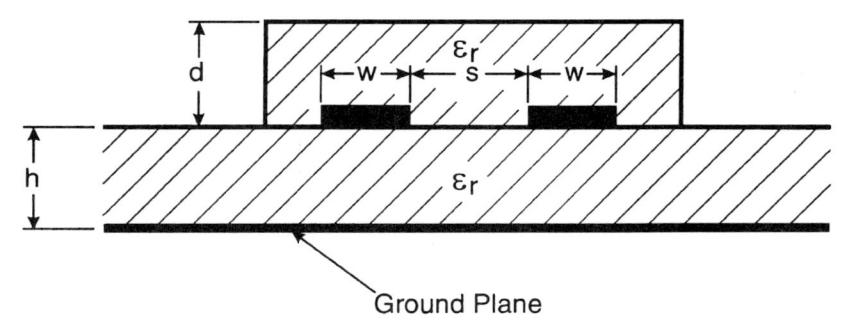

Figure 6.13 Parallel-coupled microstrip coupler with dielectric overlay compensation.

Broadband directional couplers with high directivity have been demonstrated [14].

Figure 6.14 provides design curves for coupled microstrip lines on alumina substrate using alumina for the overlay [15]. The top alumina layer is assumed to cover the bottom alumina layer completely. Here the overlay thickness is the same as the thickness of the substrate. The strip conductor thickness has been assumed to be zero. Simulated and measured data have shown about 10-dB improvement in the directivity of the 8.34-dB coupler designed using the curves in Figure 6.14. Similar improvements have been demonstrated [17] using low-dielectric constant substrate having $\epsilon_r = 2.48$. For a 10-dB coupler at S-band, the design parameters are $\epsilon_r = 2.48$, $W = 3.2$ mm, $h = d = 1.42$ mm, $S = 0.4$ mm, and coupled length $L = 20.5$ mm.

6.4.3 Use of Wiggly Lines

Although use of lumped capacitances or dielectric overlay structures lead to an improved directivity of microstrip directional couplers, both techniques complicate fabrication and may undermine the advantages of MICs. Another technique to equalize the phase velocities of even- and odd-mode signals that is compatible with MIC technology is to use wiggly lines instead of straight lines [18–20]. A top view of wiggly-coupled lines is shown in Figure 6.15b. It is assumed that by wiggling the lines, the odd-mode phase velocity is slowed down, whereas the even-mode phase velocity is not affected. Further, wiggling affects only the mutual capacitance between the coupled lines. Although these approximations are not strictly valid, it has been found that these give practically useful results [19].

The geometrical parameters of straight-coupled and wiggly-coupled lines are defined in Figure 6.15. To consider the effect of wiggling, let us consider the

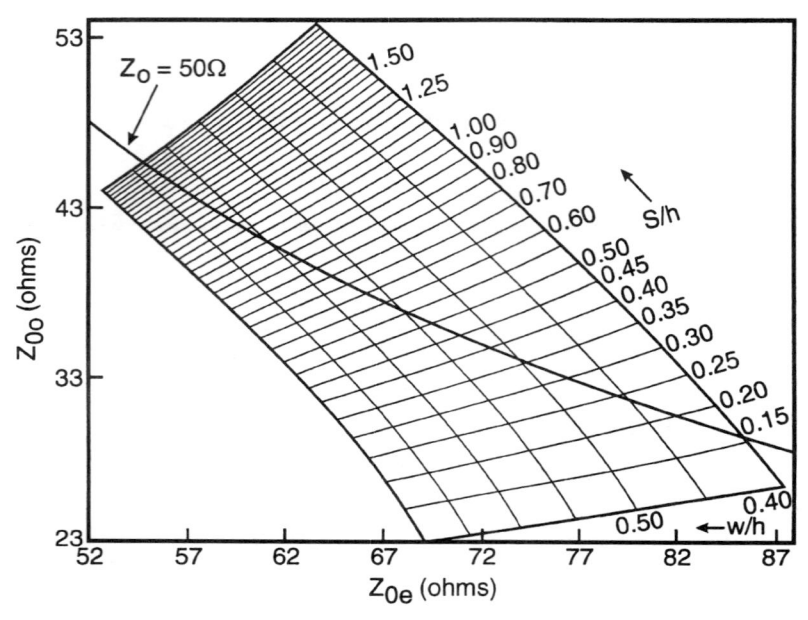

Figure 6.14 Design curves for coupled microstrip lines covered with dielectric overlay $\varepsilon_r = 10.1$, $d/h = 1.0$. (From [15], ©1978 IEEE. Reprinted with permission.)

capacitance of a section of length ΔL (between reference planes AA and BB) of straight- and wiggly-coupled sections. For the straight-coupled section, the odd-mode capacitance between reference planes AA and BB is given by

$$C_o = (C_f + C_p + C_{fo})\Delta L \tag{6.34}$$

In the above equations, C_f, C_p, and C_{fo} denote per-unit length capacitances as defined in Figure 3.9.

On the other hand, the odd-mode capacitance of wiggly-coupled lines is given by

$$C_{ow} = (C_f + C_p)\Delta L + C_{fo}L_w \tag{6.35}$$

The above relation results because in the case of wiggly-coupled lines, the effective length seen by the capacitance C_f and C_p between the reference planes AA and BB is ΔL, which is the same as that for the straight-coupled lines. The effective length seen by the odd-mode fringing capacitance C_{fo} in this case is L_w, however, which is achieved by wiggling the lines.

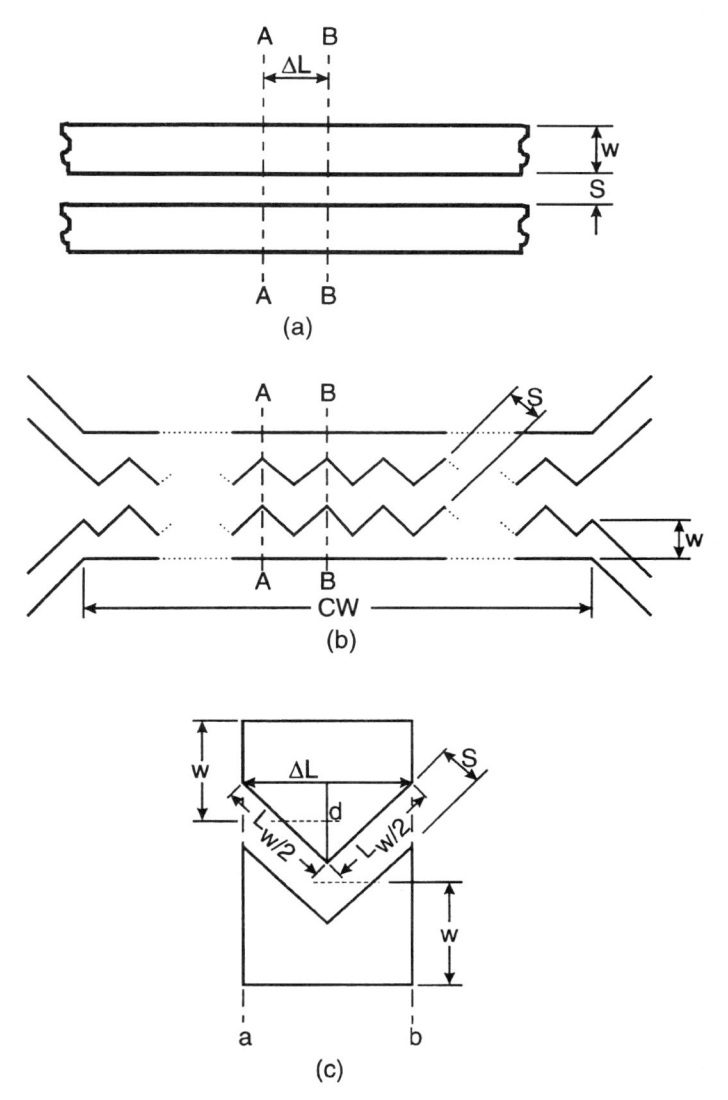

Figure 6.15 Top view of (a) parallel-coupled straight lines, (b) wiggly-coupled lines, and (c) exploded view of wiggly-coupled lines between planes AA and BB.

To equalize the odd- and even-mode phase velocities, the following relation should be satisfied:

$$C_{ow} = \frac{\epsilon_{ree}}{\epsilon_{reo}} C_o \tag{6.36}$$

where ϵ_{ree} and ϵ_{reo} denote the effective dielectric constants for the even and odd modes, respectively. Using (6.34) to (6.36), we find that the length L_w of wiggly-coupled lines should be chosen as

$$L_w = \Delta L \frac{C'_{fo}}{C_{fo}} \tag{6.37}$$

where

$$C'_{fo} = (C_p + C_f)\left(\frac{\epsilon_{ree}}{\epsilon_{reo}} - 1\right) + \frac{\epsilon_{ree}}{\epsilon_{reo}} C_{fo}$$

To obtain the value of L_w as given by (6.37), the wiggle depth d should be chosen as

$$d = \frac{\Delta L}{2}\sqrt{\left(\frac{C'_{fo}}{C_{fo}}\right)^2 - 1} \tag{6.38}$$

The capacitance parameters C_p, C_f, and C_{fo} are defined in Section 3.2.2. For microstrip lines, the capacitances C_p and C_f can be determined as follows:

$$C_p = \frac{\epsilon_0 \epsilon_r W}{h}$$

$$2C_f = \frac{\sqrt{\epsilon_{re}}}{c Z_0} - C_p$$

where c is the velocity of light in free space, and Z_0 and ϵ_{re} denote the quasistatic characterisstic impedance and effective dielectric constant, respectively, of a single microstrip line of width W. Futher, the odd-mode fringing capacitance C_{fo} can be determined using

$$C_{fo} = C_o - C_p - C_f$$

where

$$C_o = \frac{\sqrt{\epsilon_{reo}}}{c Z_{0o}}$$

In the above equation, C_o denotes the odd-mode capacitance, and Z_{0o} and ϵ_{reo} denote, respectively, the characteristic impedance and effective dielectric constant of the odd mode.

References

[1] Levy, R., "General Synthesis of Asymmetric Multielement Coupled-Transmission-Line Directional Couplers," *IRE Trans.*, Vol. MTT-11, July 1963, pp. 226–237.

[2] Levy, R., "Tables for Asymmetric Multielement Coupled-Transmission-Line Directional Couplers," *IRE Trans.*, Vol. MTT-12, May 1964, pp. 275–279.

[3] Cristal, E. G., and L. Young, "Theory and Tables of Optimum Symmetrical TEM-Mode Coupled-Transmission-Line Directional Couplers," *IEEE Trans.*, Vol. MTT-13, Sept. 1965, pp. 544–558.

[4] Arai, S., et al., "A 900-MHz 90-Degree Hybrid for QPSK," *IEEE MTT-S Int. Microwave Symp. Dig.*, 1991, pp. 857–860.

[5] Tanaka, H., et al., "2-GHz One-Octave-Band 90-Degree Hybrid Coupler Using Coupled Meander Line Optimized by 3-D FEM," *IEEE MTT-S Int. Microwave Symp. Dig.*, 1994, pp. 903–906.

[6] Tanaka, H., et al., "Miniaturized 90-Degree Hybrid Coupler Using High Dielectric Substrate for QPSK Modulator," *IEEE MTT-S Int. Microwave Symp. Dig.*, 1996, pp. 793–796.

[7] Young, L., "Stepped Impedance Transformers and Filter Prototypes," *IRE Trans.*, Vol. PGMTT-10, Sept. 1962, pp. 339–359.

[8] Seidel, H., and J. Rosen, "Multiplicity in Cascade Transmission Line Synthesis—Part I," *IEEE Trans.*, Vol. MTT-13, May 1965, pp. 275–283; and Part II, July 1965, pp. 398–407.

[9] Touplios, P. P., and A. C. Todd, "Synthesis of Symmetrical TEM-Mode Directional Couplers," *IEEE Trans.*, Vol. MTT-13, Sept. 1965, pp. 536–544.

[10] March, S. L., "Phase Velocity Compensation in Parallel-Coupled Microstrip Line," *IEEE MTT-S Int. Microwave Symp. Dig.*, 1982, pp. 410–412.

[11] Schaller, G., "Optimization of Microstrip Directional Couplers With Lumped Capacitors," *AEU*, Vol. 31, July–August 1977, pp. 301–307.

[12] Kajfez, D., "Raise Coupler Directivity With Lumped Compensation," *Microwaves*, Vol. 27, Mar. 1978, pp. 64–70.

[13] Dydyk, M., "Accurate Design of Microstrip Directional Couplers With Capacitive Compensation," *IEEE Int. Microwave MTT-S Int. Microwave Symp. Dig.*, 1990, pp. 581–584.

[14] Sheleg, B., and B. E. Spielman, "Broadband Directional Couplers Using Microstrip With Dielectric Overlays," *IEEE Trans.*, Vol. MTT-22, Dec. 1974. pp. 1,216–1,220.

[15] Paolino, D. D., "MIC Overlay Coupler Design Using Spectral Domain Techniques," *IEEE Trans. Microwave Theory Tech.*, Vol. MTT-26, Sept. 1978, pp. 646–649.

[16] Klein, J. L., and K. Chang, "Optimum Dielectric Overlay Thickness for Equal Even- and Odd-mode Phase Velocities in Coupled Microstrip Circuits," *Electronics Letters*, Vol. 26, 1990, pp. 274–276.

[17] Su, L., T. Itoh and J. Rivera, "Design of an Overlay Directional Coupler by a Full-Wave Analysis," *IEEE Trans.*, Vol. MTT-31, Dec. 1983, pp. 1,017–1,022.

[18] Podell, A., "A High-Directivity Microstrip Coupler Technique," *IEEE MTT-S Int. Microwave Symp. Dig.*, 1970, pp. 33–36.

[19] Uysal, S., and H. Aghvami, "Synthesis, Design and Construction of Ultra-Wideband Nonuniform Directional Couplers in Inhomogeneous Media," *IEEE Trans.*, Vol. MTT-37, June 1989, pp. 969–976.

[20] Uysal, S. *Nonuniform Line Microstrip Directional Couplers and Filters*, Artech House, Norwood, MA, 1993.

7

Nonuniform Broadband TEM Directional Couplers

A major disadvantage of using multisection directional couplers to obtain broadband coupling is that there is an abrupt change in the transverse dimensions of the coupler (e.g., width of the lines and spacing between them) at the start and end of each section. The physical discontinuities that occur from a change in the dimensions lead to poor match and directivity of the coupler. These effects increase in severity with increasing frequency. To avoid the abrupt change, a continuous variation in the coupling between the lines along the length of the coupler can be implemented. This requires a continuous variation in the transverse dimensions of the coupler, and these type of couplers are called *nonuniform couplers*.

In this chapter, we discuss the theory and design of symmetrical and asymmetrical nonuniform TEM couplers. Symmetrical couplers have end-to-end symmetry and have a property that the phase difference between the coupled ports is 90 degrees. On the other hand, asymmetrical couplers do not have end-to-end symmetry. By proper design, asymmetrical couplers can be designed to have the phase property of a magic-T. The design of symmetrical and asymmetrical couplers reported here is based largely on the work of Tresselt [1] and DuHamel and Armstrong [2].

7.1 Symmetrical Couplers

Figure 7.1 shows a symmetrical nonuniform TEM coupler. The width of the lines and the spacing between them vary continuously along the length of the structure. This leads to a continuous variation in the coupling along the length

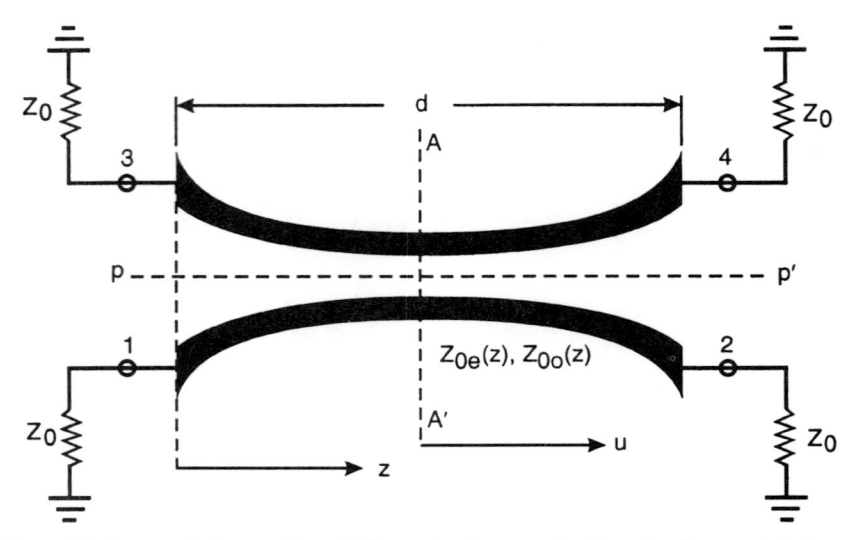

Figure 7.1 Symmetrical nonuniform TEM coupler. Even- and odd-mode characteristic impedances vary continuously along the length of the structure.

of the coupler. The coupler is symmetrical about the planes PP' and AA' which ensures that the signal coupled to the backward port is 90 degrees out of phase with that coupled to the direct port.

Because the coupler is assumed to be symmetrical about the plane PP', we can analyze it in terms of even- and odd-mode parameters. The scattering parameters of a symmetrical four-port network (symmetrical about plane PP') are given by (4.22) as follows:

$$S_{11} = \frac{S_{11e} + S_{11o}}{2}$$
$$S_{21} = \frac{S_{21e} + S_{21o}}{2}$$
$$S_{31} = \frac{S_{11e} - S_{11o}}{2}$$
$$S_{41} = \frac{S_{21e} - S_{21o}}{2}$$

(7.1)

where S_{11e} and S_{11o} denote the reflection coefficients of the even- and odd-mode signals respectively and S_{21e} and S_{21o} denote the transmission coefficients of the even- and odd-mode signals, respectively.

The nonuniform coupling is achieved basically by varying the even- and odd-mode characteristic impedances of the coupled lines along the length of the structure. If the dimensions of the nonuniform coupler are chosen such that at any cross section, the values of the even- and odd-mode impedances satisfy the following condition:

$$Z_{0e}(z) Z_{0o}(z) = Z_0^2 \tag{7.2}$$

then, as in the case of uniform couplers discussed in Section 4.2.2, the scattering parameters of a nonuniform coupler reduce to

$$\begin{aligned} S_{11} &= S_{41} = 0 \\ S_{31} &= S_{11e} \\ S_{21} &= S_{21e} \end{aligned} \tag{7.3}$$

The equivalent circuit for obtaining the scattering parameters S_{11e} and S_{21e} of a nonuniform coupler is shown in Figure 7.2(a) where $Z_{0e}(z)$ denotes the characteristic impedance of the nonuniform transmission line as a function of longitudinal coordinate z.

7.1.1 Coupling in Terms of Even-Mode Characteristic Impedance

We now proceed to find the reflection coefficient S_{11e} at port 1 of the circuit shown in Figure 7.2(a) in terms of the even-mode characteristic impedance of the coupled lines. Because the even-mode characteristic impedance of the coupled lines varies continuously along the length of the structure, it can be represented as shown in Figure 7.2(b). (The odd-mode characteristic impedance also varies along the length of the structure, but because it is always related to the even-mode impedance by (7.2), it is possible to analyze the network in terms of even-mode impedances only.) The incident wave is partially reflected at every step because of a mismatch of the impedance. The total reflection coefficient at port 1 can be found by using the small-signal reflection [3]. Using this theory, the total reflection coefficient at the input is found by summing differential contributions of the reflection coefficient from each step in proper phase. The differential reflection coefficient at the step defined by plane BB' is given by

$$\begin{aligned} d\Gamma &= \frac{Z_{0e} + dZ_{0e} - Z_{0e}}{Z_{0e} + dZ_{0e} + Z_{0e}} \approx \frac{dZ_{0e}}{2Z_{0e}} \\ &= \frac{1}{2} d(\ln Z_{0e}) \\ &= \frac{1}{2} \frac{d}{dz}(\ln Z_{0e}) dz \end{aligned} \tag{7.4}$$

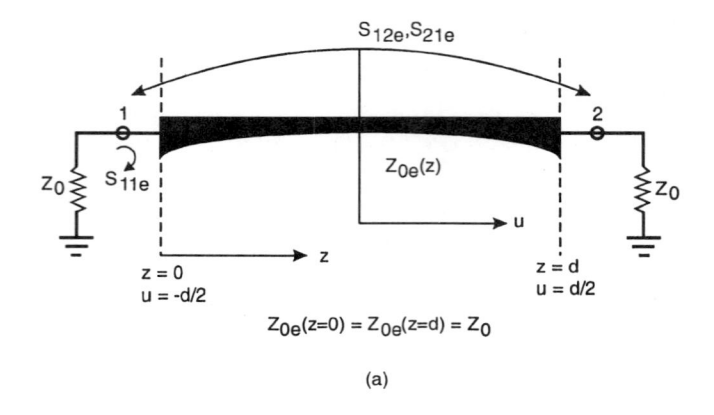

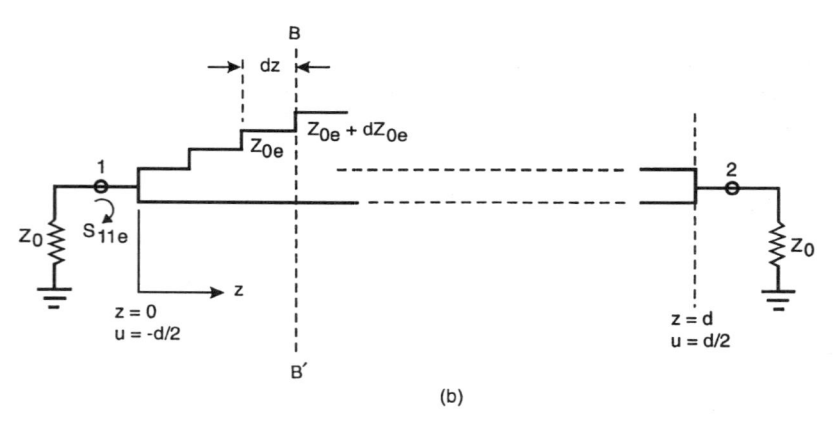

Figure 7.2 (a) Equivalent transmission line circuit of nonuniform TEM coupler shown in Figure 7.1; (b) equivalent circuit for determining coupling.

where Z_{0e} is a function of z, and it is assumed that $2Z_{0e} \gg dZ_{0e}$. The contribution to the total reflection coefficient at port 1 ($z = 0$) from the differential reflection coefficient $d\Gamma$ at plane BB' is obtained by multiplying (7.4) by the phase term $e^{-2j\beta z}$, or

$$dS_{11e} = d\Gamma\, e^{-2j\beta z} = \frac{1}{2} e^{-2j\beta z} \frac{d}{dz}(\ln Z_{0e})dz \qquad (7.5)$$

where β is the phase constant of the wave along the direction of propagation.

Summing reflections from $z = 0$ to $z = d$, we obtain

$$S_{11e} = \frac{1}{2} \int_0^d e^{-2j\beta z} \frac{d}{dz}(\ln Z_{0e})dz \qquad (7.6)$$

Further, using (7.3), the coupling between ports 1 and 3 of the configuration shown in Figure 7.1 is given by

$$S_{31} = S_{11e} = \frac{1}{2}\int_0^d e^{-2j\beta z}\frac{d}{dz}(\ln Z_{0e})dz \qquad (7.7)$$

The above equation is valid if it is assumed that the reflection at any step is quite small compared with unity.

The coupler is assumed to be symmetrical about the plane $z = d/2$. It is useful to define another coordinate u as

$$u = z - \frac{d}{2} \qquad (7.8)$$

such that the structure is symmetrical with respect to $u = 0$. Equation 7.7 then becomes

$$S_{31} = \frac{1}{2}e^{-j\beta d}\int_{-d/2}^{d/2} e^{-2j\beta u}\frac{d}{du}(\ln Z_{0e})du$$

$$= e^{-j\beta d}\int_{-d/2}^{d/2} e^{-2j\beta u} p(u)du \qquad (7.9)$$

where

$$p(u) = \frac{1}{2}\frac{d}{du}(\ln Z_{0e}) \qquad (7.10)$$

The function $p(u)$ is an odd function of u and hence (7.9) reduces to

$$S_{31} = -je^{-j\beta d}\int_{-d/2}^{d/2} \sin(2\beta u)p(u)du \qquad (7.11)$$

Inspection of (7.11) shows that the amplitude of the coupled signal is symmetrical about $\beta = 0$; that is, $|S_{31}(\beta)| = |S_{31}(-\beta)|$. Further, the phase of the coupled signal is given by

$$\angle S_{31} = \frac{\pi}{2} - \beta d \text{ rad, for } \beta > 0$$

$$= -\left(\frac{\pi}{2} + \beta d\right) \text{ rad, for } \beta < 0 \qquad (7.12)$$

Information on the amplitude and phase response for negative values of β are required for the synthesis of the coupler.

Using (7.9) or (7.11), we can determine the coupling of a symmetrical nonuniform coupler, if the distribution of the even-mode characteristic impedance is known, and (7.9) and (7.11) can be easily evaluated numerically.

7.1.2 Synthesis

For a coupler of finite length d, the even-mode impedance varies only over the length of the coupler ($|u| \leq d/2$). The function $p(u)$ given by (7.10) is therefore zero for $|u| \geq d/2$. The lower and upper limits of integration in (7.9) or (7.11) can therefore be changed to $u = -\infty$ and $u = \infty$, respectively, without affecting the value of the integral. In that case, (7.9) becomes

$$S_{31} = e^{-j\beta d} \int_{-\infty}^{\infty} e^{-2j\beta u} p(u) du$$

or

$$S_{31} e^{j\beta d} = \int_{-\infty}^{\infty} e^{-2j\beta u} p(u) du \tag{7.13}$$

It is interesting to note that the quantity on the right-hand side of (7.13) denotes the Fourier transform of the function $p(u)$. The coupling is therefore given by the Fourier transform of the function $p(u)$. In a synthesis problem, however, it is required to find the function $p(u)$ for a given coupling response S_{31}, and this can be done by using the inverse Fourier transform. Thus, we obtain

$$p(u) = \frac{1}{2\pi} \int_{-\infty}^{\infty} S_{31} e^{j\beta d} e^{j2\beta u} d(2\beta) \tag{7.14}$$

where $d(2\beta)$ denotes that the integration is with respect to 2β. Equation 7.14 can be used for the synthesis of a nonuniform TEM directional coupler.

Ideal Directional Coupler

We now discuss the synthesis of an ideal directional coupler having a flat amplitude response in the range $2\beta = 0$ to $2\beta = 1$ as shown in Figure 7.3. The synthesis can be easily extended to a coupler in a different frequency range by scaling the length of the coupler as discussed later in the chapter.

Figure 7.3 shows the amplitude of the coupling of an ideal coupler only in the range $2\beta > 0$. As shown by (7.14), however, we need to know the amplitude and phase response of the coupler in the range $-\infty < 2\beta < \infty$ to synthesize a coupler. Because the amplitude response of the coupler is symmetrical about $\beta = 0$ as discussed earlier, we have

$$|S_{31}| = R, \quad \text{for } |2\beta| < 1$$
$$= 0, \quad \text{for } |2\beta| > 1 \tag{7.15}$$

If the coupling is specified in decibels, then

$$R = 10^{-C/20} \tag{7.16}$$

where C is the coupling in decibels, and is a positive quantity.

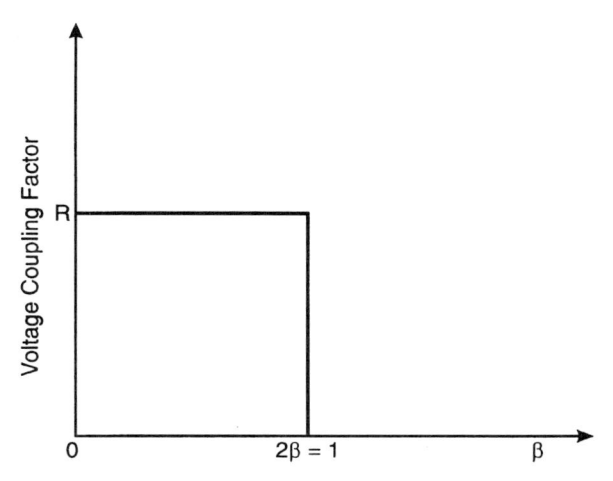

Figure 7.3 Coupling response of an ideal directional coupler.

The phase constant β is related to the frequency f by the relation

$$\beta = \frac{2\pi f}{v} \qquad (7.17)$$

where v is the phase velocity along the direction of propagation.

The phase of S_{31} is given by (7.12). Using (7.12) and (7.15), we obtain

$$\begin{aligned} S_{31} &= Re^{j(\pi/2-\beta d)}, \quad \text{for } 0 < 2\beta < 1 \\ &= Re^{-j(\pi/2+\beta d)}, \quad \text{for } -1 < 2\beta < 0 \\ &= 0 \quad \text{otherwise} \end{aligned} \qquad (7.18)$$

Substituting S_{31} from the above equation in (7.14) we obtain

$$p(u) = -\frac{R}{\pi} \frac{\sin^2(u/2)}{u/2} \qquad (7.19)$$

which is plotted in Figure 7.4 for an arbitrary assumed value of coupling R. Note from (7.19) that the function $p(u)$ extends from $u = -\infty$ to $u = \infty$. In other words, if it is desired to obtain the coupling response shown in Figure 7.3, then the coupler will have to be infinitely long. However, it is also seen from Figure 7.4 that the function $p(u)$ becomes quite small when $|u|$ becomes large. Therefore, if the length d of the coupler is chosen as equal to $4n\pi$ (corresponding to n positive and n negative lobes of Figure 7.4), the coupling response is not expected to deviate much from the ideal response, if the value of n is sufficiently high. Figures 7.5(a) to (e) show the variation in coupling for different lengths of the

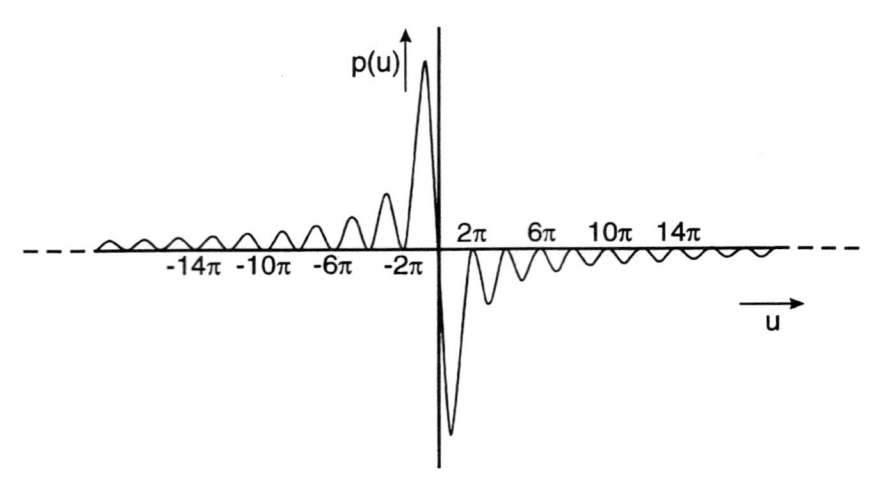

Figure 7.4 The function $p(u)$ required to obtain the coupling response shown in Figure 7.3.

coupler, for $n = 1, \ldots, 5$; that is, $d = 4\pi$, $d = 8\pi$, $d = 12\pi$, $d = 16\pi$, and $d = 20\pi$. From these figures we see that as the length of the coupler is increased, the response approaches the ideal response. We also see that for a coupler of length $d = 4n\pi$, the number of ripples in the frequency band of interest is equal to n. It may be emphasized that the stated length corresponds to a coupler designed to operate in the range $\beta = 0$ to $\beta = 1/2$ where β is related to frequency by (7.17).

Although increasing the length of the coupler brings the response of the coupler closer to the ideal response, it may be noted that there is an overshoot always present near $\beta = 0$. The value of the overshoot does not decrease in amplitude with increasing the number of lobes used. This phenomenon is known as *Gibb's phenomenon* and occurs when a step function such as shown in Figure 7.3 is approximated by a finite number of Fourier terms. The problem is resolved by employing weighting functions to remove the problem of overshoot and to make the level of ripples equal. In this technique, instead of using function $p(u)$ for the synthesis as given by (7.19), we use the weighted function

$$p_w(u) = w(u) p(u) \tag{7.20}$$

where $w(u)$ is the weighting function and $p(u)$ is given by (7.19).

A common form of the weighting function is shown in Figure 7.6. The value of the weighting function remains constant over a length equal to 4π of the coupler (e.g., the value of the weighting function is w_2 in the intervals $-4\pi < u < -2\pi$ and $2\pi < u < 4\pi$). The weighting function can be determined

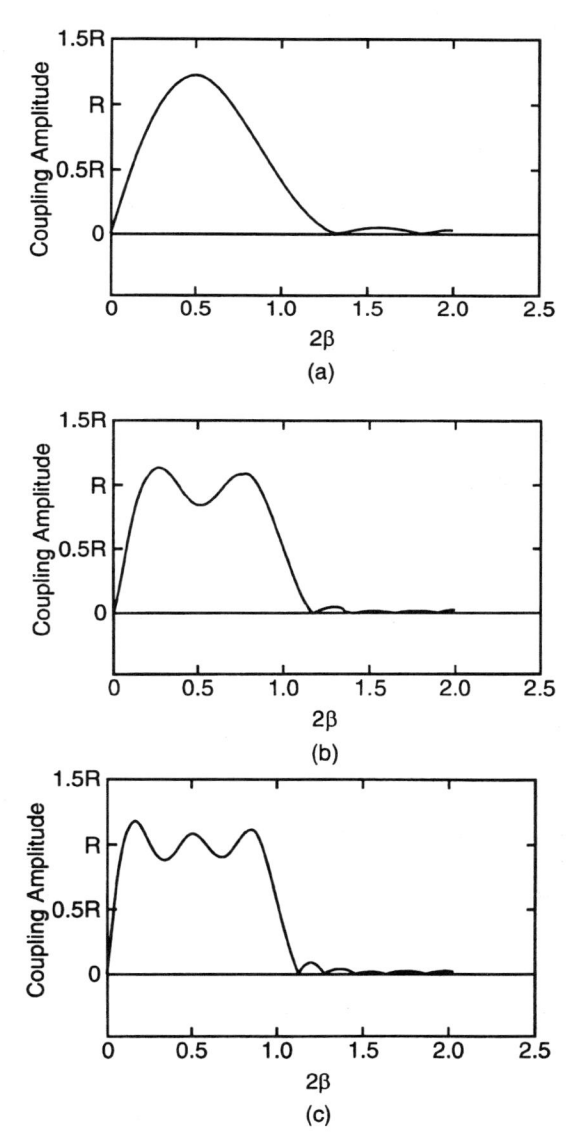

Figure 7.5 Coupling response of a nonuniform coupler of length (a) 4π, (b) 8π, (c) 12π, (d) 16π, and (e) 20π.

(d)

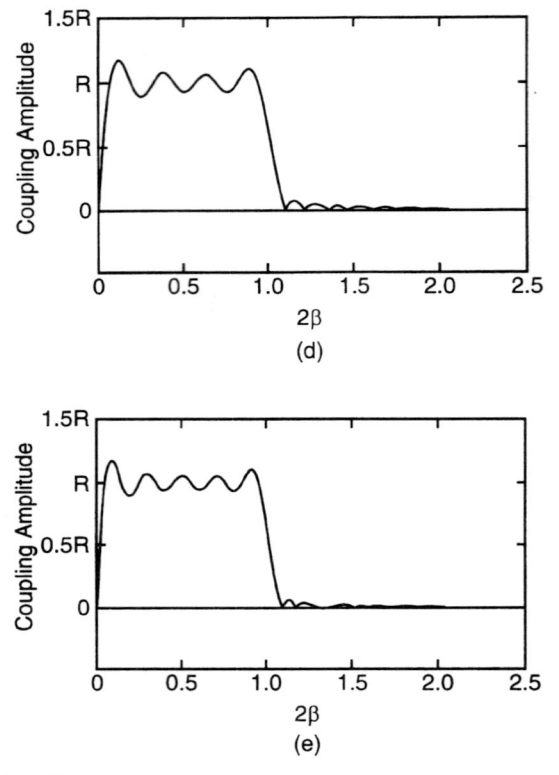

(e)

Figure 7.5 *(continued)*

by a term-by-term comparison of a known equal ripple-level function with a function containing a few weighted Fourier series terms. A technique for determining weighting functions is discussed later in Section 7.1.3. Before that, we discuss the final step for the synthesis of a nonuniform coupler.

Computing $Z_{0e}(u)$ From $p(u)$ or $p_w(u)$

Once the value of $p(u)$ or $p_w(u)$ is known, the corresponding value of $Z_{0e}(u)$ needs to be determined for the realization of the physical circuit. Integrating (7.10) with respect to u, we obtain

$$\frac{1}{2} \ln Z_{0e}(u) = \frac{1}{2} \ln Z_{0e} \bigg|_{u=-d/2} + \int_{-d/2}^{u} p(u) du \qquad (7.21)$$

where $p(u)$ is given by (7.19).

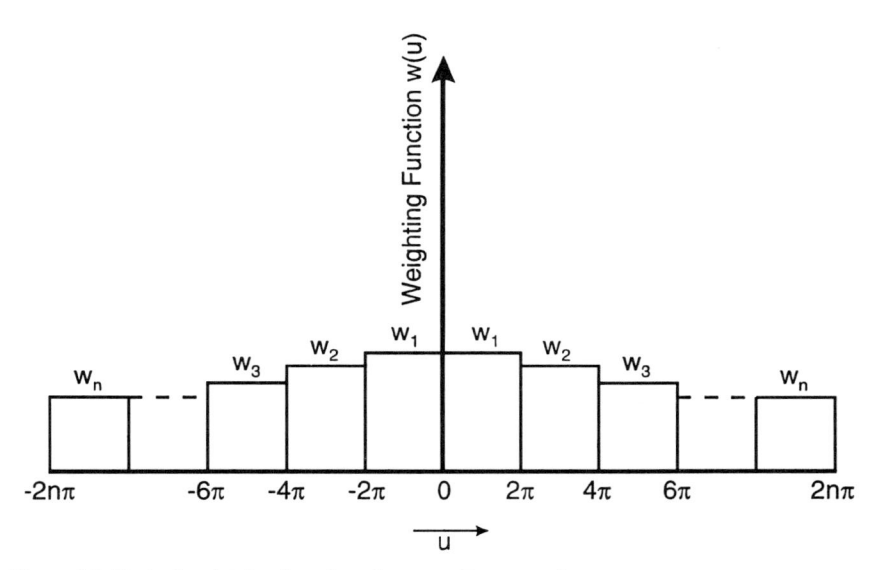

Figure 7.6 Typical weighting function of a nonuniform coupler.

Further, because

$$Z_{0e}|_{u=-d/2} = Z_0$$

(7.21) can also be expressed as

$$\frac{1}{2}\ln\frac{Z_{0e}(u)}{Z_0} = \int_{-d/2}^{u} p(u)\,du = -\frac{R}{\pi}\int_{-d/2}^{u}\frac{\sin^2(u/2)}{u/2}\,du \quad (7.22)$$

Unfortunately, the integral in (7.22) cannot be evaluated in closed form. It is, however, quite simple to evaluate the above integrals numerically using available computer software programs such as *MATHCAD*™ [3] or *MATLAB*™ [4]. If the weighted function $p_w(u)$ given by (7.20) is used for the synthesis of the coupler, the even-mode impedance is then given by

$$\frac{1}{2}\ln\frac{Z_{0e}(u)}{Z_0} = \frac{R}{\pi}\int_{-d/2}^{u} w(u)p(u)\,du = -\frac{R}{\pi}\int_{-d/2}^{u} w(u)\frac{\sin^2(u/2)}{u/2}\,du \quad (7.23)$$

Evaluation of (7.23)

Let us assume that in a particular case, the length d of the coupler is chosen as 20π (i.e., it extends from $u = -10\pi$ to $u = 10\pi$) and it is desired to find the even-mode impedance at $u = -\pi$. Assuming that the weighting function is of the form as shown in Figure 7.6, the value of Z_{0e} at $u = -\pi$ can be found using (7.23) as follows:

$$\frac{1}{2} \ln \frac{Z_{0e}|_{u=-\pi}}{Z_0} = -\frac{R}{\pi} \left[w_5 \int_{-10\pi}^{-8\pi} \frac{\sin^2(u/2)}{u/2} du \right.$$

$$+ w_4 \int_{-8\pi}^{-6\pi} \frac{\sin^2(u/2)}{u/2} du + w_3 \int_{-6\pi}^{-4\pi} \frac{\sin^2(u/2)}{u/2} du$$

$$\left. + w_2 \int_{-4\pi}^{-2\pi} \frac{\sin^2(u/2)}{u/2} du + w_1 \int_{-2\pi}^{-\pi} \frac{\sin^2(u/2)}{u/2} du \right]$$

(7.24)

Each integral in (7.24) can now be easily evaluated numerically.

7.1.3 Technique for Determining Weighting Functions

As discussed earlier, (7.7) is valid if the reflection at any step is small compared with unity. The synthesis technique described earlier is thus strictly valid for realization of couplers having coupling of less than about 10 dB. However, (7.23) can also be used to synthesize a "tight coupler" if the weighting function terms w_i are chosen properly. A general technique for deriving the weighting terms valid for "tight" as well as "loose" couplers is given by Tressel[1]. In this technique, a symmetrical multisection coupler as discussed in the last chapter is first designed for given coupler specifications. Let the multisection coupler employ $2n-1$ sections with impedance levels denoted as shown in Figure 7.7. Such a coupler will have n ripples in the frequency band of interest. Similarly, a nonuniform coupler that has a length equal to $4n\pi$ [1] will have n ripples in the frequency band. Therefore, the response of a nonuniform coupler can be made identical to the response of a multisection coupler of $2n-1$ sections if the length of the nonuniform coupler is chosen equal to $4n\pi$. The nonuniform coupler can then be synthesized using (7.23) where the weighting function terms w_i are determined by the following method.

Consider the function shown in Figure 7.8. The value of the function for $0 \le \theta \le \pi$ represents the value of desired coupling R. The Fourier series representation of the function in the range $0 < \theta < \pi$ is

$$g(\theta) = R \frac{4}{\pi} \left(\sin \theta + \frac{\sin 3\theta}{3} + \frac{\sin 5\theta}{5} + \frac{\sin 7\theta}{7} + \frac{\sin 9\theta}{9} + \cdots \right) \quad (7.25)$$

---

1. This length is for a coupler designed to operate from $2\beta = 0$ to $2\beta = 1$. For a coupler in a different frequency range, the length of the coupler is different.

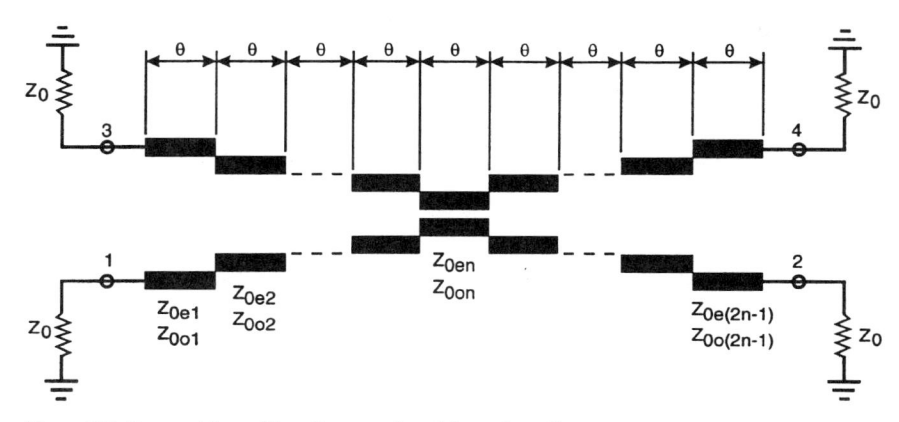

Figure 7.7 Symmetric multisection coupler of $2n - 1$ sections.

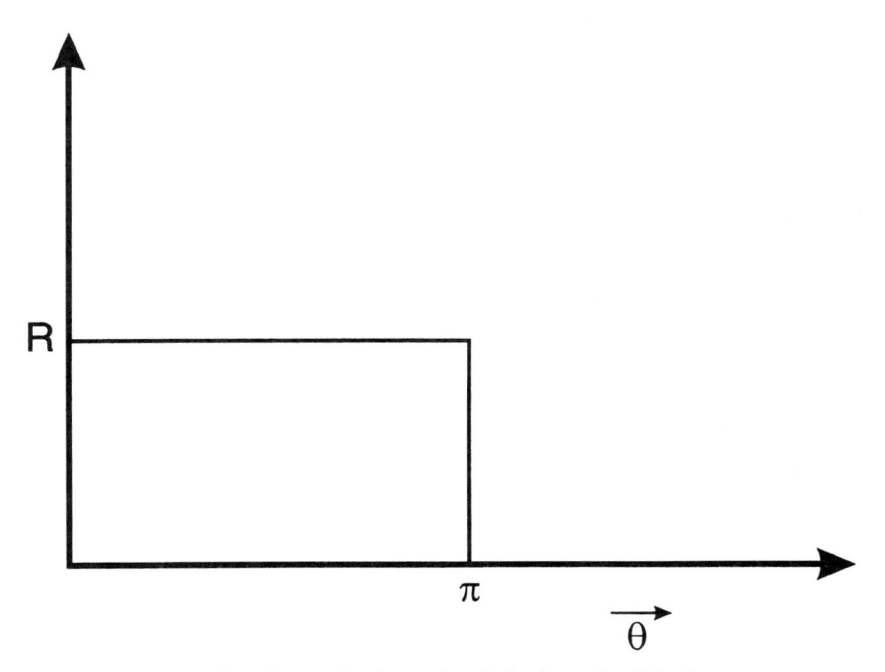

Figure 7.8 Gotte function whose value is equal to desired coupling R for $0 < \theta < \pi$.

With a few terms from the above series used to describe the function, the response would be similar to that shown in Figures 7.5(a) to (e); that is, there will exist an overshoot near $\theta = 0$ and the ripples would be unequal. If only n terms of the series of (7.25) are used, it is more appropriate to construct a new function $f(\theta)$ as follows:

$$f(\theta) = R\frac{4}{\pi}\left[w_1 \sin\theta + w_2\frac{\sin 3\theta}{3} + \cdots + w_n\frac{\sin(2n-1)\theta}{(2n-1)}\right] \quad (7.26)$$

where $w_1, \ldots, w_n$ are the weighting function terms. With a suitable choice of these terms, the function $f(\theta)$ can be made an equal ripple.

A suitable equal ripple function can be easily determined by using the design tables of symmetrical multisection couplers. If (7.11) is used to find the coupling of a symmetrical coupler of $2n - 1$ sections shown in Figure 7.7, the coupling (excluding the phase factor) is given by

$$h(\theta) = \sum_{r=1}^{n} \ln\frac{Z_{0e(n+1-r)}}{Z_{0e(n-r)}} \sin(2r - 1)\theta \quad (7.27)$$

In (7.27), θ denotes the electrical length of each section and is given by

$$\theta = \frac{\beta d}{2n - 1} \quad (7.28)$$

where β is the phase constant, d denotes the total physical length of the coupler, and

$$Z_{0e(n-r)}|_{r=n} = Z_0 \quad (7.29)$$

If the various impedances in (7.27) correspond to that of an equal ripple symmetrical coupler, the function $h(\theta)$ is also equal ripple. Comparing term by term (7.26) and (7.27), we can determine the weighting function terms w_i.

Example 7.1

Determine the weighting function terms for the design of a nonuniform coupler having a mean coupling $C = 3.01$ dB, a frequency bandwidth ratio $(B) = 8$, and a ripple tolerance $\delta = \pm 0.20$ dB.

From (7.16), we find that

$$R = 10^{-.15005} = 0.707$$

Further, from Table 6.1(k) in last chapter, we find that a symmetrical multisection coupler with the above specifications requires seven sections. This gives

$$N = 2n - 1 = 7$$

or

$$n = 4$$

Also from the same table, the even-mode impedances of the various sections are

$$\frac{Z_{oe1}}{Z_0} = 1.07950$$

$$\frac{Z_{oe2}}{Z_0} = 1.23581$$

$$\frac{Z_{oe3}}{Z_0} = 1.65795 \qquad (7.30)$$

$$\frac{Z_{oe4}}{Z_0} = 4.90662$$

(7.27) then becomes

$$h(\theta) = 1.085 \sin\theta + 0.29385 \sin 3\theta + 0.13522 \sin 5\theta + 0.076497 \sin 7\theta \qquad (7.31)$$

Substituting $R = 0.707$ and $n = 4$ in (7.26), we obtain

$$f(\theta) = 0.90034 w_1 \sin\theta + 0.30011 w_2 \sin 3\theta$$
$$+ 0.18006 w_3 \sin 5\theta + 0.12862 w_4 \sin 7\theta \qquad (7.32)$$

Comparing term by term (7.31) and (7.32), the weighting function terms are found as follows:

$$\begin{aligned} w_1 &= 1.205 \\ w_2 &= 0.979 \\ w_3 &= 0.751 \\ w_4 &= 0.595 \end{aligned} \qquad (7.33)$$

7.1.4 Electrical and Physical Length of a Coupler

The synthesis described so far is for a coupler having an equal-ripple response in the range $2\beta = 0$ to $2\beta = 1$ (the value of β at the center frequency is $1/4$). For such a coupler, the length of the coupler corresponds to those of $2n$ lobes (n positive and n negative) and is therefore equal to $4n\pi$. The number of ripples for such a coupler is equal to n in the frequency band of interest. The *total* electrical length of the nonuniform coupler at the center frequency is therefore

$$\theta_c = \frac{1}{4} 4n\pi = n\pi \qquad (7.34)$$

If β_c denotes the phase constant of the wave in the medium of the coupler corresponding to the center frequency of design f_0, then

$$\theta_c = \beta_c d = n\pi \tag{7.35}$$

where d is the total physical length of the coupler, or

$$d = \frac{n\pi}{\beta_c} \tag{7.36}$$

In (7.36), $\beta_c = 2\pi f_0/v$, where v is the velocity of propagation in the medium of the coupler.

Length of Multisection Coupler

The electrical length of each section of a multisection coupler is $\pi/2$ rad at the center frequency of design. The electrical length of a multisection coupler of $2n-1$ sections is therefore

$$\beta_c d = (2n-1)\frac{\pi}{2} \tag{7.37}$$

Comparison of (7.34) and (7.37) shows that a nonuniform coupler is a quarter-wave longer than a multisection coupler with the identical response.

7.1.5 Design Procedure

Based on the discussions so far, the design of a nonuniform symmetrical TEM mode coupler can be summarized as below:

- Step 1
  1. Specify coupling level $C(dB)$, or voltage coupling factor R. Find one from the other using (7.16).
  2. Specify ripple level δ in dB.
  3. Specify frequency bandwidth ratio B, and center frequency of design f_0 (or frequency range of operation f_1 to f_2). If the frequency range of operation of the coupler is specified, find B and f_0 using (6.11) to (6.13), respectively.
- Step 2
  Using Table 6.1 from Chapter 6, find the minimum number of sections required for a multisection symmetrical coupler for the given values of C (coupling), δ (coupling ripple), and B (bandwidth ratio). Let the number of sections required be N, then

  $$n = \frac{N+1}{2}$$

- Step 3
  Using the same tables, find $Z_{0e1}, Z_{0e2}, \ldots, Z_{0en}$.
- Step 4
  Using computed values of $Z_{0e1}, Z_{0e2}, \ldots, Z_{0en}$, construct the function $h(\theta)$ using (7.27).
- Step 5
  Construct the function $f(\theta)$ using (7.26).
- Step 6
  Comparing term by term the functions $h(\theta)$ and $f(\theta)$ computed using steps 4 and 5, respectively, find the weighting function terms $w_1, \ldots, w_n$.
- Step 7
  Evaluate $Z_{0e}(u)$ numerically using (7.23) for discrete values of u in the range $-2n\pi \leq u \leq 2n\pi$, and by substituting $d/2 = 2n\pi$.
- Step 8
  The values of u for which computations are made in step 7 correspond to the design of the coupler having a mean value of $\beta = 1/4$. To determine the values of u for $\beta = \beta_c$, divide the values of u computed in step 7 by a factor of $4\beta_c$, where β_c is the phase constant in the medium of the coupler at the center frequency of the design.

Example 7.2

Design a nonuniform coupler in a homogeneous dielectric medium ($\epsilon_r = 2.32$) with the following specifications:
Mean coupling $(C) = 8.34$ dB
Frequency of operation $= 1-11$ GHz
Ripple level $(\delta) = \pm 0.3$ dB

- Step 1
  From the above specifications, we have
  $$R = 10^{-C/20} = 0.3828$$
  $$f_0 = \frac{f_1 + f_2}{2} = \frac{1 + 11}{2} = 6.0 \, GHz$$
  $$B = \frac{f_2}{f_1} = 11$$
  At 6.0 GHz, the phase constant in free space is given by
  $$\beta_0 = \frac{2\pi f_0}{0.3} = 125.67 \text{ rad/m}$$

or, the phase constant in the medium of the coupler at the center frequency of design is given by

$$\beta_c = \sqrt{\epsilon_r}\beta_0 = 1.524\beta_0 = 191.51 \text{ rad/m}$$

- Step 2
From Table 6.1(r) in Chapter 6, we find that with a mean coupling level of $C = 8.34$ dB, ripple level $= \pm 0.3$ dB, a nine-section symmetrical coupler will exhibit a frequency bandwidth ratio of $B = 10.96$, which is very close to the specified value. We thus have

$$n = \frac{N+1}{2} = \frac{9+1}{2} = 5 \tag{7.38}$$

- Step 3
Using the same table, the impedances of the different sections of the nine-section coupler is

$$\frac{Z_{oe1}}{Z_0} = 1.03134$$

$$\frac{Z_{oe2}}{Z_0} = 1.07697$$

$$\frac{Z_{oe3}}{Z_0} = 1.16469 \tag{7.39}$$

$$\frac{Z_{oe4}}{Z_0} = 1.35771$$

$$\frac{Z_{oe5}}{Z_0} = 2.25315$$

- Step 4
Substituting the above values in (7.27), we obtain

$$h(\theta) = 0.5071 \sin\theta + 0.1532 \sin 3\theta + 0.0785 \sin 5\theta$$
$$+ 0.0433 \sin 7\theta + 0.0309 \sin 9\theta \tag{7.40}$$

- Step 5
Substituting $R = .3827$ in (7.26):

$$f(\theta) = 0.4873 w_1 \sin\theta + 0.1624 w_2 \sin 3\theta + 0.09745 w_3 \sin 5\theta$$
$$+ 0.06961 w_4 \sin 7\theta + 0.05414 w_5 \sin 9\theta \tag{7.41}$$

- Step 6

  Comparing (7.40) and (7.41) term by term, we obtain the following values for the weighting functions:

  $$w_1 = 1.040$$
  $$w_2 = 0.943$$
  $$w_3 = 0.805 \qquad (7.42)$$
  $$w_4 = 0.623$$
  $$w_5 = 0.569$$

- Step 7

  The values of $Z_{0e}(u)$ were computed for values of u in the range $-10\pi \le u \le 10\pi$ using (7.23) by numerical evaluation of the integral using software package $MATHCAD^{TM}$. The obtained values of $Z_{0e}(u)$ are shown in Table 7.1. The computed values of u are valid for a coupler having $\beta_c = \frac{1}{4}$.

- Step 8

  The values of u for the specified coupler ($\beta_c = 191.51$) were determined by dividing the values of u derived in step 7 by a factor $4\beta_c$. These values are also shown in Table 7.1.

Because the coupler is symmetrical about $u = 0$, it is sufficient to find the structure parameters for $0 \le u \le d/2$. The overall length of the coupler is $d = 2 \times 4.1 = 8.2$ cm.

Once we determine the even-mode characteristic impedances of the coupler at various cross sections of the coupler, we can find the corresponding odd-mode characteristic impedances using (7.2). Further, the physical dimensions of the lines (i.e., their width and the spacing between them) can be determined. The design and performance of nonuniform couplers using microstrip lines is described in [5, 6].

The design procedure described above requires data of symmetrical multi-section filters, which is available only for some specific cases as discussed in the last chapter. A design procedure of nonuniform couplers that is based on optimization methods is described in [7].

7.2 Asymmetrical Couplers

A nonuniform asymmetric coupler which exhibits a broadband coupling and the phase properties of a magic-T is shown in Figure 7.9. The structure does not have an end to end symmetry but has a symmetry with respect to plane PP'.

Table 7.1
Design procedure for a nonuniform TEM coupler

u (m) Normalized coupler $\beta_c = 1/4$ rad/m	u (cm) Actual coupler $\beta_c = 191.51$ rad/m	$\ln\dfrac{Z_{0e}}{Z_0}$	$\dfrac{Z_{0e}}{Z_0}$
10π	4.10	0.00	1.00
9.5π	3.89	0.00268	1.002
9π	3.69	0.01486	1.0149
8.5π	3.48	0.0277	1.028
8.0π	3.28	0.0309	1.031
7.5π	3.08	0.0346	1.035
7.0π	2.87	0.517	1.053
6.5π	2.67	0.702	1.073
6.0π	2.46	0.743	1.077
5.5π	2.25	0.806	1.084
5.0π	2.05	0.1113	1.117
4.5π	1.85	0.1449	1.156
4.0π	1.64	0.1532	1.165
3.5π	1.44	0.1647	1.1790
3.0π	1.23	0.2231	1.2500
2.5π	1.03	0.2911	1.3379
2.0π	0.82	0.3087	1.3616
1.5π	0.62	0.3369	1.400
1.0π	0.41	0.5087	1.663
0.5π	0.21	0.7852	2.1929
0π	0	0.9263	2.525

The coupler consists of a section of coupled lines of length θ extending from $z = -d$ to $z = 0$ and a section of uncoupled lines of length θ. The coupling between lines increases from $z = -d$ to $z = 0$. However it reduces abruptly to zero at $z = 0$.

The scattering parameters of the coupler can be determined using the even-mode analysis (as discussed earlier in Section 7.1) if the even- and odd-mode characteristic impedances of the coupled line section at any cross section are chosen to satisfy

$$Z_{0e}(z) Z_{0o}(z) = Z_0^2 \qquad (7.43)$$

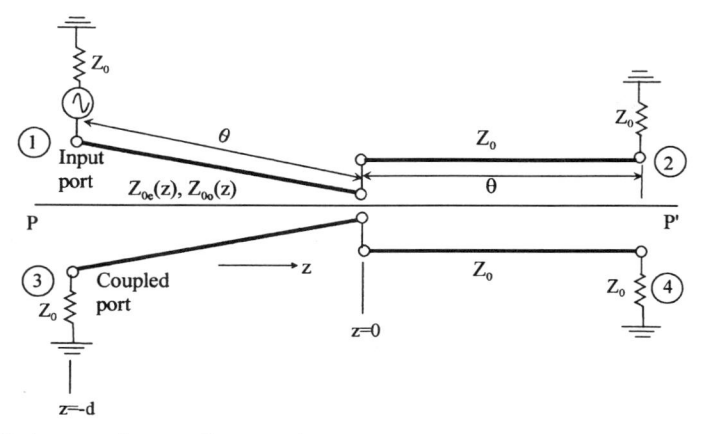

Figure 7.9 Asymmetric nonuniform coupler.

The even-mode equivalent circuit of the coupler is shown in Figure 7.10. In terms of even-mode scattering parameters, the scattering parameters of the coupler are given by

$$S_{11} = S_{22} = S_{33} = S_{44} = 0 \tag{7.44}$$

$$S_{41} = S_{14} = S_{23} = S_{32} = 0 \tag{7.45}$$

$$S_{21} = S_{12} = S_{34} = S_{43} = S_{21e} = S_{12e} \tag{7.46}$$

$$S_{31} = S_{13} = S_{11e}, \quad S_{24} = S_{42} = S_{22e} \tag{7.47}$$

The even-mode reflection coefficient S_{11e} at port 1 can be determined by summing reflections in proper phase as discussed in Section 7.1. If the even-mode impedance of the coupled line section from $z = -d$ to $z = 0^-$ is so tapered ($z = 0^-$ denotes location just to the left of abrupt discontinuity at $z = 0$) such that the contribution of the taper section to the overall reflection coefficient S_{11e} is zero, the reflection coefficient at port 1 results solely from the abrupt impedance discontinuity at $z = 0$. The reflection coefficient S_{11e} is given by

$$S_{11e} = \frac{Z_0 - Z_0/a}{Z_0 + Z_0/a} e^{-2j\theta} = \frac{a-1}{1+a} e^{-2j\theta} \tag{7.48}$$

where it is assumed that the even-mode characteristic impedance of coupled lines is Z_0/a at $z = 0^-$. Similarly,

$$S_{22e} = \frac{Z_0/a - Z_0}{Z_0 + Z_0/a} e^{-2j\theta} = \frac{1-a}{1+a} e^{-2j\theta} \tag{7.49}$$

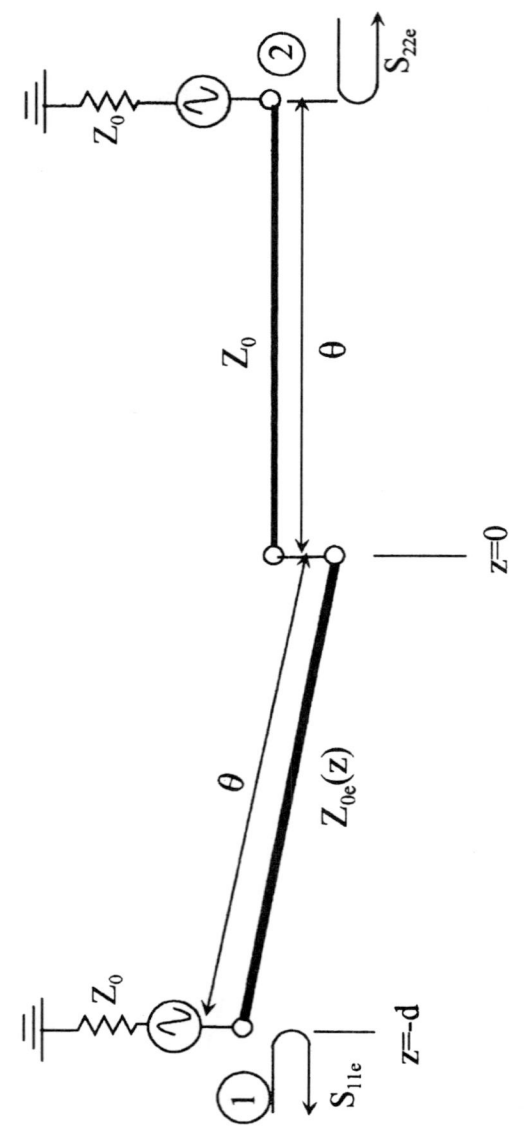

Figure 7.10 Equivalent circuit to determine coupling of the asymmetric nonuniform coupler shown in Figure 7.9.

Nonuniform Broadband TEM Directional Couplers

The properties of lossless two-ports were discussed in Chapter 2. Using (2.77), the transmission coefficient between ports 1 and 2 of the network shown in Figure 7.10 is found as

$$S_{21e} = S_{12e} = \frac{2\sqrt{a}}{1+a} e^{-2j\theta} \quad (7.50)$$

Furthermore, using (7.44)–(7.47), the scattering matrix of the coupler can be expressed as

$$[\mathbf{S}] = \begin{bmatrix} 0 & \beta & -\alpha & 0 \\ \beta & 0 & 0 & \alpha \\ -\alpha & 0 & 0 & \beta \\ 0 & \alpha & \beta & 0 \end{bmatrix} \quad (7.51)$$

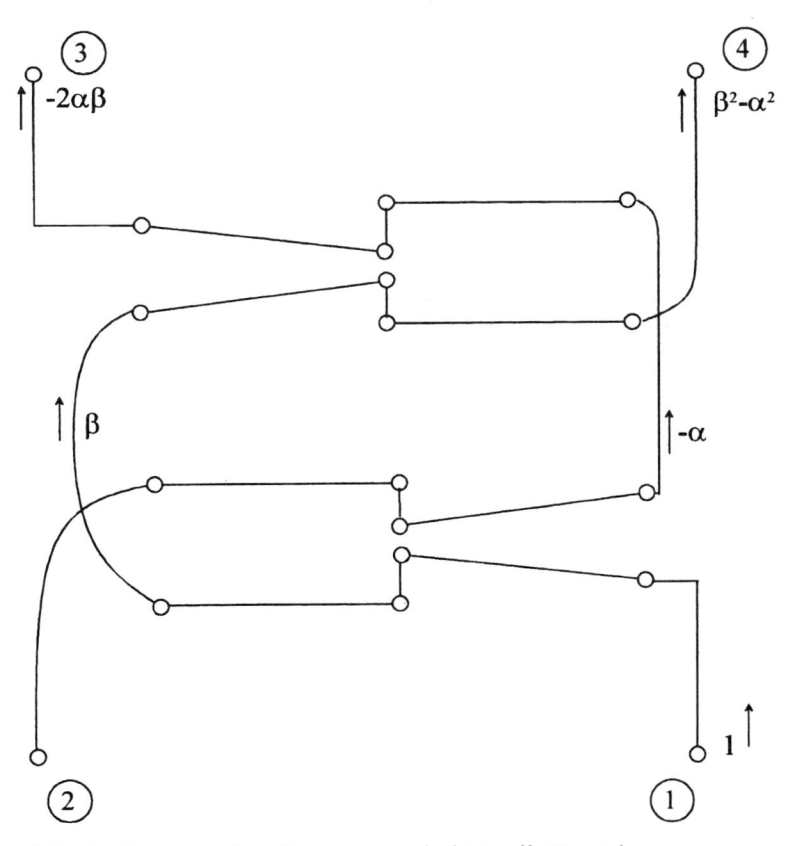

Figure 7.11 Tandem connection of two asymmetrical nonuniform couplers.

where

$$\alpha = \frac{1-a}{1+a}e^{-2j\theta}, \beta = \frac{2\sqrt{a}}{1+a}e^{-2j\theta} \qquad (7.52)$$

For a 3-dB coupler, it is found that $a = 0.1717$ or $Z_{0e} = 5.83Z_0$ and $Z_{0o} = 0.1717Z_0$ at $z = 0^-$. It is very difficult to obtain coupled lines with such extreme values of even- and odd-mode characteristic impedances. However, two 8.36-dB asymmetric couplers can be connected in tandem as shown in Figure 7.11 to achieve a 3-dB coupler. For a 8.36-dB coupler, $a = 0.446$ or $Z_{0e} = 2.24Z_0$ and $Z_{0o} = 0.44Z_0$. These values are relatively more easily obtained in practice using planar transmission lines such as broadside-coupled offset striplines discussed in Section 3.7.3.

As discussed above, the design of nonuniform asymmetric coupler requires a reflectionless taper from $z = -d$ to $z = 0^-$. There are many possible designs. One frequently employed taper design is due to Klopfenstein [8]. In practice, it is not possible to design a completely reflectionless taper using a finite length. Furthermore, it is not possible to layout a structure in which the coupling changes abruptly (at $z = 0$ in Fig. 7.9). However, it is still possible to obtain performance which is quite close to the ideal performance [2].

References

[1] Tresselt, C. P., "The Design and Construction of Broadband, High-Directivity, 90-Degree Couplers Using Nonuniform Techniques," *IEEE Trans.*, Vol. MTT-14, Dec. 1966, pp. 647–656.

[2] DuHamel, R. H., and M. E. Armstrong, "A Wideband Monopulse Antenna Utilizing the Tapered-Line Magic-T," *USAF Anenna Research and Development Program 15th Symp.*, University of Illinois, 1965, pp. 1–30.

[3] MATHCAD, is a trademark of Mathsoft Inc., MA.

[4] MATLAB, is a trademark of The Mathworks Inc., MA.

[5] Uysal, S., and H. Aghvami, "Synthesis, Design, and Construction of Ultra-Wide-Band Nonuniform Quadrature Directional Couplers in Inhomogeneous Media," *IEEE Trans.*, Vol. MTT-37, June 1989, pp. 969–976.

[6] Uysal, S., *Nonuniform Line Microstrip Directional Couplers*, Artech House, Boston, MA 1993.

[7] Kammler, D. W., "The Design of Discrete N-Section and Continuously Tapered Symmetrical Microwave TEM Directional Couplers," *IEEE Trans.*, Vol. MTT-17, Aug. 1969, pp. 577–590.

[8] Klopfenstein, R. W., "A Transmission Line Taper of Improved Design," *Proc. IEEE.*, Vol. 44, Jan 1954, pp. 31–35.

8

Tight Couplers

8.1 Introduction

In planar quasi-TEM transmission line media such as microstrip, it is difficult to obtain tight coupling between lines because they require a very small spacing between them. Tight couplers, especially 3-dB couplers, are used in many practical circuits such as balanced mixers, amplifiers, and so forth. In this chapter, we concentrate on discussing directional couplers specifically suited for obtaining tight coupling values. These include branch-line couplers, rat-race couplers, Lange couplers, tandem couplers, and several other structures using new concepts and multilayer dielectric configurations. Although conventional branch-line and rat-race couplers do not use coupled lines, many of their recent modifications employ coupled structures to enhance their bandwidths or use folded coupled lines or inductors to make them compact.

Branch-line and rat-race couplers are easily analyzed using the even- and odd-mode approach [1–4]. Only the final design equations for branch-line and rat-race couplers are given in this chapter. The equations given, however, are valid for arbitrary division of power between the ports of a branch-line or a rat-race coupler. These couplers are inherently narrowband (<20% bandwidth) circuits; some techniques used to enhance their bandwidth are described briefly.

Although branch-line and rat-race couplers are more suitable for obtaining tight coupling values (such as 3 dB), they can also be useful for obtaining loose coupling values in certain applications. For a branch-line coupler designed for loose coupling, the impedance of the shunt branches becomes very high and cannot be easily realized using planar transmission lines. A modified branch-line coupler in which the high-impedance shunt branches are replaced by coupled lines is described. The modified branch-line coupler can be easily implemented in microstrip configuration.

The size of conventional branch-line and rat-race couplers becomes quite large at low frequencies (below 2 GHz). Further, their size is "too large" for their implementation in MMICs even at frequencies as high as 10 GHz. The realization of branch-line and rat-race couplers using lumped inductive and capacitive elements is also described. The size of lumped-element couplers is very small compared with that of conventional couplers. Reduced-size branch-line and rat-race couplers that use only lumped capacitors and small sections of transmission lines (smaller than $\lambda_g/4$) are also described. These couplers are quite suitable for realization in MMICs using coplanar waveguide as the transmission line.

Tight coupling between edge-coupled lines can also be achieved by connecting a number of lines in an interdigital manner. This is essentially the basis of the Lange coupler, which is widely used in microwave circuits. Equations for the design of a general N-conductor interdigital coupler are given. Next, the operation of a tandem coupler is explained. We show that connecting two loose couplers in tandem result in a tight coupler. Finally, compact couplers for wireless applications and tight couplers using multilayer dielectric structures are described.

8.2 Branch-Line Couplers

Figure 8.1 shows a branch-line coupler. It consists of two quarter-wave-long transmission line sections of characteristic impedance Z_{0s} each, connected by two shunt branches. The shunt branches are quarter-wave-long transmission line sections of characteristic impedance Z_{0p} each. By properly choosing the values of Z_{0s} and Z_{0p}, the circuit can be made to operate like a directional coupler. At the center frequency the scattering parameters of a branch-line coupler are given by

$$S_{21} = -j\frac{Z_{0s}}{Z_0} \qquad (8.1)$$

$$S_{31} = -\frac{Z_{0s}}{Z_{0p}} \qquad (8.2)$$

$$S_{11} = 0 \qquad (8.3)$$

$$S_{41} = 0 \qquad (8.4)$$

where Z_0 denotes the impedance of various ports of a branch-line coupler. The scattering parameters of a branch-line coupler also satisfy the following

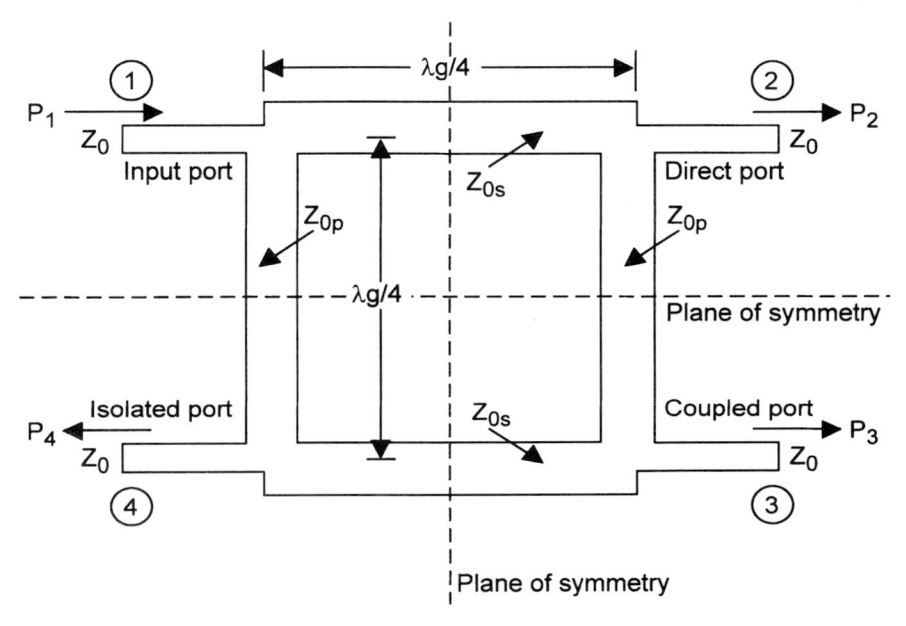

Figure 8.1 Layout of a branch-line coupler in planar circuit configuration.

condition, which follows from the principle of conservation of energy:

$$|S_{21}|^2 + |S_{31}|^2 = 1 \tag{8.5}$$

where it is assumed that $S_{11} = S_{41} = 0$. Substituting values of scattering parameters from (8.1) and (8.2) in (8.5), it is found that Z_{0s} and Z_{0p} should satisfy the following condition:

$$\frac{Z_{0s}^2}{Z_0^2} + \frac{Z_{0s}^2}{Z_{0p}^2} = 1 \tag{8.6}$$

A branch-line coupler as shown in Figure 8.1 has two planes of symmetry. The scattering matrix of a branch-line coupler can therefore be expressed as

$$[S] = \begin{bmatrix} 0 & -j\frac{Z_{0s}}{Z_0} & -\frac{Z_{0s}}{Z_{0p}} & 0 \\ -j\frac{Z_{0s}}{Z_0} & 0 & 0 & -\frac{Z_{0s}}{Z_{0p}} \\ -\frac{Z_{0s}}{Z_{0p}} & 0 & 0 & -j\frac{Z_{0s}}{Z_0} \\ 0 & -\frac{Z_{0s}}{Z_{0p}} & -j\frac{Z_{0s}}{Z_0} & 0 \end{bmatrix} \tag{8.7}$$

The characteristic impedance of the main-line and shunt branches of a branch-line coupler can be computed using (8.1) and (8.2) as

$$Z_{0s} = Z_0|S_{21}| = Z_0\sqrt{1 - |S_{31}|^2} \qquad (8.8)$$

and

$$Z_{0p} = \frac{Z_{0s}}{|S_{31}|} = \frac{Z_{0s}}{\sqrt{1 - |S_{21}|^2}} \qquad (8.9)$$

Example 8.1

Compute the characteristic impedances of the main-line and shunt branches of a branch-line coupler given $Z_0 = 50\ \Omega$ and coupling from port 1 to ports 2 and 3 is equal ($S_{21} = S_{31} = -3.01$ dB).

Given that coupling from port 1 to port 2 is 3.01 dB, therefore

$$|S_{21}| = 10^{-3.01/20} = 0.707$$

Substituting the value of $|S_{21}|$ in (8.8), we obtain

$$Z_{0s} = 0.707\, Z_0 = 35.4\ \Omega$$

Further, substituting $Z_{0s} = 35.4\ \Omega$ and $|S_{21}| = 0.707$ in (8.9), the value of Z_{0p} is found as

$$Z_{0p} = 50\ \Omega$$

The transmission line sections having characteristic impedances of 35 and 50 Ω can be easily realized using planar transmission lines such as microstrip line. A 3-dB branch-line coupler is also known as a 90-deg hybrid.

Properties of the Branch-Line Coupler

In a branch-line coupler, the following performance is obtained only at the center frequency (i.e., the frequency at which the main-line and shunt branches are each a quarter-wave long):

$$S_{11} = S_{22} = S_{33} = S_{44} = 0 \qquad (8.10)$$
$$S_{14} = S_{41} = S_{23} = S_{32} = 0 \qquad (8.11)$$

A branch-line coupler is thus perfectly matched only at its center frequency, at which there is also complete isolation between "decoupled" ports. In the case of backward parallel-coupled TEM couplers, these properties are satisfied independent of the frequency (as discussed in Section 6.4). In parallel-coupled TEM couplers, the relationship of phase quadrature between the signals at direct and coupled ports is satisfied independent of the operating frequency. In a branch-line coupler, the relationship of phase quadrature between the signals at direct and coupled ports is satisfied only at the center frequency.

The VSWR, coupling, and isolation of a 3-dB branch-line coupler as a function of frequency are plotted in Figure 8.2. It is seen that the input VSWR is unity (equivalent to $|S_{11}| = 0$) at the center frequency, but increases rapidly away from the center frequency and reaches a value of 2 (equivalent to $|S_{11}| = -9.54$ dB) about 20% away from the center frequency. Similarly, the isolation between "decoupled" ports falls to about 15 dB at frequencies about 10% away from the center frequency. The branch-line couplers are useful in applications requiring less than about 20% frequency bandwidth only.

Physical Implementation of Branch-Line Couplers

The layout of a branch-line coupler in microstrip form is shown in Figure 8.1. It is seen that there is a physical discontinuity present at each junction. The effect of these discontinuities is to add extra reactances that alter the response of the physical coupler compared with the ideal one, and should be considered in evaluating the performance of the circuit, especially at high frequencies. This can be done using the equivalent circuit models of discontinuities, which are available in the literature [5] and also in many commercially available microwave CAD software. A more accurate analysis of such couplers is generally performed using electromagnetic (EM) simulators.

8.2.1 Modified Branch-Line Coupler

Branch-line couplers are generally used for equal power division. For loose coupling, parallel-coupled TEM couplers are preferred. If parallel-coupled directional couplers are realized in microstrip configuration, the directivity of the coupler may not be very high because of unequal even- and odd-mode phase velocities. A branch-line coupler may be a better choice under this situation [6]. When a branch-line coupler is designed for loose coupling ($|S_{31}| < -10$ dB), the characteristic impedance of the shunt branches becomes quite high. The impedances of the main-line and shunt branches of a 20-dB coupler are found to be nearly 50 and 500 Ω, respectively, using (8.8) and (8.9). However, a high impedance line of 500 Ω cannot be realized using microstrip. A similar problem is faced in the design of planar multisection branch-line couplers. The

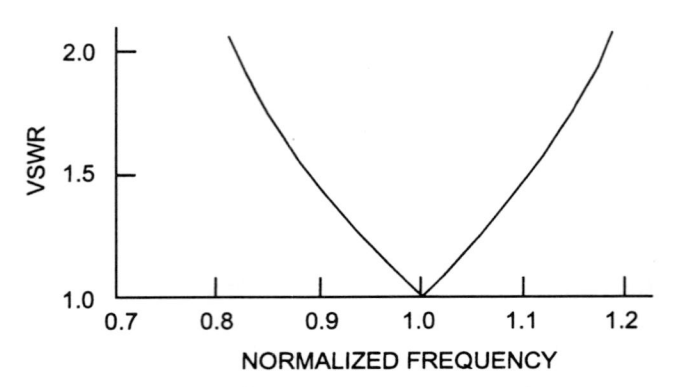

Figure 8.2 (a) Variation of VSWR with frequency of a 3-dB branch-line coupler.

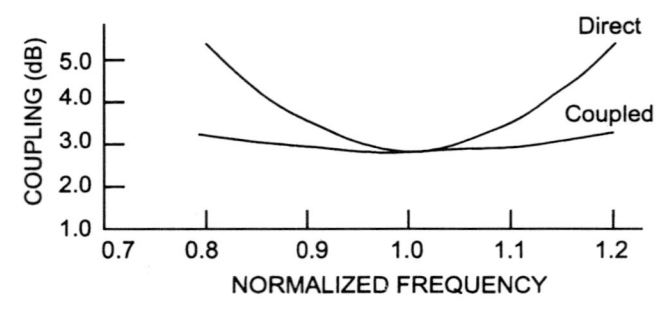

Figure 8.2 (b) Variation of coupling with frequency of a 3-dB branch-line coupler.

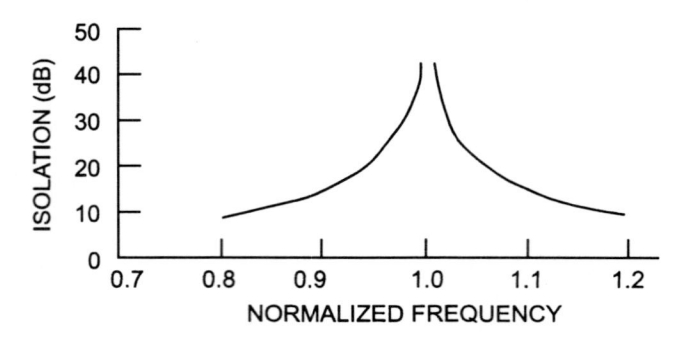

Figure 8.2 (c) Variation of isolation with frequency of a 3-dB branch-line coupler.

Tight Couplers 249

impedances of the end sections of a multisection branch-line coupler tend to be quite high.

A high-impedance transmission line section can be realized using coupled lines as shown in Figure 8.3(a). The electrical length of the coupled line is θ and each line is shorted at one of its ends. The equivalent circuit of the coupled lines between ports 1 and 2 is shown in Figure 8.3(b). When $\theta = 90$ degrees (which is equivalent to lines being $\lambda_g/4$ long), the shorting stubs of characteristic admittance Y_{0e} appear as open circuit across the main transmission line. Under this condition, the equivalent circuit further reduces to that as shown in Figure 8.3(c) where the characteristic impedance Z_c of the line is given by

$$Z_c = \frac{2 Z_{0e} Z_{0o}}{Z_{0e} - Z_{0o}} \tag{8.12}$$

Note, however, that the length of the equivalent transmission line is 270 degrees compared with 90 degrees for the coupled lines, where an additional 180-degrees phase is introduced by the short-circuited ends of the coupled-line section.

Equation 8.12 shows that very high values of Z_c can easily be obtained using coupled lines (by choosing $Z_{0e} \approx Z_{0o}$). In practice, it is feasible to obtain impedances above a certain value only. For example, for coupled lines spaced 1 mil on a 25-mil substrate, the lowest impedance Z_c that can be realized is about 115 to 120 Ω for a dielectric constant of 2. Similarly, on a substrate of dielectric constant 10, the lowest value of impedance Z_c that can be realized using coupled lines is about 70 Ω.

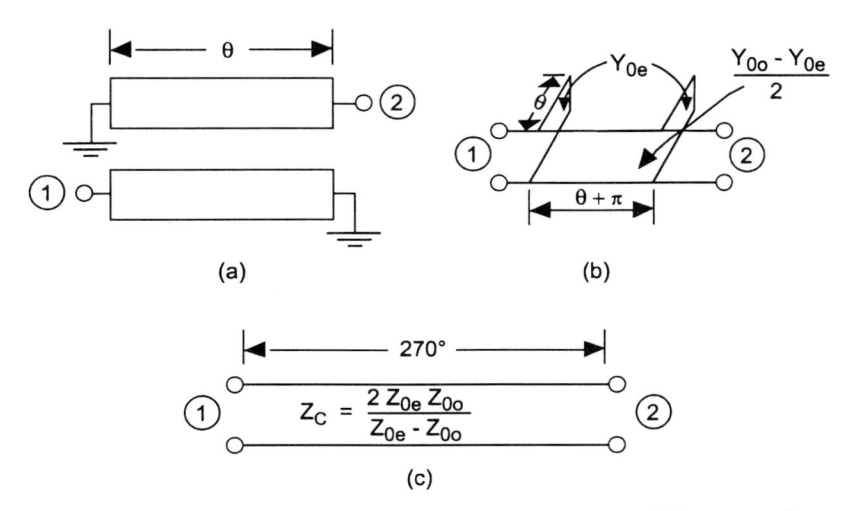

Figure 8.3 (a) Shorted coupled-line pair and (b) equivalent circuit. (c) Equivalent circuit of coupled-line pair of Figure 8.3(a) when $\theta = 90$ degrees.

A coupled section as shown in Figure 8.3(a) can therefore be used to replace the shunt branches of high-impedance values in a branch-line coupler designed for loose coupling. The modified branch-line coupler is shown in Figure 8.4. Although the equivalent length of the line realized using coupled lines is 180 degrees longer than that of the coupled lines, the performance of a modified branch-line coupler at the midband frequency is not affected because of this reason, in comparison to a conventional branch-line coupler. The frequency response of a modified branch-line coupler will be somewhat different from that of a conventional branch-line coupler, however, because the equivalent circuit of the coupled lines shown in Figure 8.3(c) will differ from that shown in Figure 8.3(b) when the value of θ is different from 90 degrees.

The simulated response of a 20-dB conventional branch-line coupler is shown in Figure 8.5(a). The isolation of an ideal conventional branch-line coupler is greater than 40 dB and 35 dB over frequency bandwidths of 12% and 20%, respectively. The directivity of the conventional design is, therefore, greater than 20 dB and 15 dB over a bandwidth of 12% and 20%, respectively. Figure 8.5(b) shows the simulated response of a modified branch-line coupler. In the modified design, the shunt branches of 500 Ω impedance have been replaced by coupled sections of even- and odd-mode impedances of 100 Ω and 71 Ω, respectively. When these values are substituted in (8.12), a characteristic impedance of 490 Ω is obtained, which is quite close to the desired value. The directivity of the modified branch-line coupler is greater than 20 dB and 15 dB over a bandwidth of 8% and 14%, respectively. Although, the bandwidth of a modified branch-line coupler is narrower than that of a conventional coupler, the modified design

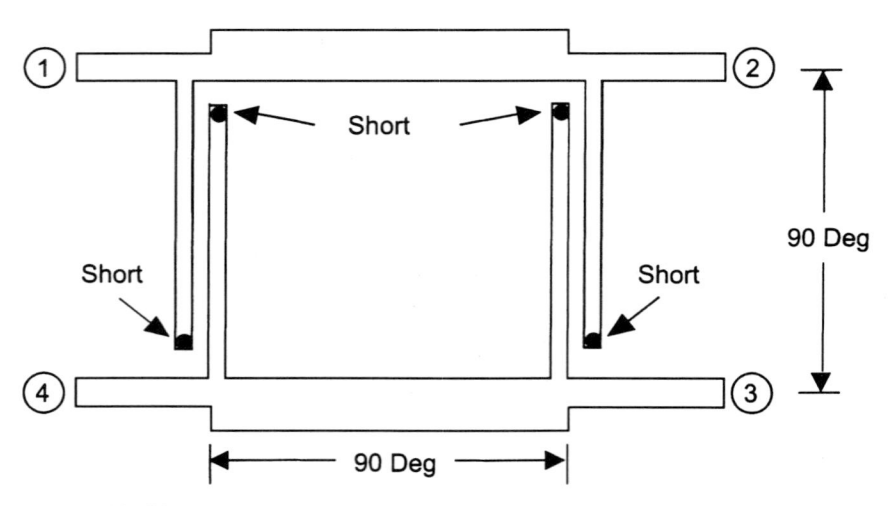

Figure 8.4 Modified branch-line coupler.

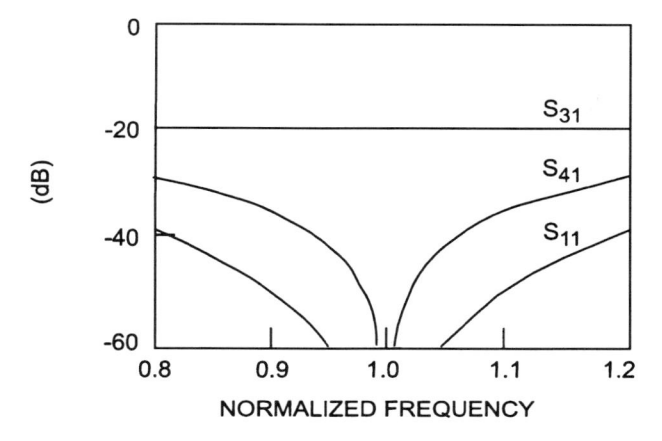

Figure 8.5 (a) Frequency response of a conventional 20-dB branch-line coupler shown in Figure 8.1 [6].

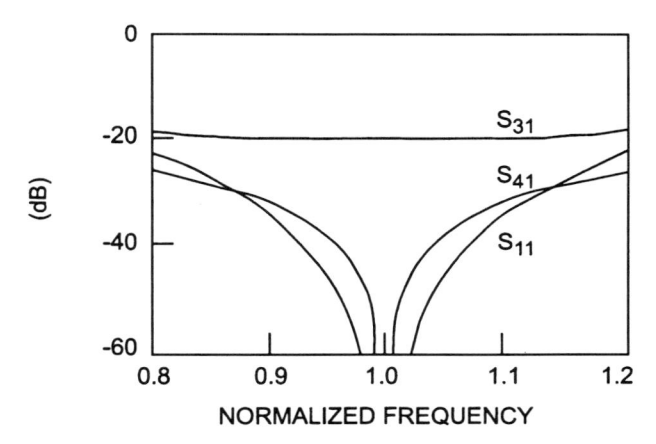

Figure 8.5 (b) Frequency response of modified 20-dB branch-line coupler shown in Figure 8.4 [6].

can be realized in microstrip configuration. For example, the coupled section can be realized by printing 5-mil-wide lines spaced 24 mil apart on a 25-mil alumina substrate. It has been reported that a simulated response of a modified branch-line coupler matches closely with experimental results [6].

8.2.2 Reduced-Size Branch-Line Coupler

In MMICs, lumped capacitors can be easily realized and have become attractive in reducing the size of passive components. Reduced-size branch-line hybrids that use only lumped capacitors and small sections of transmission lines (smaller than $\lambda_g/4$) have also been reported [7]. The size of these hybrids is about 80% smaller than those for conventional hybrids and is therefore quite suitable for MMICs. Reduced-size branch-line couplers are discussed in the following.

A transmission line section of impedance Z_c and electrical length of 90 degrees is shown in Figure 8.6(a). This section serves as a basic building block of a branch-line coupler. Its ABCD matrix is given by (8.13). The quarter-wave section can be replaced by a section shown in Figure 8.6(b), which comprises a transmission line of characteristic impedance Z and electrical length θ and shunt capacitances C at either end. By choosing $Z > Z_c$, the electrical length θ can be shorter than 90 degrees. The ABCD matrix of the circuit shown in Figure 8.6(a) is given by

$$\begin{bmatrix} A & B \\ C & D \end{bmatrix} = \begin{bmatrix} 0 & jZ_c \\ \frac{j}{Z_c} & 0 \end{bmatrix} \qquad (8.13)$$

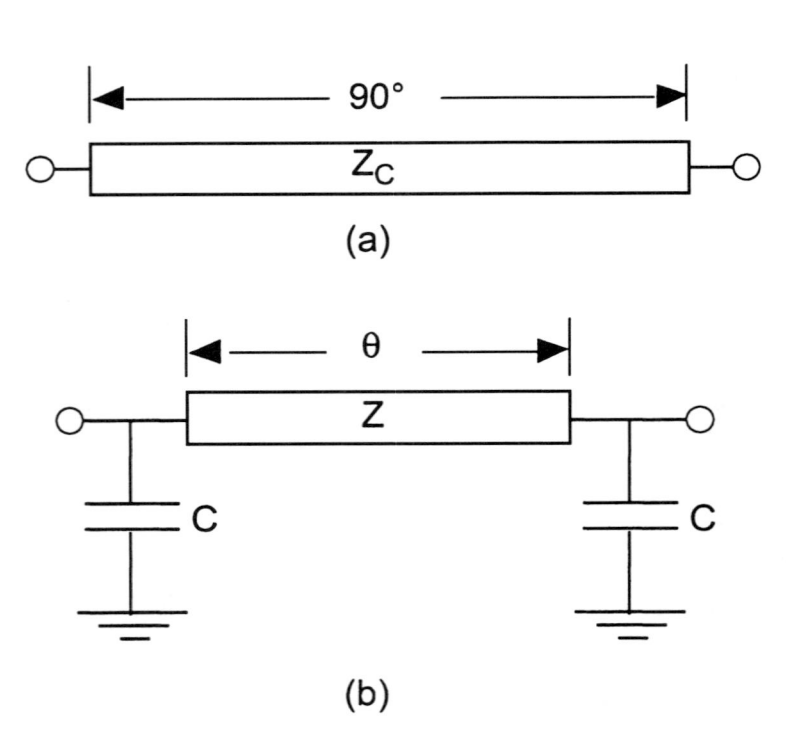

Figure 8.6 (a) Quarter-wave-long transmission section. (b) Reduced-size circuit equivalent to quarter-wave section.

On the other hand, the ABCD matrix of a lumped-element circuit as shown in Figure 8.6(b) is given by

$$\begin{bmatrix} A & B \\ C & D \end{bmatrix} = \begin{bmatrix} 1 & 0 \\ j\omega C & 1 \end{bmatrix} \begin{bmatrix} \cos\theta & jZ\sin\theta \\ j\frac{\sin\theta}{Z} & \cos\theta \end{bmatrix} \begin{bmatrix} 1 & 0 \\ j\omega C & 1 \end{bmatrix}$$

$$= \begin{bmatrix} \cos\theta - \omega CZ\sin\theta & jZ\sin\theta \\ j\frac{\sin\theta}{Z} + 2j\omega C\cos\theta - j(\omega C)^2 Z\sin\theta & \cos\theta - \omega CZ\sin\theta \end{bmatrix}$$
(8.14)

It can be easily shown that the above matrix becomes identical to (8.13) if the values of Z and C are chosen as follows:

$$Z = \frac{Z_c}{\sin\theta} \tag{8.15}$$

and

$$\omega C = \frac{\cos\theta}{Z_c} \tag{8.16}$$

where θ denotes the electrical length of the shortened transmission line shown in Figure 8.6(b).

For example, if the value of Z/Z_c is chosen to be equal to 2, then using (8.15), we obtain

$$\sin\theta = \frac{1}{2}, \quad \text{or } \theta = 30 \text{ degrees}$$

In other words, the length of the shortened transmission line becomes equal to $\lambda_g/12$, which is one-third of the usual quarter-wave section. The relationship between θ, Z/Z_c, and C is shown in Figure 8.7.

The characteristic impedances of the main-line and shunt branches of a conventional 3-dB branch-line coupler are $Z_0/\sqrt{2}$ and Z_0, respectively. If the characteristic impedances of the main-line and shunt branches are replaced by Z, where $Z > Z_0$, the lengths of the main-line and shunt branches become less than $\lambda_g/4$. More specifically, the lengths of the main-line (θ_1) and shunt (θ_2) branches and the value of capacitance C_b are given by

$$\theta_1 = \sin^{-1} y \tag{8.17}$$

$$\theta_2 = \sin^{-1} \frac{y}{\sqrt{2}} \tag{8.18}$$

$$\omega C_b Z_0 = \sqrt{1-y^2} + \sqrt{2-y^2} \tag{8.19}$$

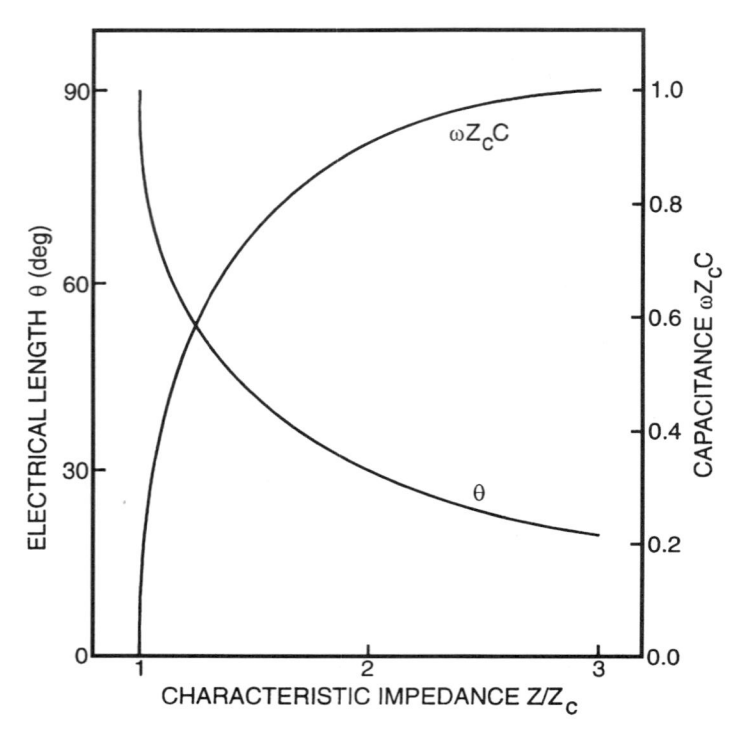

Figure 8.7 Relation between elements of Figure 8.6. (From [7], © 1990 IEEE, Reprinted with permission.)

where $y = Z_0/Z$. When $y = 1/\sqrt{2}$, or $Z = \sqrt{2}\, Z_0$, $\theta_1 = 45$ degrees and $\theta_2 = 30$ degrees. Therefore, if the main-line and shunt branch impedances of a 50 Ω branch-line coupler are chosen equal to 70.7 Ω, the lengths of the main-line and shunt branches become equal to $\lambda_g/8$ and $\lambda_g/12$, respectively. The value of C_b required can be determined using (8.19). The reduced-size branch-line hybrid is shown in Figure 8.8. The bandwidth of the reduced-size hybrid is a little wider than that of the purely lumped hybrid but is narrower than that of the conventional quarter-wavelength hybrid. The calculated phase difference between the signals at the direct (S_{21}) and coupled (S_{31}) ports is shown in Figure 8.9.

Reduced-size hybrids can be easily implemented in MMICs where the coplanar line is used as the principal transmission line medium. Because the center and ground conductors of a coplanar transmission line lie on the same side of the MMIC substrate, lumped capacitors can be easily fabricated or added.

A photograph of a reduced-size 25-GHz branch-line hybrid is shown in Figure 8.10(a). The circuit is fabricated on a GaAs substrate. All the transmission lines are 70 Ω coplanar waveguides with a 10-μm center conductor width. The

Tight Couplers 255

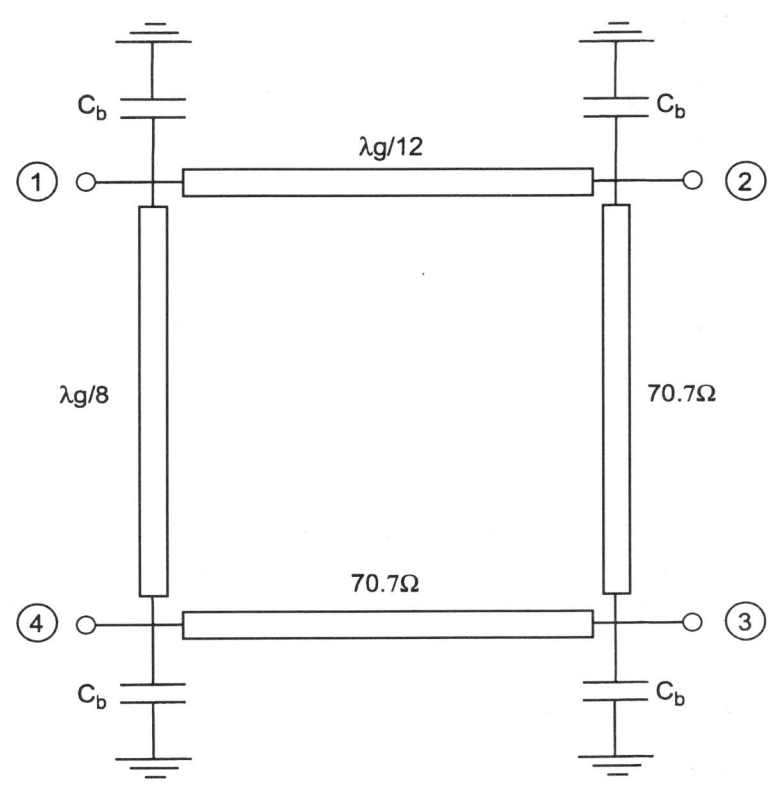

Figure 8.8 Reduced-size branch-line hybrid.

lengths of the main-line and shunt branches are $\lambda_g/8$ and $\lambda_g/12$, respectively. Metal-insulator-metal (MIM) shunt capacitors are located at the four T-junctions and between the inner conductors and the ground metal. The size of the overall hybrid is 500 μm × 500 μm, representing an 80% savings over a conventional branch line hybrid. Its measured performance is shown in Figure 8.10(b).

8.2.3 Lumped-Element Branch-Line Coupler

The size of a branch-line coupler using sections of transmission lines becomes quite large at frequencies below about 2 GHz. At these frequencies, a branch line can be realized using lumped inductor and capacitor elements. The lumped-element 90-degrees hybrid can be realized either in a "pi" or a "tee" equivalent network. In MMICs a "pi" network is preferred to "tee" because it uses fewer inductive elements which have lower Q and occupy more space.

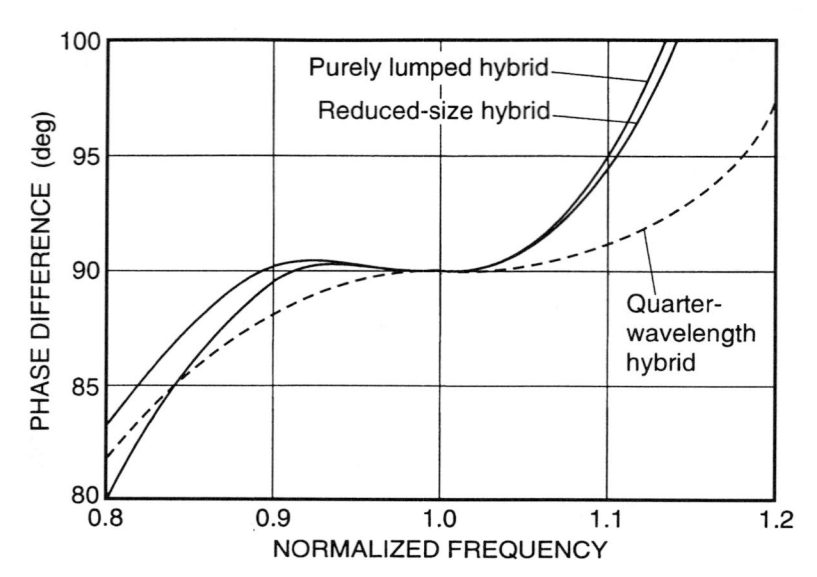

Figure 8.9 Calculated phase difference between S_{21} and S_{31} of the reduced-size hybrid, the conventional hybrid, and the purely lumped-element hybrid. (From [7], © 1990 IEEE. Reprinted with permission.)

In lumped-element implementation, basically each transmission line section shown in Figure 8.11(a) is replaced by an equivalent "pi" lumped-element network as shown in Figure 8.11(b). The values of the lumped elements are obtained by equating ABCD matrix parameters for both these structures. The ABCD matrix of a lossless transmission line section of characteristic impedance Z_c and electrical length θ is given by

$$\begin{bmatrix} A & B \\ C & D \end{bmatrix} = \begin{bmatrix} \cos\theta & jZ_c \sin\theta \\ j\frac{\sin\theta}{Z_c} & \cos\theta \end{bmatrix} \tag{8.20}$$

On the other hand, the ABCD matrix of a lumped-element circuit as shown in Figure 8.11(b) is given by

$$\begin{bmatrix} A & B \\ C & D \end{bmatrix} = \begin{bmatrix} 1 & 0 \\ j\omega C & 1 \end{bmatrix} \begin{bmatrix} 1 & j\omega L \\ 0 & 1 \end{bmatrix} \begin{bmatrix} 1 & 0 \\ j\omega C & 1 \end{bmatrix}$$

$$= \begin{bmatrix} 1 - \omega^2 LC & j\omega L \\ 2j\omega C - j\omega^3 LC^2 & 1 - \omega^2 LC \end{bmatrix} \tag{8.21}$$

Tight Couplers 257

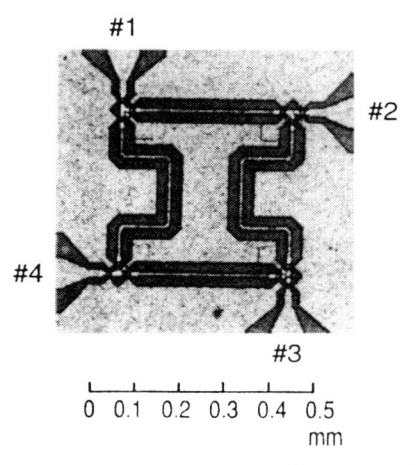

Figure 8.10 (a) Photomicrograph of the fabricated 25-GHz reduced-size branch-line hybrid. (From [7], © 1990 IEEE. Reprinted with permission.)

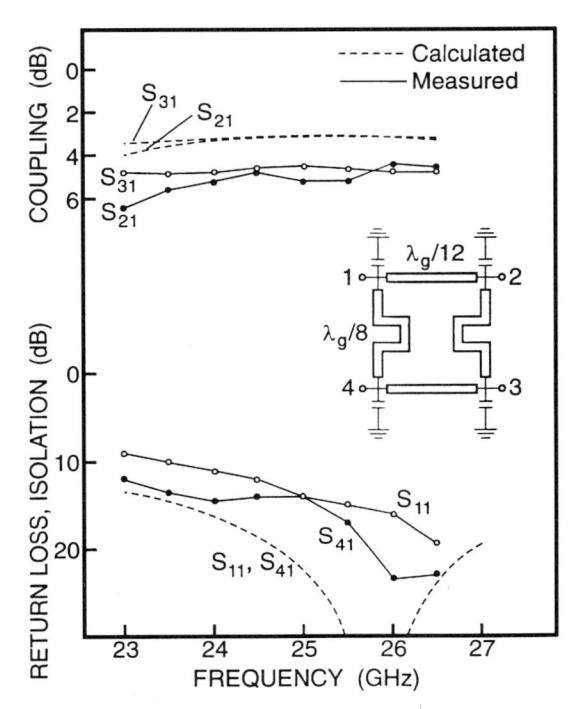

Figure 8.10 (b) Measured performance of the 25-GHz reduced-size branch-line hybrid [7].

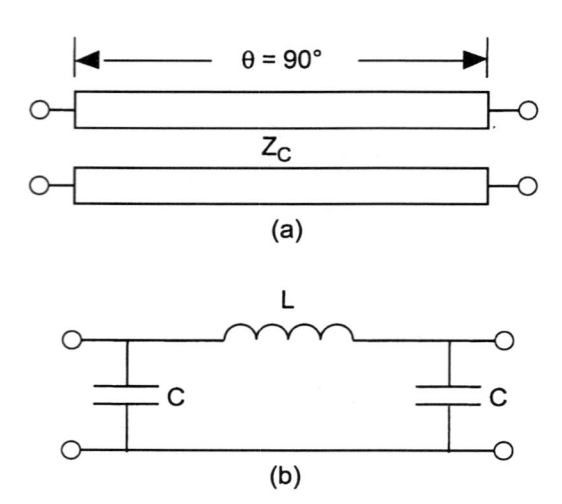

Figure 8.11 (a) Quarter-wave section and (b) its lumped equivalent.

where ω denotes the radian frequency. By equating the matrix elements in (8.20) and (8.21) and simplifying, we achieve

$$L = \frac{Z_c \sin\theta}{\omega} \qquad (8.22a)$$

$$C = \frac{1}{\omega Z_c}\sqrt{\frac{1-\cos\theta}{1+\cos\theta}} \qquad (8.22b)$$

The lumped circuit shown is therefore equivalent to a quarter-wave-long transmission line of characteristic impedance Z_c, if the elements of the lumped circuit are chosen as follows:

$$L = \frac{Z_c}{\omega}$$

$$C = \frac{1}{Z_c\omega}$$

Note that circuits shown in Figure 8.11(a) and (b) have identical response at the frequency at which the transmission line section is quarter-wave long. At other frequencies, the response of both the circuits will be different in general. Therefore, the bandwidth of circuits realized using transmission line sections will be different from those realized using lumped elements.

A branch-line coupler as shown in Figure 8.1 uses quarter-wave transmission line sections of characteristic impedances Z_{0s} and Z_{0p}. By replacing these sections with equivalent lumped elements, the circuit as shown in Figure 8.12 is

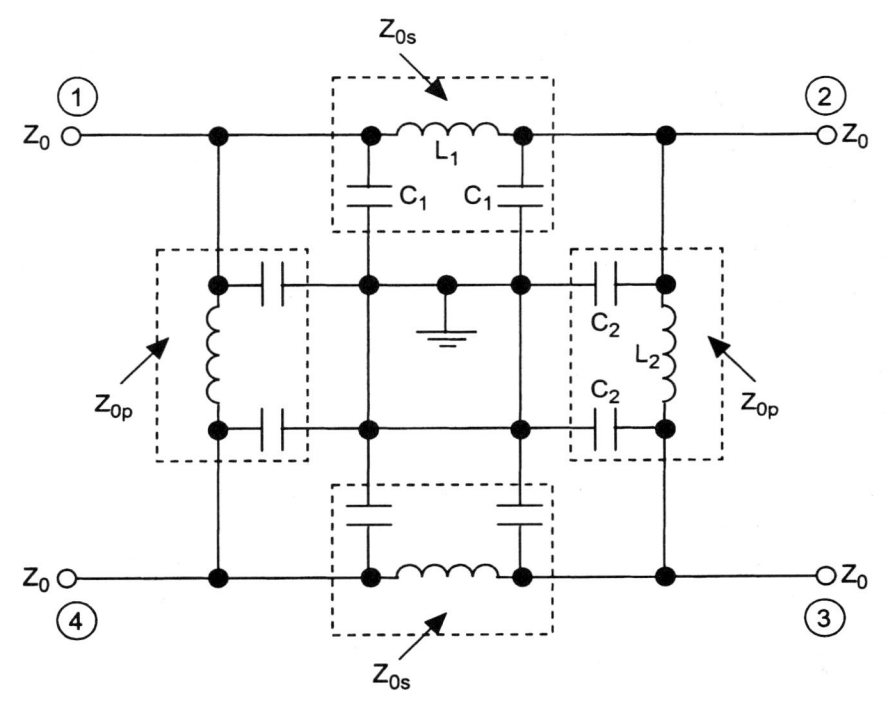

Figure 8.12 Lumped-element equivalent circuit model for the 90-deg hybrid shown in Figure 8.1.

obtained. The values of lumped inductive and capacitive elements are given by

$$L_1 = \frac{Z_{0s}}{\omega_0}, \quad C_1 = \frac{1}{Z_{0s}\omega_0} \tag{8.23}$$

and

$$L_2 = \frac{Z_{0p}}{\omega_0}, \quad C_2 = \frac{1}{Z_{0p}\omega_0} \tag{8.24}$$

where ω_0 denotes the radian frequency corresponding to the center frequency of the branch-line coupler. More specifically, the element values of a lumped 3-dB branch-line coupler are given by

$$L_1 = \frac{Z_0}{\sqrt{2}\omega_0}, \quad C_1 = \frac{\sqrt{2}}{Z_0\omega_0} \tag{8.25}$$

and

$$L_2 = \frac{Z_0}{\omega_0}, \quad C_2 = \frac{1}{Z_0\omega_0} \tag{8.26}$$

where Z_0 denotes the terminal impedance of the ports of the branch-line coupler.

Typical lumped-element values for a 900-MHz coupler designed for 50 Ω terminal impedance are $L_1 = 6.3$ nH, $L_2 = 8.8$ nH, and $C_t = 8.5$ pF, where $C_t = C_1 + C_2$. Over 900 ± 45 MHz the calculated value of amplitude unbalance and the phase difference between the output ports is ± 0.2 dB and 90 ± 2 degrees, respectively. The bandwidth of these couplers can be increased by using more sections of "pi" or "tee" equivalent networks; that is, two sections of 45 degrees or three sections of 30 degrees to realize a 90-degree section, and so on, or by properly selecting highpass and lowpass networks [8, 9]. In general, two to three sections are sufficient to realize a broadband 90-degree hybrid.

8.2.4 Broadband Branch-Line Coupler

As already discussed, the bandwidth of a branch-line coupler is quite narrow. To increase its bandwidth, a multisection branch-line coupler can be used. The synthesis of multisection branch-line couplers has been described by Levy and Lind [10]. The synthesis has been described for both maximally flat and Chebyshev type of response of VSWR and directivity of the coupler. The similar techniques can also be applied to reduced-size and lumped-element couplers [11].

8.3 Rat-Race Coupler

The strip conductor layout of a rat-race coupler in microstrip form is shown in Figure 8.13. In the case of a branch-line coupler as shown in Figure 8.1, the spacing between all adjacent ports is $\lambda_g/4$. In the case of a rat-race coupler, however, the spacing between two of the adjacent ports is $3\lambda_g/4$ and $\lambda_g/4$ between all other adjacent ports. By properly choosing the values of Z_{01} and Z_{02}, the circuit can be made to operate like a directional coupler. At the center frequency the scattering parameters of a rat-race coupler are given by

$$S_{21} = -j\frac{Z_0}{Z_{02}} \tag{8.27}$$

$$S_{41} = j\frac{Z_0}{Z_{01}} \tag{8.28}$$

$$S_{11} = S_{31} = 0 \tag{8.29}$$

Further,

$$S_{32} = -j\frac{Z_0}{Z_{01}} \tag{8.30}$$

and

$$S_{42} = 0 \tag{8.31}$$

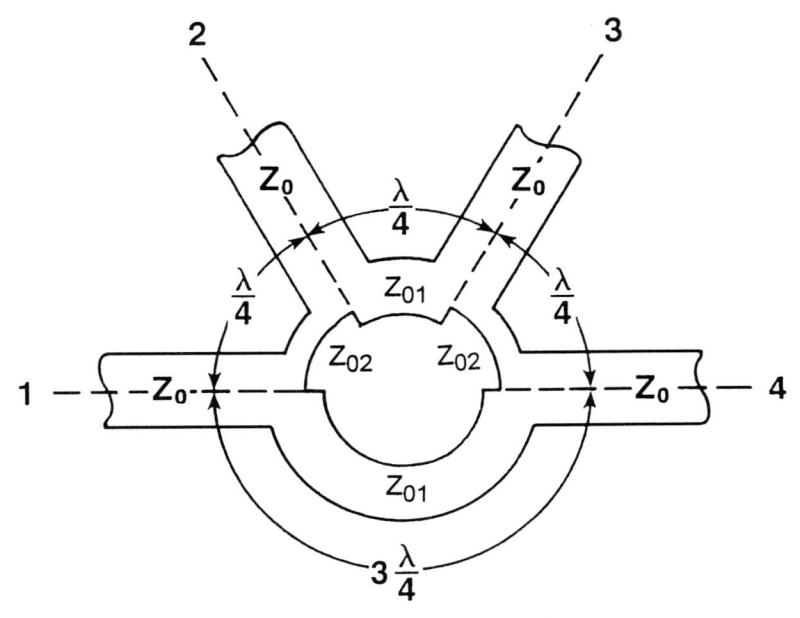

Figure 8.13 Layout of a rat-race coupler in planar circuit configuration.

The impedances Z_{01} and Z_{02} should satisfy the condition:

$$\frac{Z_0^2}{Z_{01}^2} + \frac{Z_0^2}{Z_{02}^2} = 1 \tag{8.32}$$

The above equation follows from the following condition that needs to be satisfied because of the principle of conservation of power:

$$|S_{21}|^2 + |S_{41}|^2 = 1 \tag{8.33}$$

where it is assumed that $S_{11} = S_{31} = 0$. The complete scattering matrix of a rat-race coupler at the midband frequency can be expressed as

$$[S] = \begin{bmatrix} 0 & -j\frac{Z_0}{Z_{02}} & 0 & j\frac{Z_0}{Z_{01}} \\ -j\frac{Z_0}{Z_{02}} & 0 & -j\frac{Z_0}{Z_{01}} & 0 \\ 0 & -j\frac{Z_0}{Z_{01}} & 0 & -j\frac{Z_0}{Z_{02}} \\ j\frac{Z_0}{Z_{01}} & 0 & -j\frac{Z_0}{Z_{02}} & 0 \end{bmatrix} \tag{8.34}$$

The scattering matrix of a rat-race coupler makes it useful in many applications. For example, consider a rat-race coupler designed for equal power division between two coupled ports. When an input signal is incident at port 2, the power is equally divided between ports 1 and 3 and no power reaches port 4. The signals arriving at ports 1 and 3 are in phase. When power is incident at port 1, no power is coupled to port 3 and the power is equally divided between ports 2 and 4 as in the previous case. In this case, however, the signals arriving at ports 2 and 4 are out of phase (i.e., 180-degree phase difference). This special property makes the rat-race useful in applications, such as in balanced mixers, where the effect of local oscillator noise can be canceled by connecting the rf signal to port 3 and local oscillator to port 1. A 3-dB rat-race coupler is also known as a 180-deg hybrid.

Design of a Rat-Race Coupler

From the scattering matrix of (8.34), we have

$$Z_{02} = \frac{Z_0}{|S_{21}|} = \frac{Z_0}{\sqrt{1-|S_{41}|^2}} \tag{8.35}$$

and

$$Z_{01} = \frac{Z_0}{|S_{41}|} = \frac{Z_0}{\sqrt{1-|S_{21}|^2}}$$

Example 8.2

Compute the characteristic impedances of the various sections of a rat-race coupler shown in Figure 8.13, given $Z_0 = 50\ \Omega$ and the coupling coefficient from port 1 to ports 2 and 3 equals ($S_{21} = S_{41} = -3.01$ dB).

Given

$$|S_{21}| = |S_{41}| = 10^{-3.01/20} = 0.707 \tag{8.36}$$

Using (8.35) and (8.36), we then obtain

$$Z_{01} = Z_0/0.707 = 70.7\ \Omega$$

and

$$Z_{02} = Z_0/0.707 = 70.7\ \Omega$$

Properties of the Rat-Race Coupler

A rat-race coupler has disadvantages similar to that of a branch-line coupler. For example, the ports of a rat-race are matched only at the center frequency. Further, there is complete isolation between decoupled ports (such as between port 1 and port 4 in Figure 8.13) at the center frequency only. Furthermore, the phase difference of 180 deg between signals arriving at ports 2 and 3 exists only at the center frequency. The scattering parameters of a 3-dB rat-race coupler as a function of frequency are shown in Figure. 8.14. Here, port 1 is the input, and ports 2 and 4 are the direct and coupled ports, respectively. The rat-race coupler finds use in applications where the frequency bandwidth requirement is less than about 20%.

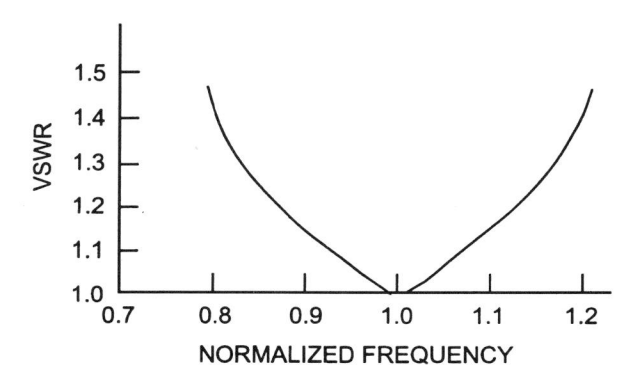

Figure 8.14 (a) Variation of VSWR with frequency of a 3-dB rat-race coupler.

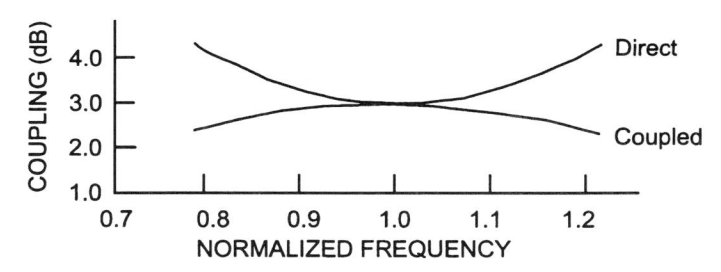

Figure 8.14 (b) Variation of coupling with frequency of a 3-dB rat-race coupler.

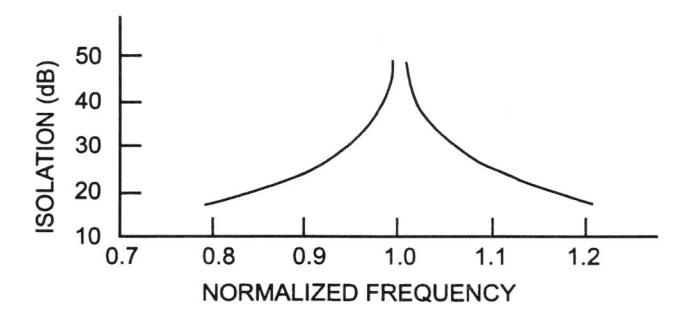

Figure 8.14 (c) Variation of isolation with frequency of a 3-dB rat-race coupler.

Because of physical discontinuities at the locations of junctions of a rat race, its performance deviates from that of an ideal rat race. The effect of these discontinuities must be considered in the design, especially at high frequencies.

8.3.1 Modified Rat-Race Coupler

The bandwidth of a rat-race coupler is only about 20% but can be increased by using multisections or significantly increased (to about an octave) by replacing the $3\lambda_g/4$ section by an equivalent coupled-line section such as shown in Figure 8.15. The coupled-line section of electrical length 90 deg is equivalent to a single transmission line of length 270 deg as described in Section 8.2.1. The three-quarter-wave-long section of a rat-race coupler (of impedance Z_{01}) can therefore be replaced by a quarter-wave-long coupled-line section. In the modified coupler, an equivalent impedance of Z_{01} can be obtained using coupled lines if the even-

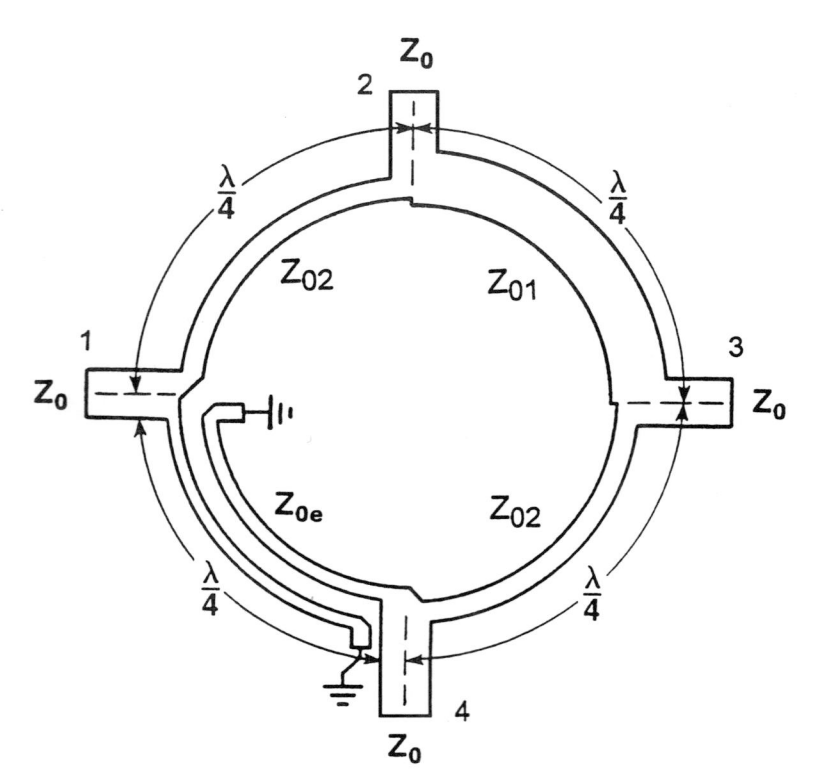

Figure 8.15 Broadband hybrid ring using a shorted parallel-coupled quarter-wave section.

and odd-mode impedances of the coupled section are chosen as follows:

$$Z_{0e} = 2.414 Z_{01} \tag{8.37}$$

$$Z_{0o} = \frac{Z_{01}}{2.414} \tag{8.38}$$

Using the above information, a broadband rat race was simulated. The performance of a broadband rat race is compared with a conventional rat race in Figure 8.16. Bandwidth improvement is quite obvious in these plots. Another advantage of the broadband rat race is that its size is smaller compared to that of a conventional rat race.

8.3.2 Reduced-Size Rat-Race Coupler

Rat-race hybrids (3-dB rat-race couplers) have unique applications in microwave circuits because they can divide an input signal into signals that are either in phase or out of phase. They can also be used to combine two signals (say, A and B) to obtain a "sum" (A + B) signal or a "difference" (A − B) signal. The size of a rat-race hybrid can also be reduced by employing techniques discussed in Section 8.2.2. A conventional 3-dB rat-race coupler is shown in Figure 8.13. It uses a transmission line section of length $3\lambda_g/4$ and three sections of length $\lambda_g/4$ each. The characteristic impedance of each section is $\sqrt{2}Z_0$. The transmission line section of length $3\lambda_g/4$ can be replaced by a lumped-equivalent circuit as shown in Figure 8.17(a). Further, the $\lambda_g/4$ sections of characteristic impedance $\sqrt{2}Z_0$ can be replaced by transmission line sections of characteristic impedance $2Z_0$ and length $\lambda_g/8$ with shunt capacitances at the two ends as shown in Figure 8.17(b). The resulting reduced-size hybrid is shown in Figure 8.18(a). (In this figure, the port labeling is different from that shown in Figure 8.13.)

In the reduced-size rat-race hybrid of Figure 8.18(a), note that parallel LC elements (L_a and C_b) between port 1 and the ground (and similarly between port 2 and ground) offer a relatively high shunt impedance at the center frequency. These elements can therefore be removed as shown in Figure 8.18(b) without significantly affecting the response of the circuit. Alternatively, to compensate for the effect of removal of these elements, the impedances of the transmission line sections, their length, and the values of other capacitances can be optimized using CAD tools.

Another advantage of the reduced-size rat-race hybrid is its flexible port arrangement. The usual port arrangement is shown in Figure 8.19(a). In this arrangement, the right edge of the bottom plate of the metal-insulator-metal (MIM) capacitor is connected to port 1 and the left edge of the top plate is

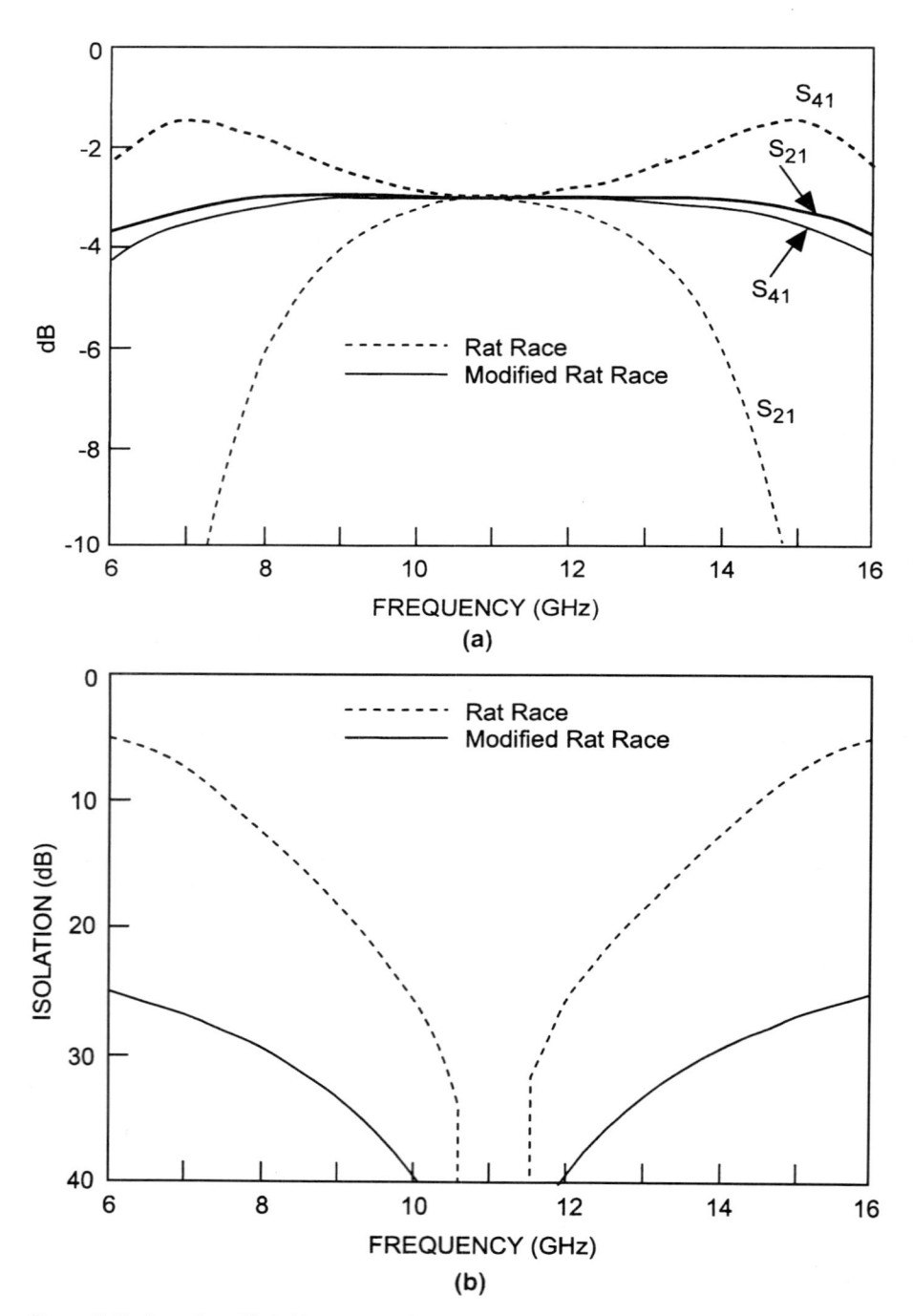

Figure 8.16 Broadband hybrid response: (a) coupling, (b) isolation, and (c) return loss.

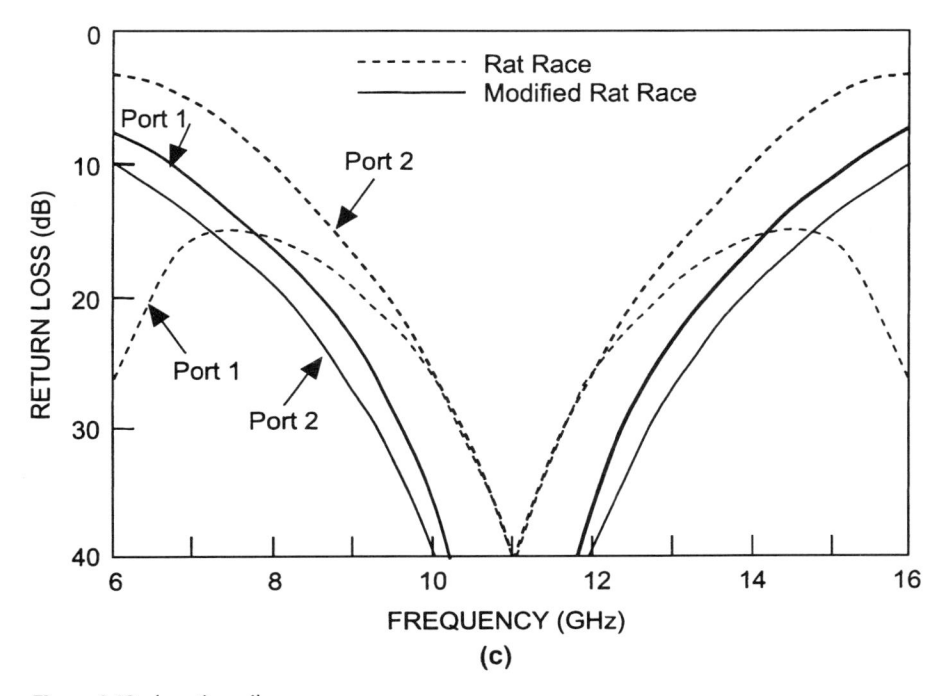

Figure 8.16 *(continued)*

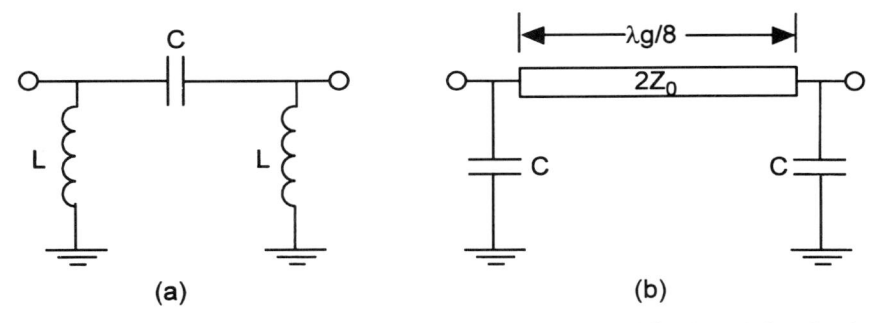

Figure 8.17 (a) Lumped-element circuit to replace $3\lambda_g/4$ section; (b) reduced-size circuit to replace $\lambda_g/4$ section of a rat-race hybrid.

connected to port 2. Port 1 can be connected to the left edge of the bottom plate without affecting the performance because the whole bottom plate is at the same potential. Similarly, port 2 can be connected to the right edge of the top plate. The resulting layout is shown in Figure 8.19(b). This layout is more convenient for mixer applications.

Figure 8.20(a) shows a photograph of a 25-GHz reduced-size rat-race hybrid [7]. The circuit consists of a 100 Ω coplanar waveguide and MIM capacitors.

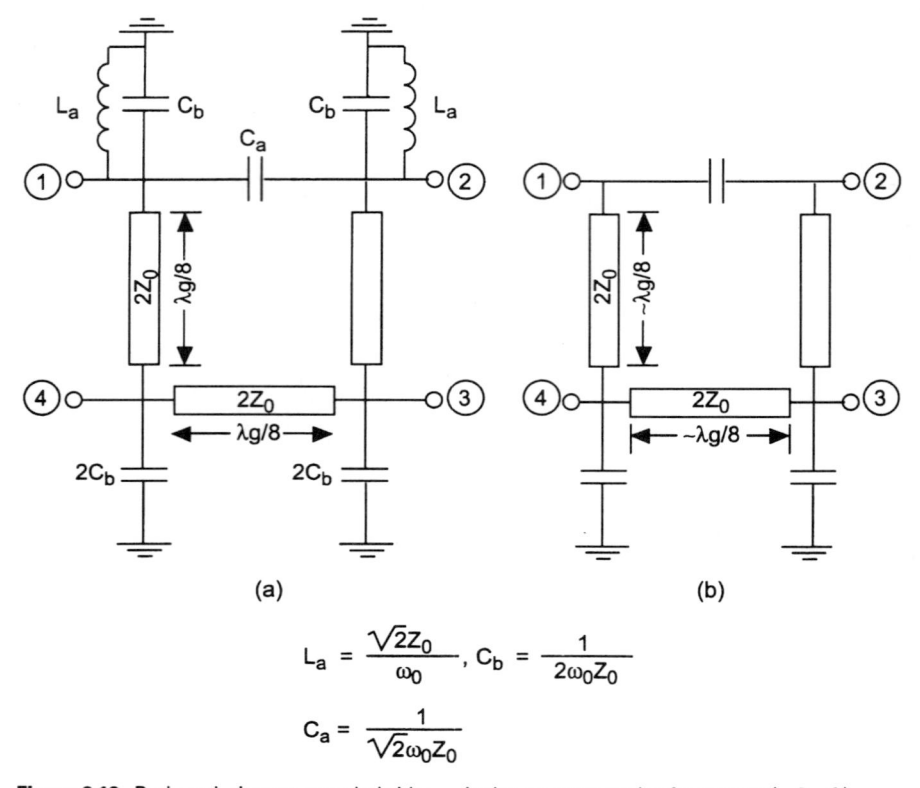

Figure 8.18 Reduced-size rat-race hybrids; ω_0 is the center angular frequency ($=2\pi f_0$).

$$L_a = \frac{\sqrt{2}Z_0}{\omega_0}, \quad C_b = \frac{1}{2\omega_0 Z_0}$$

$$C_a = \frac{1}{\sqrt{2}\omega_0 Z_0}$$

The center conductor width of the coplanar waveguide is 10 μm. The measured results of this hybrid are shown in Figure 8.20(b). Note that the insertion loss of the rat-race hybrid is much smaller than that of the branch-line coupler shown in Figure 8.10(b).

A very small, low-loss MMIC rat-race hybrid using elevated coplanar waveguides has been reported by Kamitsuna [12]. A 15-GHz rat race was developed on a chip size of 0.5 × 0.55 mm.

8.3.3 Lumped-Element Rat-Race Coupler

The design of lumped-element rat-race hybrids (Figure 8.21(a)) is similar to that of a lumped-element 90-degree hybrid described in the previous section. A lumped element equivalent circuit model for the 180-deg hybrid is shown in Figure 8.21(b). The 90-degree sections are replaced with a lowpass "pi" network as shown in Figure 8.11 and the 270-degree (or 90-degree) section is replaced with an equivalent highpass "tee" network [13] shown in Figure 8.21(a).

Tight Couplers 269

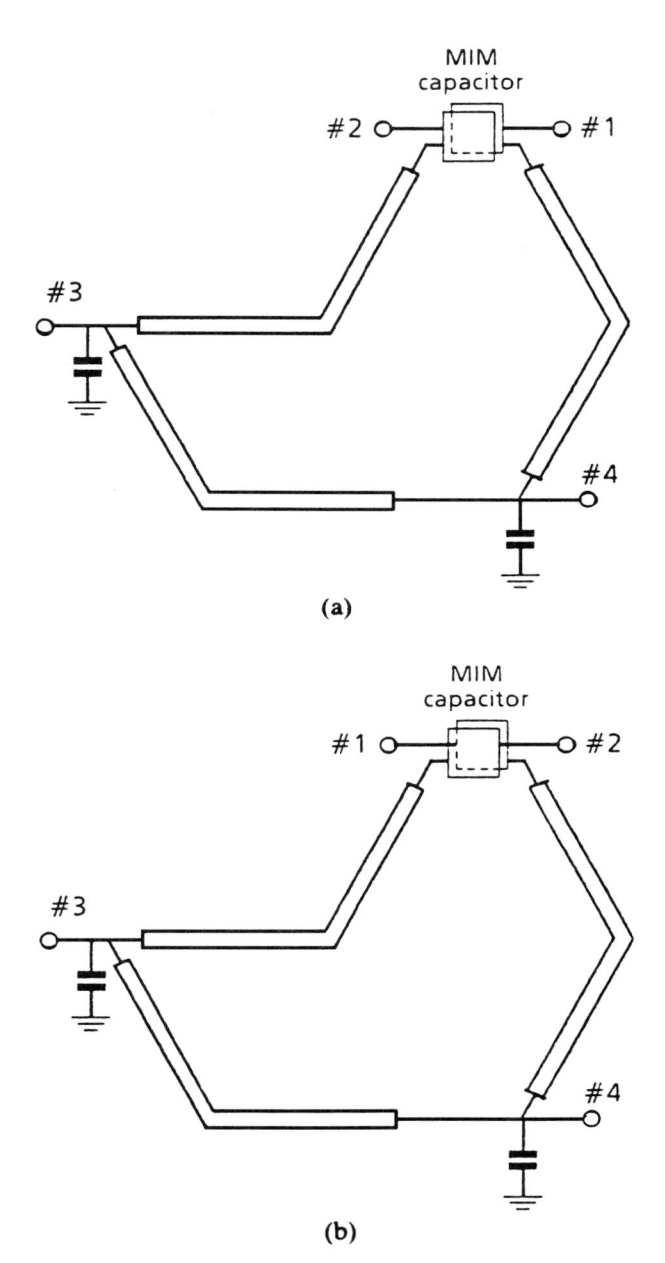

Figure 8.19 Port interchange in the reduced-size rat-race hybrid: (a) usual port layout; (b) port layout convenient for mixer applications. (From [7], © 1990 IEEE. Reprinted with permission.)

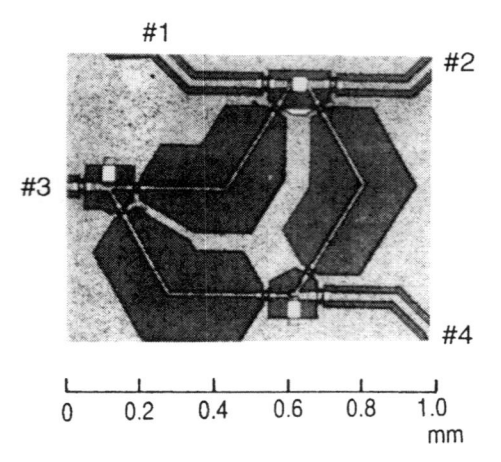

Figure 8.20 (a) Photomicrograph of the fabricated 25-GHz reduced-size rat-race hybrid. (From [7], © 1990 IEEE. Reprinted with permission.)

Following the same procedure as described for the 90-degree hybrid, the lumped elements for the "pi" section and the "tee" can be expressed as

$$L_1 = \frac{\sqrt{2}Z_0 \sin \theta_1}{\omega} \tag{8.39a}$$

$$C_1 = \frac{1}{\sqrt{2}Z_0\omega}\sqrt{\frac{1-\cos\theta_1}{1+\cos\theta_1}} \tag{8.39b}$$

$$L_2 = -\frac{\sqrt{2}Z_0}{\omega \sin \theta_2} \tag{8.40a}$$

$$C_2 = \frac{1}{\sqrt{2}Z_0\omega}\sqrt{\frac{1+\cos\theta_2}{1-\cos\theta_2}} \tag{8.40b}$$

When $\theta_1 = 90$ degree and $\theta_2 = 270$ degree or -90 degree, element values for a 50 Ω system become $L_1 = L_2 = 11.25/f$ nH and $C_1 = C_2 = 2.25/f$ pF, where f is in GHz.

8.4 Multiconductor Directional Couplers

The coupling between two TEM lines increases as the mutual capacitance between them is increased. The mutual capacitance between the lines can be increased by decreasing the spacing between them. However, to obtain tight coupling values such as 3 dB between planar transmission lines such as microstrip lines, the spacing between the lines becomes too small to be realized in

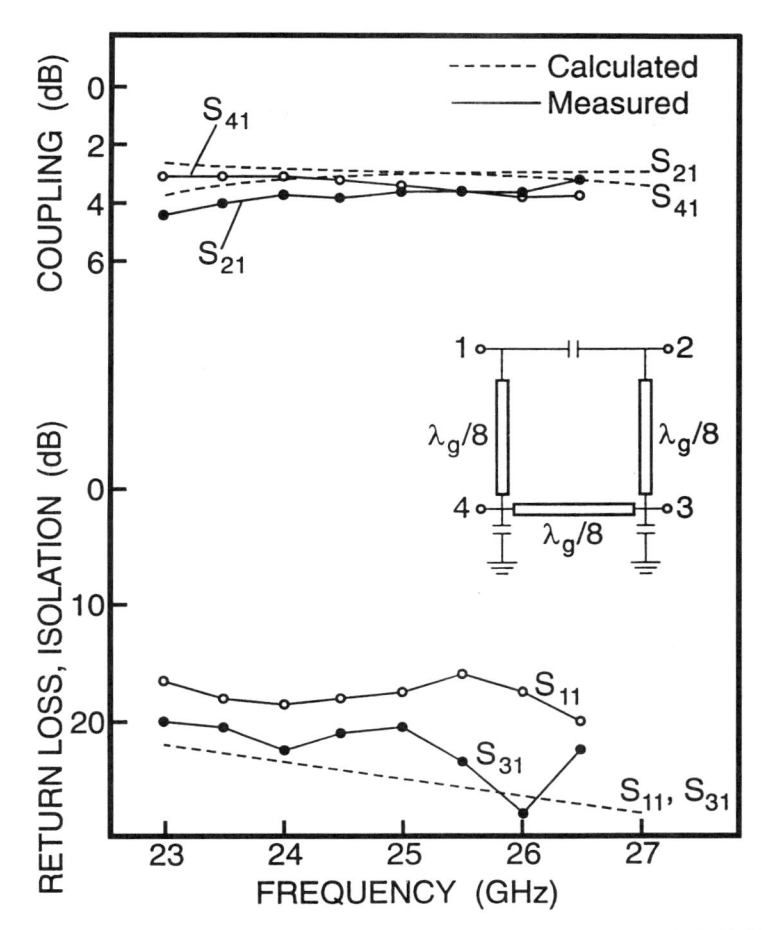

Figure 8.20 (b) Measured performance of the 25-GHz reduced-size rat-race hybrid. (From [7], © 1990 IEEE. Reprinted with permission.)

a convenient manner using commonly used photolithographic techniques. Further, very small transverse dimensions lead to increased conductor loss. Lange [14] described a scheme using multiconductors in interdigital configuration in which the mutual capacitance between the lines can be increased without the need for a small spacing between them. This section describes the theory and design of interdigital couplers [14–19].

8.4.1 Theory of Interdigital Couplers

Figure 8.22 shows two coupled lines and their capacitances. C_s and C_{mu} denote the capacitances of lines (per-unit length) with respect to the ground and the

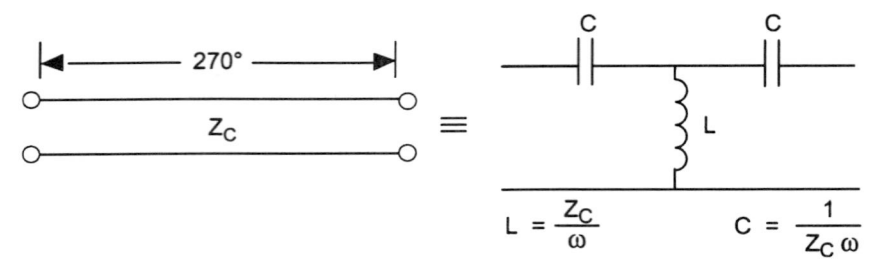

Figure 8.21 (a) Equivalent lumped circuit of a transmission line section of length $3\lambda_g/4$.

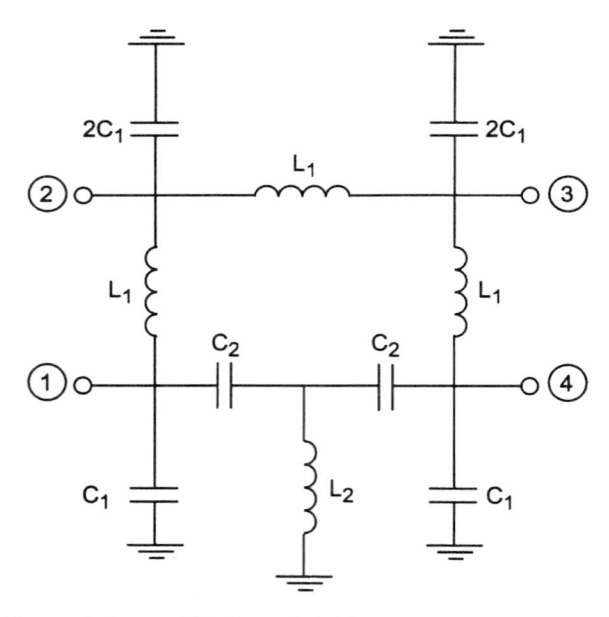

$$L = \frac{Z_C}{\omega} \qquad C = \frac{1}{Z_C \omega}$$

Figure 8.21 (b) Lumped-element 180-degree hybrid.

mutual capacitance between them, respectively. To increase the mutual capacitance, the two lines can be divided into two lines of half-width each and arranged in an interdigital manner as shown in Figure 8.23(a) with alternate conductors connected together at the ends. The spacing between adjacent conductors remains the same as in the undivided case shown in Figure 8.22(a). The capacitance of each divided line with respect to the ground is approximately $C_s/2$ (because of half-width), whereas the mutual capacitance between adjacent lines is somewhat smaller than between the lines in the configuration shown in Figure 8.22(a). The total capacitance of each line with respect to the ground is therefore approximately C_s as in the original case. However, the mutual capacitances between neighboring conductors add together to give an overall value

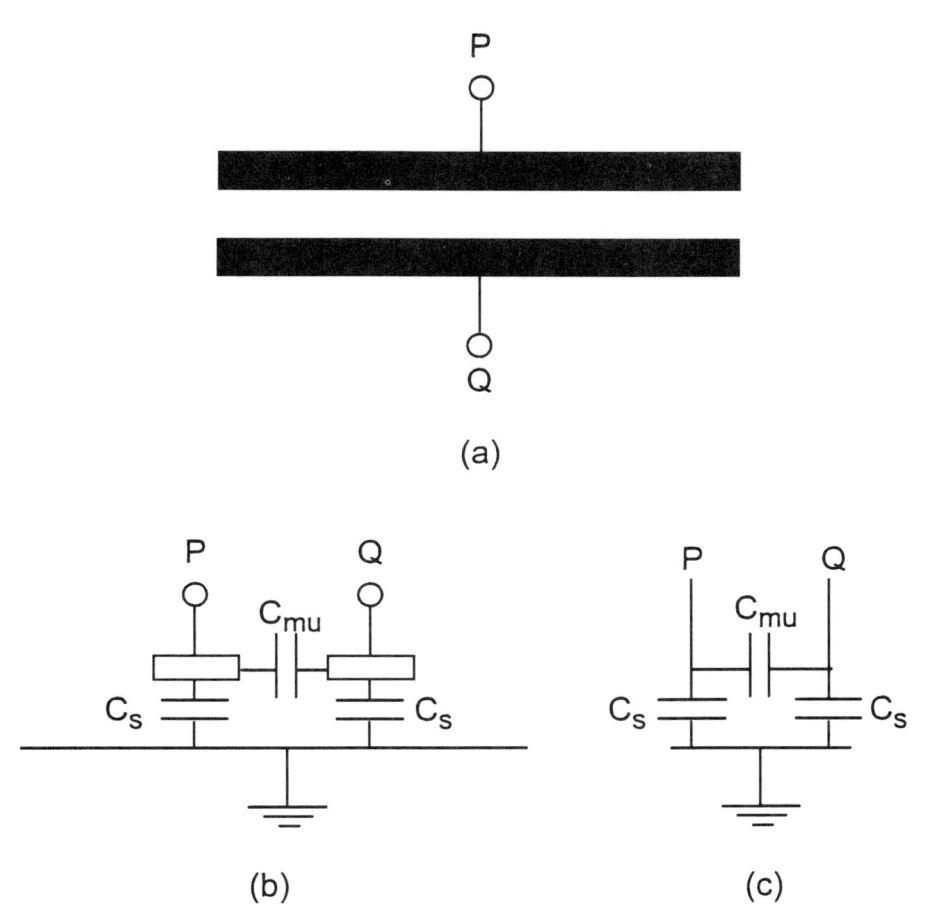

Figure 8.22 (a) Top view of coupling between parallel TEM lines, (b) side view, and (c) equivalent capacitance network showing static per-unit-length self-capacitance of the lines and mutual capacitance between them.

of mutual capacitance that is much larger than between the lines in the configuration shown in Figure 8.22(a). This is essentially the basis of a Lange coupler. The equivalent capacitances of the interdigital configuration are shown in Figure 8.23(b) and (c).

Figure 8.24 shows how a coupler consisting of two parallel-coupled lines can be rearranged. For the same spacing between adjacent lines, the coupler configurations shown in Figure 8.24(b) and (c) offer larger coupling than the configuration shown in Figure 8.24(a). The configurations shown in Figure 8.24(b) and (c) are identical electrically, but the configuration shown in Figure 8.24(c), which is also known as a Lange coupler, offers an extra advantage in

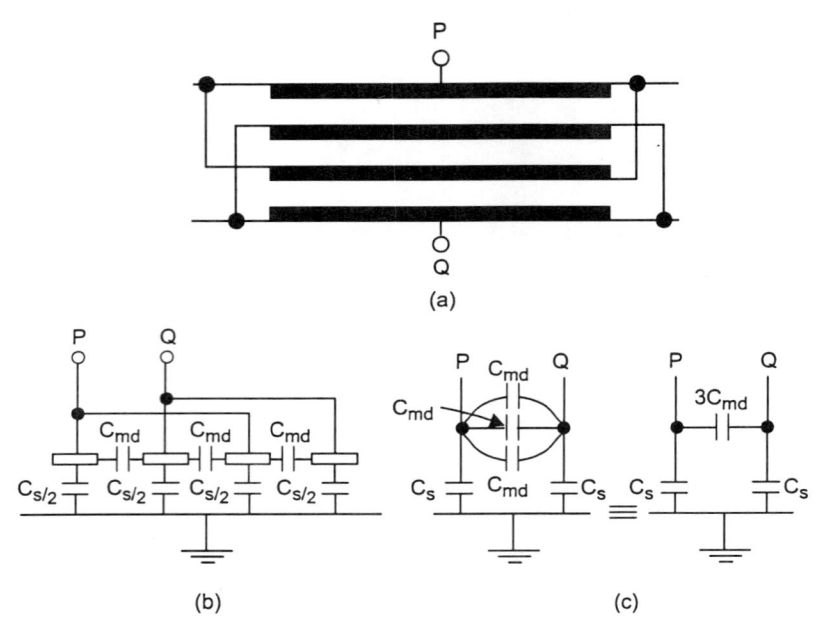

Figure 8.23 (a) Top view of an interdigital coupler, (b) side view, and (c) equivalent capacitance network showing static per-unit-length self-capacitance of the lines and mutual capacitance.

that both the direct and coupled ports are on the same side. Figure 8.24(b) is known as a unfolded Lange coupler.

8.4.2 Design of Interdigital Couplers

In this section, the design relations of an N-conductor (N-even) interdigital coupler as shown in Figure 8.25 are presented. It is assumed that the number of conductors (N) is even. The design of a Lange coupler is obtained when $N = 4$. All the lines in the configuration shown in Figure 8.25 are assumed to have the same width. The spacing between all adjacent lines is also assumed to be the same.

For given values of N, Z_0 (the impedance of various ports) and coupling, the design is obtained in terms of even- and odd-mode impedances of a pair of coupled lines as shown in Figure 8.25(c). Once the even- and odd-mode impedances of one pair of coupled lines are known, the width (W) of the lines and the spacing (S) between them can be determined. The design is obtained in terms of even- and odd-mode impedances of one pair of coupled lines because these data are available in the literature for a large class of transmission lines.

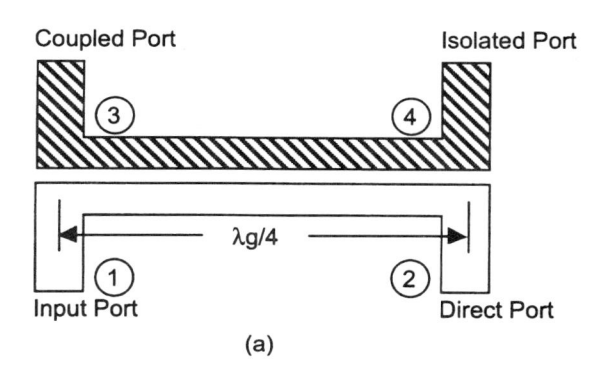

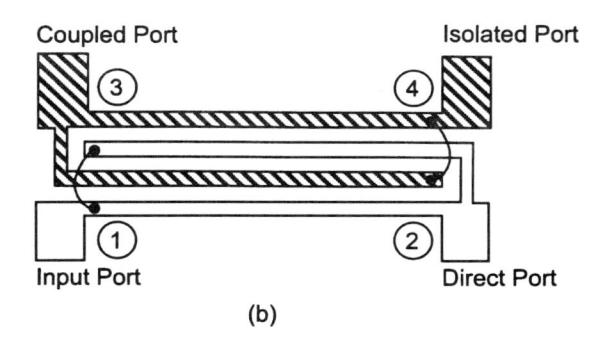

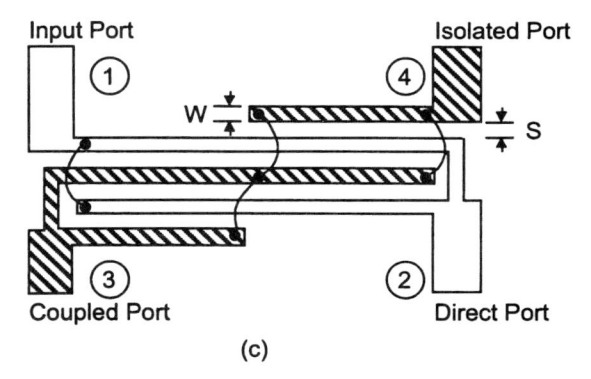

Figure 8.24 TEM couplers using (a) two parallel-coupled lines, (b) interdigital configuration known as unfolded Lange coupler, and (c) interdigital configuration known as a Lange coupler.

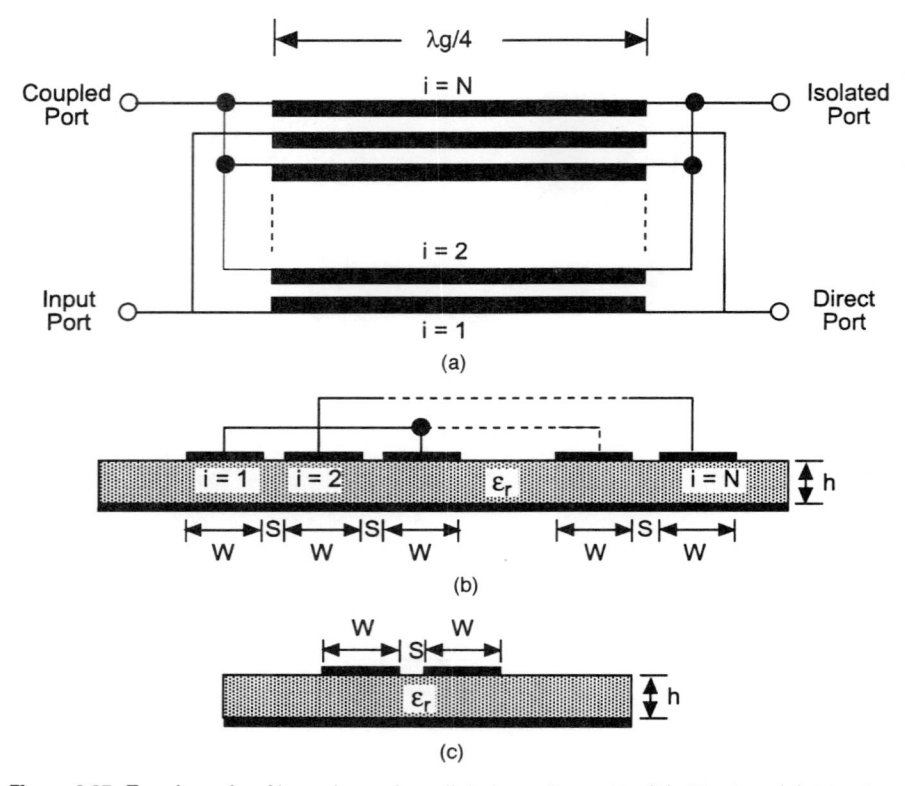

Figure 8.25 Top view of an N-conductor interdigital coupler and its (b) side view; (c) side view of two parallel-coupled lines having the same width and spacing between them as the N-conductor interdigital coupler.

The design equations of a N-conductor (N even) interdigital coupler are given by [15, 16]

$$k = \frac{(N-1)(1-R^2)}{(N-1)(1+R^2) + 2R} \tag{8.41}$$

$$Z = \frac{Z_{0o}}{Z_0} = \frac{\sqrt{R[(N-1)+R][(N-1)R+1]}}{(1+R)} \tag{8.42}$$

$$R = \frac{Z_{0o}}{Z_{0e}} \tag{8.43}$$

where k is the voltage coupling coefficient between the input and coupled ports at the center frequency of the design and N (N is even) is the total number of conductors. Z_{0e} and Z_{0o} denote, respectively, the even- and odd-mode impedances of a *pair* of coupled lines having the same width and spacing between them as any pair of N-conductor interdigital coupler. It may be worth remarking that the usual relation $Z_0 = \sqrt{Z_{0e} Z_{0o}}$ is valid in the case of a interdigital coupler only when $N = 2$. For other values of N, this relation is not satisfied.

The length of the interdigital coupler at the center frequency of design is given by

$$\ell = \frac{\lambda_g}{4} \qquad (8.44)$$

where $\lambda_g = 0.5(\lambda_{ge} + \lambda_{go})$ is the wavelength in the medium of the coupler at the center frequency of the design. Here λ_{ge} and λ_{go} are the guide wavelengths for the even- and odd-modes, respectively.

For a given value of voltage coupling factor, k, and number of conductors N, (8.41) can be used to find the value of R and, (8.42) can be used to determine the odd-mode impedance Z_{0o}. Further, the even-mode impedance Z_{0e} can be determined using (8.43). Using now the known values of the even- and odd-mode impedances, the dimensions of the lines and the spacing between them can be determined either by using available nomograms, computer programs [17] or design equations [5].

Equations 8.41 to 8.43 are not exact but are based on the following assumptions:

- The mode of propagation along the structure is TEM.
- The length of bonding wires is negligible compared with the wavelength at the frequency of operation.
- The mutual capacitance between any two neighboring conductors of the interdigital coupler is the same as for the two conductor lines shown in Figure 8.25(c).

Results

Using (8.41), the value of R has been plotted as a function of coupling in Figure 8.26 for values of $N = 2$ and 4 [15, 16]. Similarly, the normalized odd-mode impedance has been plotted as a function of coupling in Figure 8.27 for values of $N = 2$ and 4. In Table 8.1 numerical values are given for an interdigital coupler for a few values of coupling and number of conductors N. The input/output impedances of the coupler are assumed to be 50 Ω. It may be verified that for $N = 2$, the values of even- and odd-mode impedances given in Table 8.1 are the same as those given by (8.41–8.43).

Table 8.1
Impedance parameters of an
N-conductor interdigital coupler

N	Coupling (dB)	$Z_{0e}\,\Omega$	$Z_{0o}\,\Omega$
2	3	120.70	20.71
	6	86.60	28.87
	10	69.37	36.04
4	3	176.20	52.61
	6	142.50	67.96
	10	118.30	76.30
6	3	243.10	82.55
	6	204.30	105.10
	10	181.11	122.10

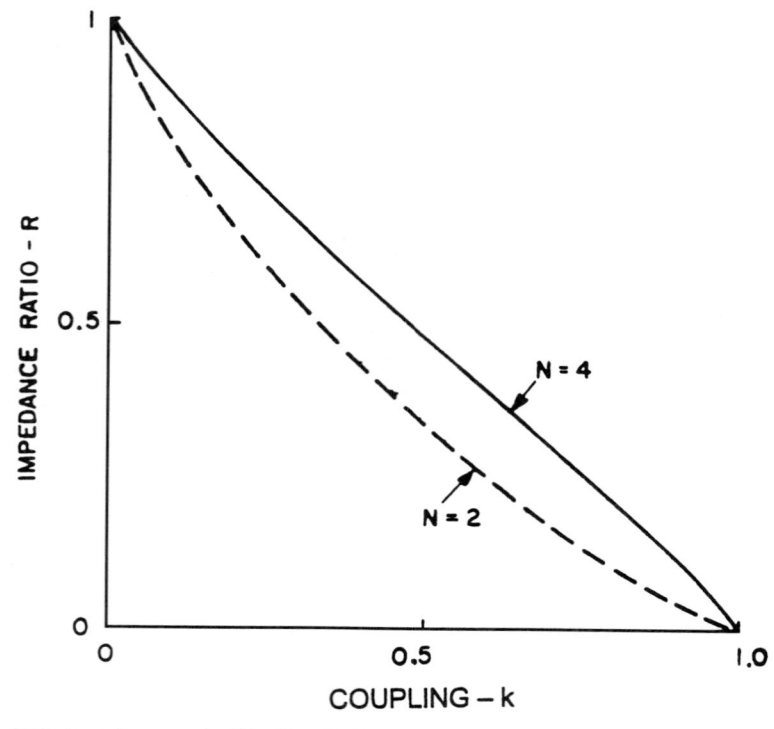

Figure 8.26 Impedance ratio $R\,(=Z_{0e}/Z_{0o})$ as a function of coupling. (From [16], © 1978 IEEE. Reprinted with permission.)

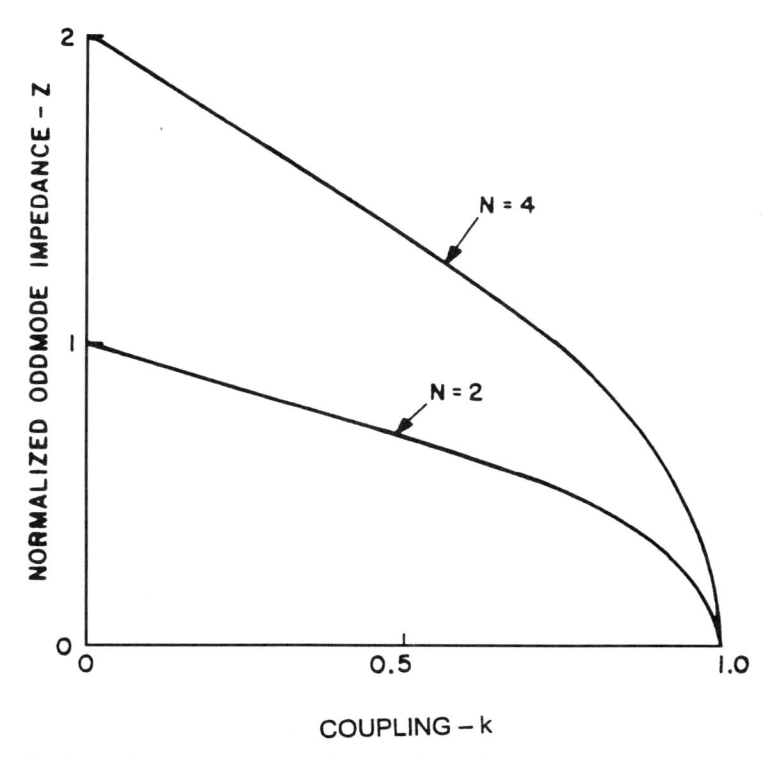

Figure 8.27 Normalized odd-mode impedance $(=Z_{oo}/Z_0)$ as a function of coupling. (From [16], © 1978 IEEE. Reprinted with permission.)

Design Data for a Lange Coupler

A Lange coupler is usually realized in microstrip configuration. In Figure 8.28, the normalized dimensions of the coupler (W/h and S/h) are given as a function of coupling for $Z_0 = 50\ \Omega$ and printed on a dielectric substrate of $\varepsilon_r = 10$ [16]. The data are based on the computation of even- and odd-mode impedance using (8.41) through (8.43). Further, using the computed values of even- and odd-mode impedances, the dimension ratios W/h and S/h of the coupler have been found using the program for coupled microstrip lines given by Bryant and Weiss [17]. Simple expressions given in reference [5] can also be used to obtain the values of W/h and S/h.

A Lange coupler is frequently designed for 3-dB coupling. In Figure 8.29, the dimensions of a 3-dB, 50 Ω Lange coupler are given as a function of the dielectric constant of the substrate material on which the coupler is printed.

The dimensions given in Figure 8.28 and 8.29 are valid for zero-thickness conductors. In practice, however, the conductors will always have a finite

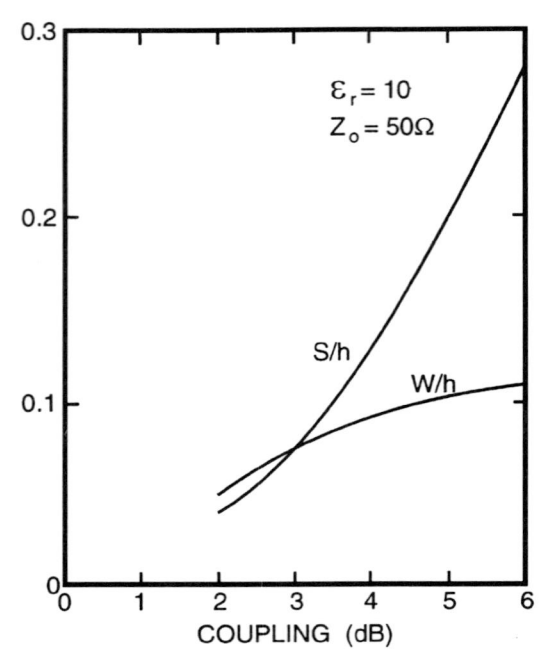

Figure 8.28 The dimensional ratios of a Lange coupler printed on a dielectric substrate ($\varepsilon_r = 10$) and $Z_0 = 50\,\Omega$ as a function of coupling. (From [16], © 1978 IEEE. Reprinted with permission.)

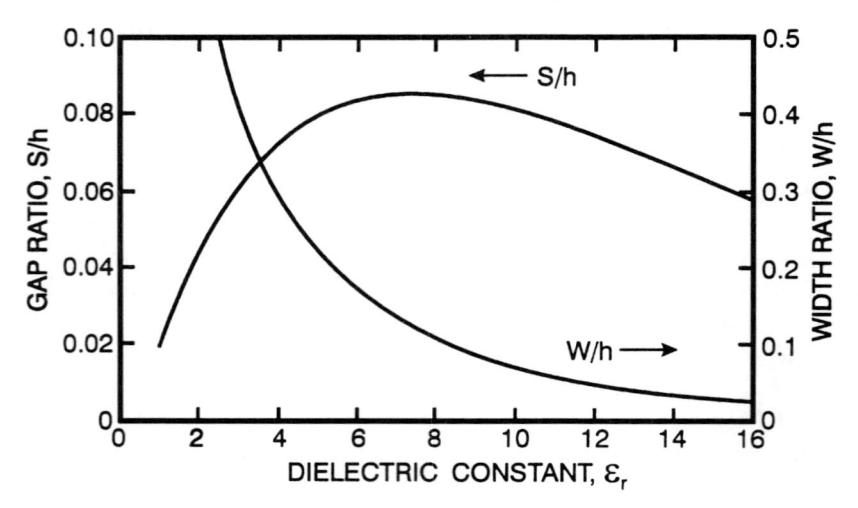

Figure 8.29 The dimensional ratios of a 3-dB Lange coupler as a function of dielectric constant of the substrate. (From [16], © 1978 IEEE. Reprinted with permission.)

thickness. Presser [16] has given an empirical formula for the correction factor that was found by performing a large number of experiments. The effect of thickness can be taken into account by increasing the separation between adjacent lines and decreasing the width of the lines as

$$\frac{S}{h} = \frac{S_0}{h} + \frac{\Delta S}{h} \qquad (8.45a)$$

and

$$\frac{W}{h} = \frac{W_0}{h} - \frac{\Delta S}{h} \qquad (8.45b)$$

where W_0 and S_0 denote the width and the separation between lines, respectively, for a coupler employing zero-thickness conductors. The value of $\Delta S/h$ in the above equations is given by

$$\frac{\Delta S}{h} = \frac{t/h}{\pi \sqrt{\varepsilon_{re}}} \left(1 + \ln \frac{4\pi W_0/h}{t/h}\right) \qquad (8.46)$$

where t denotes the thickness of the metalization and ε_{re} can be assumed to be the effective dielectric constant of a single, uncoupled microstrip line having width W.

Further, Presser [16] found from sensitivity analysis that the most sensitive parameters in the design of a Lange coupler are the gap dimensions and the metalization thickness, whereas 10% changes in the width and the dielectric constant cause practically insignificant deviations in the performance.

Many computer programs are now available, which give the impedance data for coupled microstrip lines taking the thickness of the conductors into account. With such a program, we can directly compute the dimensions of the interdigital structure using the values of even- and odd-mode impedances computed using (8.42) and (8.43). Commercial CAD tools can also be used to design such couplers. More accurate design can also be performed using EM simulators.

8.5 Tandem Couplers

A tight coupler can also be obtained by connecting two loose couplers in tandem [20] as shown in Figure 8.30. In this arrangement, the direct and coupled ports of the first coupler are connected to the isolated and input ports of the second

282 RF and Microwave Coupled-Line Circuits

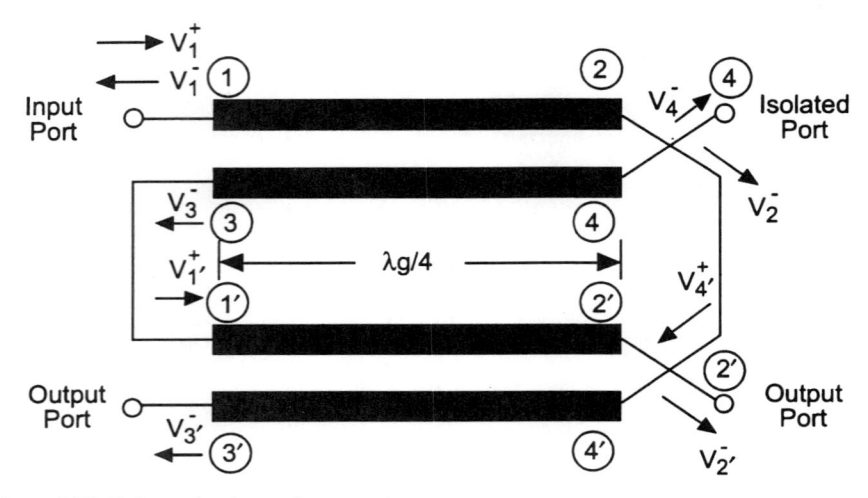

Figure 8.30 Schematic of a tandem coupler.

coupler, respectively. Both the couplers are of the TEM type, whose scattering parameters are of the form given by (8.47) and (8.48). Let the voltage coupling factors for the two couplers be k_1 and k_2, respectively. Then the scattering matrix of the first coupler at the center frequency (the center frequency is the one at which the electrical length of the TEM coupler is 90 degrees) is given by

$$[S_1] = \begin{bmatrix} 0 & -j\cos\alpha_1 & \sin\alpha_1 & 0 \\ -j\cos\alpha_1 & 0 & 0 & \sin\alpha_1 \\ \sin\alpha_1 & 0 & 0 & -j\cos\alpha_1 \\ 0 & \sin\alpha_1 & -j\cos\alpha_1 & 0 \end{bmatrix} \quad (8.47)$$

where $\sin\alpha_1 = k_1$ is the voltage coupling coefficient for coupler 1. Similarly, the scattering matrix of the second coupler can be expressed as

$$[S_2] = \begin{bmatrix} 0 & -j\cos\alpha_2 & \sin\alpha_2 & 0 \\ -j\cos\alpha_2 & 0 & 0 & \sin\alpha_2 \\ \sin\alpha_2 & 0 & 0 & -j\cos\alpha_2 \\ 0 & \sin\alpha_2 & -j\cos\alpha_2 & 0 \end{bmatrix} \quad (8.48)$$

where $\sin\alpha_2 = k_2$ denotes the voltage coupling coefficient for the second coupler. Let a wave with voltage amplitude of unity ($V_1^+ = 1$) be incident on port

1 of coupler 1, then the reflected voltages V_1^-, V_2^-, V_3^-, and V_4^- are given by

$$\begin{bmatrix} V_1^- \\ V_2^- \\ V_3^- \\ V_4^- \end{bmatrix} = \begin{bmatrix} 0 & -j\cos\alpha_1 & \sin\alpha_1 & 0 \\ -j\cos\alpha_1 & 0 & 0 & \sin\alpha_1 \\ \sin\alpha_1 & 0 & 0 & -j\cos\alpha_1 \\ 0 & \sin\alpha_1 & -j\cos\alpha_1 & 0 \end{bmatrix} \begin{bmatrix} 1 \\ 0 \\ 0 \\ 0 \end{bmatrix} \quad (8.49)$$

which gives

$$\begin{aligned} V_1^- &= 0, \\ V_2^- &= -j\cos\alpha_1, \\ V_3^- &= \sin\alpha_1, \text{ and} \\ V_4^- &= 0 \end{aligned} \quad (8.50)$$

Further, the reflected voltages at ports 2 and 3 of coupler 1 are, respectively, the incident voltages for the ports 4' and 1' of the second coupler. Therefore

$$V_{1'}^+ = V_3^- = \sin\alpha_1 \quad (8.51a)$$
$$V_{4'}^+ = V_2^- = -j\cos\alpha_1 \quad (8.51b)$$

The output at the second coupler can then be found using

$$\begin{bmatrix} V_{1'}^- \\ V_{2'}^- \\ V_{3'}^- \\ V_{4'}^- \end{bmatrix} = \begin{bmatrix} 0 & -j\cos\alpha_2 & \sin\alpha_2 & 0 \\ -j\cos\alpha_2 & 0 & 0 & \sin\alpha_2 \\ \sin\alpha_2 & 0 & 0 & -j\cos\alpha_2 \\ 0 & \sin\alpha_2 & -j\cos\alpha_2 & 0 \end{bmatrix} \begin{bmatrix} \sin\alpha_1 \\ 0 \\ 0 \\ -j\cos\alpha_1 \end{bmatrix}$$
$$(8.52)$$

which gives

$$V_{1'}^- = V_{4'}^- = 0 \quad (8.53a)$$
$$V_{2'}^- = -j\cos\alpha_2 \sin\alpha_1 - j\sin\alpha_2 \cos\alpha_1 = -j\sin(\alpha_1 + \alpha_2) \quad (8.53b)$$

and

$$V_{3'}^- = \sin\alpha_2 \sin\alpha_1 - \cos\alpha_2 \cos\alpha_1 = -\cos(\alpha_1 + \alpha_2) \quad (8.53c)$$

Now let us choose

$$\alpha_1 = \alpha_2 = \frac{\pi}{8} \quad (8.54)$$

or

$$k_1 = k_2 = \sin\left(\frac{\pi}{8}\right) = 0.3827$$

which is also equivalent to coupling in decibels for the two couplers as C_1 and C_2, respectively, where

$$C_1 = C_2 = -20\log(0.3827) = 8.34 \text{ dB}$$

Substituting the values of α_1 and α_2 from (8.54) in (8.53), we obtain

$$V_{2'}^- = -j\sin\left(\frac{\pi}{4}\right) = -\frac{j}{\sqrt{2}} \quad (8.55a)$$

and

$$V_{3'}^- = -j\cos\left(\frac{\pi}{4}\right) = -\frac{1}{\sqrt{2}} \quad (8.55b)$$

The fractional power coupled from port 1 of the first coupler to port 3' of the second coupler is therefore given by

$$\frac{P_{3'}^-}{P_1^+} = \frac{|V_{3'}^-|^2}{|V_1^+|^2} = \frac{1}{2} \quad (8.56)$$

which is equivalent to 3-dB coupling. Therefore, by connecting two 8.34-dB couplers in tandem, a 3-dB coupler is obtained.

For physical realization of tandem couplers, a scheme such as shown in Figure 8.31 is used. Note that crossovers are required in tandem couplers to achieve proper interconnections between the couplers as required by the connection scheme symbolically shown in Figure 8.30.

8.6 Multilayer Tight Couplers

A recent upsurge in MMIC-based system demands and wireless applications has led to new configurations for tight directional couplers. These are broadside, embedded microstrip, re-entrant, and compact directional couplers. Compact couplers include lumped-element, spiral, and meander line structures, which are described briefly in Section 8.7.

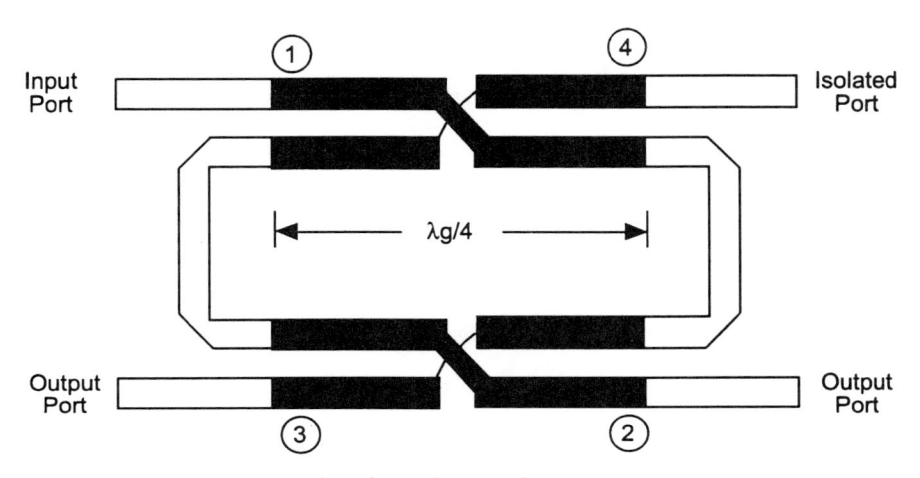

Figure 8.31 Physical configuration of a tandem coupler.

8.6.1 Broadside Couplers

The asymmetrical broadside-coupled microstrip lines configuration is the simplest technique to realize tight coupling. Several different configurations and fabrication technologies to design 3-dB couplers have been reported in the literature [21–26]. A basic configuration of a MIC/MMIC asymmetric broadside coupler, which consists of two conductors separated by a thin layer of polyimide dielectric ($\varepsilon_{r2} = 3.2$), is shown in Figure 8.32(a). This requires multilayer MMIC technology. The 3-dB coupler design is usually done using an EM simulator and involves an optimal solution for polyimide dielectric thickness and conductor width for given GaAs substrate thickness. The conductors can be folded or meandered to reduce the overall chip size.

Recently a coupled-multilayer microstrip line structure as shown in Figure 8.32(b), which works similar to symmetrical broadside-coupled striplines structure, was reported by Okazaki and Hirota [26]. The coupler was fabricated using multilayer thin-film microstrip line technology and consists of four parallel-strip conductors. The ground plane, which is 1-μm thick, for these conductors is placed on the top surface of the GaAs. Diagonal conductors as shown in the figure are connected at the ends, and the structure behaves like symmetrical broadside-coupled striplines, and the couplers can be designed using broadside-coupled stripline formulas given in Sec. 3.7. Various dimensions for the coupler reported in [26] are as follows:

Conductors A and C are buried in polyimide layers ($\varepsilon_r \cong 3.3$) and have widths (W_1) of 3 μm. The gap (S_1) between them is 5 μm and are placed above the ground plane level (h_1) at 6.5 μm. The other two conductors B and D have

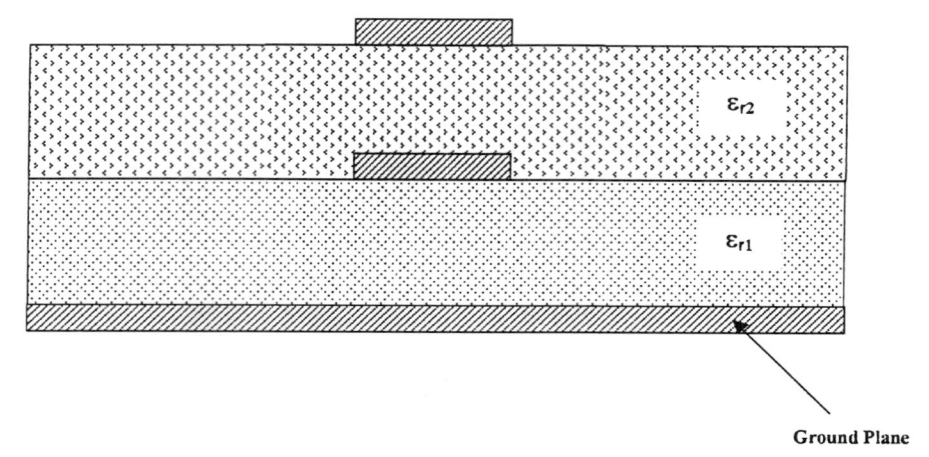

Figure 8.32 (a) Cross section of an asymmetric broadside-coupled microstrip line coupler.

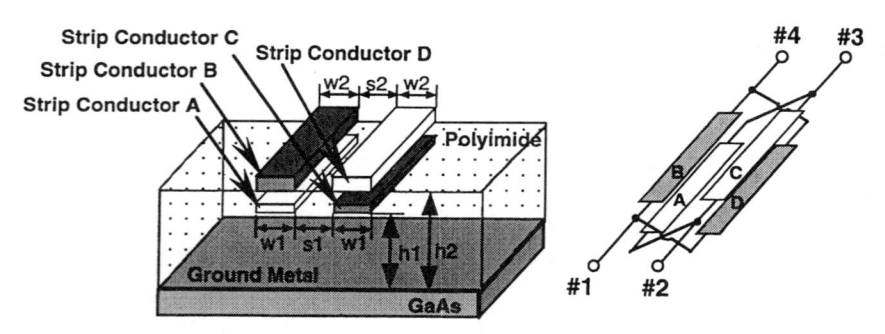

Figure 8.32 (b) Structure of the multilayer symmetric broadside coupler. (From [26], © 1997 IEEE. Reprinted with permission.)

widths (W_2) of 5 μm and the gap between them (S_2) is also 5 μm. They are placed above the ground plane level (h_2) at 9 μm. Conductors A and D, and B and D, are connected at both ends. The width of the entire coupled line is only 15 μm, which allows layout in meander shape (Figure 8.33) to reduce the chip size, which measures only 1.3 × 0.4 mm for the X/Ku-band coupler design.

Figure 8.34 shows the measured performance of this coupler; coupling is 4.2 ± 0.4 dB, return loss and isolation better than 20 and 15 dB, respectively, over the 10 to 17.5 GHz frequency range. The measured phase difference between the direct and coupled ports was about 91 ± 5 degree over the 10 to 17.5-GHz frequency range.

Figure 8.33 Photograph of the MMIC symmetric broadside coupler. (From [26], © 1997 IEEE. Reprinted with permission.)

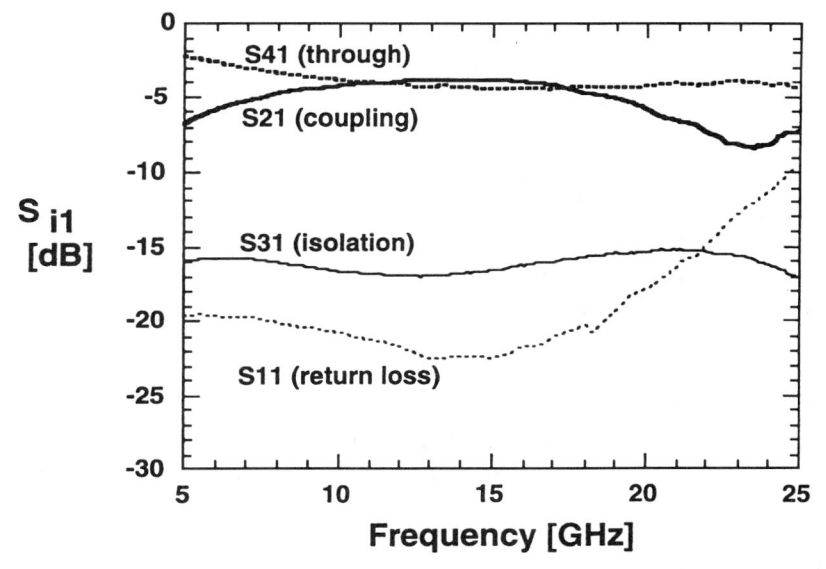

Figure 8.34 Amplitude characteristics of the symmetric, broadside microstrip coupler. (From [26], © 1997 IEEE. Reprinted with permission.)

8.6.2 Embedded Microstrip Couplers

The coupling factor of multiconductor couplers such as a Lange coupler can be increased by decreasing the spacing between the coupler's conductors. Typically, MMIC metalization processes use a plate-up technique that can achieve low loss and uniform spacings. The conductors, which are plated 4 to 5 μm thick, must have spacings between the conductors greater than 8 μm to achieve high yields. Unfortunately, dimensions of half this size are required for the realization of a 3-dB coupler on a 75-μm-thick GaAs substrate. The embedded microstrip coupled structure [27] mitigates the limitations of the photolithographic and plating processes. The embedded microstrip line coupler consists of two parallel strip conductors placed in close proximity, in which one strip is embedded

Table 8.2
The physical dimensions of 3-dB couplers designed on GaAs substrate

Parameter	6-21 GHz	2-7 GHz
Conductor width (μm)	30	30
Conductor length (μm)	1,900	5,700
Conductor overlap (μm)	0	0
Plating thickness (μm)	4.5	4.5
Dielectric layer's thickness (μm)	0.2	0.2
Dielectric constant of dielectric layer	6.7	6.7
Unplated metal thickness (μm)	0.6	0.6
Substrate thickness (μm)	125	125
Dielectric constant of substrate	12.9	12.9

in a dielectric. In this case, the parallel conductors can be placed very close or overlapped to each other. Figure 8.35 shows a cross sectional representation of an embedded microstrip coupled-line structure that provides a coupling factor of 2 dB, and that can be used to make couplers having a bandwidth of several octaves.

The analysis of these coupled-line geometries is difficult because the lines are asymmetrically edge-coupled, similar to an offset-coupled stripline. Accurate solutions of these structures can be obtained by using a 3D electromagnetic simulator. In this case, "em"[TM] by Sonnet Software was used to analyze this structure. The design parameters for two couplers are summarized in Table 8.2, and their layout is shown in Figure 8.36(a). A 3-dB coupler was also designed and fabricated with coupled and direct ports on the same side using a crossover, a configuration commonly used for the construction of balanced amplifiers. The crossover, shown in Figure 8.36(b) by a physical layout, uses an airbridge. The crossover at the midpoint switches the embedded microstrip line to the plated microstrip line and vice versa. In the coupler, the length of the crossover is only 60 μm and has negligible effect on the coupler's performance.

The couplers were tested on wafer using RF probes Thru-Reflect-Line (TRL) de-embedding techniques. Each coupler had one port terminated, by connecting a 50 Ω resistor to the ground through a hole on the chip itself, forming a three-port test structure. Testing a three-port is much simpler than testing a four-port and yet, the coupler can be completely characterized because it is a reciprocal device. Figure 8.37 shows the measured performance (coupling and return loss at the three ports) of nine 6-21 GHz broadband couplers tested in the 3 × 3 array on a 125-μm-thick GaAs wafer. A maximal amplitude variation of

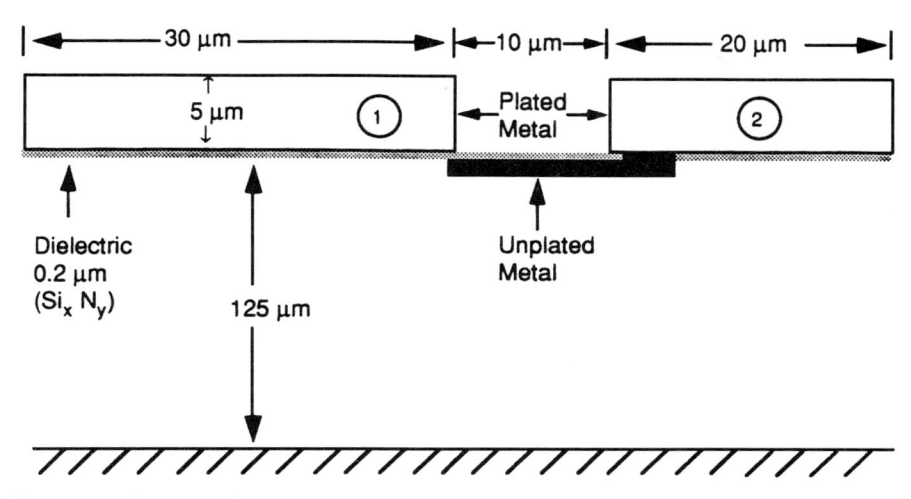

Figure 8.35 Cross sectional representation of the coupled line structure used to fabricate a 6-21 GHz coupler on a 125-μm-thick GaAs substrate.

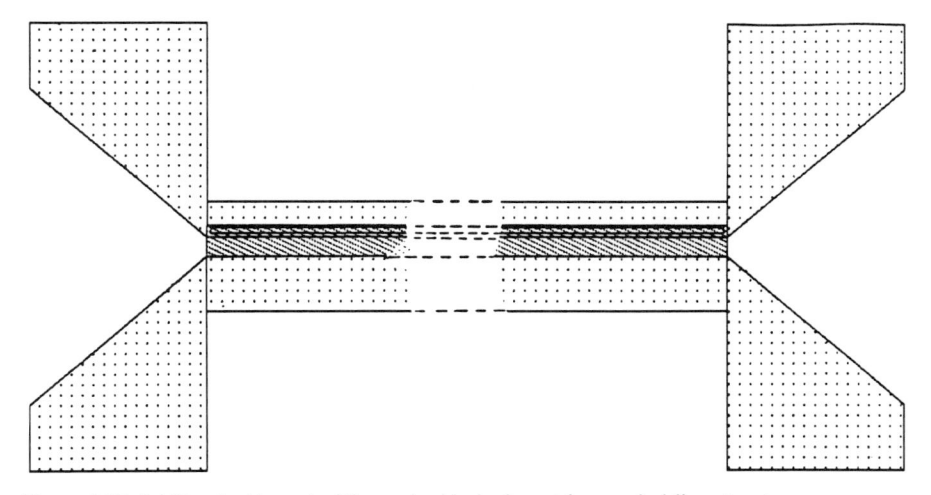

Figure 8.36 (a) Physical layout of the embedded microstrip coupled-line structure.

±1.5 dB was achieved between the coupled and direct ports over a 16-GHz bandwidth. The measured return loss at all ports was greater than 13 dB from 0.5 to 24 GHz. The typical measured phase difference between the direct and coupled ports, and the isolation between the input and isolated ports of this coupler, as a function of frequency, are plotted in Figure 8.38. The phase difference is 93 ± 9 degrees, and the isolation is better than 10 dB across the 0.5 to 20-GHz frequency range. Similar performance was measured on other wafers in different wafer lots.

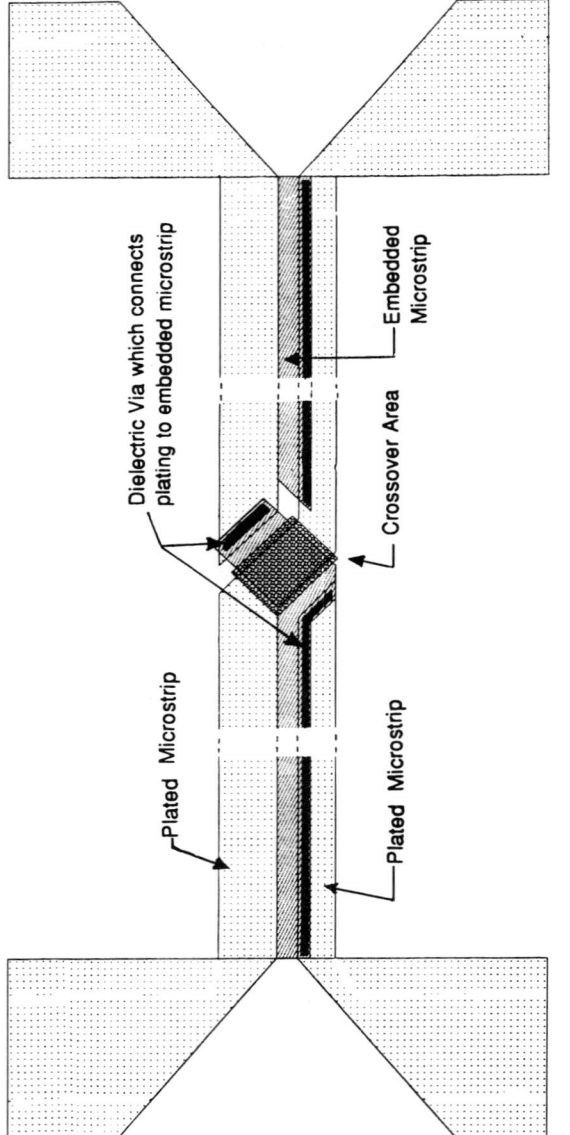

Figure 8.36 *(continued)* (b) Top view of a crossover for broadband embedded microstrip coupler to bring the coupled and direct ports on the same side.

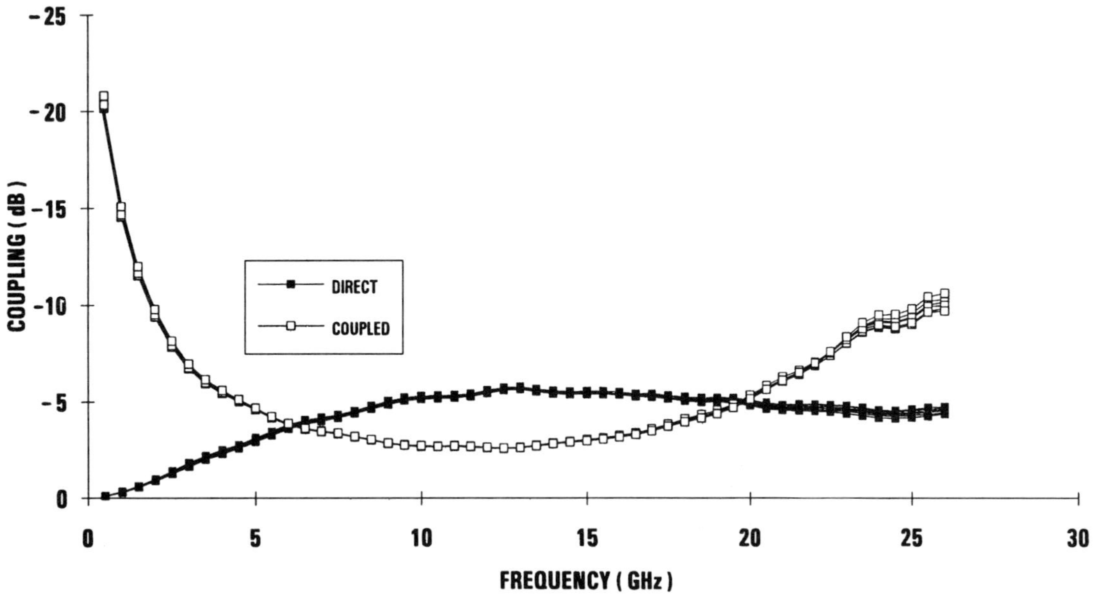

Figure 8.37 Measured performance of the 5–21 GHz coupler fabricated on a 125-μm-thick GaAs substrate. (a) Coupling at coupled and direct ports, (b) return loss at input, coupled and direct port.

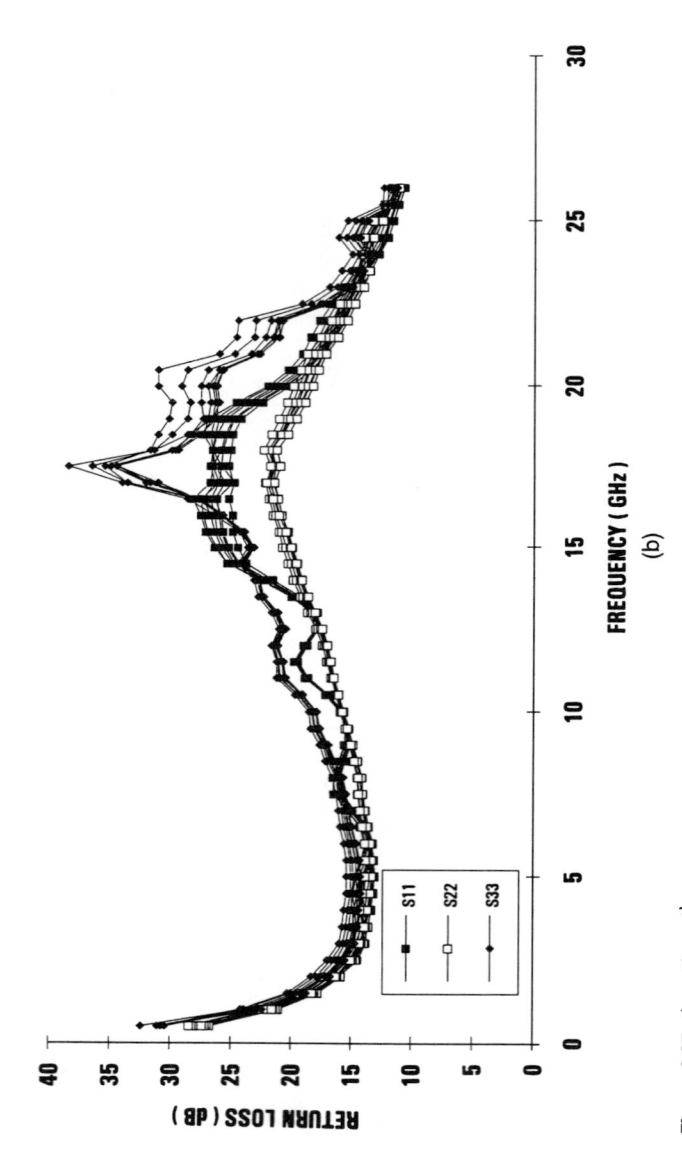

Figure 8.37 (*continued*)

Tight Couplers

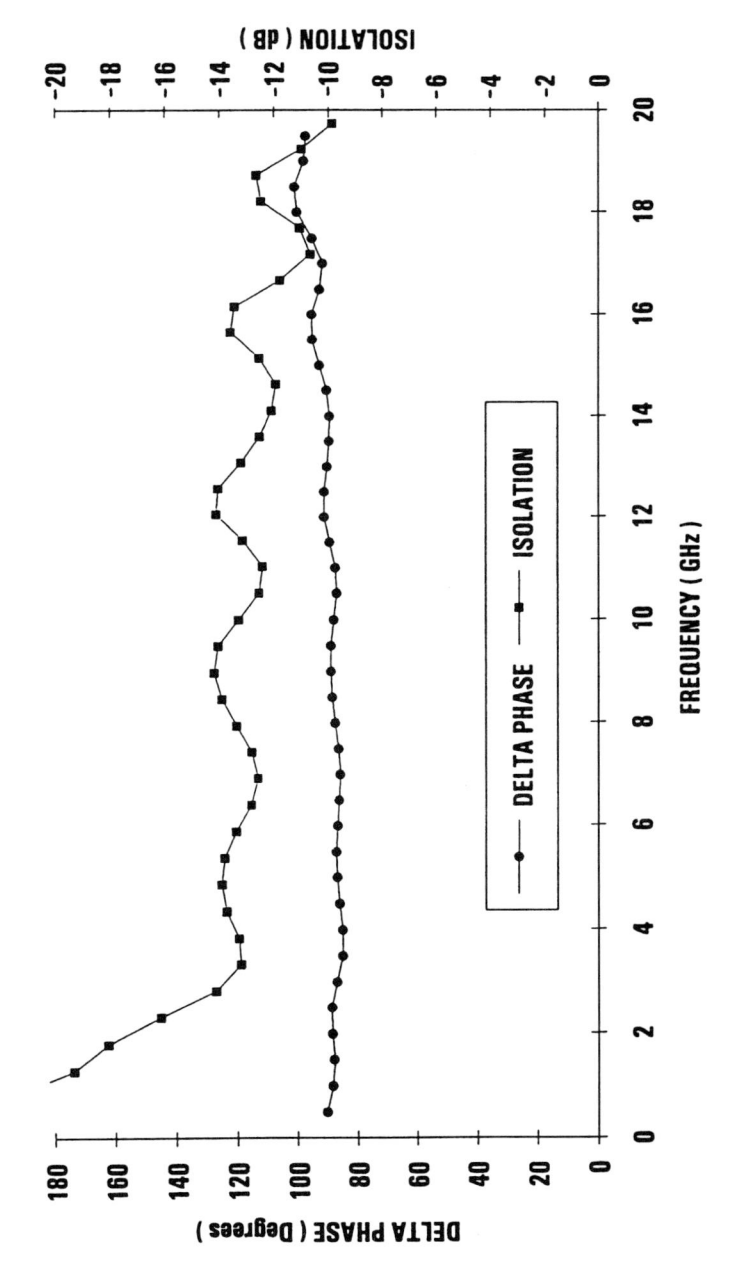

Figure 8.38 Measured insertion phase difference between coupled and direct ports and isolation of the 5-21 GHz coupler fabricated on a 125-μm-thick GaAs substrate.

8.6.3 Re-Entrant Mode Couplers

Re-entrant mode couplers have been designed to obtain tight coupling in coaxial [28], stripline [29], and microstrip [30, 31] media. Design procedures of semi-re-entrant [32, 33] and re-entrant [34] have also been discussed. The basic theory of such couplers is very simple and can be described by referring to Figures 8.39 and 8.40(a). In Figure 8.39, the top conductor 1 is floating while in Figure 8.40(a), the underneath conductor is floating. The latter coupler consists of a parallel-coupled microstrip line (conductors A and B) with another conductor C floating underneath. This is again a multilayer configuration, the dielectric constant and the thickness of the dielectric layer between conductors A or B and C is determined to achieve the required coupling coefficient.

The even- and odd-mode impedances, like any other symmetrical edge-coupled lines, can be determined by placing magnetic and electric walls at the plane of symmetry $P-P'$ as shown in Figure 8.40. In the case of even mode shown in Figure 8.40(b), the impedance of floating conductor C (having characteristic impedance of Z_{01}) is in series with the transmission lines $A-C$ and $B-C$, each having the characteristic impedance of Z_{02}. By placing a magnetic wall

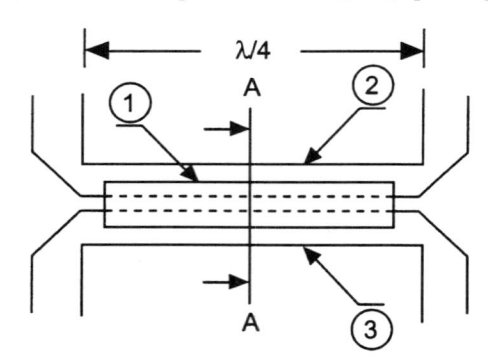

Figure 8.39 (a) A schematic view of a single-section semi-re-entrant coupler.

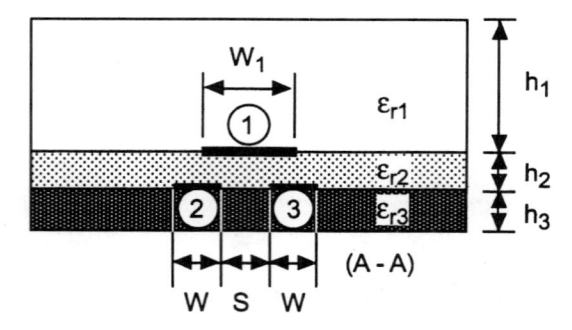

Figure 8.39 (b) A cross sectional view of a semi-re-entrant coupled section.

Tight Couplers 295

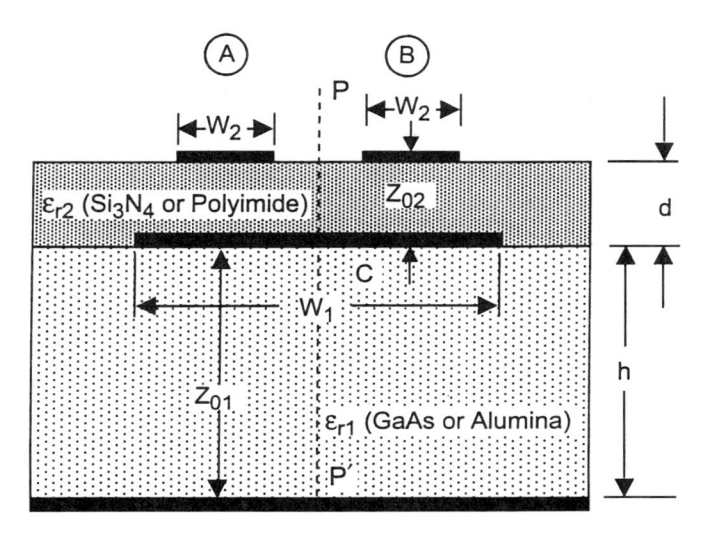

Figure 8.40 (a) Microstrip re-entrant mode coupler cross section. (From [31], © 1990 IEEE. Reprinted with permission.)

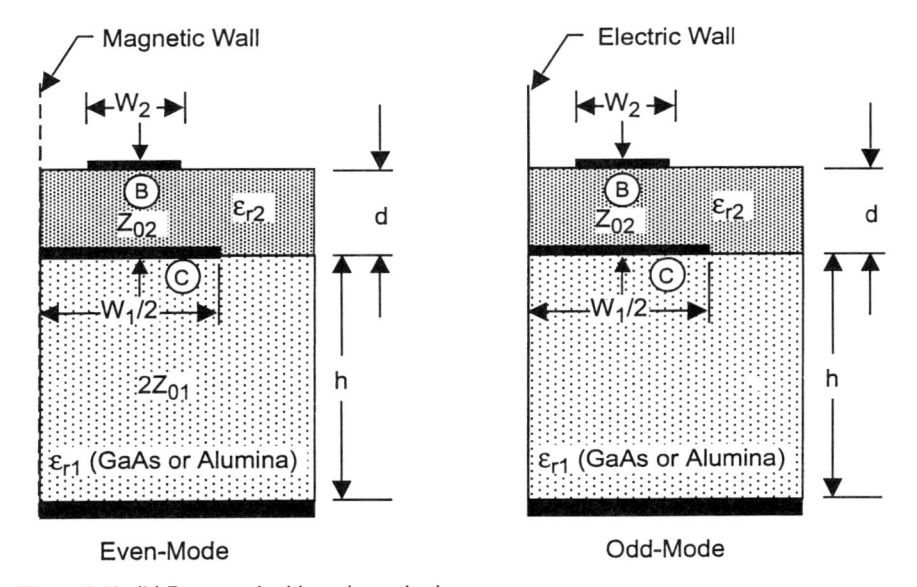

Figure 8.40 (b) Even- and odd-mode excitations.

that bisects the coupler's cross section, the even-mode characteristic impedance becomes

$$Z_{0e} = Z_{02} + 2Z_{01} \tag{8.57}$$

In the case of the odd mode, Figure 8.40(b), the electric wall passing through the middle of conductor C sets the conductor at ground plane reference, and the odd-mode characteristic impedance of the coupled structure is

$$Z_{0o} = Z_{02}$$

Therefore, the coupling coefficient in this case can be expressed as

$$k = \frac{Z_{0e} - Z_{0o}}{Z_{0e} + Z_{0o}} = \frac{Z_{01}}{Z_{01} + Z_{02}} \tag{8.58}$$

and the termination impedance $Z_0 = \sqrt{Z_{0e}Z_{0o}}$. Equation 8.58 shows that the coupling coefficient, in this case, does not depend upon the spacing between the conductors A and B, but mostly depends upon the Z_{02} value, which can easily be controlled by the parameters ε_{r2} and d. Table 8.3 lists typical design parameters for several 3-dB couplers matched approximately to 50 Ω.

Figure 8.41 shows the measured performance of a re-entrant coupler fabricated on 0.381-mm-thick alumina substrate. The top gold conductors were about 3 to 4 μm thick and coupler worked over the 5 to 19-GHz range.

8.7 Compact Couplers

In cellular wireless microwave applications, quadrature 3-dB couplers are required to determine the phase error of a transmitter using the QPSK modulation scheme. The basic requirements for such couplers include small size, low cost, tight amplitude balance, and 90-degree phase difference between the coupled and direct ports. At the L-band, the distributed couplers are big in

Table 8.3
Typical dimensions for various re-entrant microstrip couplers

Substrate	W_1 (mm)	W_2 (mm)	d (mm)	h (mm)	Z_{01} (Ω)	Z_{02} (Ω)
Alumina ($\varepsilon_{r1} = 9.9$; $\varepsilon_{r2} = 3.7$)	0.254	0.0685	0.0075	0.381	59.83	18.14
GaAs ($\varepsilon_{r1} = 12.9$; $\varepsilon_{r2} = 6.8$)	0.075	0.015	0.0023	0.100	49.8	18.00
	0.115	0.015	0.0025	0.150	49.8	18.00
GaAs ($\varepsilon_{r1} = 12.9$; $\varepsilon_{r2} = 3.7$)	0.115	0.015	0.006	0.150	49.8	18.00

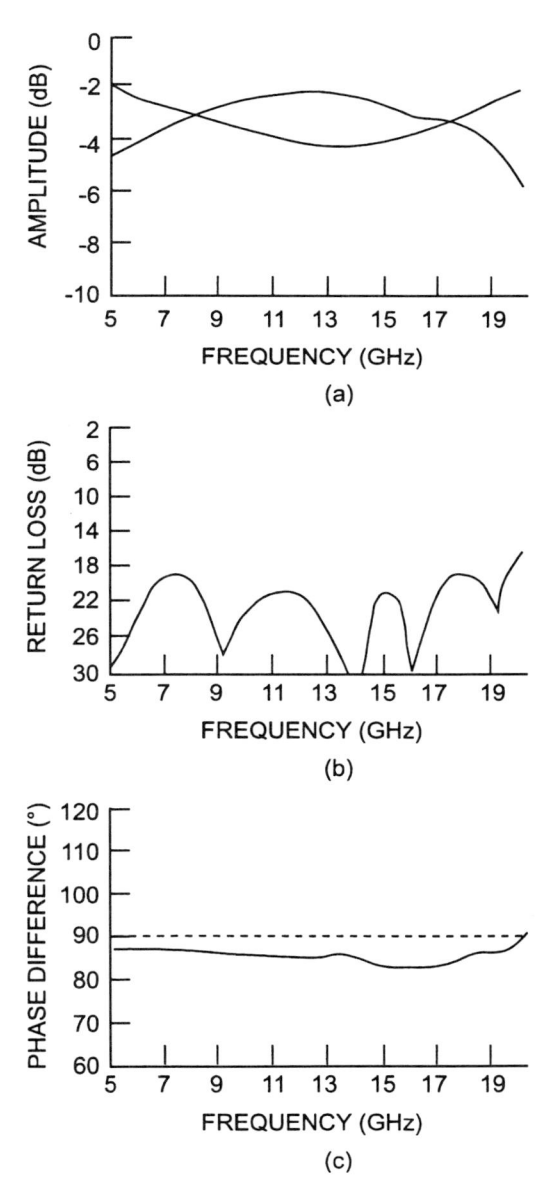

Figure 8.41 Coupler performance: (a) coupled and direct power, (b) return loss, and (c) phase difference between the coupled and direct ports. (From [31], © 1990 IEEE. Reprinted with permission.)

size and also expensive. An equivalent lumped-element implementation is compact in size and has the potential to be low cost. For example, the size of a monolithic coupler on GaAs substrate has to be of the order of about 1 to 2 mm² to be cost effective for low-frequency wireless applications. Several existing coupler configurations have been transformed into new layouts to meet size target values. Some of these new configurations such as lumped-element couplers [35], the spiral directional coupler [36], and meander line couplers [37–39] are described briefly in this section.

8.7.1 Lumped-Element Couplers

The coupler shown in Figure 8.42(a) can be modeled as a lumped-element equivalent circuit as shown in Figure 8.42(b). The values for L, M, C_g and C_c in terms of Z_{0e}, Z_{0o}, and θ are given as follows [35]:

$$L = \frac{(Z_{0e} + Z_{0o}) \sin \theta}{4\pi f_o} \qquad C_g = \frac{\tan(\theta/2)}{Z_{0e} 2\pi f_o} \qquad (8.59a)$$

$$M = \frac{(Z_{0e} - Z_{0o}) \sin \theta}{4\pi f_o} \qquad C_c = \left(\frac{1}{Z_{0o}} - \frac{1}{Z_{0e}}\right) \frac{\tan(\theta/2)}{4\pi f_o} \qquad (8.59b)$$

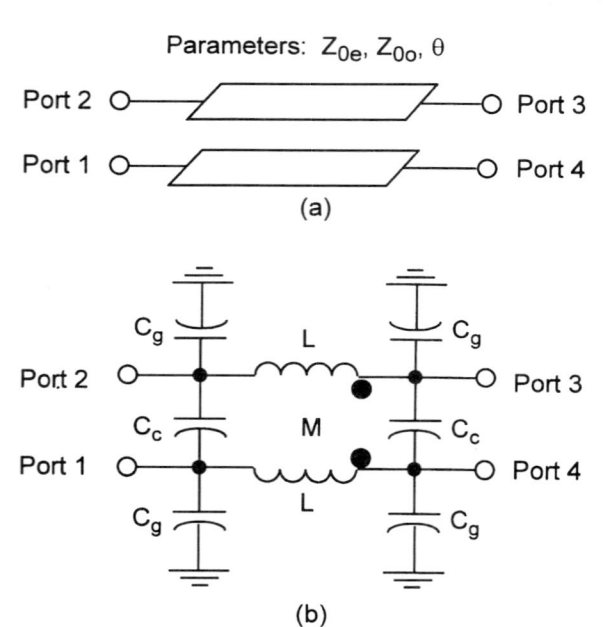

Figure 8.42 Two-conductor coupled microstrip line models: (a) distributed elements, (b) lumped elements.

where f_o is the center frequency and $\theta = 90$ degree at f_o. For a given coupling, using (4.59), the values of Z_{0e} and Z_{0o} are determined and then lumped-element values are calculated using (8.59). The self- and mutual inductors are realized using a spiral inductor transformer and the capacitors C_g and C_c are of MIM type and their partial value is also included in the transformers.

8.7.2 Spiral Directional Couplers

To obtain miniature directional couplers with tight coupling, a coupled structure in a spiral shape (also known as "spiral coupler") is realized. Printing the spiral conductor on high dielectric constant materials further reduces the size of the coupler. In this case, tight coupling is achieved by using loosely coupled parallel-coupled microstrip lines placed in a close-proximity spiral configuration. This structure as shown in Figure 8.43 uses two turns and resembles a multiconductor structure. Design details of such couplers and their modifications are given in [36] and are briefly summarized below. An accurate design of such structures, however, is only possible by using EM simulators.

As reported in [36], the total length of the coupled line, on the alumina substrate, along its track is $\lambda_0/8$, where λ_0 is the free-space wavelength at the center frequency and $D \cong \lambda_0/64 + 4W + 4.5S$. Longer lengths result in tighter couplings. Typical line widths and spacings are approximately 500 μm

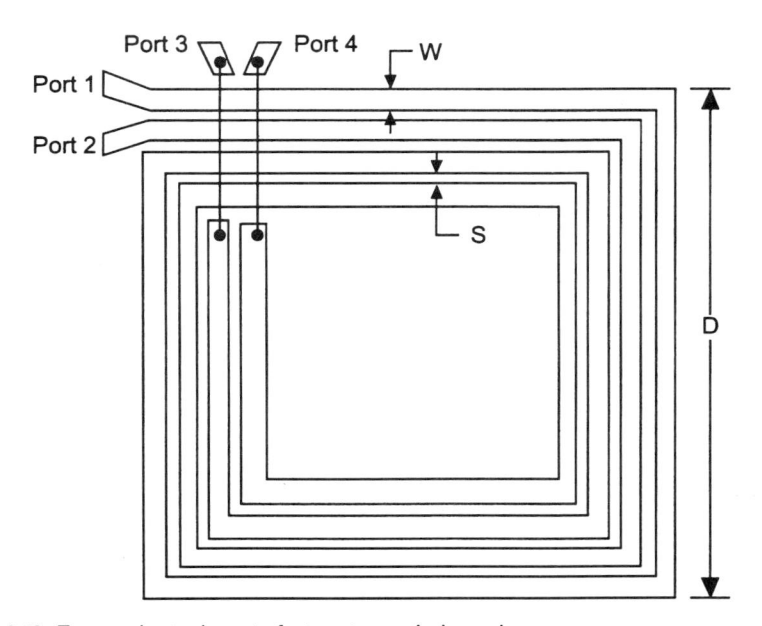

Figure 8.43 Top conductor layout of a two-turn spiral coupler.

and 40 μm, respectively, for a 0.635-mm alumina ($\varepsilon_r = 9.6$) substrate. In the spiral configuration, coupling is not a strong function of spacing between the conductors. The conductors were about 5 μm thick. Measured coupled power, direct power, return loss, and isolation at 800 MHz for the two-turn spiral coupler, were approximately −3.5 dB, −3.5 dB, 22 dB, and 18 dB, respectively.

8.7.3 Meander Line Directional Coupler

A compact coupler can also be realized by meandered line edge-coupled microstrip lines [37–39] using high dielectric constant substrates. Figure 8.44

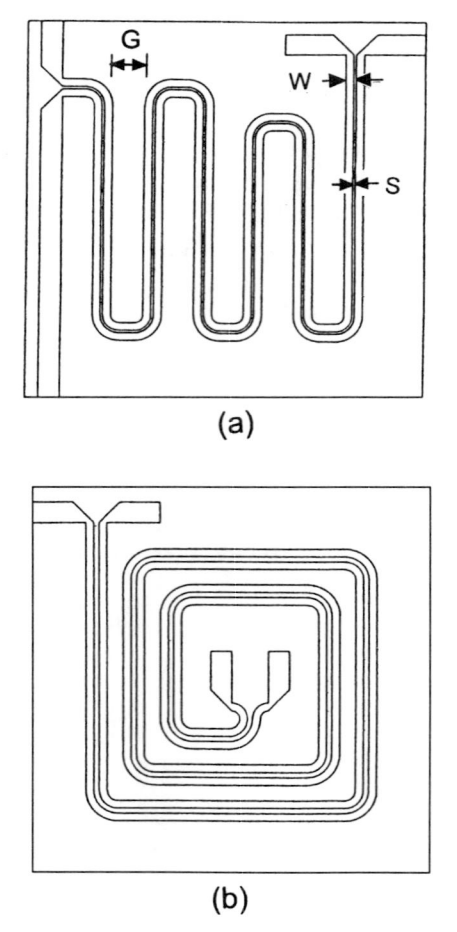

Figure 8.44 Compact coupled-line directional couplers: (a) meander type and (b) spiral type. (From [39], © 1990 IEEE. Reprinted with permission.)

Table 8.4
Physical dimensions of two compact couplers

Parameters	Meander type	Spiral type
W (μm)	30	30
S (μm)	10	20
G (μm)	120	60
Line length (mm)	6.5	7.5
Substrate thickness (μm)	300	300
Chip size (mm)	1.5 × 1.5	1.5 × 1.5

Table 8.5
Measured electrical performance of two compact couplers

Configuration	Meander type	Spiral type
Frequency (GHz)	1.8–3.8	1.8–3.8
Insertion loss (dB)	4.5–7.5	3.8–5.0
Amplitude balance (dB)	±0.7	±0.5
Phase difference (deg)	93 ± 2	90 ± 2
Return loss (dB)	>15	>15

shows a meander-type coupler and a spiral-type 90-degree coupler for comparison. In these couplers, tight coupling is achieved by placing coupled pair sections in close proximity, which also results in compact size. These couplers are normally designed using EM simulators. The physical dimensions of meander- and spiral-type couplers, designed to work over the S-band, are given in Table 8.4, where W, S, and G are the conductor width, separation between the conductors, and gap between the neighboring pair lines. Measured performance of these couplers is summarized in Table 8.5. The spiral-type configuration provides tighter coupling than the meander type.

8.8 Other Tight Couplers

There are several other types of tight couplers described in the literature [40–54]. These include braided microstrip [40, 41], vertically installed [42, 43], slot-coupled [44, 45], combline [46, 47], finline [48, 49], coplanar waveguide [50], wiggly two-line [51], and dielectric waveguide [52–54] directional couplers. Because it is not within the scope of this book to go into such detail, readers are referred to the above-mentioned references.

References

[1] Reed, J., and G. Wheeler, "A Method of Analysis of Symmetrical Four-Port Networks," *IRE Trans.*, Vol. PGMTT-6, Oct. 1958, pp. 398–403.

[2] Howe, H., Jr., *Stripline Circuit Design*, Artech House, Norwood, MA, 1974.

[3] Sazanov, D. M., A. N. Gridin, and B. A. Mishustin, *Microwave Circuits*, Mir Publishers, Moscow, CIS, 1982.

[4] Pozar, D. M., *Microwave Engineering*, Addison-Wesley Publishing Company, Reading, MA, 1990.

[5] Gupta, K. C., et al., *Microstrip Lines and Slotlines*, Sec. Ed. Artech House, Norwood, MA, 1996.

[6] Gipprich, J. W., *A New Class of Branch-Line Directional Couplers*, IEEE MTT-S Int. Microwave Symp. Dig., 1993, pp. 589–592.

[7] Hirota, T., A. Minakawa, and M. Muraguchi, "Reduced-Size Branch-Line and Rat-Race Hybrids for Uniplanar MMICs," *IEEE Trans. Microwave Theory Tech.*, Vol. MTT-38, Mar. 1990, pp. 270–275.

[8] Caulton, M., et al., "Status of Lumped Elements in Microwave Integrated Circuit—Present and Future," *IEEE Trans., Microwave Theory Tech.*, Vol. MTT-19, July 1991, pp. 558–599.

[9] Fusco, V. F., and S. B. D. O'Caireallain, "Lumped Element Hybrid Networks for GaAs MMICs," *Microwave and Optical Tech. Lett.*, Vol. 2, Jan. 1989, pp. 19–23.

[10] Levy, R., and L. F. Lind, "Synthesis of Symmetrical Branch-Guide Directional Couplers," *IEEE Trans. Microwave Theory Tech.*, Vol. MTT-16, Feb. 1968, pp. 80–89.

[11] Ali, F., and A. Podell, "Design and Application of a 3:1 Bandwidth GaAs Monolithic Spiral Quadrature Hybrid," *IEEE GaAs IC Symp. Dig.*, 1990, pp. 279–282.

[12] Kamitsuna, H., "A Very Small, Low-Loss MMIC Rat-Race Hybrid Using Elevated Coplanar Waveguides," *IEEE Microwave Guided Wave Letts.*, Vol. 2, Aug. 1992, pp. 337–339.

[13] Parisi, S. J., "A Lumped Element Rat-Race Coupler," *Applied Microwave*, Vol 1, Aug/Sep. 1989, pp. 84–93.

[14] Lange, J., "Interdigital Stripline Quadrature Hybrid," *IEEE Trans., Microwave Theory Tech.*, Vol. MTT-17, Dec. 1969, pp. 1,150–1,151.

[15] Ou, W. P., "Design Equations for an Interdigitated Directional Coupler," *IEEE Trans., Microwave Theory Tech.*, Vol. MTT-23, Feb. 1975, pp. 253–255.

[16] Presser, A., "Interdigitated Microstrip Coupler Design," *IEEE Trans., Microwave Theory Tech.*, Vol. MTT-26, Oct. 1978, pp. 801–805.

[17] Bryant, T. G., and J. A. Weiss, "MSTRIP: Computer Program Descriptions," *IEEE Trans., Microwave Theory Tech.*, Vol. MTT-19, Apr. 1971, pp. 418–419.

[18] Paolino, D. D., Design More Accurate Interdigitated Couplers, *Microwaves*, Vol. 15, May 1976, pp. 34–38.

[19] Kajfez, D., Z. Paunovic, and S. Pavlin, "Simplified Design of Lange Coupler," *IEEE Trans., Microwave Theory Tech.*, Vol. MTT-26, Oct. 1978, pp. 806–808.

[20] Shelton, J. P., J. Wolfe, and R. Von Wagoner, "Tandem Couplers and Phase Shifters for Multi-Octave Bandwidth," *Microwaves*, Vol. 4, Apr. 1965, pp. 14–19.

[21] Robertson, I. D., and A. H. Aghvami, "Novel Coupler for Gallium Arsenide Monolithic Microwave Integrated Circuit Applications," *Electronics Letts.*, Vol. 24, Dec. 1988. pp. 1,577–1,578.

[22] Izadian, J. S., "A New 6-18 GHz, -3 dB Multisection Hybrid Coupler Using Asymmetric Broadside, and Edge Coupled Lines," *IEEE MTT-S Int. Microwave Symp. Dig.*, 1989, pp. 243–247.

[23] Toyoda, I., T. Hirota, T. Hiraoka, and T. Tokumitsu, "Multilayer MMIC Branch-Line Coupler and Broad-Side Coupler," IEEE Microwave and Millimeter-Wave Monolithic Circ. Symp. Dig., 1992, pp. 79–82.

[24] Bamba, S., and H. Ogawa, "Multilayer MMIC Directional Couplers Using Thin Dielectric Layers," *IEEE Trans., Microwave Theory Tech.*, Vol. 43, June 1995, pp. 1,270–1,275.

[25] Gokdemir, T., et al., "K/Ka-Band Coplanar Waveguide Directional Couplers Using a Three-Metal-Level MMIC Process," *IEEE Microwave Guided Wave Lett.*, Vol. 6, Feb. 1996, pp. 76–78.

[26] Okazaki, H., and T. Hirota, "Multilayer MMIC Broad-Side Coupler With a Symmetric Structure," *IEEE Microwave Guided Wave Lett.*, Vol. 7, June 1997, pp. 145–146.

[27] Willems, D., and I. Bahl, "An MMIC-Compatible Tightly Coupled Line Structure Using Embedded Microstrip," *IEEE Trans. Microwave Theory Tech.*, Vol. MTT- 41, Dec. 1993, pp. 2,303–2,310.

[28] Seymour, S.B., "The Re-Entrant Cross Section and Wide-Band 3-dB Hybrid Couplers," *IEEE Trans. Microwave Theory Tech.*, Vol. MTT-11, Jul. 1963, pp. 254–258.

[29] Lavendol, L., and J. J. Taub, "Re-Entrant Directional Coupler Using Strip Transmission Line," *IEEE Trans. Microwave theory Tech.*, Vol. MTT-13, Sept. 1965, pp. 700–701.

[30] Malherbe, J. A. G., and I. E. Losch, "Directional Couplers Using Semi-Re-entrant Coupled Lines," *Microwave J.*, Vol. 30, Nov. 1987, pp. 121–128.

[31] Pavio, A. M., and S. K. Sutton, "A Microstrip Re-entrant Mode Quadrature Coupler for Hybrid and Monolithic Circuit Applications," *IEEE MTT-S Int. Microwave Symp. Dig.*, 1990, pp. 573–576.

[32] Nakajima, M., E. Yamashita, and M. Asa, "New Broad-Band 5-Section Microstrip-Line Directional Coupler," *IEEE MTT-S Int. Microwave Symp. Dig.*, 1990, pp. 383–386.

[33] Nakajima, M., and E. Yamashita, "A Quasi-TEM Design Method for 3 dB Hybrid Couplers Using a Semi-Re-entrant Coupling Section," *IEEE Trans. Microwave Theory Tech.*, Vol. 38, Nov. 1990, pp. 1731–1733.

[34] Tsai, C.M., and K. C. Gupta, "CAD Procedures for Planar Re-entrant Type Couplers and Three-Line Baluns," *IEEE MTT-S Int. Microwave Symp. Dig.*, 1993, pp. 1013–1016.

[35] Hagerheiden, J., M. Ciminera, and G. Jue, "Improved Planar Spiral Transformer Theory Applied to a Miniature Lumped Element quadrature Hybrid," *IEEE Trans. Microwave Theory Tech.*, Vol. 45, Apr. 1997, pp. 543–545.

[36] Shibata, K., et. al., "Microstrip Spiral Directional Coupler," *IEEE Trans. Microwave Theory Tech.*, Vol., MTT-29, July, 1981, pp. 680–689.

[37] Arai, S., et al., "A 900 MHz 90 Degrees Hybrid for QPSK Modulator," *IEEE MTT-S Int. Microwave Symp. Dig.*, 1991, pp. 857–860.

[38] Tanaka, H., et al., "2 GHz One Octave-band 90 Degree Hybrid Coupler Using Coupled Meandered Line Optimized by 3-D FEM," *IEEE MTT-S Int. Microwave Symp. Dig.*, 1994, pp. 903–906.

[39] Tanaka, H., et al., "Miniaturized 90 Degree Hybrid Coupler Using High Dielectric Substrate for QPSK Modulator," *IEEE MTT-S Int. Microwave Symp. Dig.*, 1996, pp. 793–796.

[40] Tajima, Y., and A. Platzker, "Monolithic Hybrid Quadrature Couplers" (Braided Structures), *IEEE GaAs IC Symposium Digest*, 1982, pp. 154–155.

[41] Willems, D. A., "A. Broadband MMIC Quadrature Coupler Using a Braided Microstrip Structure," *IEEE MTT-S Int. Microwave Symp. Dig.*, 1994, pp. 889–902.

[42] Dongtien, L., "New types of 3-dB Directional Couplers of Microstrip Transmission Lines," *IEEE MTT-S Int. Microwave Symp. Dig.*, 1986, pp. 265–266.

[43] Konishi, Y., et al., "A Directional Coupler of a Vertically Installed Planar Circuit Structure," *IEEE Trans. Microwave Theory Tech.*, Vol. 36, Jun. 1988, pp. 1057–1063.

[44] Wong, M. F., et al., "Analysis and Design of Slot-Coupled Directional Couplers Between Double-Sided Substrate Microstrip Lines," *IEEE Trans. Microwave Theory Tech.*, Vol. 39, Dec. 1991, pp. 2123–2128.

[45] Gillick, M., I. D. Robertson, and J. S. Joshi, "Design Analysis of Novel Coupling Structures for Multilayer MMICs," *IEEE Trans. Microwave Theory Tech.*, Vol. 41, Feb. 1993, pp. 346–349.

[46] Islam, S., "Multiway Uniform Combline Directional Couplers for Microwave Frequencies," *IEEE Trans. Microwave Theory Tech.*, Vol. 36, Jun. 1988, pp. 985–993.

[47] Islam, S., "A New Analytic Design Technique for Two- and Three-Way Warped Mode Combline Directional Couplers," *IEEE Trans. Microwave Theory Tech.*, Vol. 37, Jan. 1989, pp. 34–42.

[48] Beyer, A., D. Kother, and I. Wolff, "Development of a Coupler in Fin Line Technique," *IEEE MTT-S Int. Microwave Symp. Dig.*, 1985, pp. 139–142.

[49] Labonte, S., and W. J. R. Hoefer, Analysis and Optimization of E-Plane Directional Couplers, *IEEE MTT-S Int. Microwave Symp. Dig.*, 1988, pp. 721–724.

[50] Rius, E., et al., A. Broadband High Directivity 3-dB Coupler Using Coplanar Waveguide Technology, *IEEE MTT-S Int. Microwave Symp. Dig.*, 1995, pp. 671–674.

[51] Uysal, S., *Nonuniform Line Microstrip Directional Couplers and Filters*, Artech House, Norwood (MA), 1993.

[52] Kim, D. I., "Directly Connected Image Guide 3-dB Couplers with Very Flat Couplings," *IEEE Trans. Microwave Theory Tech.*, Vol. MTT-32, Jun. 1984, pp. 621–627.

[53] Ikalainen, P. K., and G. L. Matthaei, "Design of Broadband Dielectric Waveguide 3-dB Couplers," *IEEE Trans. Microwave Theory Tech.*, Vol. MTT-35, Jul. 1987, pp. 621–628.

[54] Rodriguez, J., and A. Prieto, "Wide-Band Directional Couplers in Dielectric Waveguide," *IEEE Trans. Microwave Theory Tech.*, Vol, 35, Aug. 1987, pp. 681–686.

9

Coupled-Line Filters

9.1 Introduction

Typically, a microwave circuit consists of a number of components, or parts, the functions of which depend on the specific application in mind. Engineering these components for a desired frequency response is often difficult and cost prohibitive, and usually the required frequency response may be obtained by the use of filters. Filters can be fabricated from lumped or distributed elements or a combination of both and can usually be designed for the precise frequency response required, at low cost. Thus, they have been used for a very long time and are popular microwave components, present in virtually every microwave subsystem.

The primary parameters of interest in a filter are the frequency range, bandwidth, insertion loss, stopband attenuation and frequencies, input and output impedance, group delay, and transient response. Consider Figure 9.1, where P_{in} is the incident power, P_R the power reflected back to the generator, P_A the power absorbed by the filter, and P_L power transmitted to the load:

$$P_{in} = P_R + P_A \tag{9.1}$$

and if the filter is lossless and there are no reflections, $P_L = P_A$ and $P_L = P_{in}$. The insertion loss (in decibels) at a particular frequency can be defined as

$$IL = -10 \log(P_L/P_{in}) \tag{9.2}$$

while the return loss is given by

$$RL = -10 \log(P_R/P_{in}) = -10 \log\left[\frac{VSWR-1}{VSWR+1}\right]^2 \tag{9.3}$$

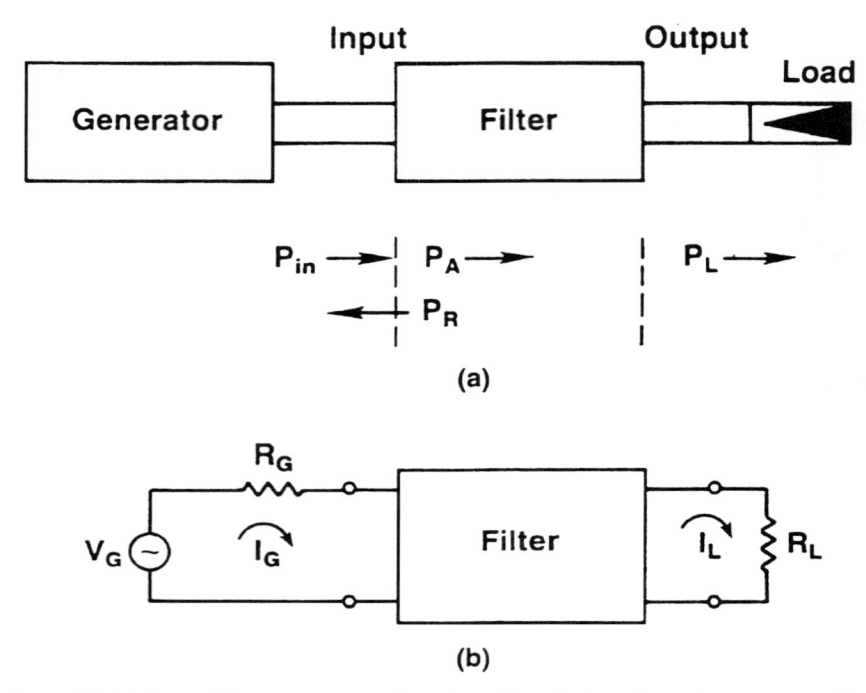

Figure 9.1 (a) General filter network configuration; (b) equivalent circuit for power-transfer calculations.

The group delay (τ_D), which is a measure of the time taken by a signal to propagate through the filter, is given by

$$\tau_D = -\frac{1}{2\pi}\frac{d\Phi_T}{df} \tag{9.4a}$$

where

$$\phi_T = arg(IL) \tag{9.4b}$$

For no frequency dispersion, the group delay should be constant over the required frequency band.

Finally, the transient and steady-state response of a filter may be different. This feature is an important consideration for certain applications. In general, transient effects can be ignored if pulsewidths are longer than the group delay.

9.1.1 Types of Filters

Filters may be classified in a number of ways. An example of one such classification is reflective versus dissipative. In a reflective filter, signal rejection is achieved by reflecting the incident power, while in a dissipative filter, the rejected signal is dissipated internally in the filter. In practice, reflecting filters are used in most applications. The most conventional description of a filter is by its frequency characteristic such as lowpass, bandpass, bandstop, or highpass. Typical frequency responses for these different types are shown in Figure 9.2. In addition, an ideal filter displays zero insertion loss, constant group delay over the desired passband, and infinite rejection elsewhere. However, in practice, filters deviate from these characteristics and the parameters in the introduction above are a good measure of performance.

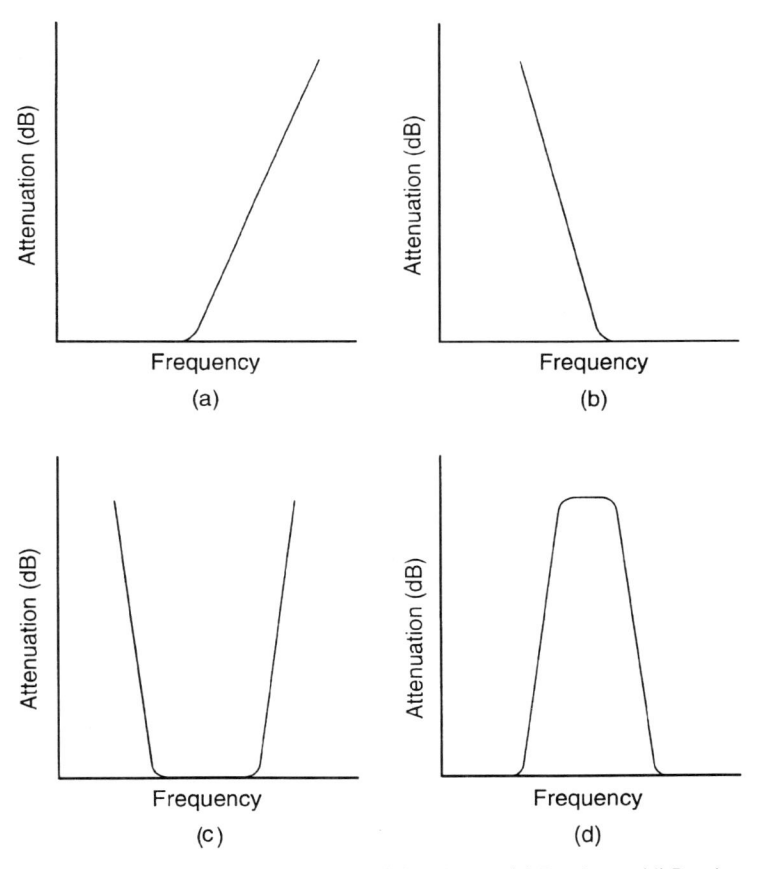

Figure 9.2 Basic filter responses (a) Lowpass, (b) Highpass, (c) Bandpass, (d) Bandstop.

9.1.2 Applications

As mentioned above, virtually all microwave receivers, transmitters, and so forth require filters. Typical commonly used circuits that require filters include mixers, transmitters, multiplexers, and the like. Multiplexers are essential for channelized receivers. System applications of filters include radars, communications, surveillance, ESM receivers, Satellite Communications (SATCOM), mobile communications, direct broadcast satellite systems, personal communication systems (PCS), and microwave FM multiplexers. In many instances, such as PCS, miniature filters are a key to realizing the required reduction in size. There is, however, a significant reduction in power handling capacity and an increase in the insertion loss. The former is not a severe limitation in such systems, however, and the latter can be compensated for by subsequent power amplification.

In this chapter we constrain ourselves to dealing mostly with coupled-line filters. In addition, a small section on computer-aided design and synthesis software is included. Finally, because of the importance of filter miniaturization for some applications, we discuss some issues related to this.

9.2 Theory and Design of Filters

An ideal bandpass filter with no attenuation or phase shift of the passband frequencies and total attenuation of all out-of-band frequencies is impractical to realize. In practice, a polynomial transfer function such as Butterworth, Chebyshev, and Bessel is used to model the filter response. A combination of inductors and capacitors, as shown in Figure 9.3, will obviously result in a lowpass filter, and we can develop a prototype normalized to 1Ω and a 1-rad cutoff frequency. From here, it is simply a matter of scaling the g values to obtain the desired frequency response and insertion loss. In addition, other filter types such as highpass, bandpass, and bandstop merely require a transformation in addition to the scaling to obtain the desired characteristics.

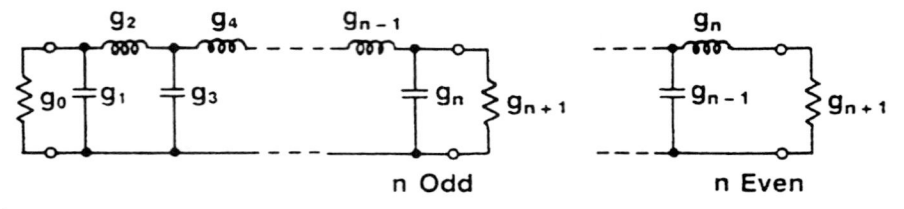

Figure 9.3 Lowpass filter prototype.

9.2.1 Maximally Flat or Butterworth Prototype

In the Butterworth lowpass prototype, the insertion loss should be as flat as possible at zero frequency and rise monotonically as fast as possible with increasing frequency. With n as the order of the filter (i.e., the number of reactive elements required to obtain the desired response), f_1 the defined 3-dB band-edge point, and f the frequency of interest, the insertion loss is given by

$$IL = 10\log[1 + (f/f_1)^{2n}] \quad (9.5)$$

Nomographs, as shown in Figure 9.4, can be used to determine the stopband attenuation versus number of sections for the desired bandwidth. For

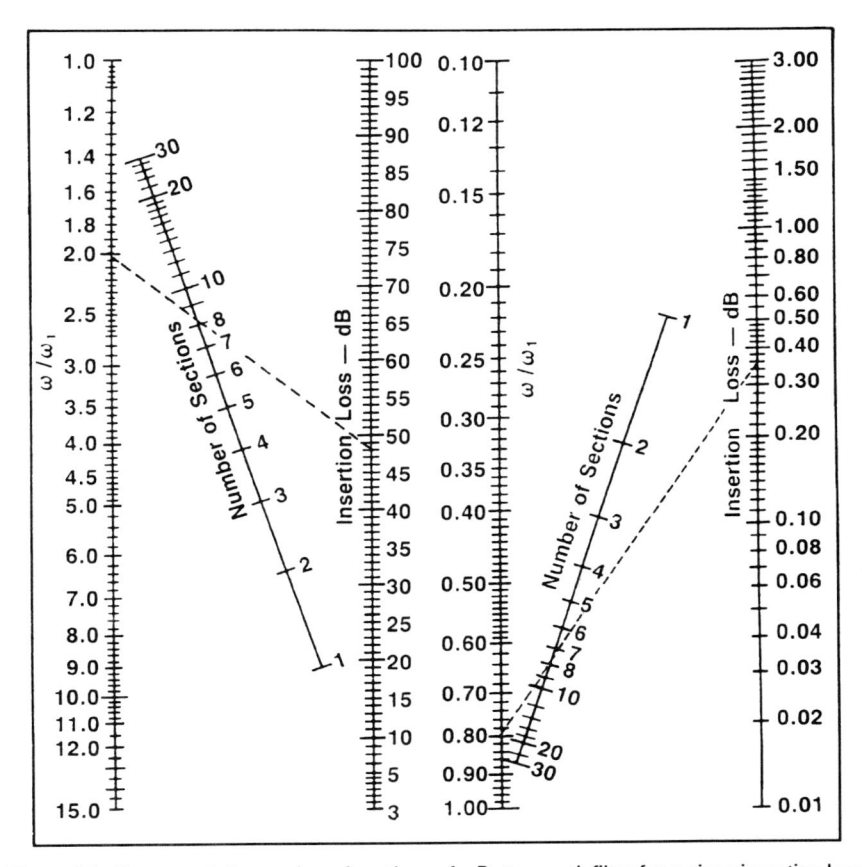

Figure 9.4 Nomograph for number of sections of a Butterworth filter for a given insertion loss in the stopband. (From [6], © 1985 Microwave and RF. Reprinted with permission.)

Table 9.1
Element values for a Butterworth filter with $g_0 = 1$, $\omega_1 = 1$, and $n = 1$ to 10

Value of n	g_1	g_2	g_3	g_4	g_5	g_6	g_7	g_8	g_9	g_{10}	g_{11}
1	2.000	1.000									
2	1.414	1.414	1.000								
3	1.000	2.000	1.000	1.000							
4	0.7654	1.848	1.848	0.7654	1.000						
5	0.6180	1.618	2.000	1.618	0.618	1.000					
6	0.5176	1.414	1.932	1.932	1.414	0.5176	1.000				
7	0.4450	1.247	1.802	2.000	1.802	1.247	0.445	1.000			
8	0.3902	1.111	1.663	1.962	1.962	1.663	1.111	0.3902	1.000		
9	0.3473	1.000	1.532	1.879	2.000	1.879	1.532	1.000	0.3473	1.000	
10	0.3129	0.908	1.414	1.782	1.975	1.975	1.782	1.414	0.908	0.3129	1.000

example, an eight-section filter gives an attenuation of approximately 48 dB at $f/f_1 = 2.0$ in the stopband, while it results in an attenuation of 0.35 dB at $f/f_1 = 0.8$ in the passband.

The Butterworth prototype values can be calculated from the equations below and are tabulated in Table 9.1 for filters with $n = 1$ to 10 reactive elements:

$$g_0 = 1$$

$$g_l = 2\sin\left(\frac{(2l-1)\pi}{2n}\right), \quad l = 1, 2, \ldots n \quad (9.6)$$

$$g_{n+1} = 1 \text{ for all } n$$

These g values can be scaled for the desired filter input termination resistance R and cutoff frequency $\omega_1 = 2\pi f_1$ as

$$L = gR/\omega_1 \quad (9.7)$$

$$C = g/(\omega_1 R) \quad (9.8)$$

9.2.2 Chebyshev Response

In the Chebyshev response filter, the insertion loss remains less than a specified level A_c, up to a specified frequency ω_1, and then rises quickly and monotonically

Coupled-Line Filters 311

with frequency. For an nth order filter, with A_c the ripple magnitude (in decibels) and ω_1, the bandwidth over which the insertion loss has maximum ripple, the insertion loss is given by

$$IL = 10\log[1 + (10^{A_c/10} - 1)\cos^2(n\cos^{-1}\omega/\omega_1)] \quad (9.9)$$

As in the case of the Butterworth response, a nomogram as shown in Figure 9.5 can be used to determine the filter characteristics. With the cutoff

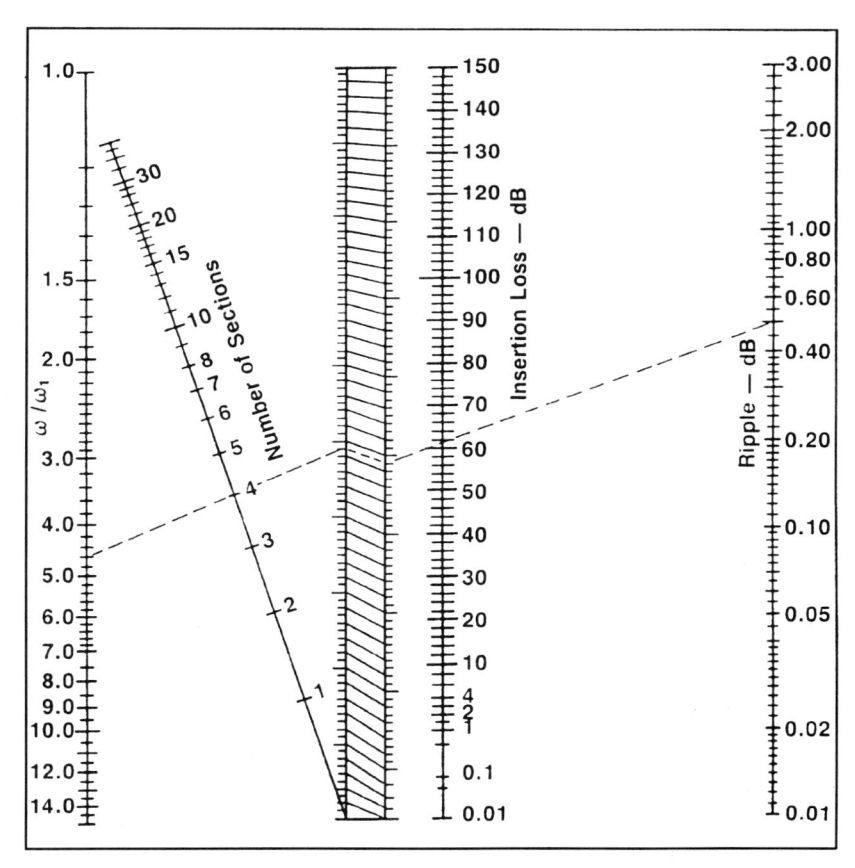

Figure 9.5 Nomograph for number of sections of a Chebyshev filter for a given ripple and stopband insertion loss. (From [6], © 1985 Microwaves and RF. Reprinted with permission.)

defined as the ripple value, the lowpass prototype g values are

$$g_0 = 1$$
$$g_1 = \frac{2a_1}{\gamma}$$
$$g_k = \frac{4a_{k-1}a_k}{b_{k-1}g_{k-1}} \qquad k = 2, 3, \ldots n \qquad (9.10)$$
$$g_{n+1} = 1 \qquad \text{for } n \text{ odd}$$
$$g_{n+1} = \coth^2(\beta/4) \qquad \text{for } n \text{ even}$$

The passband $VSWR$ maximum is related to the ripple level A_c by

$$VSWR_{\max} = \frac{1+A}{1-A}$$

where

$$A = 10\left[1 - 10^{-A_c/10}\right]^{1/2}$$

and

$$a_k = \sin\left[\frac{(2k-1)\pi}{2n}\right] \qquad k = 1, 2, \ldots n$$
$$b_k = \gamma^2 + \sin^2\left(\frac{k\pi}{n}\right) \qquad k = 1, 2, \ldots n \qquad (9.11)$$
$$\beta = \ln\left(\coth\frac{A_c}{17.37}\right)$$
$$\gamma = \sinh\left(\frac{\beta}{2n}\right).$$

Notice that for n even, the terminating impedances are not equal. The g values are tabulated in Table 9.2 for $g_0 = 1$, $\omega_1 = 1$, and $n = 1$ to 10 for various ripple values. In general, a ripple value in the 0.01- to 0.2-dB range is used.

9.2.3 Other Response-Type Filters

Some other response-type filters are also commonly used, including the elliptic function response, the Bessel response, and the generalized Chebyshev response [1–9]. The elliptic function response is a popular type, and some characteristics of this are discussed below.

Here the stopband has a series of peaks and a minimal attenuation level L_m. However, no simple equation for the insertion loss is possible. These filters are treated in detail in [1–4], with specific element values for different n values

Table 9.2
Element values for a Chebyshev lowpass prototype with $g_0 = 1$, $\omega_1 = 1$, and $n = 1$ to 10 for different ripple values

Value of n	g_1	g_2	g_3	g_4	g_5	g_6	g_7	g_8	g_9	g_{10}	g_{11}
					0.01-db ripple						
1	0.0960	1.0000									
2	0.4488	0.4077	1.1007								
3	0.6291	0.9702	0.6291	1.0000							
4	0.7128	1.2003	1.3212	0.6476	1.1007						
5	0.7563	1.3049	1.5773	1.3049	0.7563	1.0000					
6	0.7813	1.3600	1.6896	1.5350	1.4970	0.7098	1.1007				
7	0.7969	1.3924	1.7481	1.6331	1.7481	1.3924	0.7969	1.0000			
8	0.8072	1.4130	1.7824	1.6833	1.8529	1.6193	1.5554	0.7333	1.1007		
9	0.8144	1.4270	1.8043	1.7125	1.9057	1.7125	1.8043	1.4270	0.8144	1.0000	
10	0.8196	1.4369	1.8192	1.7311	1.9362	1.7590	1.9055	1.6527	1.5817	0.7446	1.1007
					0.1-dB ripple						
1	0.3052	1.0000									
2	0.8430	0.6220	1.3554								
3	1.0315	1.1474	1.0315	1.0000							
4	1.1088	1.3061	1.7703	0.8180	1.3554						
5	1.1468	1.3712	1.9750	1.3712	1.1468	1.0000					
6	1.1681	1.4039	2.0562	1.5170	1.9029	0.8618	1.3554				
7	1.1811	1.4228	2.0966	1.5733	2.0966	1.4228	1.1811	1.0000			
8	1.1897	1.4346	2.1199	1.6010	2.1699	1.5640	1.9444	0.8778	1.3554		
9	1.1956	1.4425	2.1345	1.6167	2.2053	1.6167	2.1345	1.4425	1.1956	1.0000	
10	1.1999	1.4481	2.1444	1.6265	2.2253	1.6418	2.2046	1.5821	1.9628	0.8853	1.3554
					0.2-dB ripple						
1	0.4342	1.0000									
2	1.0378	0.6745	1.5386								
3	1.2275	1.1525	1.2275	1.0000							
4	1.3028	1.2844	1.9761	0.8468	1.5386						
5	1.3394	1.3370	2.1660	1.3370	1.3394	1.0000					
6	1.3598	1.3632	2.2394	1.4555	2.0974	0.8838	1.5386				
7	1.3722	1.3781	2.2756	1.5001	2.2756	1.3781	1.3722	1.0000			
8	1.3804	1.3875	2.2963	1.5217	2.3413	1.4925	2.1349	0.8972	1.5386		

(Continued)

Table 9.2 *(Continued)*

Value of n	g_1	g_2	g_3	g_4	g_5	g_6	g_7	g_8	g_9	g_{10}	g_{11}
9	1.3860	1.3938	2.3093	1.5340	2.3728	1.5340	2.3093	1.3938	1.3860	1.0000	
10	1.3901	1.3983	2.3181	1.5417	2.3904	1.5536	2.3720	1.5066	2.1514	0.9034	1.5386
					0.5-dB ripple						
1	0.6986	1.0000									
2	1.4029	0.7071	1.9841								
3	1.5963	1.0967	1.5963	1.0000							
4	1.6703	1.1926	2.3661	0.8419	1.9841						
5	1.7058	1.2296	2.5408	1.2296	1.7058	1.0000					
6	1.7254	1.2479	2.6064	1.3137	2.4758	0.8696	1.9841				
7	1.7372	1.2583	2.6381	1.3444	2.6381	1.2583	1.7372	1.0000			
8	1.7451	1.2647	2.6564	1.3590	2.6964	1.3389	2.5093	0.8796	1.9841		
9	1.7504	1.2690	2.6678	1.3673	2.7239	1.3673	2.6678	1.2690	1.7504	1.0000	
10	1.7543	1.2721	2.6754	1.3725	2.7392	1.3806	2.7231	1.3485	2.5239	0.8842	1.9841

from 3 to 9 given in [7]. This type of filter provides a much steeper stopband skirt for a given n and passband/stopband insertion loss than either the Butterworth or Chebyshev response filters.

9.2.4 LC Filter Transformation

As mentioned before, highpass, bandpass or bandstop filters require a transformation in addition to scaling, and these transformations are discussed below. For lowpass filters, scaling to the desired frequency band and impedance level is accomplished by using the equations given below for the series inductors and shunt capacitors. In this case

$$\omega/\omega_1 = \omega/\omega_{Lpb}$$

$$L_k = g_k(Z_0/\omega_{Lpb}) \qquad (9.12)$$

$$C_k = g_k(1/\omega_{Lpb} Z_0) \qquad (9.13)$$

where ω_{Lpb} is the required lowpass cutoff radian frequency and Z_0 is the input and output termination impedance.

Highpass Transformation. Transposing the series inductances into series capacitances and the shunt capacitances into shunt inductances transforms the lowpass prototype into a highpass filter (see Figure 9.6). Thus

$$\frac{\omega}{\omega_1} = -\frac{\omega_{Hpb}}{\omega}$$

Coupled-Line Filters

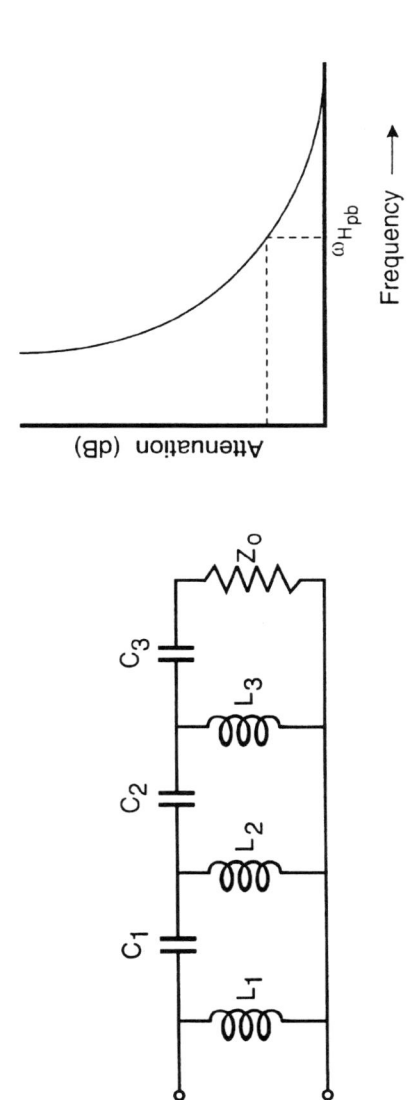

Figure 9.6 Highpass filter schematic and typical frequency response.

and
$$C_{hp} = \frac{1}{L_{lp}}, \quad L_{hp} = \frac{1}{C_{lp}} \tag{9.14}$$

where ω_{Hpb} is the band-edge frequency. The element values obtained are

$$C_k = \frac{1}{g_k \omega_{Hpb} Z_0} \tag{9.15a}$$

$$L_k = \frac{Z_0}{g_k \omega_{Hpb}} \tag{9.15b}$$

Bandpass Transformation. Bandpass filters also require transformation and scaling. Series inductors of the lowpass prototype are transformed into a series combination of an inductor and a capacitor, while the shunt capacitors are transformed into a parallel combination of an inductor and a capacitor (see Figure 9.7). Hence the bandpass filter has twice the number of elements. With a lower cutoff frequency f_l and upper cutoff frequency f_u defined for the bandpass, the center frequency f_o, bandwidth BW, and fractional bandwidth f_b are defined by

$$f_0 = \sqrt{f_\ell f_u} \tag{9.16}$$

$$BW = f_u - f_\ell \tag{9.17}$$

$$f_b = BW/f_0 \tag{9.18}$$

The transformed series elements are

$$L_{fbse} = L_{lp}/f_b; \quad C_{fbse} = 1/L_{fbse} \tag{9.19}$$

while the transformed shunt element values are

$$C_{fbsh} = C_{\ell p}/f_b, \quad L_{fbsh} = 1/C_{fbsh} \tag{9.20}$$

where C_{lp} is the lowpass prototype capacitor g value and L_{lp} the inductor g value. A typical filter structure that results is shown in Figure 9.7. In this case

$$\frac{\omega}{\omega_1} = \frac{f_0}{BW}\left(\frac{f}{f_0} - \frac{f_0}{f}\right)$$

and

$$L_k = g_k \frac{Z_0}{2\pi BW}; \quad C_k = \frac{2\pi BW}{g_k Z_0 \omega_0^2} \quad \text{(Series elements)} \tag{9.21a}$$

$$L_k = \frac{2\pi BW Z_0}{g_k \omega_0^2}; \quad C_k = \frac{g_k}{2\pi BW Z_0} \quad \text{(Shunt elements)} \tag{9.21b}$$

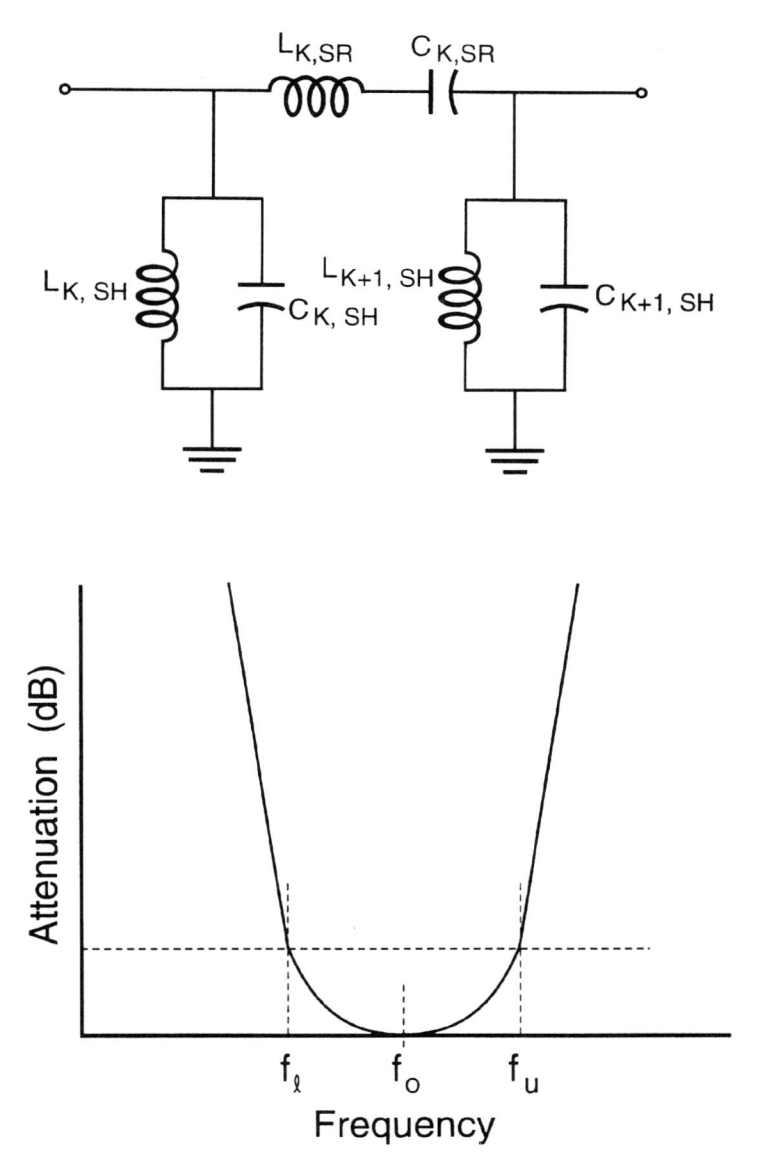

Figure 9.7 Bandpass filter structure with typical frequency response.

where

$$\omega_0^2 = \frac{1}{L_k C_k}$$

Bandstop Transformation. Here, the shunt capacitor of the lowpass prototype is transformed into a series inductor and capacitor in shunt to ground and the series inductor is replaced with a parallel inductor-capacitor in series as shown in Figure 9.8. The shunt element values are

$$L_{bssh} = 1/(C_{\ell p} \cdot f_b); \qquad C_{bssh} = 1/L_{bssh} \qquad (9.22)$$

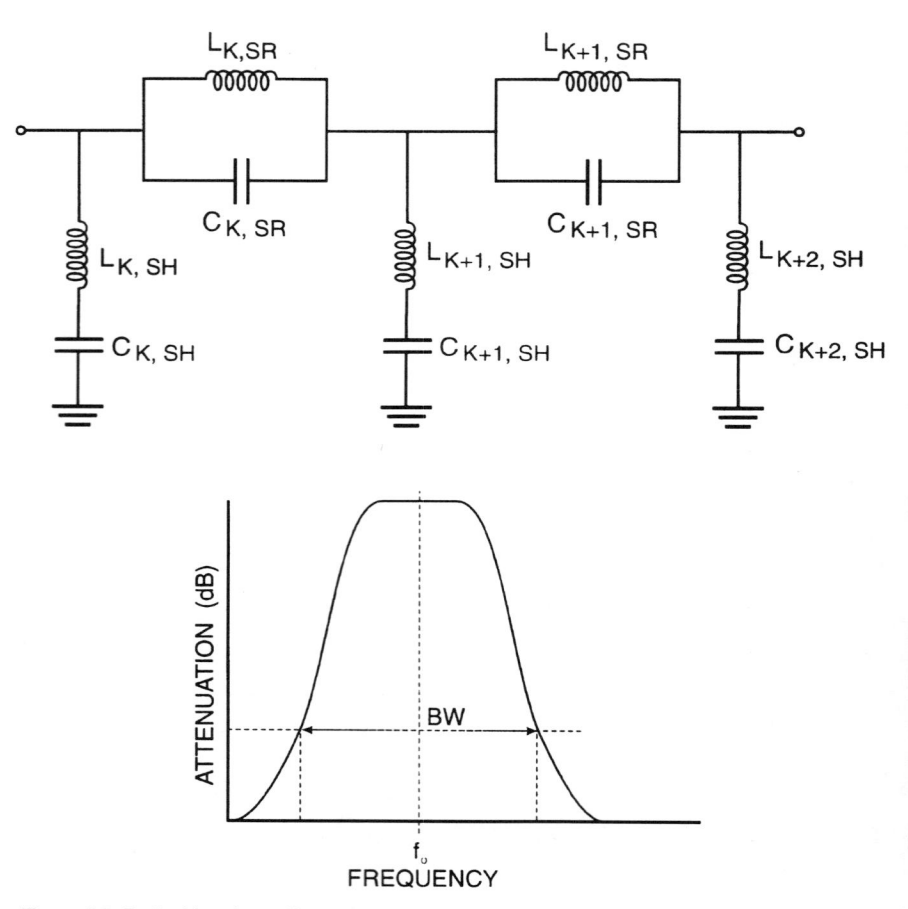

Figure 9.8 Typical bandstop filter structure and its frequency response.

whereas the series elements are given by

$$C_{bsse} = 1/L_{\ell pfl}; \qquad L_{bsse} = 1/C_{fbse} \qquad (9.23)$$

In this case

$$\frac{\omega_1}{\omega} = \frac{f_0}{BW}\left(\frac{f}{f_0} - \frac{f_0}{f}\right)$$

with parallel-tuned circuit element values

$$\omega_0 C_k = \frac{1}{\omega_0 L_k} = \frac{\omega_0}{2\pi BW Z_0 g_k} \qquad (9.24a)$$

and series-tuned circuit element values

$$\omega_0 L_k = \frac{1}{\omega_0 C_k} = \frac{\omega_0 Z_0}{2\pi BW g_k} \qquad (9.24b)$$

9.2.5 Filter Analysis and CAD Methods

Filter Analysis. While the discussion above has covered some analytical aspects of various types of filters, in general to account for phase characteristics and finite Q of circuit elements, we resort to the use of either ABCD matrices or Kirchhoff's equations. The ABCD matrix method is limited to ladder networks, while Kirchhoff's equations can be applied in general to any network. Both of these techniques are well covered in [10]. Knowing the ABCD matrices for a variety of elements as given in Chapter 2, we can obtain the overall matrix of the circuit by simple matrix multiplication, taking care to perform the multiplication in the right order. This process is considerably simplified by the use of computers. The ABCD matrix method, however, cannot handle re-entrant combinations (i.e., nonladder networks) such as are present in high-performance bandpass filters. This limitation is readily overcome through the use of Kirchhoff's equations.

Computer-Assisted Design. Computers can be used effectively to simplify and speed up the design of filters, and in some cases are the only means to render practical the synthesis of filter transfer approximations. Software packages are available to simulate performance before prototype construction, thus permitting fine tuning of parameters for optimization, taking into account practical realities and fabrication constraints. Some of these packages include LINMIC + [11], which uses a full-wave field solution and planar-field simulators such as em, IE3D, and SFPMIC [12–15]. These codes can deal with multiple dielectric layers and multiple conductors and are excellent for design verification. For three-dimensional circuits, other packages, such as the HP High-Frequency Structure Simulator (HFSS) [16], which handles multiport structures with unrestricted dielectric and conductor geometry, can be used.

In general, we commence the design process by selecting the appropriate circuit construction, modeling the filter response against the physical parameters, and then using accurate models for the structure to determine the response and optimize parameters for the application at hand. Most of these aspects can be performed iteratively on a computer using an optimization routine.

One excellent software package for performing the above operation is the Genesys software suite from Eagleware Corp [17]. The tools are available in this package for the design of a wide set of LC filter topologies including conventional, narrowband, flat-delay, symmetric, elliptic, zig-zag, bandpass, lowpass, bandstop, and highpass structures. It also supports a large number of prototype transfer functions including the most commonly used, such as Butterworth, Chebyshev, and Bessel. The advantage of this package is that there are other files included that permit the design of an overall circuit network including the filter, oscillators, matching networks, and equalizers. The filters can be designed in the media of interest here (e.g., stripline, microstrip) in most configurations such as hairpin, interdigital and combline.

9.2.6 Some Practical Considerations

In the previous sections we started with the lumped-element filters. From this we can move into distributed-line filters and from there to coupled-line filters. The distributed lines can be realized in any desired form such as waveguide, coaxial lines, or planar transmission lines. Kuroda's identities [18, 19] allow one to realize lowpass structures using shunt elements with the identical response and with Richards transformations [19, 20] one can establish the distributed line parameters. Table 9.3 provides a listing of transmission line lengths with the equivalent RLC network, which can simplify design considerably. However, as the passband for distributed filters can reoccur at frequencies of twice or thrice the initial pass band frequency, stopband attenuation is severely compromised. In addition, discontinuities such as open ends, steps in linewidth, and T and cross-junction effects can degrade filter performance.

A number of practical considerations and limitations often determine the actual filter construction. For example, for satellite systems applications, naturally, size, weight, and the like are major considerations. In other cases, such as for PCS, while size and weight are again major factors, cost is critical to maintaining overall system costs at acceptable levels. Other aspects, such as finite Q, group delay, temperature effects, power-handling capacity, and tunability may also be important factors in determining filter configuration and design [10].

Table 9.3
Equivalent RLC networks for transmission line lengths

Tx Line Configuration	Equivalent Circuit	Element Values
Z_0, $\theta < 90°$ (shorted)	L	$\omega L = Z_0 \tan\theta$
Z_0 Open, $\theta < 90°$	C	$\omega C = \dfrac{1}{Z_0 \cot\theta}$
$Z_0 = \dfrac{1}{Y_0}$, $\theta = 90°$ (shorted)	L, G, C parallel	$\omega C = \dfrac{\pi}{4} Y_0$ $\omega L = \dfrac{4 Z_0}{\pi}$ $G = Y_0 \alpha \theta$ $Q = \dfrac{\omega C}{G} = \dfrac{\pi}{4\alpha\theta}$
$Z_0 = \dfrac{1}{Y_0}$, $\theta = 90°$ Open	R, L, C series	$R = Z_0 \alpha \theta$ $\omega L = \dfrac{\pi}{4} Z_0$ $\omega C = \dfrac{4}{\pi Z_0}$ $Q = \dfrac{\omega L}{R} = \dfrac{\pi}{4\alpha\theta}$
Z_0, $\theta = \dfrac{\pi}{2}$ (shorted)	R, L, C	$\omega L = \dfrac{\pi Z_0}{2}$ $\omega C = \dfrac{2}{\pi Z_0}$ $R = Z_0 \alpha \theta$ $Q = \dfrac{\pi}{2\alpha\theta}$

(Continued)

Table 9.3 *(Continued)*

Tx Line Configuration	Equivalent Circuit	Element Values
$Z_o = \frac{1}{Y_o}$, $\theta = \frac{\pi}{2}$, Open	L, G, C (parallel)	$\omega_0 C = \frac{\pi}{2} Y_0$ $\omega_0 L = \frac{2}{\pi Y_0}$ $R = Y_0 \alpha \theta$ $Q = \frac{\pi}{2\alpha\theta}$
Z_o, $\theta < 45°$	L	$\omega L = Z_0 \sin\theta$
Z_o, $\theta > 45°$	C	$\omega C = \frac{\sin\theta}{Z_0}$

For low selectivity, wideband applications, striplines, and microstrip filters are ideal. They do suffer from temperature effects however, and are difficult to tune. Suspended-substrate filters give a higher Q over microstrip, resulting in lower filter loss, sharper band edges, and better temperature stability. Dielectric resonator filters have been developed over a wide range of frequencies, and high dielectric constant materials (e.g., barium tetratitanate) can be used to decrease the size.

When one considers the losses in filters (e.g., finite Q of resonators) the passband insertion loss is increased or, as in the case of equal-ripple response filters, ripples are suppressed. The loaded Q of the resonators normally determines the bandwidth of the filter. The loaded Q in turn depends on the losses and the external circuit. When the Q of the external circuit is much less than the unloaded Q of the filter, the filter bandwidth is almost independent of the unloaded Q but as both these Q values become comparable, the circuit becomes lossy and the bandwidth is broadened, insertion loss is increased, and stop band rejection is reduced.

Printed circuit filters can handle a few hundred watts of power. The specific power-handling capability depends on the filter topology and will depend on the transmission media used. In general, the type of media discussed in this text (i.e., stripline, microstrip) are good for low-power levels but can handle high-pulse power levels (up to a few kW) and approximately 50W average power levels. Environmental temperature changes result in changes to the physical dimensions of the filter structure and therefore changes to the electrical characteristics. These effects are analyzed and discussed in [10].

Finally, the group delay of a filter depends on the selected prototype and number of sections with both band center group delay and group delay variation increasing with an increase in the number of sections. In many applications, a constant group delay over the desired bandwidth is required. The design of these filters is discussed in the literature [21].

9.3 Types of Coupled Filters

Direct-coupled resonator filters tend to have excessive lengths and generally can be reduced significantly in size by the use of parallel-coupled lines. Since parallel coupling results in much tighter coupling than with the use of end-coupled structures, greater bandwidths are possible. The first spurious response occurs at two to three times the center frequency, depending upon the media, with a larger gap being permitted between adjacent strips. A broader bandwidth is also obtained for a given gap tolerance. For compactness, resonator sections are placed side-by-side. Some commonly used designs using this concept include the parallel-coupled, interdigital, combline and hairpin line filters, and these configurations are shown in Figure 1.15 (Chapter 1). These parallel-coupled line filters will be discussed in greater detail below. The tapped input-output structures shown in the figure have a relatively narrow bandwidth.

9.3.1 Parallel-Coupled Line Filters

The design of parallel-coupled line filters was formulated by Cohn [22] and has been refined or modified by others for specific design situations or operating conditions [23–29]. For a typical n section parallel-coupled filter, one starts with computing the $g_1, g_2, \ldots g_{n+1}$ values of the lowpass prototype for either maximally flat or equal-ripple response, using (9.6) or (9.10). As shown in Figure 9.9(a), the filter is assembled from $n+1$ sections of equal length ($\lambda/4$) at the center frequency, giving a structure of n resonators. The electrical design is specified by the even- and odd-mode characteristic impedances Z_{0o}, Z_{0e} on the parallel conductors. Using these impedance values, the transmission line dimensions of each section can be determined.

A filter section and its equivalent are shown in Figures 9.9(b) and 9.9(c), respectively. The ABCD matrix of the ideal impedance inverter can be obtained

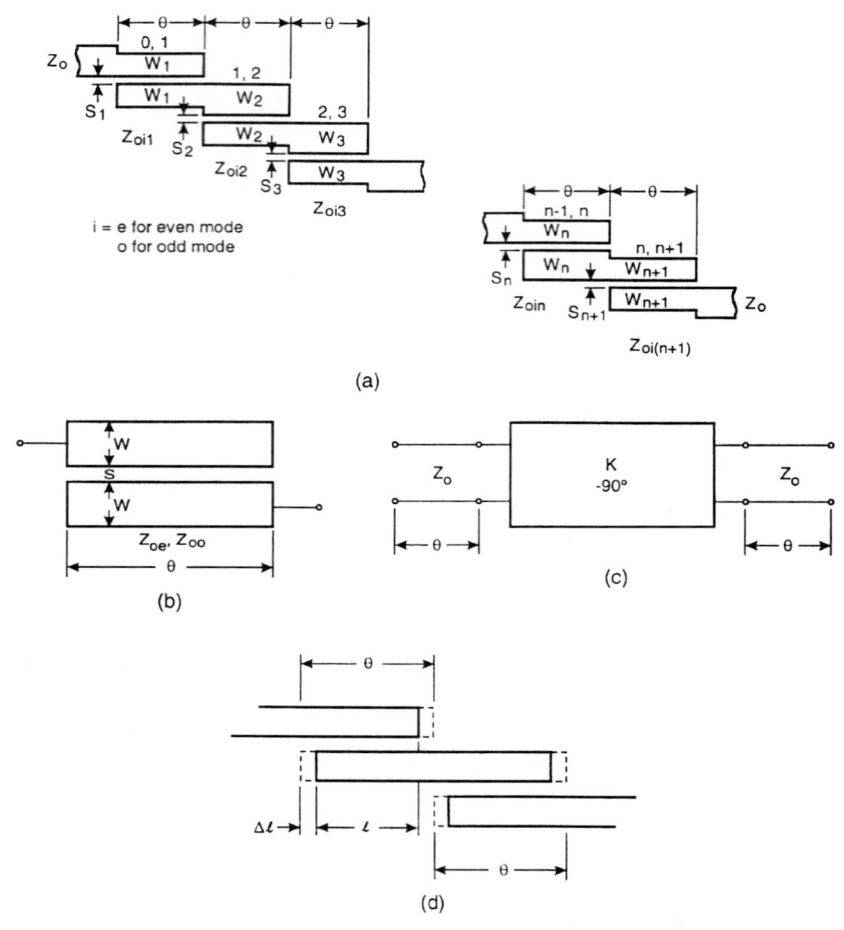

Figure 9.9 (a) Parallel-coupled transmission line resonator filter; (b) parallel-coupled line resonator; (c) K inverter-type equivalent circuit of the coupled line resonator; and (d) resonators with end effects.

by substituting $\theta = -90$ deg and $Z_0 = K$ in the ABCD matrix of a transmission line of electrical length θ and characteristic impedance Z_0. Therefore, the ABCD matrix of the complete filter section is

$$\begin{bmatrix} A & B \\ C & D \end{bmatrix} = \begin{bmatrix} \cos\theta & jZ_0\sin\theta \\ \frac{j\sin\theta}{Z_0} & \cos\theta \end{bmatrix} \begin{bmatrix} 0 & -jK \\ -\frac{j}{K} & 0 \end{bmatrix} \begin{bmatrix} \cos\theta & jZ_0\sin\theta \\ \frac{j\sin\theta}{Z_0} & \cos\theta \end{bmatrix} \quad (9.25)$$

The K inverters for the various sections are defined in terms of the $g_0, g_1, \ldots$

elements as follows [22]:

$$\frac{Z_0}{K_{01}} = \sqrt{\frac{\pi \Delta f}{2\omega_1 g_0 g_1}} \quad (9.26a)$$

$$\frac{Z_0}{K_{j,j+1}} = \frac{\pi \Delta f}{2\omega_1 \sqrt{g_j g_{j+1}}}, \quad j = 1 \text{ to } n-1 \quad (9.26b)$$

$$\frac{Z_0}{K_{n,n+1}} = \sqrt{\frac{\pi \Delta f}{2\omega_1 g_n g_{n+1}}} \quad (9.26c)$$

where Δf is the fractional bandwidth and ω_1 has been defined previously. After calculating K, the even- and odd-mode impedances of the coupled lines are calculated from the following relations:

$$\frac{(Z_{0e})_{j+1}}{Z_0} = 1 + \frac{Z_0}{K_{j,j+1}} + \left(\frac{Z_0}{K_{j,j+1}}\right)^2, \quad j = 0 \text{ to } n \quad (9.27)$$

$$\frac{(Z_{0o})_{j+1}}{Z_0} = 1 - \frac{Z_0}{K_{j,j+1}} + \left(\frac{Z_0}{K_{j,j+1}}\right)^2, \quad j = 0 \text{ to } n \quad (9.28)$$

The physical dimensions of the filter sections are then calculated [30] for the desired Z_{0e} and Z_{0o}. An approximate value for the physical length is obtained from the average value of the even- and odd-mode velocities, that is:

$$\theta = \frac{2\pi}{\lambda} l = \frac{2\pi}{\lambda_0} \frac{\sqrt{\epsilon_{ree}} + \sqrt{\epsilon_{reo}}}{2} l = \frac{\pi}{2} \quad (9.29)$$

A more accurate value for the length of the resonator is obtained by taking into account the open-end discontinuity capacitance [30], which gives rise to an additional line length Δl as shown in Figure 9.9(d). In this case, the line length calculated from (9.29) is shortened at each end by Δl:

$$\Delta l = 0.6 C_o Z_{01} / \left(\sqrt{\epsilon_{ree}} + \sqrt{\epsilon_{reo}}\right) \quad (9.30)$$

where Δl is in millimeters, the open-end capacitance C_o in picofarads [30], and $Z_{01} = \sqrt{Z_{0e} Z_{0o}}$.

Figure 9.10 depicts the bandpass characteristics of a six-section, 1-dB ripple filter designed using microstrip as a transmission media for 44 to 48 GHz [31].

In parallel-coupled microstrip filters, the physical lengths of the coupled sections are the same for both the even and odd modes, and we obtain

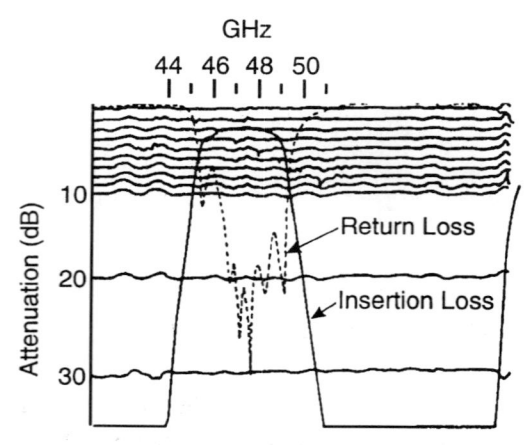

Figure 9.10 Performance of a six-section parallel-coupled microstrip line filter. (From [31], © 1978 IEEE. Reprinted with permission.)

an asymmetrical passband response with deterioration of the upper stopband from differences in phase velocities in these two modes. To improve the stopband performance, the phase velocities of the two modes should be equalized. Various techniques can be used to do this, including overcoupling the resonators, suspending the substrate, using parallel-coupled stepped impedance resonators or using capacitors at the end of the coupled sections [29]. We can also introduce wiggles in the coupled lines to accomplish this [26]. Some of these techniques are described in Chapter 6. Similarly, we can use other methods such as half-wavelength broadside-coupled microstrip lines to enhance bandwidth [23]. These techniques are useful for MIC and MMIC applications.

9.3.2 Interdigital Filters

The interdigital filter is popular because it is compact and uses the available space efficiently. It can be designed for both narrow (2–30%) and wide (30–70%) bandwidths. A typical interdigital bandpass structure is shown in Figure 1.15(b). Accurate design procedures are available for these filters [32] together with an exact design theory [33]. This filter is not as compact as the combline filter but has higher unloaded Q, thus making it a good choice for narrow-bandwidth filter applications. For large bandwidths, the capacitively-loaded parallel-coupled line filter is ideal. This also reduces the filter size. In addition, the tolerances required in their manufacture can be relaxed, spurious responses are not present, the rates of cutoff and strength of the stopbands can be increased by multiple poles of attenuation at dc and at even multiples of the center frequency of the first passband, and these filters can be fabricated without dielectrics, thereby eliminating dielectric loss.

The design equations for parallel-coupled line interdigital filters are given in [32, 33] and will not be repeated here. Essentially these equations yield the various line capacitances per unit length of the line and these can be used to determine the physical dimensions of the line. Alternately, narrow-to-moderate-bandwidth bandpass filters using resonators can be designed by calculating or measuring the coupling coefficient between resonators and the external quality factor of the input and output resonators [10]. These coupling values are then related to a normalized lowpass prototype value and can be used to realize all possible response shapes. This procedure is the most practical design method when the filter structure is complex or its equivalent circuit model is not readily available. The procedure described below is applicable to all types of coupled resonator filters, whether realized in microstrip or in any other medium.

The first step in the design procedure is finding the necessary normalized coupling coefficients in terms of lowpass prototype element values and design frequencies, as follows:

$$K_{n,n+1} = \frac{BW}{f_0\sqrt{g_n g_{n+1}}} \qquad (9.31)$$

The next step is to determine physical dimensions of the coupled resonators, depending on the transmission medium used. The final step is to determine the loaded Q of the first and the last resonators required to connect the coupled resonators to input and output terminals.

In a filter design, the singly-loaded Q is calculated from

$$Q_L = \frac{f_0}{BW}g_1 = \frac{f_0}{BW}g_{n+1} \qquad (9.32)$$

To illustrate the above-described design method, the following microstrip example is considered:

Center frequency	4 GHz
Response	Chebyshev with 0.2-dB ripple
Bandwidth	0.4 GHz
35-dB attenuation points	3.6 and 4.4 GHz
Substrate	$\epsilon_r = 9.8$, $h = 1.27$mm

From the nomograph (Figure 9.5), we find that the required number of resonators is 5. The prototype values are $g_0 = 1.0$, $g_1 = g_5 = 1.3394$, $g_2 = g_4 = 1.337$, and $g_3 = 2.166$. From (9.31) the coupling parameters are determined to be $K_{12} = K_{45} = 0.0747$, $K_{23} = K_{34} = 0.0588$, and from (9.32),

$Q_L = 13.4$. The filter can be realized using coupled microstrip medium on RT/duroid substrate or alumina substrate. Figure 9.11 shows [25] measured coupling coefficients as a function of S/h for $\epsilon_r = 2.22$ and $W/h = 1.8$, which corresponds to a single-strip impedance Z_{0I} of approximately 70 Ω, and for alumina substrate ($\epsilon_r = 9.8$) and $W/h = 0.7$ ($Z_{0I} = 58\ \Omega$). Similar curves can be obtained for other line widths and dielectric substrates. In these filters, the final step in the design is to determine the tap-point location for a given Q_L. The tap point ℓ/L (Figure 1.15(b)) can be calculated from the following equation [25]:

$$\frac{Q_L}{Z_0/Z_{0I}} = \frac{\pi}{4\sin^2(\pi\ell/2L)} \tag{9.33}$$

Thus, the filter dimensions in Figure 1.15(b) on a 1.27-mm-thick alumina substrate become

$$\begin{aligned} h &= 1.27\ \text{mm} \\ W_0 &= 1.27\ \text{mm}\ (Z_0 = 50\ \Omega) \\ W &= 0.889\ \text{mm} \\ S_{12} &= 2.16\ \text{mm} \\ S_{23} &= 2.54\ \text{mm} \\ \ell &= 2.25\ \text{mm} \\ L &= 7.5\ \text{mm} \end{aligned}$$

where S_{12} and S_{23} are the spacings between lines 1 and 2 and lines 2 and 3, respectively.

9.3.3 Combline Filters

A typical combline filter configuration is shown in Figure 1.15(c). Together with the capacitively loaded interdigital filter, the combline filter is one of the most commonly used bandpass structures. This filter is more compact than the interdigital, is generally easier to fabricate, and is an attractive alternative to other filter types, especially for narrow bandwidths. Here the resonators consist of TEM mode transmission line elements that are short-circuited at one end and have a lumped capacitance between the other end of each resonator line element and ground. Without these capacitances, the resonator lines would be exactly $\lambda/4$ at resonance and the structure would have no passband. Hence the lines must be less than 90 degrees long and capacitively loaded to achieve resonance. The capacitances are made relatively large and the lines 45 degrees or less in length, thereby resulting in a very compact and efficient coupling structure. With the line elements $\lambda/8$ at the primary passband, the second passband is far removed and the attenuation is infinite at a frequency where the line length

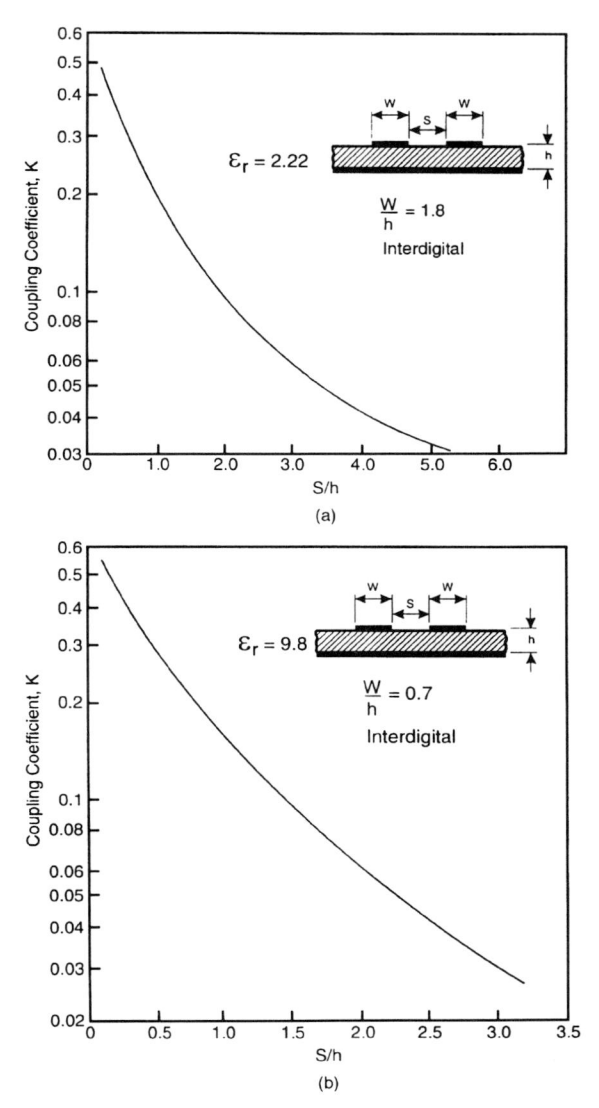

Figure 9.11 Measured coupling coefficients versus S/h for (a) RT/duroid and (b) alumina. (From [25], © 1979. IEEE. Reprinted with permission.)

is $\lambda/4$. So beyond the passband, the attenuation is very high and cutoff at the upper side of the passband can be very steep. Further, adequate coupling can be maintained between resonator elements with sizeable spacing, thereby providing more margin in tolerance requirements.

Design procedures for combline filters are discussed in [34–36]. The process is parallel to the design process of the interdigital filter discussed above and therefore is not covered in any detail here. Refer to [34–36] for the actual design equations to calculate the distributed line capacitances and from there the dimensions of the structure.

9.3.4 The Hairpin-Line Filter

The interdigital and combline filters described above require ground connections, which may be difficult to achieve when using microstrip lines on ceramic substrates. When striplines or microstrip is used, the hairpin-line filter is one of the preferred configurations. This is particularly useful when one is interested in MIC or MMIC circuits. The hairpin-line filter can be considered basically to be a folded version of a half-wave parallel coupled-line filter. It is much more compact, though, and gives approximately the same performance. A typical schematic of this type of filter is shown in Figure 1.15(d). As the frequency increases, the length-to-width ratio is smaller for a given substrate thickness, so that folding the resonator becomes impractical. Hence, this type of resonator is more suitable at lower frequencies. In general, the hairpin-line filter is larger than the combline or interdigital filter. But because no grounding is required, it is amenable to mass production as a large number of filters can be simultaneously printed on a single substrate, thereby lowering production costs.

The design of hairpin-line filters has been discussed by various researchers [37–41]. Although one could normally use the design expressions for parallel-coupled filters, here, the line between the resonators decreases the length of the coupling sections, so that the coupled sections are less than a quarter-wavelength. In addition, the bend discontinuities are difficult to handle. Using the parallel-coupled bandpass design and compensating for the end capacitance, however, dissimilar propagation velocities, shortened coupling elements, and bends [36] through an optimization routine on a computer permits rapid design of such a filter.

We can use the methods given above for the design of hairpin-line filters, so details are not provided here. Keep in mind a few points, however. The interarm spacings in the hairpin-line filter design should be kept large, so that coupling (≥ 15 dB) between the arms can be neglected. However, this leads to a larger filter size. Cristal and Frankel's [37] unified design method takes the interarm coupling into account, but assumes negligible phase shift over the line joining the arms of a hairpin. Thus, we must exercise caution when high-dielectric constant substrates are used, as small physical length variations will result in large phase shifts. Cristal and Frankel's equations are also not applicable for inhomogeneous media and do not account for the effects of right-angled bends and corners. A good comparison of several types of hairpin filters is given by Matthaei [42].

Hairpin-line filters can also be designed by calculating the inter-resonator coupling as a function of the spacing between resonators for a given set of substrate parameters, frequency, and microstrip width. The microstrip width is selected to obtain maximum Q for the resonators. Figure 9.12 shows measured coupling coefficients for hairpin resonators on alumina ($h = 1.27$ mm, $W/h = 0.7$) and high dielectric constant substrates. In these filters, the tap point (ℓ/L) is calculated using the following relation [25]:

$$\frac{Q_L}{(Z_0/Z_{01})} = \frac{\pi}{2\sin^2(\pi\ell/2L)} \tag{9.34}$$

The physical layout of a low-frequency hairpin-line filter designed using microstrip on a high-dielectric constant material is shown in Figure 9.13, while

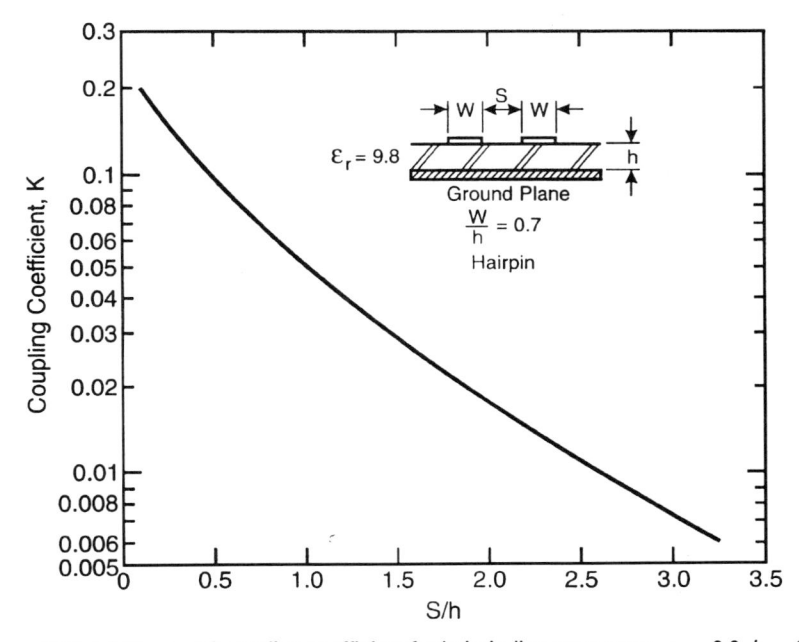

Figure 9.12 (a) Measured coupling coefficient for hairpin-line resonators, $\epsilon_r = 9.8$, $h = 1.27$ $W/h = 0.7$. (From [25], © 1979 IEEE. Reprinted with permission.)

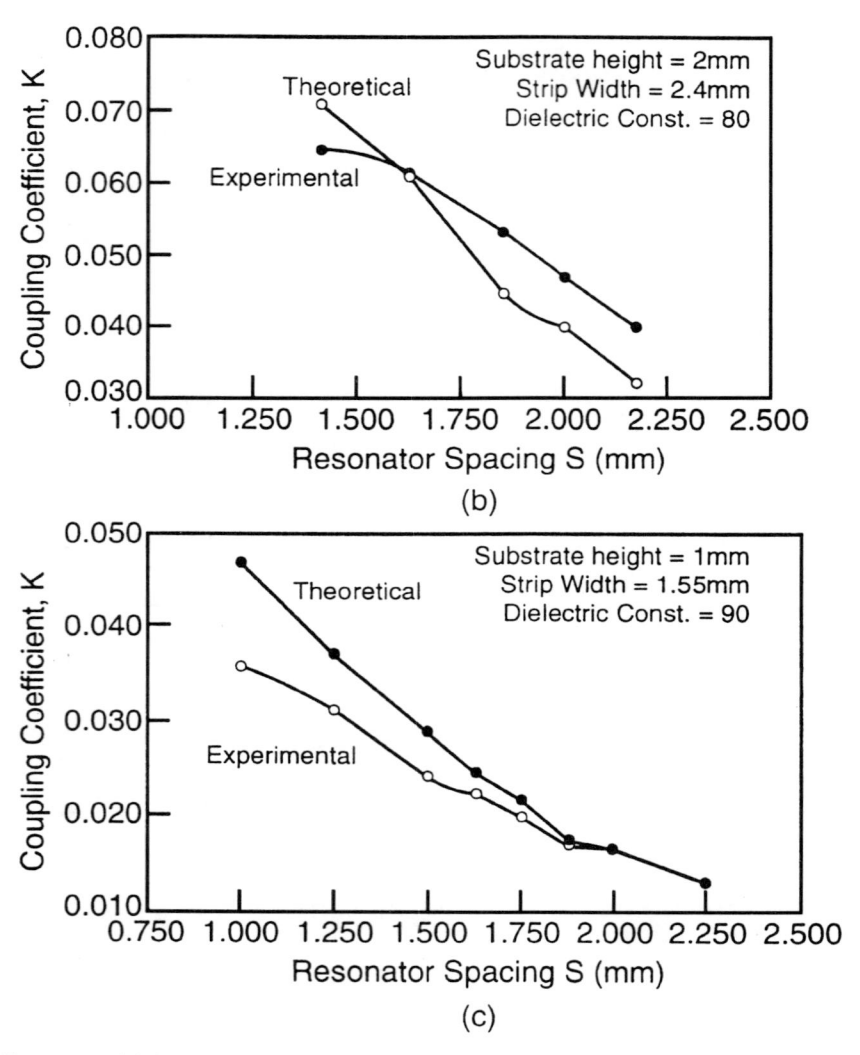

Figure 9.12 (b) Computed and measured coupling for high dielectric constant resonators. $\epsilon_r = 80$, $h = 2$ mm, $f_0 = 905$ MHz; (c) $\epsilon_r = 90$, $h = 1$ mm, $f_0 = 854$ MHz. (From [40], © 1994 Int. J. Microwave Millimeter Wave Computer Aided Engineering. Reprinted with permission.)

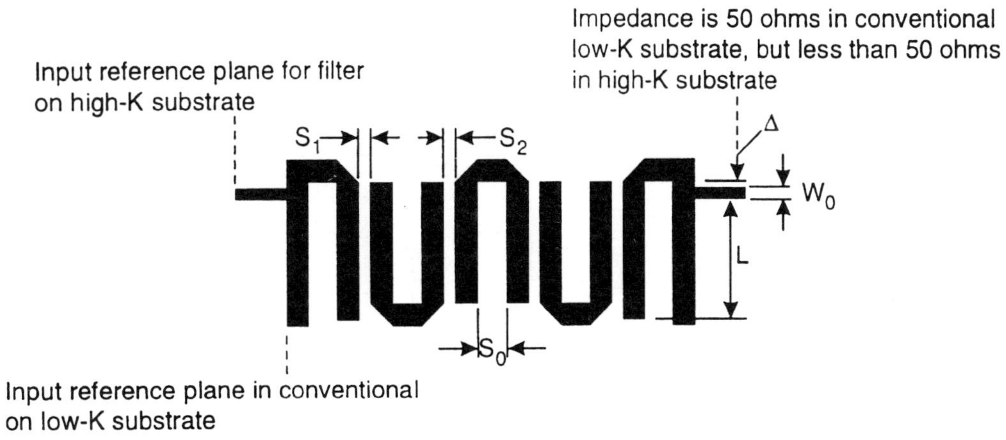

Coupled-Line Filters 333

Impedance is 50 ohms in conventional low-K substrate, but less than 50 ohms in high-K substrate

Input reference plane for filter on high-K substrate

Input reference plane in conventional on low-K substrate

Filter 1:

$\varepsilon_r = 80.00$ $S_0 = 3.00$ mm $L = 9.60$ mm
$h = 2.00$ mm $S_1 = 1.85$ mm $W_0 = 0.33$ mm
Strip width = 2.40 mm $S_2 = 2.10$ mm $\Delta = 0.62$ mm

Figure 9.13 Hairpin-line filter layout. (From [40], © 1994 Int. J. Microwave Millimeter Wave Computer Aided Engineering. Reprinted with permission.)

Figure 9.14 shows the measured data. The filter was designed having five-pole Chebyshev response centered at 905 MHz with 46-MHz bandwidth and 20-dB return loss. The substrate thickness and the dielectric constant are 2 mm and 80, respectively. The substrate material is a solid mixture of barium titrate and barium zirconate.

For regular hairpin-line filters, we can use Eagleware's M/FILTER software [41] for quick results. We start with a filter topology such as edge-coupled, hairpin, interdigital or stepped Z, together with selecting the frequency response, such as Butterworth, Chebyshev, or Bessel. The transmission line format is selected among microstrip, stripline, and so on, together with the performance parameters such as lower cutoff frequency, upper cutoff frequency, and passband ripple. Finally, the substrate parameters are entered, including dielectric constant and thickness. Entering these, we obtain a layout of the filter, while the frequency response can be evaluated using Eagleware's Superstar software and the output file of M/FILTER. Should further optimization be required, it can be accomplished by an iterative process of reloading the final values into the filter program to arrive at the final dimensions of the circuit. The output plot file can be used to create another complete file to drive a numerically controlled milling machine and thereby achieve prototype fabrication.

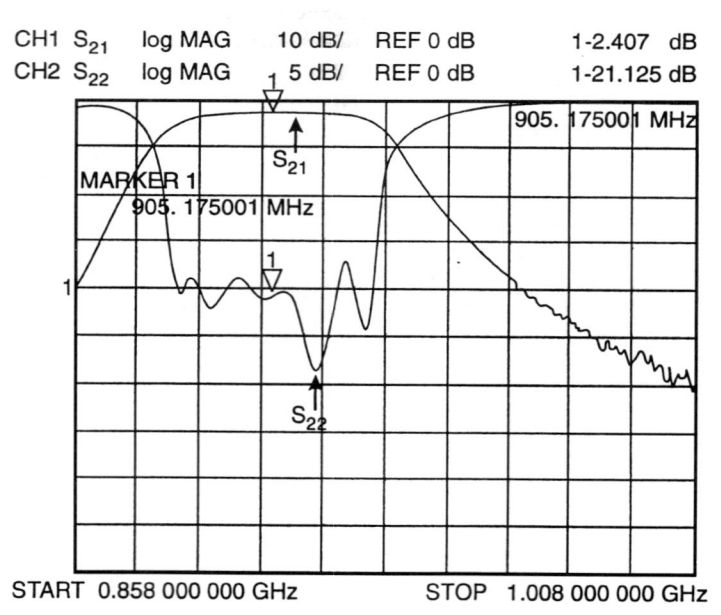

Figure 9.14 Measured response of a 905-MHz hairpin-line filter. (From [40], © 1994 Int. J. Microwave Millimeter Wave Computer Aided Engineering.)

9.3.5 Parallel-Coupled Stepped-Impedance Filters

A stepped-impedance lowpass filter is shown in Figure 9.15. In practice, this type of filter is easily realized in planar media by cascading narrow and wide transmission line segments [43–46]. Makimoto and Yamashita [46] published the details for the design of a stepped-impedance resonator stripline bandpass filter using nonuniform transmission lines. Designs in microstrip and suspended microstrip have also been reported [44]. With the impedance ratio defined as

$$K = Z_2/Z_1 \tag{9.35}$$

the fundamental resonance condition for the structure can be expressed as

$$K = \tan\theta_1 \cdot \tan\theta_2 \tag{9.36}$$

The resonator length $\theta_T = 2(\theta_1 + \theta_2)$ has maximal value for $K > 1$ and minimal value for $0 < K < 1$. For practical applications we choose $\theta_1 = \theta_2$, so that the fundamental resonance frequency is given by

$$\theta_0 = \tan^{-1}\sqrt{K} \tag{9.37}$$

The first three spurious resonance frequencies are given by

$$\frac{f_{s1}}{f_0} = \frac{\pi}{2\tan^{-1}\sqrt{K}} \tag{9.38}$$

$$\frac{f_{s2}}{f_0} = 2\left(\frac{f_{s1}}{f_0}\right) - 1 \tag{9.39}$$

$$\frac{f_{s3}}{f_0} = 2\left(\frac{f_{s1}}{f_0}\right) \tag{9.40}$$

and are all a function of K. Hence, the spurious response can be controlled by the choice of K, which is a key feature of this type of filter. The above equations neglect the physical step discontinuity effect at the junction of the two lines.

For designing bandpass filters using the stepped-impedance resonator configuration, it is necessary to calculate the even- and odd-mode impedances in the parallel-coupled sections and from these the spacings and widths of the coupled lines can be determined. The actual dimensions must account for end and junction effects, which can be done by using equations previously developed [30, 45, 46]. Design examples for stepped-impedance filters are given in the references above.

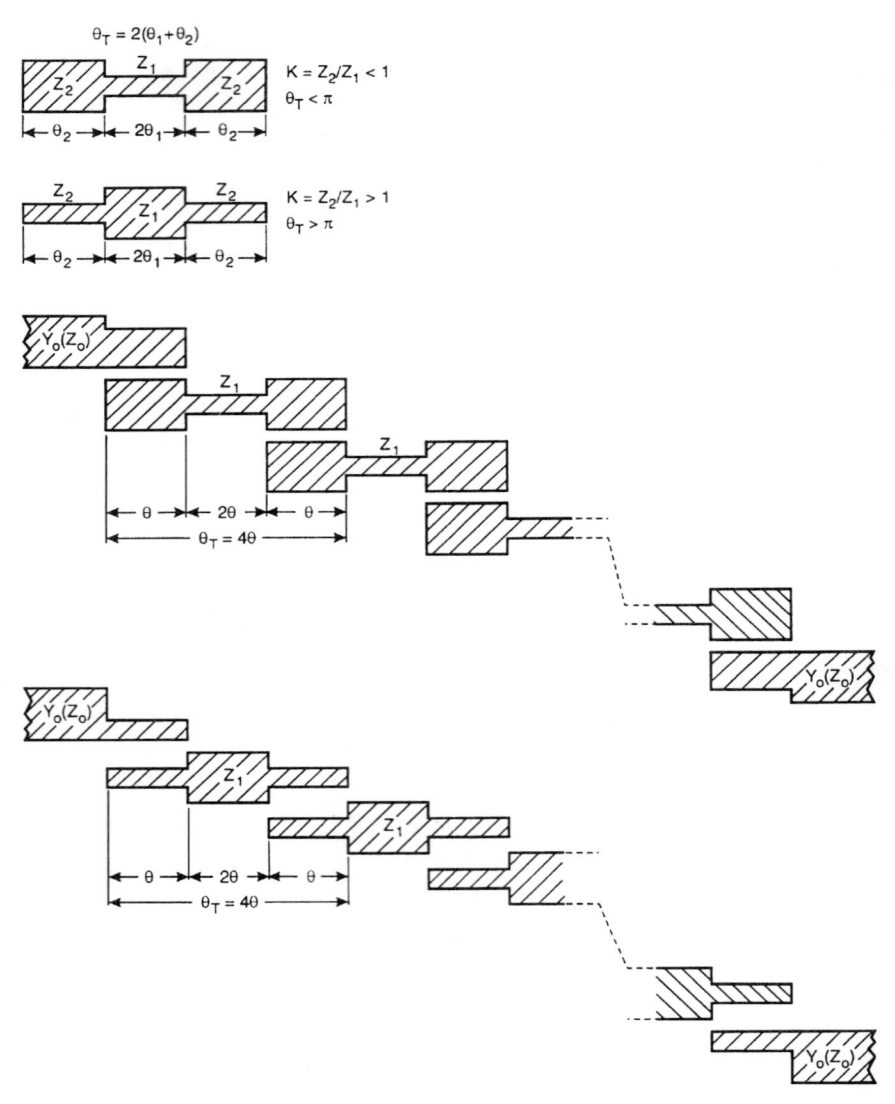

Figure 9.15 Stepped-impedance resonator and filter configurations.

9.4 Miniature Filters

Over the past decade, the explosive growth in wireless PCS and other portable receivers, such as handheld global positioning system (GPS) for position location and navigation, have generated a significant market for lightweight, ultrasmall filters that are amenable to mass manufacturing without tuning and so forth. The requirements encompass virtually all types of filters that are currently realizable using conventional lumped-element or distributed-element technology. Miniaturization is obtained with a drastic reduction in power-handling capability and increase in insertion loss. In many applications, however, particularly those cited above, power-handling limitations are not a primary concern, and the insertion loss can be compensated for by subsequent, inexpensive power amplification.

Miniaturization in microwave filter technology has been achieved in several different areas. These include lumped-element-based technology, [47] monolithic integrated circuit-based technology, high-dielectric constant ceramic-based technology, [48] and active filter-based technology. [49] In the case of lumped elements, which can be either two- or three-dimensional in nature, chip capacitors, chip inductors, and chip resistors can accomplish miniaturization. The high-dielectric constant ceramic-based technology is built on distributed-circuit technology with the reduction in size being accomplished by a reduction in wavelength in the high-dielectric constant medium.

9.4.1 Lumped-Element-Based Miniaturization

For proper operation, without excessive loss due to radiation, lumped elements must be small, relative to the operating wavelength. Thus, miniaturization is achieved at the expense of low Q and high loss. Realization of passive circuits and components such as filters, couplers, and power dividers is simple using lumped elements. Accurate models for the elements are available, requiring minimal post-production tuning. For example, Figure 9.16 shows the practical realization of a completely lumped-element diplexer for operation between 6 and 18 GHz. [50] The circuit consists of two six-pole elliptic and singly-terminated filter configurations. The overall dimension of the diplexer is 0.2 by 0.3 in, with the printed microstrip loop inductors and interdigital capacitors being realized on a 25-mil alumina substrate. The principal obstacle in realizing such printed circuits is the very narrow widths of the microstrip lines and gaps.

Besides the two-dimensional filters above, another popular filter is the Minnis-type filter, [51] which is a quasi-planar quasi-lumped-element filter. This filter uses a combination of distributed planar transmission line elements and lumped capacitors in chip form and is easily designed using commercial

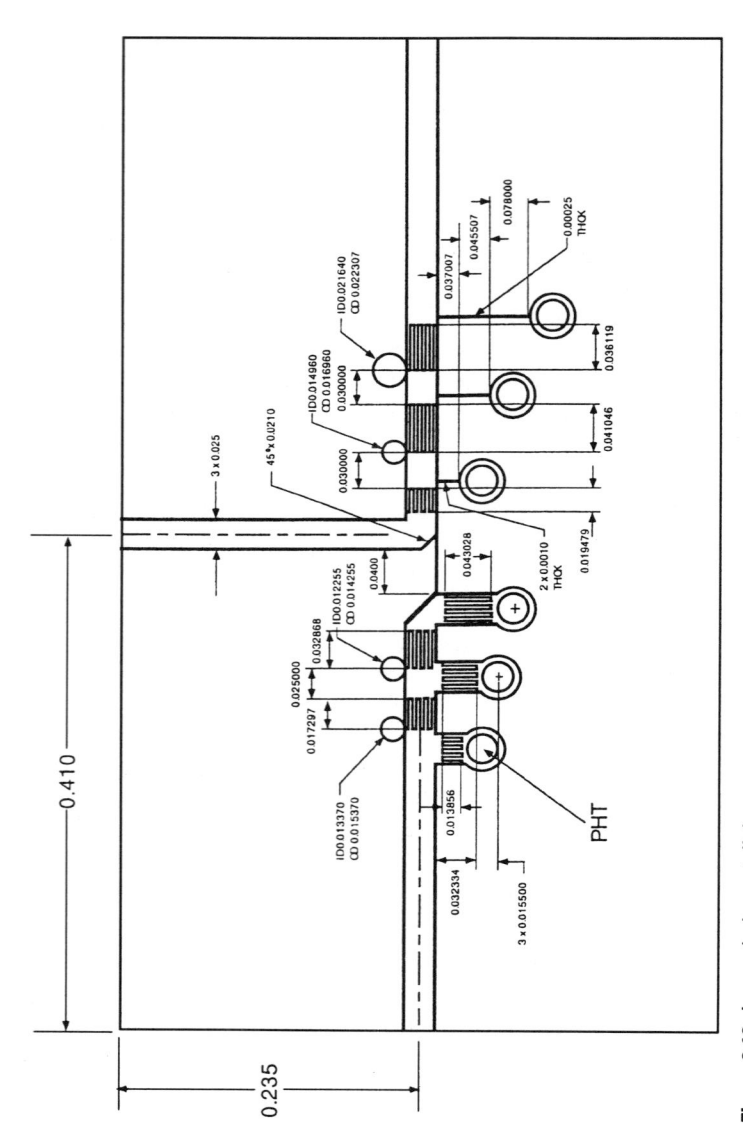

Figure 9.16 Lumped-element diplexer.

software such as FILSYN [52]. Figure 9.17 shows a Minnis-type filter centered at 3 GHz with 20% bandwidth [51].

9.4.2 Monolithic Microwave Integrated Circuit Filters

The drive for a higher level of circuit integration, lower costs, and mass production resulted in the development of MMICs in the mid-1960s. Currently, MMIC technology offers the highest level of miniaturization in microwave and millimeter-wave integrated circuits. In any typical MMIC-based circuit, the active components such as amplifiers, switches, and oscillators occupy a very small area (typically 2 to 3 mm^2) compared with that of their passive counterparts. Filters can take up to five times the area of the active circuitry. Therefore, filter realization methods are exploited in such a way as to retain and be compatible with the size-reduction benefits obtained from the active components. Because MMIC technology is mainly based on microstrip line, coplanar waveguide, and slot-line, filter designs intended for use here must be easily adaptable to microstrip, coplanar waveguide and slot-line level of integration. A typical coplanar waveguide bandpass filter structure used in MMICs is shown in Figure 9.18. The filter is based on the coupled-printed-line concept with the coupled lines acting as $\lambda/2$ or $\lambda/4$ resonators.

Currently, the principal impediment in MMIC-compatible filter design is the availability of accurate and reliable software for the CAD of MMICs. With the software packages currently available, a first-pass design in MMIC is virtually impossible as the filters are very small and extremely sensitive to dimensional tolerances. Current development using full-wave analysis with the finite element method, the boundary element method, the moment method, or the finite difference time-domain method show considerable promise.

9.4.3 Miniaturization Using High-Dielectric Constant Ceramics

It is well known that the wavelength of a signal is inversely proportional to the square root of the dielectric constant of the medium in which the signal propagates. Hence, increasing the dielectric constant of the medium a hundredfold will reduce the circuit dimensions by a factor of ten. This simple concept is being exploited extensively as distributed circuit technology is being adopted in the L-band and below for cellular telephony, GPS receivers, and mobile SATCOM [40, 53, 54].

A number of very-high-dielectric constant ceramic substrates with $\epsilon_r = 20$ to 95, very-low-dielectric loss (Q factor $= 5,000$ to $20,000$), and high temperature stability (3 ppm/°C) are currently available. They are composed of solid solutions of various titanates and are relatively inexpensive. These substrates can be

340   RF and Microwave Coupled-Line Circuits

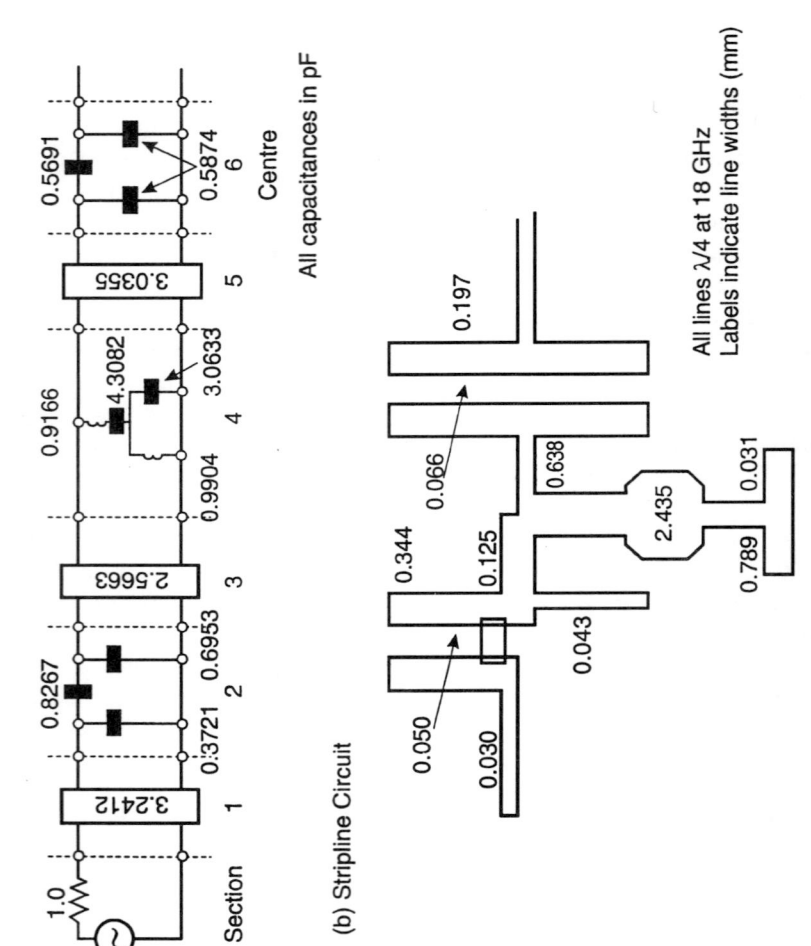

Figure 9.17 Minnis filter. (From [51], © 1982 IEEE. Reprinted with permission.)

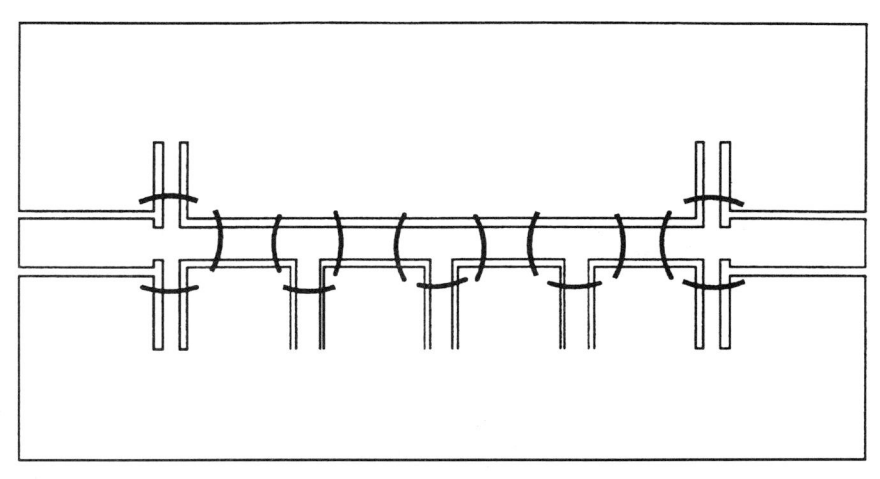

Figure 9.18 Coplanar waveguide bandpass filter.

used to realize filters in stripline, microstrip, or other configurations [54, 55]. One major obstacle in realizing components that support the TEM mode is the very narrow width and the loss of the resulting 50Ω line. Hence, in every design, the line impedances are kept sufficiently low so that the linewidths are physically realizable and an optimum Q factor is obtained. Obviously, therefore, the inputs and outputs of the circuit require impedance transformers. Various techniques are used to achieve this goal, depending on the application at hand [40].

To circumvent the low power limitations of striplines and microstrip, we can use ceramic-block waveguides. These offer not only a higher power-handling capability, but also higher Q for a comparable volume circuit in another media. Another important advantage is the reduction in component size by a factor of two that can be realized by creating a nearly perfect magnetic wall at a suitable plane of symmetry, because of the air and high-dielectric constant interface. An excellent account of waveguide ceramic-block filters is available from Konishi [48].

With the well-defined boundary conditions in a ceramic-block waveguide, the design problem is relatively easy to solve on a computer using numerical techniques. Unlike the ceramic block technology based on TEM wave propagation, however, the waveguide ceramic-block technology still requires considerable development but holds tremendous potential.

Miniaturization in microwave filter technology remains a significant challenge. Dramatic success is possible, however, with the use of new ceramic materials of higher dielectric constant and progress in making use of superconducting materials more economical.

Coaxial interdigital and combline filters (Figure 1.14) using ceramic block are commonly used in applications from 200 to 3,000 MHz. These filters have typically a 2-dB insertion loss and have bandwidth from 1 to 20%. Ceramic filters are temperature-stable, and their temperature range of operation is normally from −30°C to +85°C. They are surface-mountable, and in high volume their cost is $2.00 to $5.00. Their specific applications include cellular radio, mobile radio, wireless LAN, PCN, GPS, CATV, and ISM band.

Figure 9.19 shows a schematic of a three-resonator combline ceramic-block filter. The ceramic materials have high-dielectric constant (i.e., $\epsilon_r \cong 40 - 80$). The coupling between pairs of adjacent resonators is realized by a circular or rectangular air hole. The inhomogeneous interface between the high-dielectric constant ceramic and air hole gives rise to different phase velocities for the even- and odd-modes of the coupled lines. This difference provides the required couplings between the resonators to realize a filter. The design of such filters is straightforward but requires numerical methods, such as EM simulators, to determine coupling between the resonators. Normally, filters are designed empirically and tuned after fabrication using ceramic grinders and metal scrapers. Analysis, design, and test results for various ceramic-block filters have been discussed by many authors [56–61]. In this section, we describe briefly the design of such filters.

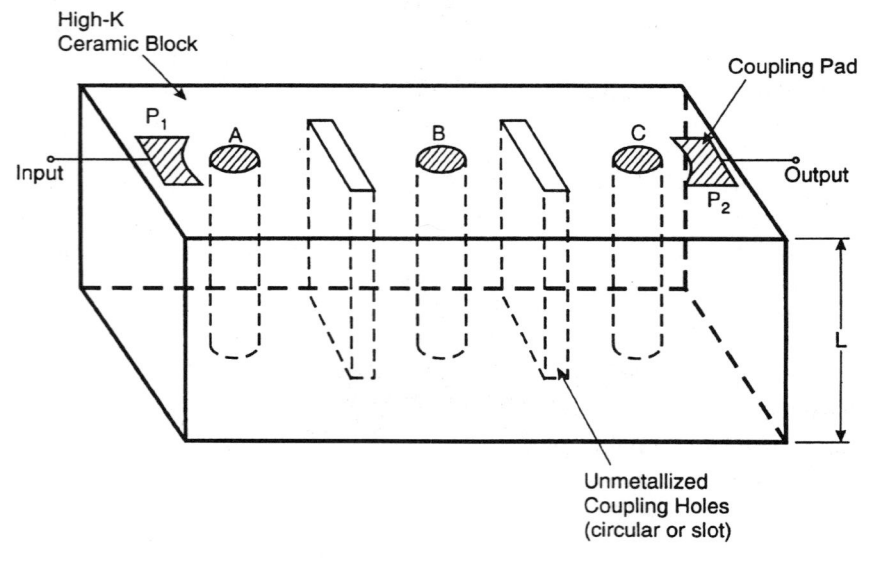

Figure 9.19 High-K ceramic back combline bandpass filter. All the surfaces are metallized except the top surface. A, B, and C are metallized coaxial resonators. Metallized sidewalls of the ceramic block act as outer conductors.

In Figure 9.19, A, B, and C are the metallized center conductors of coaxial resonators. All resonators are short-circuited at the bottom and open-circuited at the top and are designed to be $\lambda/4$ long at the operating center frequency. Resonators A, B, and C are coupled to each other for filter action through air holes between them. The first and last resonators are coupled to input and output ports, respectively, by coupling pads P_1 and P_2 located near them. The capacitive coupling between the filter and input and output is usually accomplished by a cut-and-try-method. Figure 9.20 shows a lumped-element equivalent-circuit for this filter; $\lambda/4$ resonators are represented by parallel resonant circuits (C_i, L_i). Air holes provide magnetic (inductive) coupling (L_{ij}), and the filter is connected to the input and output (usually 50 Ω) through capacitive coupling represented by C_{in} and C_{out}.

At the center frequency f_o, the length of each resonator is given by

$$L = 0.25\lambda = 0.25\frac{\lambda_0}{\sqrt{\epsilon_{re}}} = \frac{0.25c}{f_0\sqrt{\epsilon_{re}}} \tag{9.41}$$

where λ_0, c, and ϵ_{re} are the free-space wavelength, velocity of light, and the effective dielectric constant, ϵ_{re}, is obtained from

$$\epsilon_{re} = \frac{\epsilon_{ree} + \epsilon_{reo}}{2} \tag{9.42}$$

where ϵ_{ree} and ϵ_{reo} are, respectively, the even- and odd-mode effective dielectric constants of the medium in which the coaxial resonators are embedded.

The coupling coefficient is given by [58]

$$K = \frac{2(\sqrt{\epsilon_{ree}} - \sqrt{\epsilon_{reo}})}{\sqrt{\epsilon_{ree}} + \sqrt{\epsilon_{reo}}} \tag{9.43}$$

Figure 9.21 shows a cross-sectional view of an air hole coupled-line structure with dimensions, and Figure 9.22 shows the calculated values of the even- and odd-mode effective dielectric constants. Finite difference method [58] was used to analyze the structure, with $\epsilon_r = 80$, $D = 2.4$ mm, $H = 6$ mm, and $S = 0.8$ mm. Figures 9.23(a) and (b) show the coupling coefficient versus air hole radius, and separation between resonator and air hole, respectively. More extensive data for the coupling coefficient has also been published by Yao, et al. [60, 61].

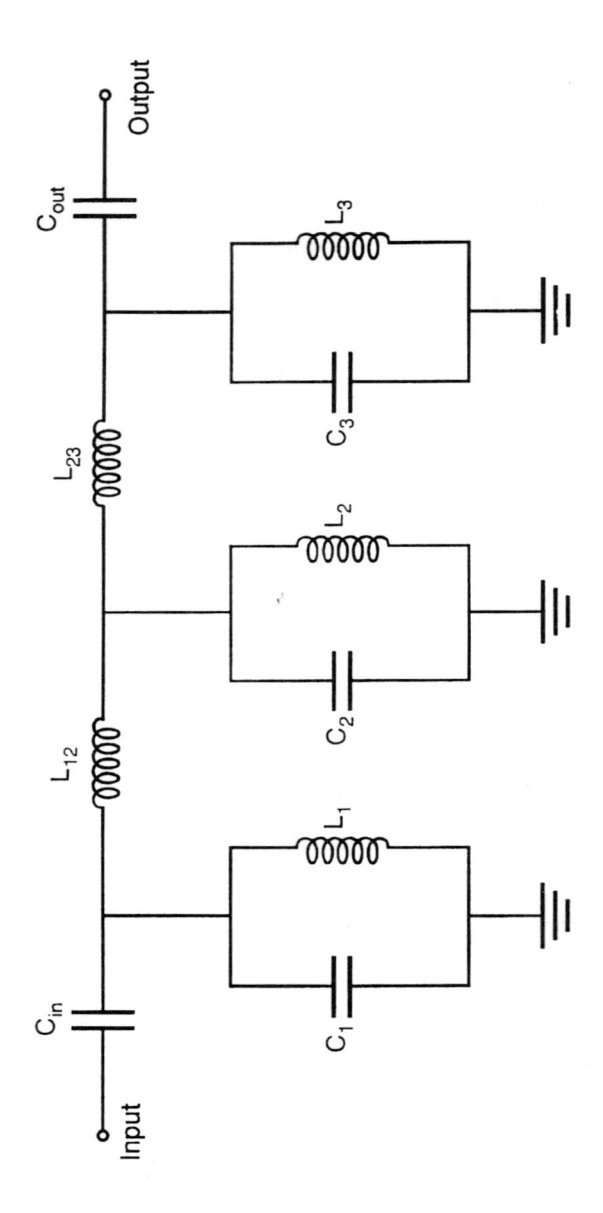

Figure 9.20 Lumped-element equivalent circuit for the three-resonator filter.

Coupled-Line Filters 345

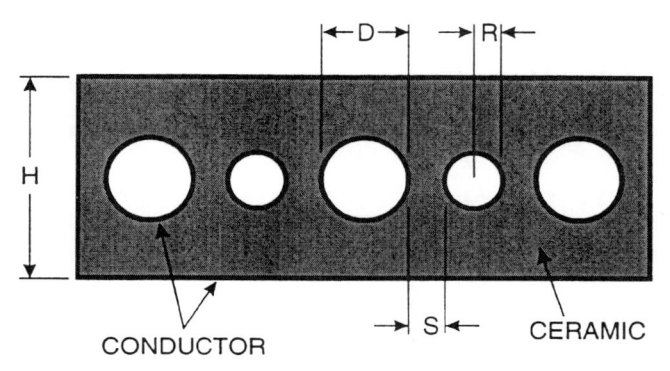

Figure 9.21 Cross-sectional view of the three-resonator ceramic-block combline filter.

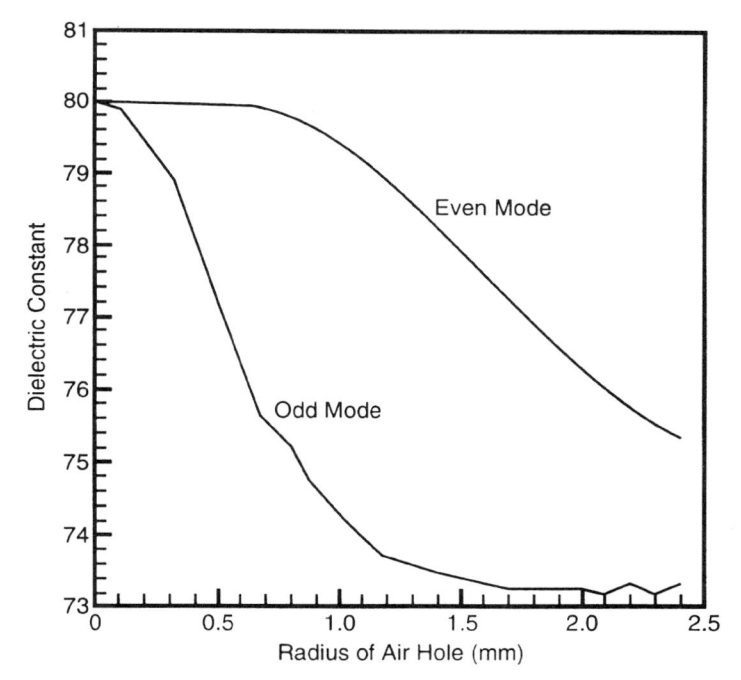

Figure 9.22 Effective dielectric constant for even- and odd-modes as a function of air hole radius. $\epsilon_r = 80$, $D = 2.4$ mm, $H = 6$ mm, and $S = 0.8$ mm.

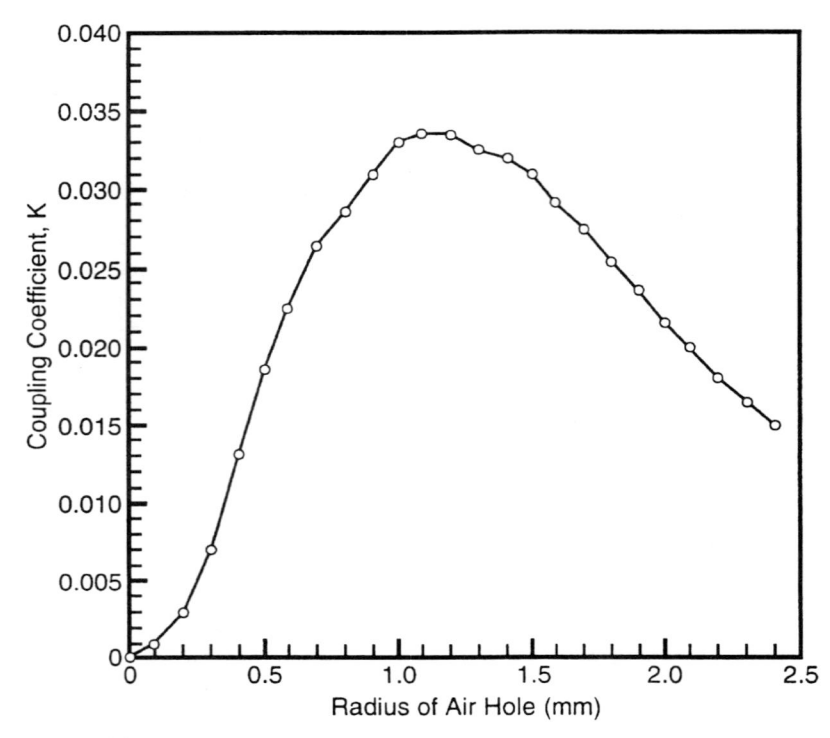

Figure 9.23 (a) Coupling coefficient as a function of air hole radius. $\epsilon_r = 80$, $D = 2.4$ mm, $H = 6$ mm, and $S = 0.8$ mm.

To illustrate a design example of a bandpass ceramic-block filter, the following specifications are chosen:

Center frequency	900 MHz
Response	Chebyshev with 0.05-dB ripple
Bandwidth	30 MHz
Number of resonators	3

This filter can be designed by using normalized coupling coefficients in terms of lowpass prototype element values and design frequencies as given in (9.31) and (9.32). The normalized bandwidth, $BW/f_0 = 30/900 = 1/30$. From Table 9.2, the lowpass prototype element values are $g_0 = g_4 = 1$, $g_1 = g_3 = 0.8794$, and $g_2 = 1.1132$. From (9.31), the coupling coefficients $K_{12} = K_{23} = 0.0339$. For the structure shown in Figure 9.21, and using data in Figure 9.23, the filter design parameters are $\epsilon_r = 80$, $H = 6$ mm, $D = 1.2$ mm, $R = 1.2$ mm, and $S = 0.7$ mm.

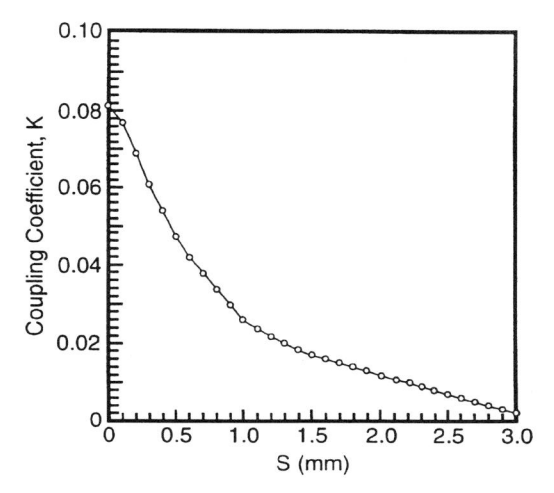

Figure 9.23 (b) Coupling coefficient as a function of separation between the metallized resonator and air hole. $\epsilon_r = 80$, $D = 2.4$ mm, $H = 6$ mm, and $R = 1.2$ mm.

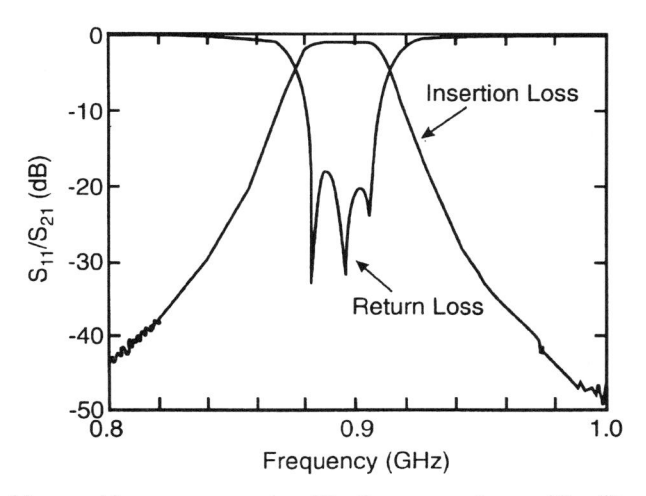

Figure 9.24 Measured frequency response of the three-resonator combline filter.

Figure 9.24 shows the measured frequency response of the three-resonator ceramic-block filter where the input and output couplings were obtained by experiments. In the 50-MHz bandwidth, the measured insertion loss was better than 1.5 dB. The ceramic materials loss tangent (tan δ) was 2.5×10^{-4} and the temperature coefficient was 3 ppm/°C.

Several other low-profile ceramic-block filters have also been reported in the literature [62, 63].

References

[1] Saal, R., "The Design of Filters Using the Catalogue of Normalized Low-Pass Filters" (in German), *Telefunken* (GMBH), Backang, W. Germany, 1961.

[2] Rhodes, J. D., *Theory of Electrical Filters*, Wiley Interscience, New York, 1976.

[3] Skwirzynski, J. K., *Design Theory and Data for Electrical Filters*, Van Nostrand, London, U.K., 1965.

[4] Zverev, A. I., *Handbook of Filter Synthesis*, John Wiley, New York, 1967.

[5] Matthaei, G. L., L. Young, and E.M.T. Jones, *Microwave Filters, Impedance, Matching Networks and Coupling Structures*, McGraw-Hill, New York, 1964.

[6] Milligan, T., "Nomographs and the Filter Designer," *Microwaves and RF*, Vol. 24, Oct. 1985, pp. 103–107.

[7] Howe, H., Jr., *Stripline Circuit Design*, Artech House, Dedham, MA, 1974.

[8] Alseyab, S. A., "A Novel Class of Generalized Chebyshev Low Pass Prototype for Suspended Substrate Stripline Filters," *IEEE Trans. Microwave Theory Tech.*, Vol. MTT-30, Sept. 1982, pp. 1,341–1,347.

[9] Mobbs, C. I., and J. D. Rhodes, "A Generalized Chebyshev Suspended Substrate Stripline Bandpass Filter," *IEEE Trans. Microwave Theory Tech.*, Vol. MTT-31, May 1983, pp. 397–402.

[10] Bahl, I. J., and P. Bhartia, *Microwave Solid-State Circuit Design*, John Wiley, New York, 1988, Ch. 6.

[11] LINMIC+ Computer Program, Jansen Microwave, Ratingen, Germany.

[12] em Computer Program, Sonnet Software, Liverpool, NY.

[13] EEpal Computer Program, Eagleware Corp., Stone Mountain, GA.

[14] IE3D, Zeland Software, Fremont, CA.

[15] SFPMIC+ Computer Program, Jansen Microwaves, Ratigen, Germany.

[16] HFSS Computer Program, Hewlett-Packard Co., Santa Clara, CA.

[17] Rhea, R., "PC Tools Simulate and Synthesize RF Circuits," *Microwaves and RF*, Vol. 33, No. 4, 1994, pp. 194–199.

[18] Ozaki, H., and J. Ishii, "Synthesis of a Class of Stripline Filters," *IRE Trans. Circuit Theory*, Vol. CT-5, June 1958, pp. 104–109.

[19] Davis, W. A., *Microwave Semiconductor Circuit Design*, Van Nostrand Reinhold Co., New York, 1984, Ch. 3.

[20] Richards, P. I., "Resistor-Transmission Line Resonator Filters," *IRE Trans. Microwave Theory Tech.*, Vol. MTT-6, Apr. 1958, pp. 223–231.

[21] Malherbe, J.A.G., *Microwave Transmission Line Filters*, Artech House, Dedham, MA, 1990.

[22] Cohn, S. B., "Parallel Coupled Transmission Line Resonator Filters," *IRE Trans. Microwave Theory Tech.*, Vol. MTT-6, No. 2, Apr. 1958, pp. 223–232.

[23] Moazzam, M. R., S. Uysal, and A. H. Aghvami, "Improved Performance Parallel Coupled Microstrip Filters," *Microwave J.*, Vol. 34, No. 11, Nov. 1991, pp. 128–135.

[24] Mara, J. F., and J. B. Schappacher, "Broadband Microstrip Parallel-Coupled Filters Using Multi-Line Sections," *Microwave J.*, Vol. 22, No. 4. Apr. 1979, pp. 97–99.

[25] Wong, J. S., "Microstrip Tapped-Line Filter Design," *IEEE Trans. Microwave Theory Tech.*, Vol. MTT-27, Jan. 1979, pp. 328–339.

[26] Tran, M., and C. Nguyen, "Wideband Bandpass Filters Employing Broadside Coupled Microstrip Lines for MIC and MMIC Applications," *Microwave J.*, Vol. 37, No. 4, Apr. 1994, pp. 210–225.

[27] Minnis, B. J., "Printed Circuit Coupled Line Filters for Bandwidths up to and Greater Than an Octave," *IEEE Trans. Microwave Theory Tech.*, Vol. MTT-29, No. 3, Mar. 1991, pp. 215–222.

[28] Ho, C. Y., and J. H. Werdman, "Improved Design of Parallel Coupled Line Filters With Tapped Input/output," *Microwave J.*, Vol. 26, No. 18, Oct. 1983, pp. 127–130.

[29] Bahl, I. J., "Capacitively Compensated High-Performance Parallel-Coupled Microstrip Filters," *IEEE MTT-S Int. Microwave Symp. Dig.* 1989, pp. 679–682.

[30] Gupta, K. C., et al., *Microstrip Lines and Slotlines*, Sec. Ed. Artech House, Norwood, MA, 1966, Ch. 3.

[31] Rubin, D., and D. Saul, "Millimeter Wave IC Band Pass Filters and Multiplexers," *IEEE MTT-S Int Microwave Symp. Dig.* 1978, pp. 208–210.

[32] Matthaei, G. L., "Interdigital Band-Pass Filters," *IRE Trans Microwave Theory Tech.*, Vol. MTT-10, No. 6, Nov. 1962, pp. 479–491.

[33] Wenzel, R. J., "Exact Theory of Interdigital Band-Pass Filters and Related Coupled Structures," *IEEE Trans Microwave Theory Tech.*, Vol. MTT-13, No. 5, Sept. 1975, pp. 559–575.

[34] Matthaei, G. L., "Comb-Line Band-Pass Filters of Narrow and Moderate Bandwidth," *Microwave J.*, Vol. 6, Aug. 1963, pp. 82–91.

[35] Wenzel, R. J., "Synthesis of Combline and Capacitively Loaded Interdigital Bandpass Filters of Arbitrary Bandwidth," *IEEE Trans Microwave Theory Tech.*, Vol. MTT-19, No. 8, Aug. 1971, pp. 678–686.

[36] Vincze, A., "Practical Design Approach to Microstrip Combine-Type Filters," *IEEE Trans. Microwave Theory Tech.*, Vol. MTT-22, No. 12, Dec. 1974, pp. 1,171–1,181.

[37] Cristal, E. G., and S. Frankel, "Hairpin-Line and Hybrid Hairpin-Line/Half-Wave Parallel-Coupled-Line Filters," *IEEE Trans. Microwave Theory Tech.*, Vol. MTT-20, No. 22, Nov. 1972, pp. 719–728.

[38] Gysel, U. H., "New Theory and Design for Hairpin-Line Filters," *IEEE Trans. Microwave Theory Tech.*, Vol. MTT-22, No. 5, May 1974, pp. 523–531.

[39] Salkhi, A., "Quick Filter Design and Construction," *Appl. Microwave and Wireless*, Vol. 6, No. 1, Jan. 1994, pp. 92–100.

[40] Pramanick, P., "Compact 900-MHz Hairpin-Line Filters Using High Dielectric Constant Microstrip Line," *Int. J. Microwave Millimetre-Wave Computer-Aided Eng.*, Vol. 4, No. 3, 1994, pp. 272–281.

[41] Rhea, R. W., "Distributed Hairpin Bandpass, RF Compute!", Eagleware Corp., Vol. 3, No. 1, 1991.

[42] Matthaei, G. L., et al., "Hairpin-Comb Filters for HTS and other Narrow-Band Applications," *IEEE Trans. Microwave Theory Tech.*, Vol. MTT-13, No. 9, Aug. 1997, pp. 1,226–1,231.

[43] Ho, C. Y., and J. H. Weidman, "Half-Wavelength and Step Impedance Resonators and Microstrip Filter Design," *Microwave Sys. News*, Vol. 13, No. 10, Oct. 1983, pp. 88–103.

[44] Rubin, D., and A. R. Hislop, "Millimeter-Wave Coupled-Line Filters—Design Techniques for Suspended Substrate and Microstrip," *Microwave J.*, Vol. 23, No. 10, Oct. 1980, pp. 67–78.

[45] Nalbandian, V., and W. Steenaart, "Discontinuities in Systematic Striplines Due to Impedance Step and their Compensations," *IEEE Trans. Microwave Theory Tech.*, Vol. MTT-20, Sept. 1972, pp. 573–578.

[46] Makimoto, M., and S. Yamashita, "Bandpass Filters Using Parallel Coupled Stripline Stepped Impedance Resonators," *IEEE Trans. Microwave Theory Tech.*, Vol. MTT-28, No. 12, Dec. 1980, pp. 1,413–1,417.

[47] Field, P. L., et al., "Asymmetric Bandpass Filter using a Ceramic Substrate," *IEEE Microwave and Guided-Wave Lett.*, Vol. 2, No. 9, Sept. 1992, pp. 361–363.

[48] Konishi, Y., "Novel Dielectric Waveguide Components-Microwave Applications of Ceramic Materials," *Proc. IEEE*, Vol. 79, No. 6, June 1981, pp. 726–740.

[49] Yamamoto, Y., et al., "A MESFET Controlled X-Band Active Bandpass Filter," *IEEE Microwave and Guided-Wave Lett.*, Vol. 1, No. 5, May 1991, pp. 110–111.

[50] Pramanick, P., University of Saskatchewan, Saskatoon, Canada (from the author's private communication).

[51] Minnis, J. B., "Classes of Subminiaturized Microwave Printed Circuit-Filters With Arbitrary Passband and Stopband Widths," *IEEE Trans Microwave Theory Tech.*, Vol. MTT-30, No. 11, pp. 1,893–1,900, Nov. 1982.

[52] FILSYN, DGS Associates, Inc., Ver. 2.7, Santa Clara, CA, 1992.

[53] Ohm, G., et al., "Miniature Microstrip Bandpass Filter on a Barium Tetratitanate Substrate," *Microwave J.*, Vol. 28, No. 11, pp. 129–134, Nov. 1985.

[54] Nishikawa, T., et al., "800-MHz Band Face Bonding Filter Using Dielectric BDLS," *IEEE MTT-S Symp. Dig.* pp. 403–406, 1986.

[55] Ishizaki, T., et al., "A Very Small Dielectric Planar Filter for Portable Telephones," *IEEE Trans. Microwave Theory Tech.*, Vol. MTT-42, No. 11, pp. 2,017–2,022, Nov. 1994.

[56] Fukasawa, A., "Analysis and Composition of a New Microwave Filter Configuration With Inhomogeneous Dielectric Medium," *IEEE Trans. Microwave Theory Tech.*, Vol. MTT-30, pp. 1,367–1,375, Sept. 1982.

[57] Levy, R., "Simplified Analysis of Inhomogeneous Dielectric Block Combline Filters," *IEEE Int. Microwave Symp. Dig.*, pp. 135–138, 1990.

[58] You, C. C., C. L. Huang, and C. C. Wei, "Single-Block Ceramic Microwave Bandpass Filters," *Microwave J.*, Vol. 37, pp. 24–35, Nov. 1994.

[59] Hano, K., H. Kohriyama, and K.-I. Sawamoto, "A Direct Coupled λ/4-Coaxial Resonator Bandpass Filter for Land Mobile Communications," *IEEE Trans. Microwave Theory Tech.*, Vol. MTT-34, pp. 972–976, Sept. 1986.

[60] Yao, H. W., et al., "Full-Wave Modeling of Conducting Posts in Rectangular Waveguide and Its Applications to Slot-Coupled Combline Filters," *IEEE Trans. Microwave Theory Tech.*, Vol. MTT-43, pp. 2,824–2,829, Dec. 1995.

[61] Yao, H. W., C. Wang, and K. A. Zaki, "Quarter-Wavelength Combline Filters," *IEEE Trans. Microwave Theory Tech.*, Vol. 44, pp. 2,673–2,679, Dec. 1996.

[62] Matsumoto, H., H. Ogura, and T. Nishikawa, "A Miniature Dielectric Monoblock Band-Pass Filter for 800-MHz Band Cordless Telephone System," *IEEE Int. Microwave Symp. Dig.*, pp. 249–252, 1994.

[63] Kobayashi, S., and K. Saito, "A Miniaturized Ceramic Bandpass Filter for the Cordless Phone," *IEEE Int. Microwave Symp. Dig.*, pp. 391–394, 1995.

10

Coupled-Line Circuit Components

DC blocks, impedance transformers, interdigital capacitors, and spiral inductors employing coupled-line sections are commonly used in microwave circuits. A coupled-line section provides the required characteristics for these components over a wide frequency range. This chapter deals with these structures and includes the basic theory, design, and circuit performance to illustrate the design principles.

10.1 DC Blocks

A series capacitor is used to isolate the bias voltages applied to various circuits as well as to block dc and low-frequency voltages while allowing the RF signal to pass through with minimal loss. At microwave frequencies, both a high-quality capacitor and a distributed coupled network are used. This section only deals with the coupled-line structures used as dc block networks.

10.1.1 Analysis

A 3-dB backward-wave coupled-line section [1–3] with open circuit terminations as shown in Figure 10.1(a) can be used instead of a series capacitor as a series dc block. The circuit used for analysis is shown in Figure 10.1(b). The scattering matrix of a microstrip backward-wave coupler, when ports are terminated in matched loads, is given by

$$\begin{bmatrix} b_1 \\ b_2 \\ b_3 \\ b_4 \end{bmatrix} = \begin{bmatrix} 0 & E_2 & 0 & E_4 \\ E_2 & 0 & E_4 & 0 \\ 0 & E_4 & 0 & E_2 \\ E_4 & 0 & E_2 & 0 \end{bmatrix} \begin{bmatrix} a_1 \\ a_2 \\ a_3 \\ a_4 \end{bmatrix} \qquad (10.1)$$

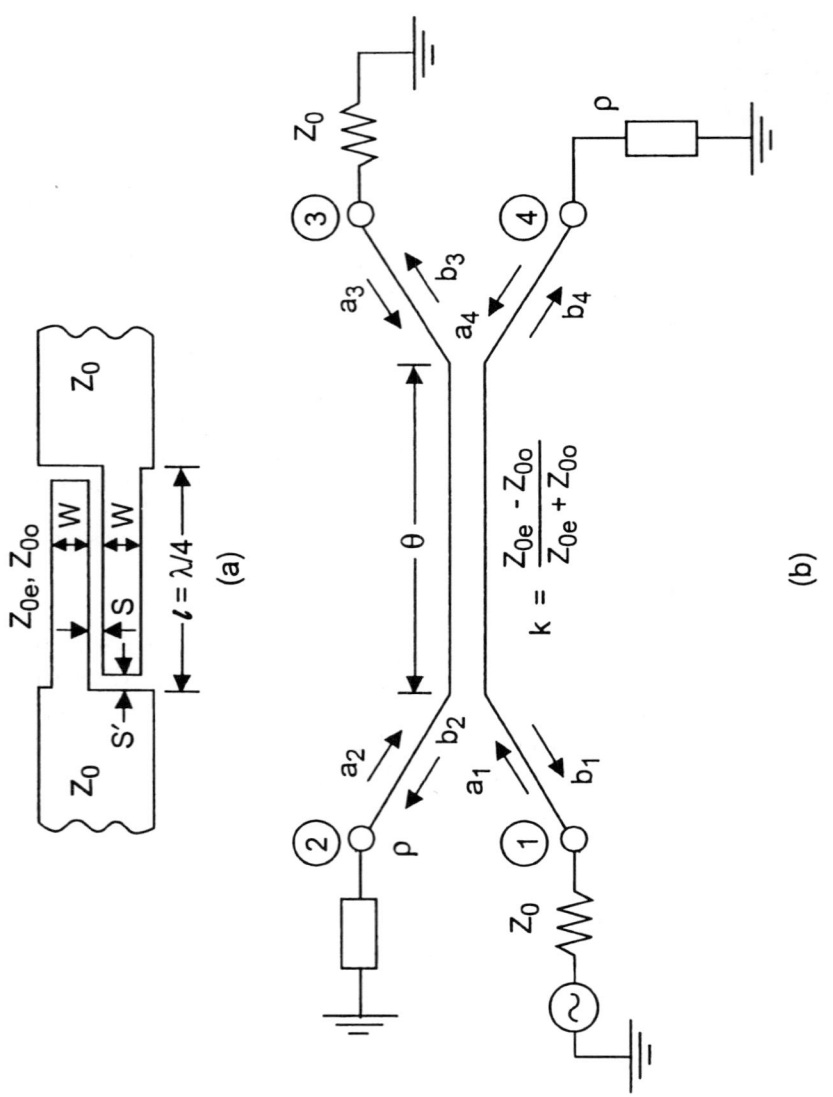

Figure 10.1 Quarter-wave coupled-line section as a dc block: (a) physical layout and (b) circuit schematic.

where

$$E_2 = \frac{jk\sin\theta}{\sqrt{1-k^2}\cos\theta + j\sin\theta}$$

$$E_4 = \frac{\sqrt{1-k^2}}{\sqrt{1-k^2}\cos\theta + j\sin\theta}$$

and

$$\theta = \frac{1}{2}(\theta_e + \theta_o)$$

$$k = \frac{Z_{0e} - Z_{0o}}{Z_{0e} + Z_{0o}}$$

where θ is the electrical line length and subscripts e and o denote even and odd modes, respectively. When ports 2 and 4 are terminated in reflecting loads, then $a_2 = \rho b_2$, $a_3 = 0$, $a_4 = \rho b_4$, where ρ is the reflection coefficient at ports 2 and 4. For an ideal coupler when $Z_{0e} Z_{0o} = Z_0^2$, the circuit ensures perfect match at the input. As there are reflecting terminations at ports 2 and 4, there is generally not a perfect match at port 1, except when the operating frequency and coupling length are such as to give exactly equal outputs at ports 2 and 4. At the input the reflected wave amplitude is given by

$$b_1 = E_2 \rho b_2 + E_4 \rho b_4 \tag{10.2}$$

Similarly, expressions for b_2 and b_4 are

$$b_2 = E_2 a_1 \tag{10.3}$$

$$b_4 = E_4 a_1 \tag{10.4}$$

From (10.2) to (10.4):

$$\rho_{in} = \frac{b_1}{a_1} = (E_2^2 + E_4^2)\rho = \frac{1 - k^2(1 + \sin^2\theta)}{[\sqrt{1-k^2}\cos\theta + j\sin\theta]^2}\rho \tag{10.5}$$

For

$$k = 1/\sqrt{2} \text{ and } \rho = 1$$

Note: $\rho = (Z_L - Z_0)/(Z_L + Z_0)$, where Z_L is the load impedance.

$$VSWR = \frac{1 + |\rho_{in}|}{1 - |\rho_{in}|}$$

$$= 1/\sin^2\theta$$

When $\theta = 90$ degrees, $VSWR = 1$.

We note that for a 3-dB backward-wave coupler, the direct and coupled waves have equal amplitude when the coupling length θ is 90° at the coupler's center frequency. The variation of VSWR with frequency for various coupling values is shown in Figure 10.2. In the calculations, ports 2 and 4 are assumed to be totally reflecting ($\rho = 1$), and perfect open circuits. From Figure 10.2 we find that for a 2.7-dB coupler ($k = 0.732$), the input VSWR is below 1.2 for $0.65 \leq f/f_o \leq 1.35$, an octave frequency bandwidth. The main disadvantage of this type of dc block is that at lower frequencies, it increases the size of the circuit. At millimeter wave frequencies, where the size becomes small, however,

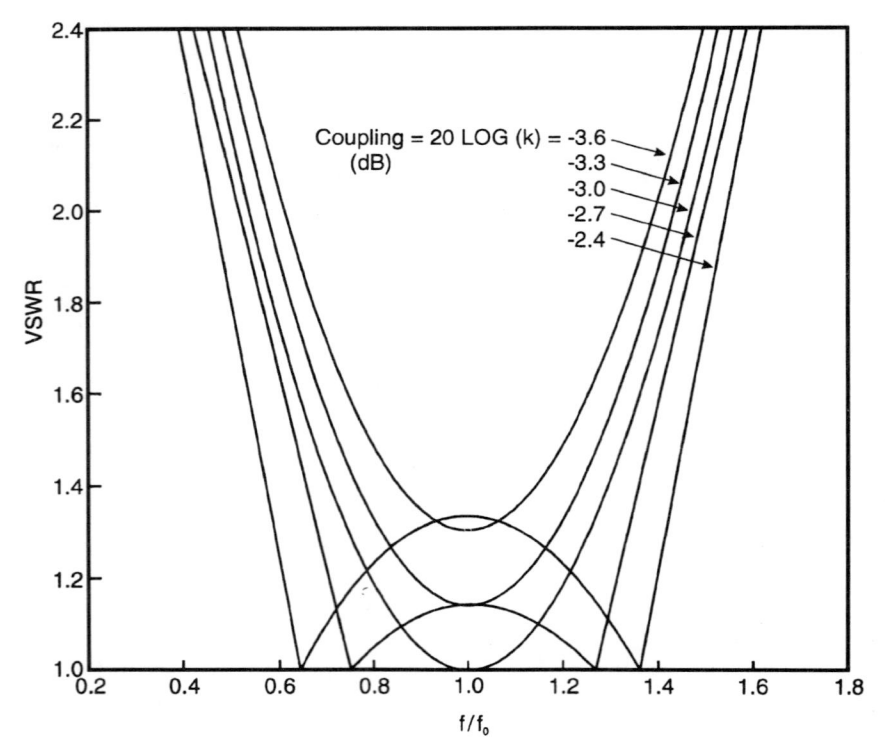

Figure 10.2 Simulated VSWR versus fractional bandwidth of a coupled-line dc block.

this structure is preferred in comparison with the chip or MIM capacitor because of lower loss.

Knowing Z_{0e} and Z_{0o}, the physical dimensions can easily be obtained as for any other coupler. [4–6] Simple expressions are also available [7] for Z_{0e} and Z_{0o} for a given VSWR and bandwidth, that is:

$$Z_{0e} = \sqrt{S}\left[1 + \sqrt{1 + \frac{1 + \sqrt{1 + \Omega^2}}{\Omega^2}\left(1 - \frac{1}{S}\right)}\right]Z_0 \quad (10.6a)$$

$$Z_{0o} = \sqrt{S}\left[-1 + \sqrt{1 + \frac{1 + \sqrt{1 + \Omega^2}}{\Omega^2}\left(1 - \frac{1}{S}\right)}\right]Z_0 \quad (10.6b)$$

where S is the voltage standing wave ratio and Ω is a normalized bandwidth given by

$$\Omega = \cot\left[\frac{\pi}{2}(1 - (f_2 - f_1)/(f_2 + f_1))\right] \quad (10.7)$$

where f_1 and f_2 are the lower and upper edge of the frequency band.

10.1.2 Broadband DC Block

A broadband dc block [1] having center frequency of 12 GHz was designed and constructed using a 25-mil-thick alumina substrate ($\varepsilon_r \cong 10$). The physical dimensions obtained for $Z_{0e} = 130\Omega$ and $Z_{0o} = 24\Omega$ were spacing $S = 1$ mil, width $W = 6.3$ mil, and length $l = 100$ mil. Measured and simulated VSWR and insertion loss responses are shown in Figure 10.3. Over an octave band, the worst case measured VSWR and insertion loss were about 1.4 and 0.2 dB, respectively. More accurate transmission phase agreement between the measured and simulated results was obtained [3] by including open-end discontinuity effects in the calculations.

10.1.3 Biasing Circuits

Solid-state circuits require low frequency biasing networks which must be separated from the RF circuit. In other words, when a bias voltage is applied to the device, the RF energy should not leak out through the bias port and also it must isolate the bias voltages applied to various devices. In the case of amplifiers and oscillators, the biasing network should not alter stability conditions. In these circuits and many others, the biasing circuitry becomes an integral part of the circuit design. There are many biasing schemes used in practice. Basically it

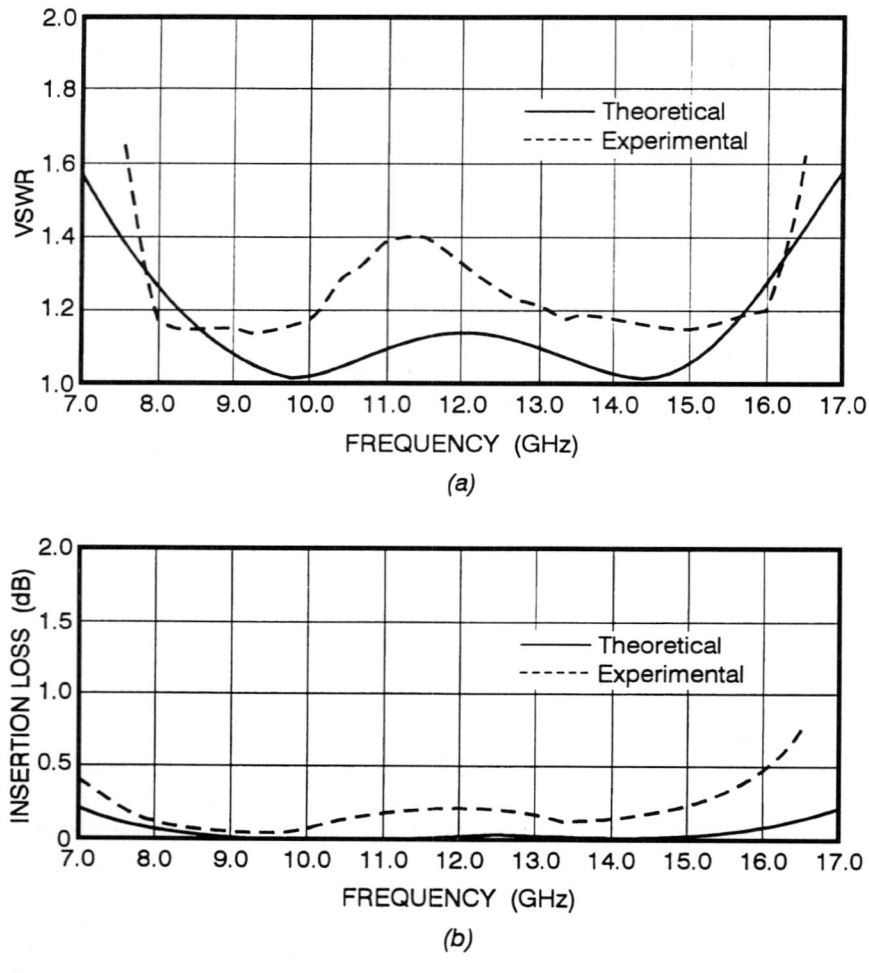

Figure 10.3 (a) VSWR versus frequency response of a microstrip interdigital dc block and (b) insertion loss versus frequency response of a microstrip interdigital dc block. (From [1], © 1972 IEEE. Reprinted with permission.)

consists of a dc block and an RF choke as shown in Figure 10.4. A dc block can be either a capacitor or a 3-dB backward-wave coupler described earlier. The RF choke at microwave frequencies is generally realized by using a high-impedance $\lambda/4$ line terminated by a RF bypass capacitor or a quarter-wave line terminated by another $\lambda/4$ open-circuited line or a radial stub as shown in Figure 10.5. For low-RF leakage through the biasing network, the ratio of the shunt stub impedance and the through-line, impedance must be much greater than unity. In this case,

Coupled-Line Circuit Components

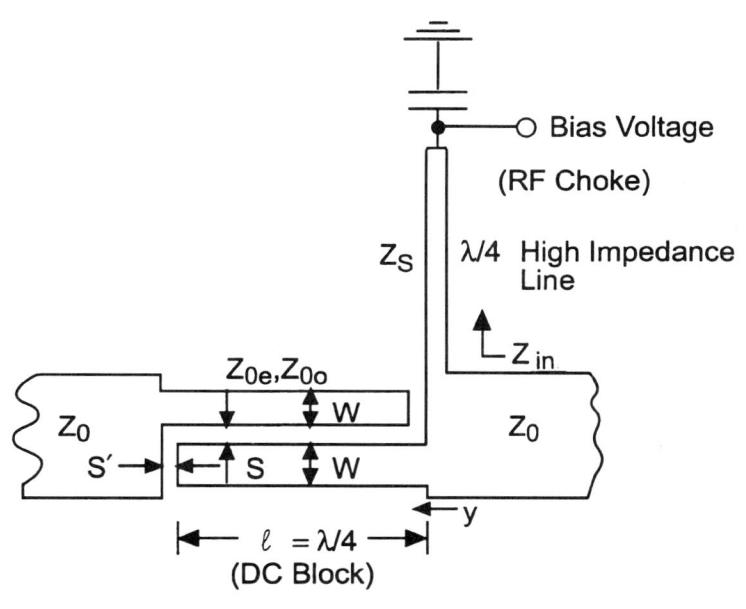

Figure 10.4 Simplified microwave biasing circuit.

the bandwidth increases when the impedance of the stub increases. For VSWR ≤ 1.05, the bandwidth for $Z_s = 100\Omega$ is about 12%. To further increase the bandwidth, two sections of quarter-wave long transmission lines are used. If an open circuit is required across the main line for RF signals, a quarter-wave high-impedance line followed by another open-circuited quarter-wave low-impedance line are connected. The configuration is shown in Figure 10.5(a). Assuming that the through line is 50Ω, the normalized admittance with the load is

$$y = 1 + 1/Z_{in} \tag{10.8}$$

where

$$Z_{in} = jZ_1 \frac{Z_1 \tan\theta_1 \tan\theta_2 - Z_2}{Z_1 \tan\theta_2 + Z_2 \tan\theta_1} \tag{10.9}$$

Here θ_1, Z_1 and θ_2, and Z_2 are the electrical line length and characteristic impedance of the first and second line sections, respectively. The VSWR response for various combinations of Z_1 and Z_2 is shown in Figure 10.6. Maximal bandwidth is obtained when Z_1/Z_2 is large. For example, with $Z_1 = 100\Omega$, $Z_2 = 10\Omega$, $Z_0 = 50\Omega$, and a $VSWR = 1.2$, the maximal bandwidth is about 40%.

A radial line section provides better bandwidth than a $\lambda/4$ open-circuit line section. A broadband biasing network structure was designed on a semi-insulating GaAs substrate with parameters listed in Table 10.1. Other structure

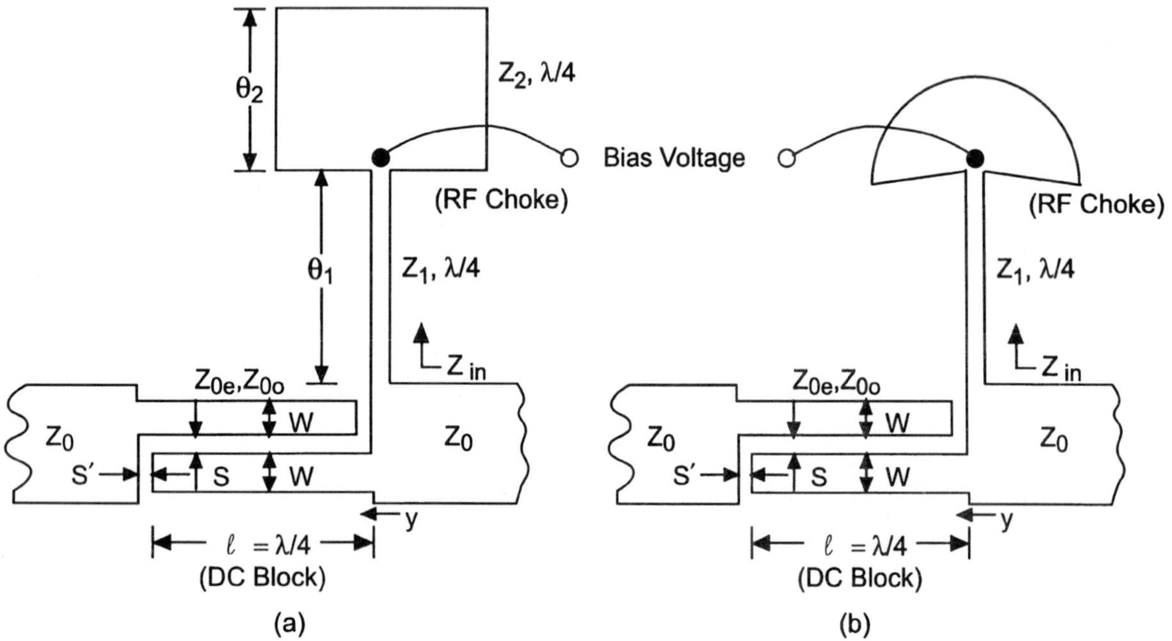

Figure 10.5 Microwave biasing circuits using multisections shunt stubs for large bandwidths. (a) Two $\lambda/4$ sections configuration and (b) a combination of $\lambda/4$ section and radial line configuration.

Coupled-Line Circuit Components

Table 10.1
Parameters for the biasing network structure

Substrate height, $h = 125\,\mu\text{m}$
Substrate dielectric constant, $\varepsilon_r = 12.9$
Conductor thickness, $t = 4.5\,\mu\text{m}$
Conductor's bulk conductivity, $\sigma = 4.9 \times 10^7$ Sm
Substrate loss tangent $\tan\delta = 0.0005$

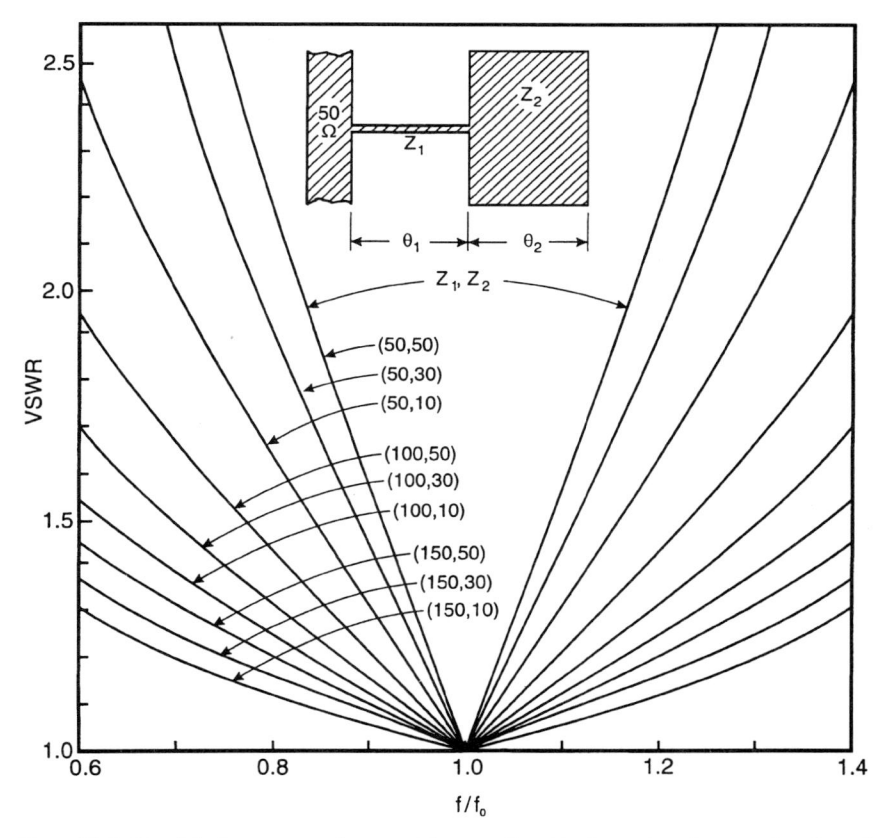

Figure 10.6 Simulated response of several two-section biasing networks as a function of normalized frequency. Impedance values are in ohms.

parameters are given in Figure 10.7. The structure was analyzed and optimized using a commercial CAD tool. The measurements were made on the wafer using RF probes and TRL calibration techniques. The measured and simulated performance (return loss and insertion loss) of this biasing network are compared in Figures 10.8 and 10.9. Over an octave bandwidth (9–18 GHz), the measured return loss and insertion loss were better than 17 dB and 0.5 dB, respectively [8].

10.1.4 Millimeter-Wave DC Block

At millimeter-wave frequencies the coupled-line dc block has distinct advantages over the conventional dc block, which generally consists of chip or MIM capacitors. These structures are broadband and cover full waveguide bandwidths with low insertion loss. Several dc blocks on RT-5880 duroid substrates were designed and fabricated with dimensions given in Table 10.2 for several millimeter-wave frequency bands [9]. The length of the coupled section is 90° at the center of the band. The measured insertion loss for these structures was between 0.2 to 0.4 dB over the full-waveguide bands.

10.1.5 High-Voltage DC Block

The above-described coupled-line dc blocks are capable of handling voltages of less than 200V, depending on the fabrication tolerances and humidity [10]. To increase the protection against voltage breakdown across the gap, an overlay of

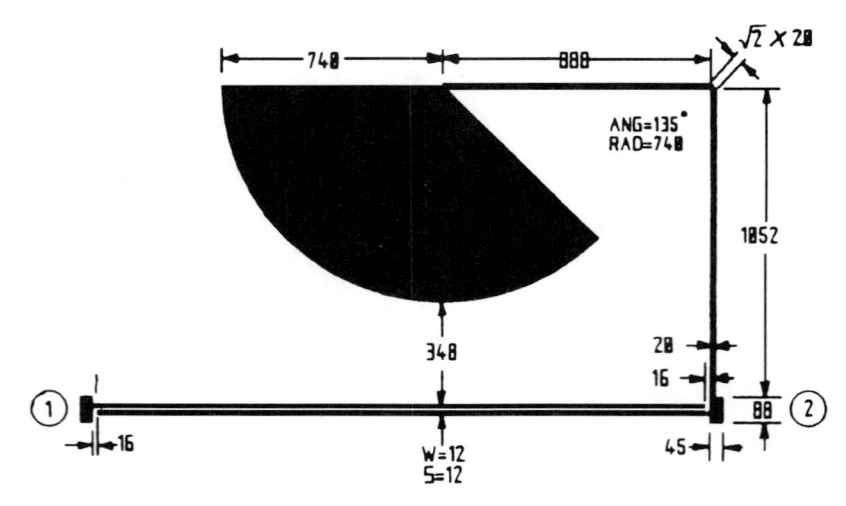

Figure 10.7 Biasing network using Figure 10.5(b) configuration. Coupled line length is 2045 μm.

Table 10.2
A comparison between designed and measured dimensions of the coupled microstrip dc blocks.

Band	h (millimeter)	ε_r	Designed dimensions (mil) W	S	Measured dimensions (mil) W	S
Ka	10	2.2	7.0	1.0	6.1	1.7
V	5	2.2	2.5	1.0	2.0	1.7
W	5	2.2	2.5	1.0	2.0	1.5

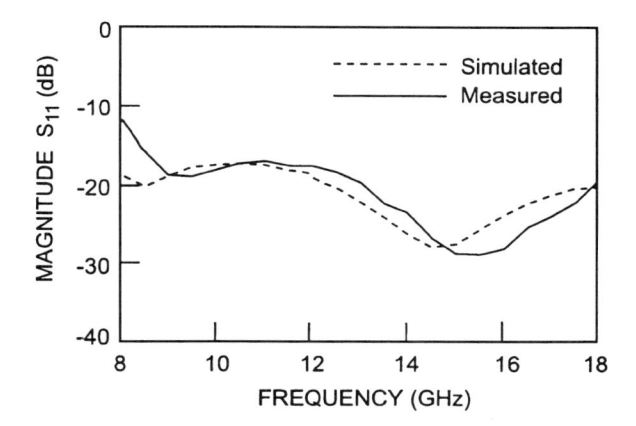

Figure 10.8 Simulated (LIBRA) and measured $|S_{11}|$ versus frequency of the biasing network.

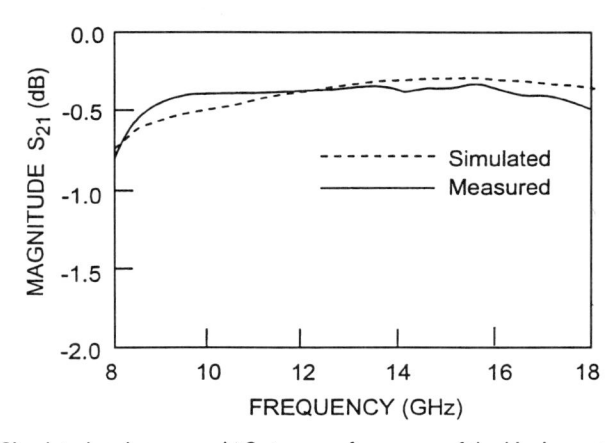

Figure 10.9 Simulated and measured $|S_{21}|$ versus frequency of the biasing network.

silicon rubber as shown in Figure 10.10 has been used. Because dielectric loading modifies the electrical characteristics of the coupled line, accurate design or simulation methods such as electromagnetic simulators are required to determine the new parameters. A dc block fabricated on RT/duroid substrate with spacing $S = 50\,\mu$m and width $W = 60\,\mu$m, achieved a breakdown voltage over 4.5 kV. In general, breakdown occurs at one of the open ends of the coupled lines. [10]

10.2 Coupled-Line Transformers

Quarter-wavelength-long tightly coupled lines have been used as filter elements, directional couplers, and dc blocks, and their other important application is in broadband impedance-matching transformers. This scheme works very well at millimeter-wave frequencies where it eliminates dc blocking capacitors and can handle large impedance transformation without transverse resonances which occurs in conventional $\lambda/4$ low-impedance microstrip single-section impedance transformers. A simple representation of a coupled-line impedance transformer is shown in Figure 10.11.

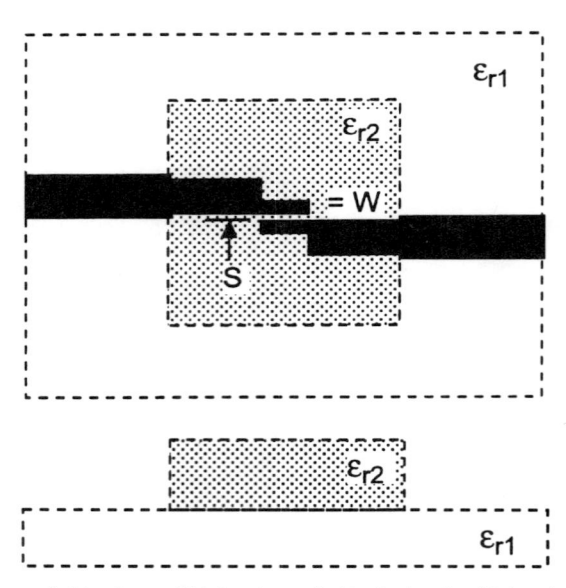

Figure 10.10 Top and side views of high voltage dc block showing high-voltage insulator dielectric. (From [10], © 1993 IEEE. Reprinted with permission.)

Coupled-Line Circuit Components

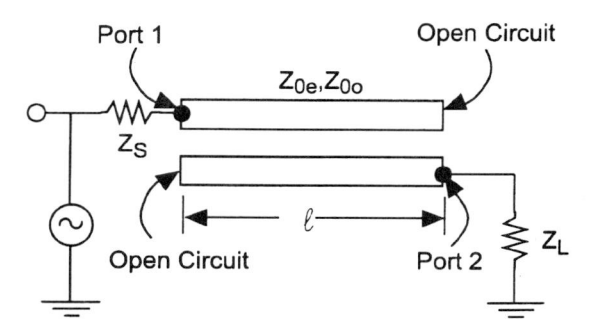

Figure 10.11 Coupled transmission-line transformer section.

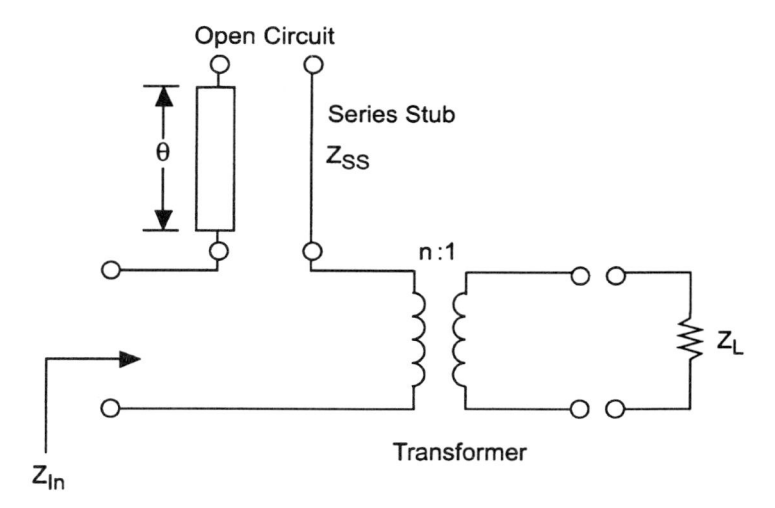

Figure 10.12 Equivalent circuit representation of the coupled-transmission-line impedance transformer.

The analysis of this structure can easily be performed by using its equivalent circuit shown in Figure 10.12, where the expressions for Z_{ss} and n are given by

$$Z_{ss} = Z_L(1 - n^2) \tag{10.10}$$

$$n = k = \frac{Z_{0e} - Z_{0o}}{Z_{0e} + Z_{0o}} \tag{10.11}$$

Under maximum power transfer from port 1 to port 2:

$$Z_{0e} = Z_{0o} + 2\sqrt{Z_{in} Z_L} \tag{10.12}$$

Suppose that if Z_L is the load impedance, the input impedance is given by
$$Z_{in} = -jZ_{ss}\cot\theta + n^2 Z_L \tag{10.13}$$
When $\theta = 90°$:
$$Z_{in} = n^2 Z_L \tag{10.14}$$
From (10.11), (10.12), and (10.14):
$$Z_L = \frac{Z_{0e} + Z_{0o}}{2} \tag{10.15}$$
If $Z_{in} = Z_S$:
$$Z_{0e} = Z_L\left(1 + \sqrt{\frac{Z_S}{Z_L}}\right) \tag{10.16a}$$

$$Z_{0o} = Z_L\left(1 - \sqrt{\frac{Z_S}{Z_L}}\right) \tag{10.16b}$$

For given Z_S and Z_L, the even- and odd-mode impedances of the coupled-line transformer can be determined from (10.16a) and (10.16b). Most of commercial CAD tools can synthesise these networks for given Z_{0o}, Z_{0e}, and substrate parameters in terms of physical dimensions W, S, and l.

If Z_S and Z_L are the source and load impedances, the transmission coefficient S_{21} is given by [7]

$$S_{21} = \frac{2(Z_{0e}\csc\theta_e - Z_{0o}\csc\theta_o)}{\left(\left(\sqrt{\frac{Z_L}{Z_S}} + \sqrt{\frac{Z_S}{Z_L}}\right)(Z_{0e}\cot\theta_e + Z_{0o}\cot\theta_o)\right.} \\ \left. + j\left[\frac{Z_{0e}^2 + Z_{0o}^2}{2\sqrt{Z_S Z_L}} - \frac{Z_{0e}Z_{0o}}{\sqrt{Z_S Z_L}}(\csc\theta_e\csc\theta_o + \cot\theta_e\cot\theta_o) + 2\sqrt{Z_S Z_L}\right]\right)$$
(10.17)

where θ_e and θ_o are the electrical lengths of the structure corresponding to the even- and odd-mode propagation constants. The electrical lengths are given as

$$\theta_e = \frac{2\pi}{\lambda_0}\ell\sqrt{\varepsilon_{ree}} \text{ and } \theta_o = \frac{2\pi}{\lambda_0}\ell\sqrt{\varepsilon_{reo}} \tag{10.18}$$

where λ_0 is the free-space wavelength, and ε_{ree} and ε_{reo} are the effective dielectric constants corresponding to the even- and odd-mode propagation, respectively.

Under first-order approximation when $\theta_e = \theta_o = \pi/2$, and $S_{21} = 1$ (matched condition for lossless network), (10.17) provides

$$Z_{0e} = Z_{0o} + 2\sqrt{Z_S Z_L} \tag{10.19}$$

which is the same as (10.12) when $Z_S = Z_{in}$.

The quarter-wavelength of the coupled section at the center frequency may be calculated from the following relation [11]:

$$\ell = \frac{\pi/2}{K + [(Z_{0e} - Z_{0o})/(Z_{0e} + Z_{0o})]\Delta K} \quad (10.20)$$

where

$$K = (\beta_e + \beta_o)/2$$
$$\Delta K = (\beta_e - \beta_o)/2$$
$$\beta_e = 2\pi\sqrt{\varepsilon_{ree}}/\lambda_0$$
$$\beta_o = 2\pi\sqrt{\varepsilon_{reo}}/\lambda_0$$

Table 10.3 summarizes the electrical and physical parameters for typical microstrip coupled-line transformers. Parameters from the table were used to compare this type of transformer with a $\lambda/4$ microstrip section to transform 10Ω impedance to 50Ω impedance. Three cases (9, 10, and 11 in Table 10.3) were considered for coupled lines with the physical line length used being 2,200 μm. For the single-section case, $W = 2,200\ \mu$m and $l = 1,850\ \mu$m were used. The substrate parameters used are thickness $h = 625\ \mu$m, conductor thickness $t = 3\ \mu$m, $\varepsilon_r = 9.8$, $\tan\delta = 0.0005$, and the conductors were of gold. Figure 10.13 shows the performance comparison. Although a $\lambda/4$ single microstrip transformer gives the best electrical performance in terms of insertion loss and bandwidth, its physical dimension (width) is too large in the transverse direction. A transverse resonance may occur when the width of the conductor is $\lambda/2$. In this example the resonance frequency is about 26 GHz.

Symmetrical coupled-line section transformers were described above. Asymmetrical coupled sections can also be used as transformers and offer better bandwidths. Coupled-line transformers are good candidates for active device impedance matching at millimeter-wave frequencies where they also serve the purpose of low-loss dc blocks. Thus, coupled-line transformers provide wide bandwidth and eliminate the use of dc blocking capacitors in active circuits.

10.3 Interdigital Capacitor

The interdigital or interdigitated capacitor is a multifinger periodic structure as shown in Figure 10.14. Interdigital capacitors use the capacitance that occurs across a gap in thin-film conductors. These gaps are essentially very long and folded to use a small amount of area. As illustrated in Figure 10.14, the

Table 10.3
Parameters for different transformation ratios: $\varepsilon_r = 9.8$, $t = 3$ to $4\ \mu m$, and $h = 635\ \mu m$

Serial Number	Transformation From: Ω	To: Ω	Desired impedances Z_{0e} Ω	Z_{0o} Ω	Actual impedances Z_{0e} Ω	Z_{0o} Ω	Velocity factors K_{0e}	K_{0o}	Physical dimensions W/h	S/h
1	50.0	75.0	162.5	40.0	162.40	40.59	0.404	0.4300	0.120	0.048
2	50.0	75.0	152.5	30.0	153.81	30.24	0.403	0.4298	0.159	0.018
3	50.0	53.0	142.96	40.0	143.06	40.47	0.401	0.4298	0.175	0.070
4	50.0	53.0	132.96	30.0	132.50	30.72	0.399	0.4296	0.240	0.030
5	50.0	35.0	123.67	40.0	123.24	39.83	0.397	0.4293	0.260	0.135
6	50.0	35.0	113.67	30.0	113.75	29.96	0.394	0.4291	0.350	0.040
7	50.0	25.0	110.71	40.0	110.28	40.03	0.394	0.4287	0.335	0.135
8	50.0	25.0	100.71	30.0	100.69	30.07	0.391	0.4284	0.455	0.055
9	50.0	10.0	84.72	40.0	84.66	39.81	0.385	0.4261	0.565	0.250
10	50.0	10.0	74.72	30.0	74.57	30.17	0.381	0.4253	0.790	0.115
11	50.0	10.0	64.72	20.0	65.11	20.14	0.376	0.4246	1.05	0.01

$K_{0e} = 1/\sqrt{\varepsilon_{ree}}$ and $K_{0o} = 1/\sqrt{\varepsilon_{reo}}$

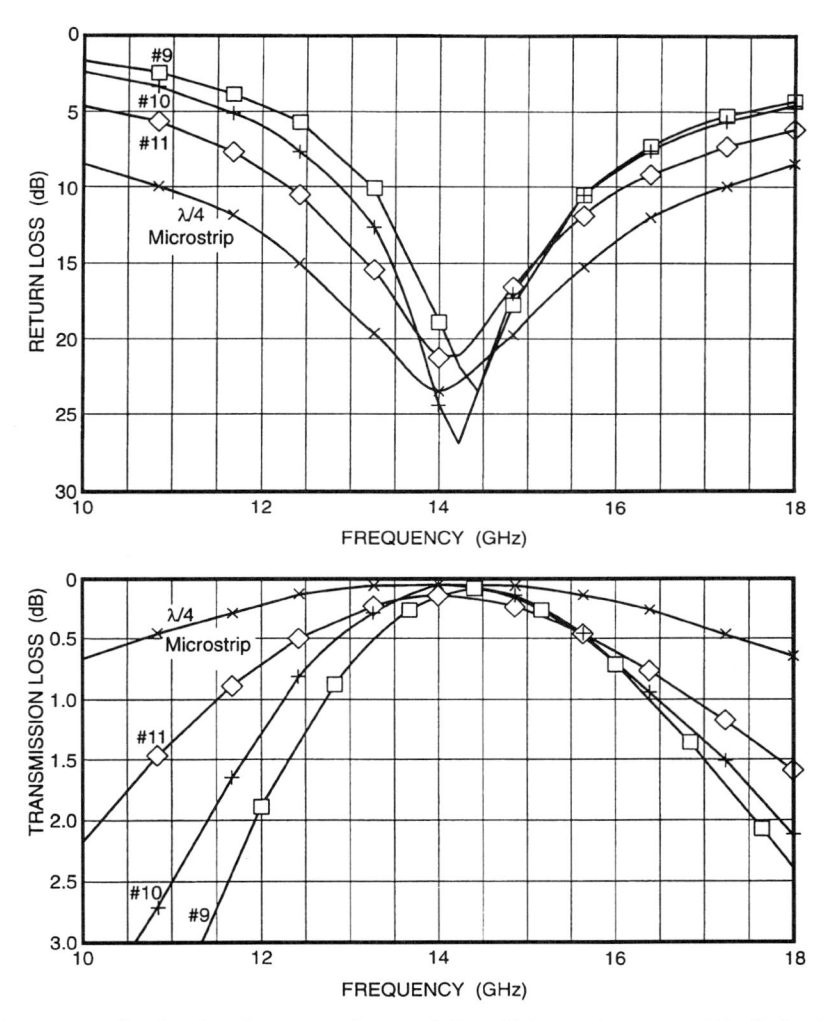

Figure 10.13 Simulated performance of several 10 to 50Ω transformers on 25-mil alumina substrates.

gap meanders back and forth in a rectangular area forming two sets of fingers, which are interdigital. By using a long gap in a small area, compact single-layer small-valued capacitors can be realized. Typically, values range from 0.05 pF to about 0.5 pF. The capacitance can be increased by increasing the number of fingers, or by putting on an overlay dielectric layer, which also acts as a protective shield. One of the important design considerations is to keep the size of the

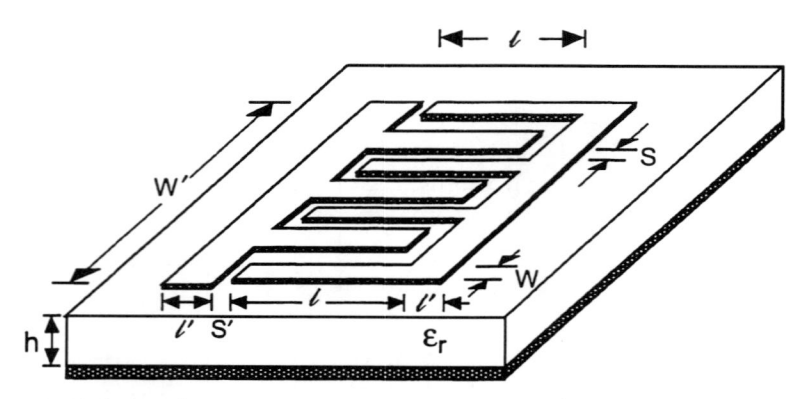

Figure 10.14 An interdigital capacitor configuration using seven fingers.

capacitor very small relative to a wavelength so that it can be treated as a lumped element. A larger total width-to-length ratio results in the desired higher shunt capacitance and lower series inductance. This type of capacitor can be fabricated by hybrid technology used in the fabrication of conventional integrated circuits or monolithic microwave integrated circuit technology and does not require any additional processing steps.

10.3.1 Approximate Analysis

Analysis and characterization of interdigital capacitors have been reported in the literature [12–17]. Earlier analyses [12–14] data when compared with the measured results showed that these analyses were inadequate to describe the capacitors accurately. The analyses were based on lossless microstrip coupled lines [12] and lossy coupled microstrip lines. [13] A more accurate characterization of these capacitors can be performed if the capacitor geometry is divided into basic microstrip sections such as the single microstrip line, coupled microstrip lines, open-end discontinuity, asymmetrical gap, 90° bend, and T-junction discontinuities [16] as shown in Figure 10.15. This model provides better accuracy than the previously reported analyses. This method still provides an approximate solution, however, because of several assumptions in the grouping of subsections and does not include interaction effects between basic microstrip sections described above.

An approximate expression for an interdigital capacitor is given by [12]

$$C = \frac{\varepsilon_r + 1}{W} \ell [(N-3)A_1 + A_2] \qquad (10.21)$$

where A_1 (the interior) and A_2 (the two exterior) are the capacitances of the fingers, N is the number of fingers, and the dimensions W and ℓ are as shown

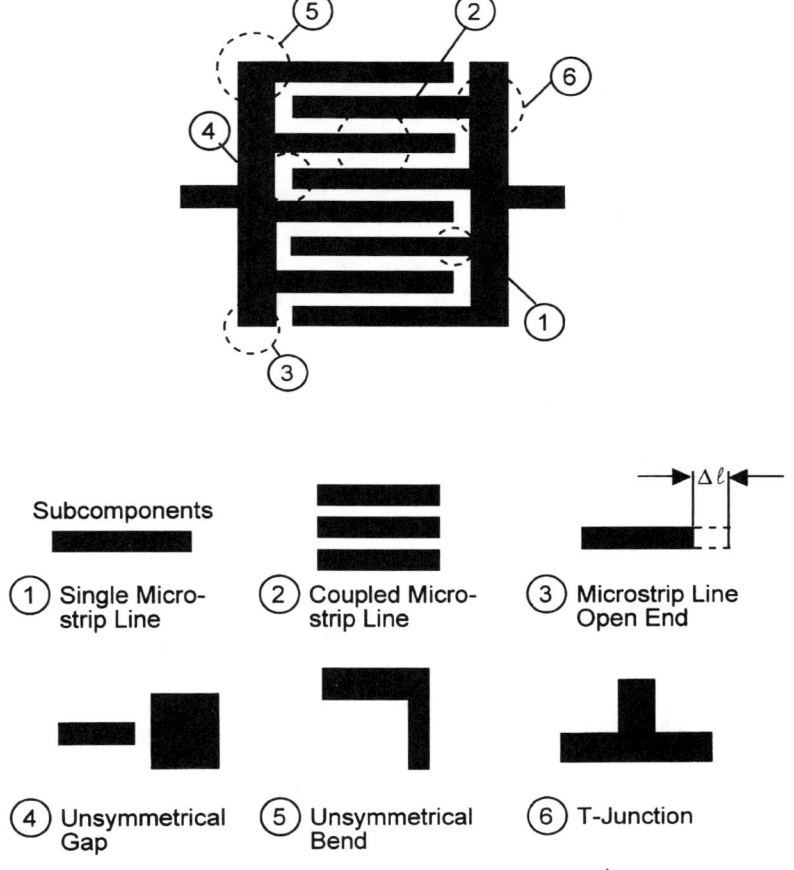

Figure 10.15 The interdigitated capacitor and its subcomponents. (From [16], © 1988 IEEE. Reprinted with permission.)

in Figure 10.14. For infinite substrate thickness (or no ground plane), $A_1 = 8.86 \times 10^{-6}$ pF/μm and $A_2 = 9.92 \times 10^{-6}$ pF/μm.

For a finite substrate, the effect of h must be included in A_1 and A_2. In the final design, usually $S = W$ and $\ell \leq \lambda/4$.

The total series capacitance can also be written as [6]

$$C = 2\varepsilon_0 \varepsilon_{re} \frac{K(k)}{K'(k)} (N-1)\ell \, \text{F}$$

$$= \frac{10^{-11}}{18\pi} \varepsilon_{re} \frac{K(k)}{K'(k)} (N-1)\ell \times 10^{-4} \text{F} \qquad (10.22a)$$

or

$$C = \frac{\varepsilon_{re} 10^{-3}}{18\pi} \frac{K(k)}{K'(k)} (N-1)\ell \text{pF} \tag{10.22b}$$

where ℓ is in μm, N is the number of fingers, ε_{re} is the effective dielectric constant of the microstrip line of width W, and

$$\frac{K(k)}{K'(k)} = \frac{1}{\pi} \ell n \left\{ 2 \frac{1+\sqrt{k}}{1-\sqrt{k}} \right\} \quad \text{for } 0.707 \leq k \leq 1 \tag{10.23a}$$

$$= \frac{\pi}{\ell n \left[2 \frac{1+\sqrt{k'}}{1-\sqrt{k'}} \right]} \quad \text{for } 0 \leq k \leq 0.707 \tag{10.23b}$$

and

$$k = \tan^2\left(\frac{a\pi}{4b}\right), \, a = W/2, \text{ and } b = (W+S)/2, \text{ and } k' = \sqrt{1-k^2} \tag{10.23c}$$

The series resistance of the interdigital capacitor is given by

$$R = \frac{4}{3} \frac{\ell}{WN} R_s \tag{10.24}$$

where R_s is the sheet resistivity (Ω/square) of the conductors used in the capacitors. The effect of metal-thickness t plays a secondary role in the calculation of capacitance. The Q of this capacitor is given by

$$Q_c = \frac{1}{\omega CR} = \frac{3WN}{\omega C 4\ell R_s} \tag{10.25}$$

10.3.2 Fullwave Analysis

Quasistatic and fullwave numerical methods have been extensively employed to analyze transmission lines and their discontinuities. The numerical data obtained from these methods have been used to develop analytical and empirical design equations along with equivalent circuit (EC) models to describe the electrical performance of planar transmission lines and their discontinuities, including microstrip, coplanar waveguide and slot lines. These equations and EC models have been exclusively used in commercial microwave CAD tools. The recent advances in workstations and user-friendly software made it possible to develop electromagnetic (EM) simulators. The emerging EM simulators [18–24] have added another dimension to computer-aided engineering (CAE) tools. These simulators play an important role in the simulation of single-layer elements such

Table 10.4
Parameters for an interdigital structure

Substrate height, $h = 100\ \mu$m
Substrate dielectric constant, $\varepsilon_r = 12.9$
Conductor thickness, $t = 0.8\ \mu$m
Conductor's bulk conductivity, $\sigma = 4.9 \times 10^7$ Sm
Substrate loss tangent, $\tan\sigma = 0.0005$
$W = 40\ \mu$m
$S = 20\ \mu$m
$\ell = 440\ \mu$m
$W' = 520\ \mu$m
$S' = 40\ \mu$m
$\ell' = 60\ \mu$m
No. of fingers = 9
Enclosure: No

as transmission lines, patches and their discontinuities, multilayer components (namely inductors, capacitors, packages, etc.), and mutual coupling between various circuit elements. Accurate evaluation of the effects of radiation, surface waves, and interaction between components on the performance of densely packed monolithic microwave integrated circuits can only be calculated using EM simulators [18–24].

Table 10.4 lists the physical parameters of a typical interdigital capacitor (Figure 10.14) analyzed using EM simulators, and Figures 10.16 and 10.17 compare the simulated S_{11} and S_{21} magnitude, and S_{11} and S_{21} phase performances [25–27], respectively.

10.4 Spiral Inductors

Spiral (rectangular or circular) inductors are used as RF chokes, matching elements, impedance transformers, and reactive terminations, and they can also be found in filters, couplers, dividers and combiners, baluns, and resonant circuits. In the low-microwave frequency monolithic approach, low-loss inductors are essential to develop compact and low-cost, low-noise amplifiers and high-power added-efficiency amplifiers.

Inductors in MICs are fabricated using standard integrated circuit processing with no additional process steps. The innermost turn of the inductor is connected to other circuitry by using a wire bond connection in the hybrid MICs and through a conductor that passes under airbridges in the monolithic MIC

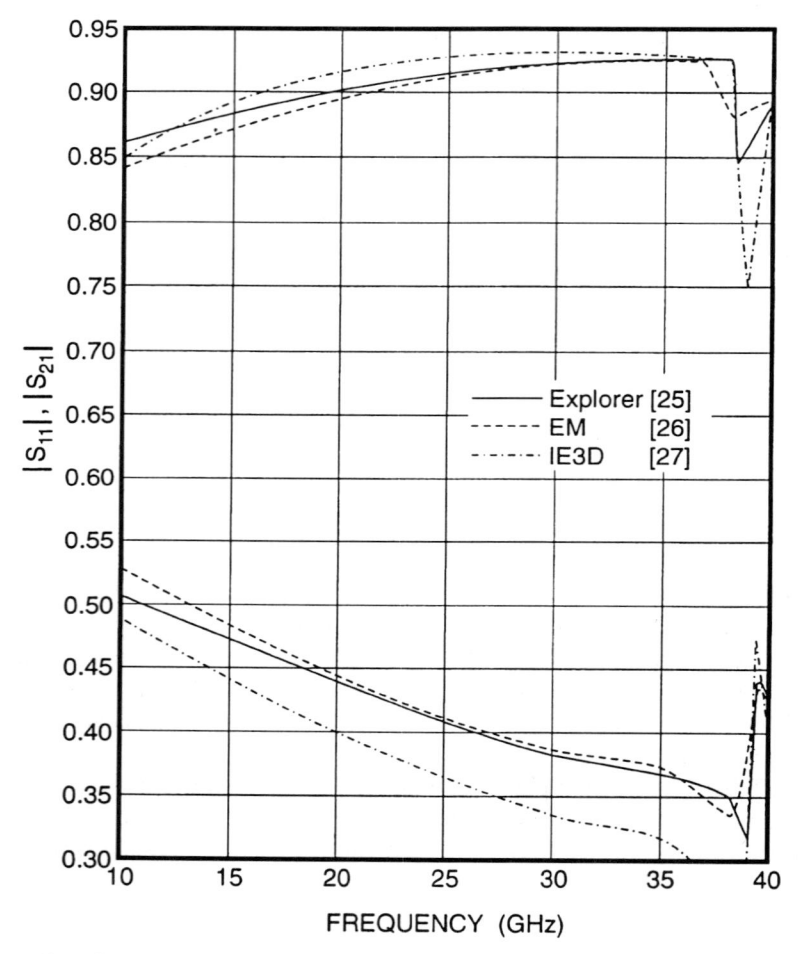

Figure 10.16 Interdigitated capacitor $|S_{11}|$ and $|S_{21}|$ responses.

technology. The width and thickness of the conductor determine the current-carrying capacity of the inductor. Typically the thickness is 0.5 to 1.0 μm, and the airbridge separates it from the upper conductors by 1.5 to 3.0 μm. In a dielectric crossover technology, the separation between the crossover conductors may be anywhere between 0.5 to 3 μm. Typical inductance values for monolithic microwave integrated circuits working above the S-band fall in the range of 0.5 to 10 nH.

The design of spiral inductors for MIC applications is usually based on two approaches: the lumped-element method and microstrip coupled-line method.

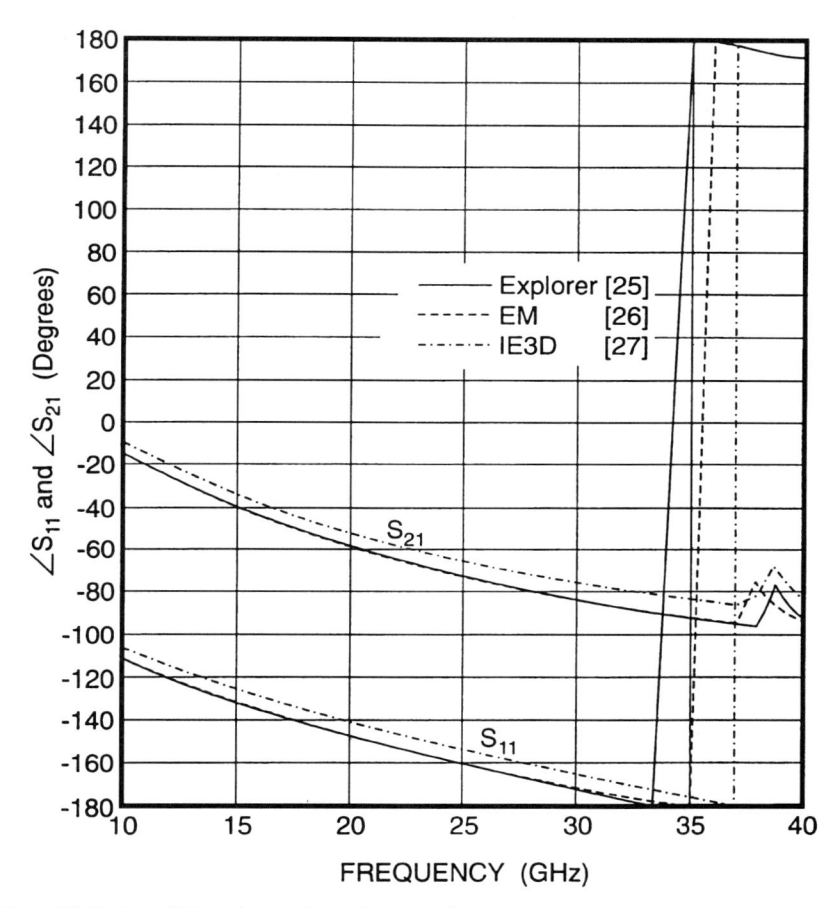

Figure 10.17 Interdigitated capacitor $\angle S_{11}$ and $\angle S_{21}$ responses.

The lumped-element approach uses frequency independent formulas for free-space inductance with ground-plane effects. These formulas are useful only when the total length of the inductor is a small fraction of the operating wavelength and when the inter-turn capacitance can be ignored. Wheeler [28] presented an approximate formula for the inductance of a circular spiral inductor, with reasonably good accuracy at lower microwave frequencies. This formula has been extensively used for the design of microwave lumped circuits [29–31]. The inductor parameters can also be obtained from two-port S-parameter measurements for the structure. This approach requires fabrication of the structure, however. In the coupled-line approach [32, 33], an inductor is analyzed using multiconductor coupled microstrip lines. This technique predicts the spiral inductor's performance reasonably well for two turns and up to 18 GHz. Inductors with their conductors supported on posts provide lower capacitances from the conductor to ground and between conductors, which result in a higher resonant frequency, thereby extending the frequency range of operation.

A 2-turn spiral microstrip inductor with opposite sides appropriately connected as shown in Figure 10.18 may be treated as a coupled-line section.

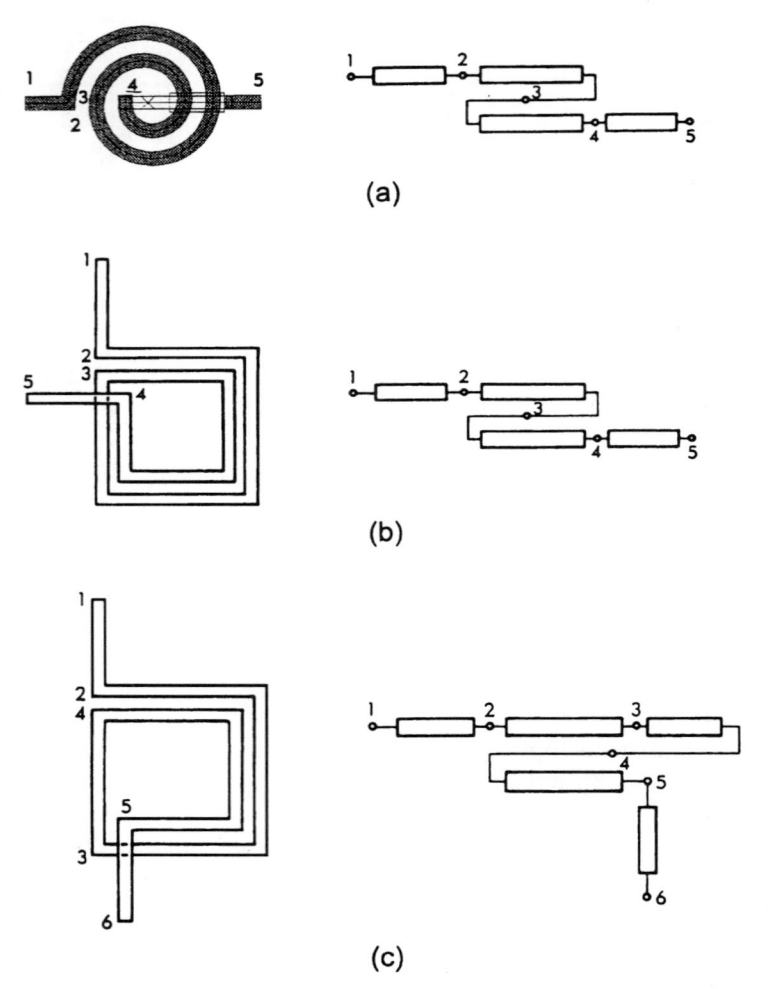

Figure 10.18 Spiral inductors and its coupled-line equivalent circuit models (a) circular 2 turns, (b) rectangular 2 turns, and (c) rectangular $1\frac{3}{4}$ turns.

This figure shows 2-turn circular and rectangular spiral inductors and 1.75-turn rectangular spiral inductor with connecting single-line section and the feed lines represented between nodes 1 and 2, and between nodes 4 and 5. In the 2-turn case, the parallel coupled-line section, which has total line length equivalent to the spiral length between nodes 2 and 4, is represented between nodes 2, 3, and 4 and constitutes the intrinsic inductor. The length is taken as the average of the outer and inner turn lengths. In the 1.75-turn inductor, an additional single line between nodes 3 and 4 is connected, whereas the coupled line is between nodes 2 and 3.

The electrical equivalent coupled-line model shown in Figure 10.18 does not include the crossover capacitance or the right-angle bend discontinuity effects. Figure 10.19 shows a modified equivalent circuit of a 1.75-turn rectangular spiral inductor that also includes the crossover capacitor. This figure also shows the physical and electrical parameters of this inductor. Figure 10.20 shows a further subdivision of the inductor that is required to evaluate more accurately its performance. In this case, the inductor is split into three sections representing elements 1, 2, and 3. The performance of these inductors can either be calculated by using commercial CAD tools or by solving cascaded ABCD or S-parameter matrices for these elements. Improved versions of these inductors include chamferred bends or compensated bends [5].

The electrical characteristics of the intrinsic 2-turn inductor can be derived from the general four-port network of a coupled-lines section as shown in Figure 10.21, where the current and voltage relationships of the pair of lines can be described by the admittance matrix equation as follows:

$$\begin{bmatrix} I_1 \\ I_2 \\ I_3 \\ I_4 \end{bmatrix} = \begin{bmatrix} Y_{11} & Y_{12} & Y_{13} & Y_{14} \\ Y_{21} & Y_{22} & Y_{23} & Y_{24} \\ Y_{31} & Y_{32} & Y_{33} & Y_{34} \\ Y_{41} & Y_{42} & Y_{43} & Y_{44} \end{bmatrix} \begin{bmatrix} V_1 \\ V_2 \\ V_3 \\ V_4 \end{bmatrix} \qquad (10.26)$$

This matrix may be reduced to two port by applying the boundary condition that ports 2 and 4 are connected, that is:

$$V_2 = V_4 \qquad (10.27a)$$

$$I_2 = -I_4 \qquad (10.27b)$$

and by rearranging the matrix elements, the two-port matrix may be written as

$$\begin{bmatrix} I_1 \\ I_3 \end{bmatrix} = \begin{bmatrix} Y'_{11} & Y'_{13} \\ Y'_{31} & Y'_{33} \end{bmatrix} \begin{bmatrix} V_1 \\ V_3 \end{bmatrix} \qquad (10.28)$$

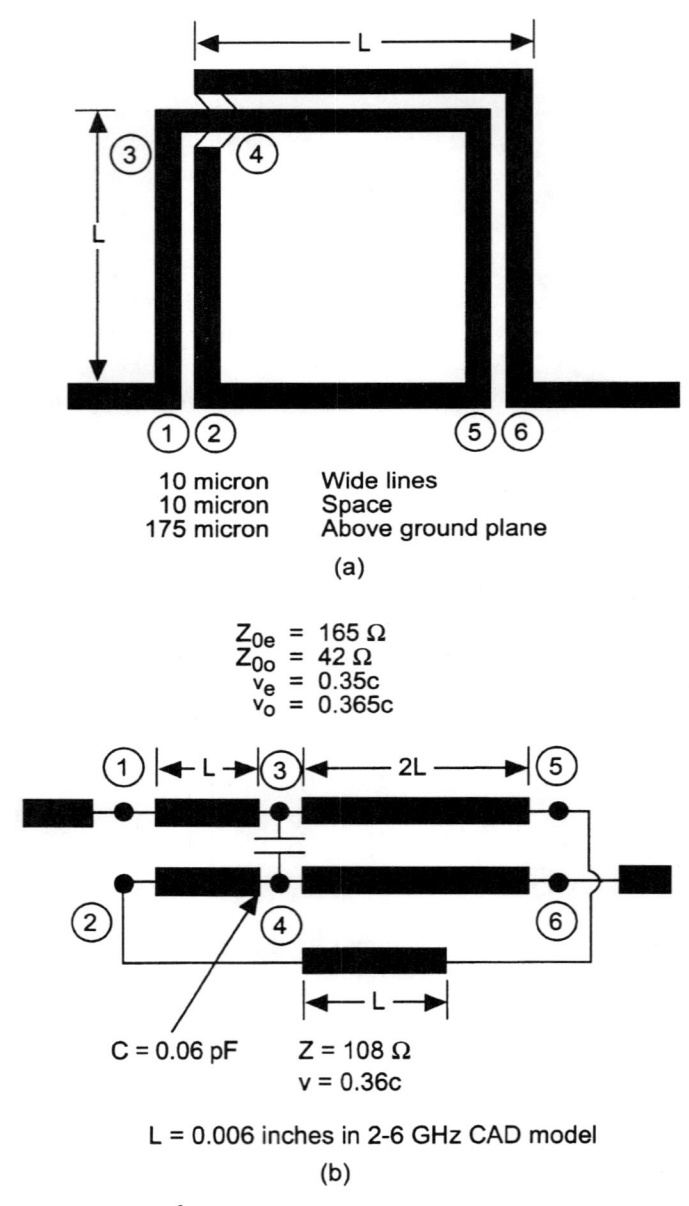

10 micron Wide lines
10 micron Space
175 micron Above ground plane

(a)

Z_{0e} = 165 Ω
Z_{0o} = 42 Ω
v_e = 0.35c
v_o = 0.365c

C = 0.06 pF   Z = 108 Ω
               v = 0.36c

L = 0.006 inches in 2-6 GHz CAD model

(b)

Figure 10.19 Rectangular $1\frac{3}{4}$-turn spiral inductor (a) physical layout and (b) coupled-line equivalent circuit model. (From [32], © 1983 IEEE. Reprinted with permission.)

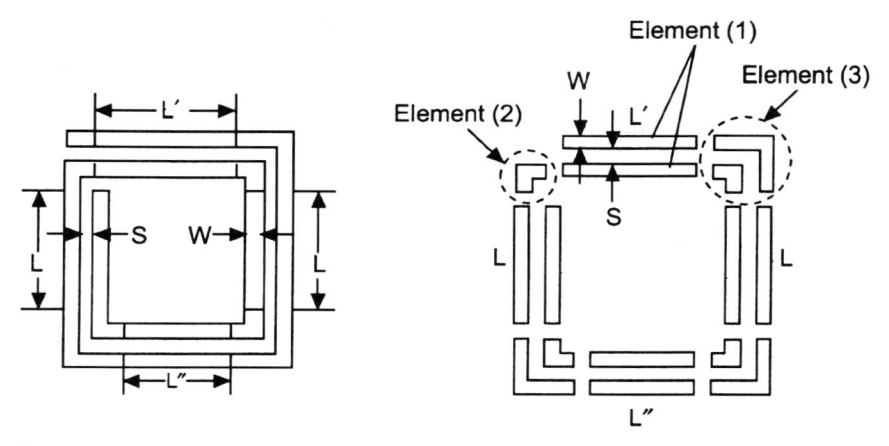

Figure 10.20 The network model for calculating the inductance of a planar rectangular spiral inductor.

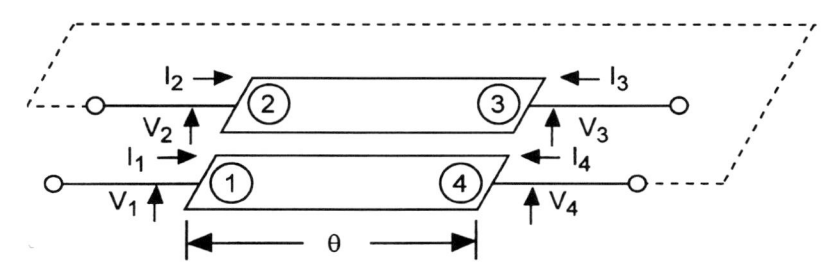

Figure 10.21 Four-port representation of a coupled-line section of an inductor.

where

$$Y'_{11} = Y_{11} - \frac{(Y_{12} + Y_{14})(Y_{21} + Y_{41})}{Y_{22} + Y_{24} + Y_{42} + Y_{44}} \tag{10.29}$$

$$Y'_{13} = Y_{13} - \frac{(Y_{12} + Y_{14})(Y_{23} + Y_{43})}{Y_{22} + Y_{24} + Y_{42} + Y_{44}} \tag{10.30}$$

and

$$Y'_{33} = Y'_{11} \tag{10.31}$$

$$Y'_{31} = Y'_{13} \tag{10.32}$$

due to symmetry.

The admittance parameters for a coupled microstrip line in inhomogenous dielectric medium are given by [34]

$$Y_{11} = Y_{22} = Y_{33} = Y_{44} = -j[Y_{0e}\cot\theta_e + Y_{0o}\cot\theta_o]/2 \quad (10.33a)$$
$$Y_{12} = Y_{21} = Y_{34} = Y_{43} = -j[Y_{0e}\cot\theta_e - Y_{0o}\cot\theta_o]/2 \quad (10.33b)$$
$$Y_{13} = Y_{31} = Y_{24} = Y_{42} = j[Y_{0e}\csc\theta_e - Y_{0o}\csc\theta_o]/2 \quad (10.33c)$$
$$Y_{14} = Y_{41} = Y_{23} = Y_{32} = j[Y_{0e}\csc\theta_e + Y_{0o}\csc\theta_o]/2 \quad (10.33d)$$

An equivalent "π" representation of a two-port network is shown in Figure 10.22, where

$$Y_A = -Y'_{13} \quad (10.34)$$
$$Y_B = Y'_{11} + Y'_{13} \quad (10.35)$$

and

$$Y_A = -j\frac{1}{2}\left\{Y_{oe}\cot\theta_e + Y_{oo}\cot\theta_o + \frac{\left[Y_{oe}\left(\frac{1-\cos\theta_e}{\sin\theta_e}\right) + Y_{oo}\left(\frac{1+\cos\theta_o}{\sin\theta_o}\right)\right]}{2\left[Y_{oe}\left(\frac{1-\cos\theta_e}{\sin\theta_e}\right) - Y_{oo}\left(\frac{1+\cos\theta_o}{\sin\theta_o}\right)\right]}\right\}$$

(10.36)

$$Y_B = \frac{2jY_{oe}Y_{oo}(1-\cos\theta_e)(1+\cos\theta_o)}{[Y_{oo}\sin\theta_e(1+\cos\theta_o) - Y_{oe}\sin\theta_o(1-\cos\theta_e)]} \quad (10.37)$$

Because the physical length of the inductor is much less than $\lambda/4$, $\sin\theta_{e,o} \cong \theta_{e,o}$ and $\cos\theta_{e,o} \cong 1 - \theta^2_{e,o}/2$. Also $Y_{oo} > Y_{oe}$. Therefore, (10.36) and (10.37) are

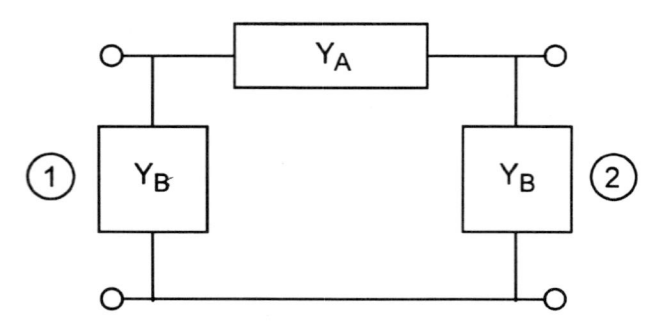

Figure 10.22 π equivalent circuit representation of the inductor.

Coupled-Line Circuit Components

approximated as follows:

$$Y_A \cong -j\frac{Y_{oe}}{2\theta_e} \tag{10.38}$$

$$Y_B \cong jY_{oe}\theta_e \tag{10.39}$$

which are independent of the odd mode. Thus the "π" equivalent circuit consists of shunt capacitance C and series inductance L as shown in Figure 10.23. The expressions for L and C may be written as

$$Y_A = \frac{1}{j\omega L} = -j\frac{Y_{oe}}{2\theta_e} \tag{10.40}$$

or

$$L = \frac{2\theta_e}{\omega Y_{oe}} \tag{10.41}$$

and

$$Y_B = j\omega C = jY_{oe}\theta_e \tag{10.42}$$

or

$$C = \frac{Y_{oe}\theta_e}{\omega} \tag{10.43}$$

If ℓ is the average length of the conductor, then

$$\theta_e = \frac{\omega \ell}{c}\sqrt{\varepsilon_{ree}} \tag{10.44}$$

where c is the velocity of light in free space and ε_{ree} is the effective dielectric constant for the even mode. When $Z_{0e} = 1/Y_{0e}$, from (10.41) and (10.43):

$$L = \frac{2\ell Z_{0e}\sqrt{\varepsilon_{ree}}}{c} \tag{10.45}$$

$$C = \frac{\ell\sqrt{\varepsilon_{ree}}}{Z_{0e}c} \tag{10.46}$$

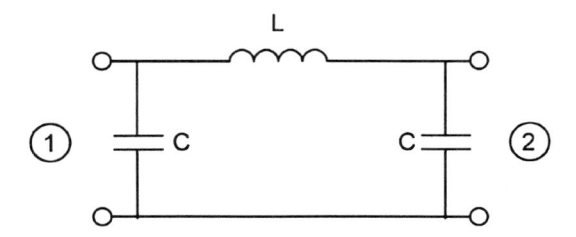

Figure 10.23 Equivalent LC circuit representation of the inductor.

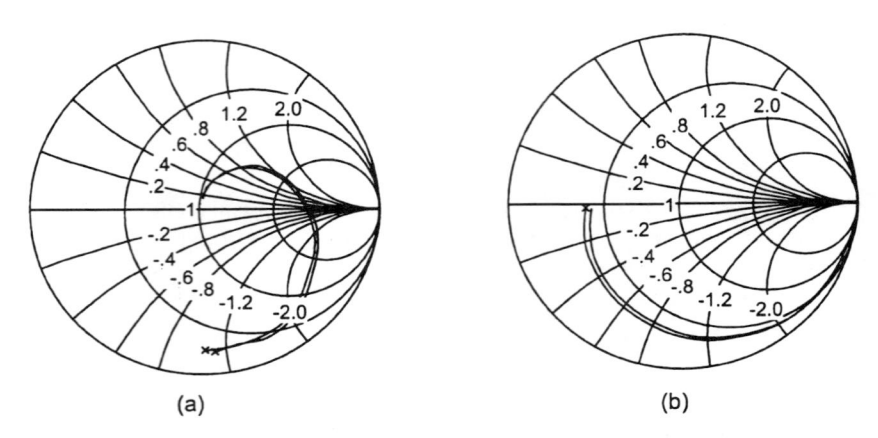

Figure 10.24 Measured and modeled (a) reflection (S_{11}) and (b) transmission (S_{21}) responses for the two-turn inductor. (From R. Plumb, workshop notes.)

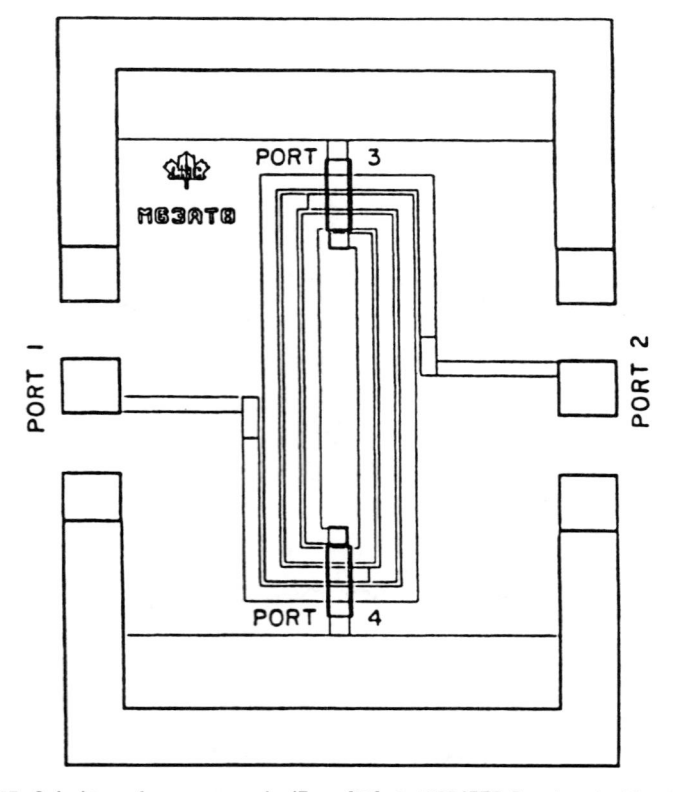

Figure 10.25 Spiral transformer example. (From [36], © 1989 IEEE. Reprinted with permission.)

In a loosely coupled inductor $Z_{oe} \cong Z_o$ and $\varepsilon_{ree} = \varepsilon_{re}$ for the single conductor microstrip line. The above equations may be used to approximately evaluate inductor performance. Figure 10.24 shows measured and modeled S_{11} and S_{21} responses of a two-turn inductor.

10.5 Spiral Transformers

The classical coil transformers used at low and radio frequencies can be realized by hybrid and monolithic techniques working in the microwave frequency range. A major challenge in the printed coil transformers is keeping the parasitic capacitances and series resistances low to operate these components at higher frequencies with low insertion loss. The transformers can be two-, three-, or four-port components. The 3-port transformers may have 0°, 90°, or 180° phase difference at the output ports.

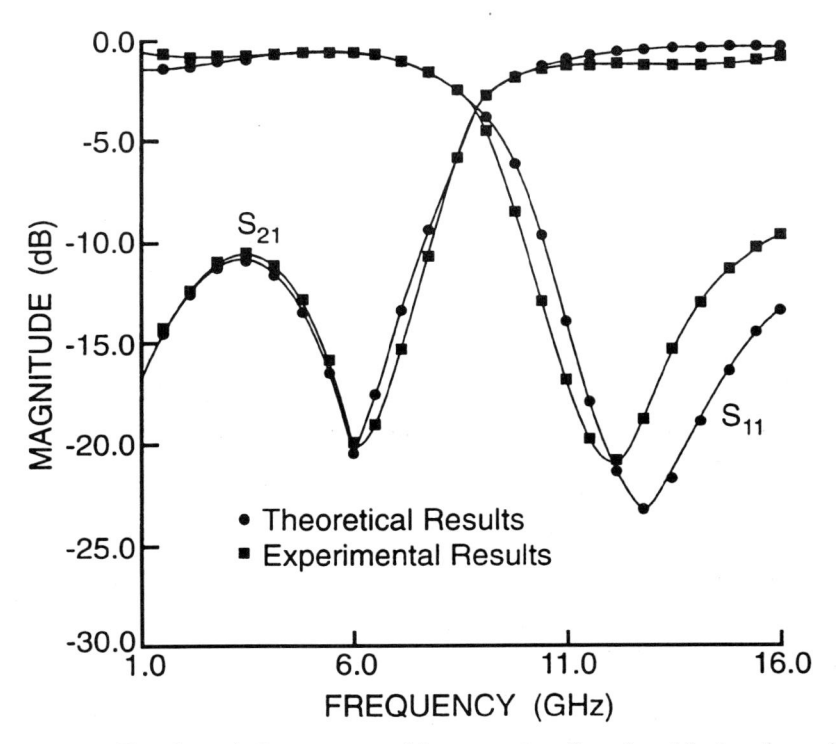

Figure 10.26 The $|S_{11}|$ and $|S_{21}|$ responses of the two-port configuration of the transformer in Figure 10.25. (From [36], © 1989 IEEE. Reprinted with permission.)

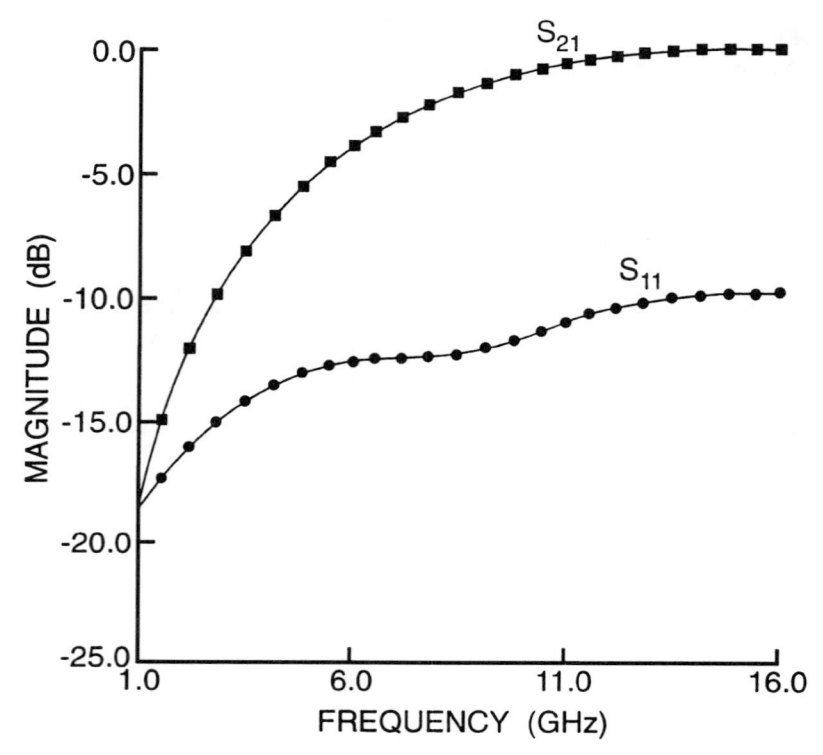

Figure 10.27 The $|S_{11}|$ and $|S_{21}|$ responses of the four-port configuration, constructed by adding 50Ω series resistors to ports 3 and 4 in Figure 10.25. (From [36], © 1989 IEEE. Reprinted with permission.)

Rectangular spiral transformers fabricated using GaAs MMIC technology have been reported in the literature [35–39]. In active circuits, their impedance transformer ratio and inductance values are used for impedance transformation and resonating out the active device's capacitance, respectively. Figure 10.25 shows the physical layout of a two-conductor transformer consisting of a series of turns of thin, metallized conductors placed on a dielectric substrate (not shown). The characterization of this structure is not straightforward; however, a multiconductor coupled microstrip line analysis [40, 41] may be used to determine its parameters approximately. A more accurate characterization of this transformer can only be achieved by using fullwave and comprehensive circuit simulators such as EM CAD tools.

The twin-coil four-port rectangular spiral transformer [36, 38] shown in Figure 10.25 has a ground ring around it. The dimensions of the transformer are outside ring = 1,020 × 710 μm, conductor thickness = 1 μm,

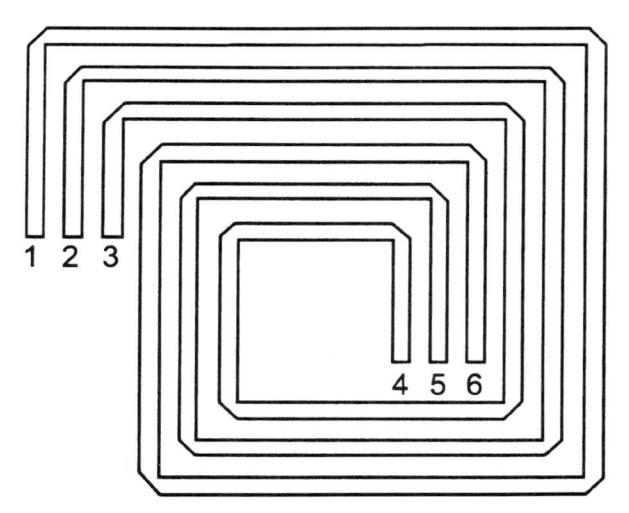

Figure 10.28 Schematic of a 1.5-turn rectangular spiral triformer.

GaAs ($\varepsilon_r = 12.9$), substrate thickness $= 250\,\mu\text{m}$, dielectric crossover height $= 1.3\,\mu\text{m}$, overlay dielectric $\varepsilon_r = 6.8$, conductor width $= 20\,\mu\text{m}$, and conductor gap $= 6\,\mu\text{m}$. Figure 10.26 shows a comparison between the measured and simulated S_{11} and S_{21} responses of this transformer when ports 3 and 4 are short-circuited as shown in Figure 10.25. The simulated performance was obtained using a WATMIC-EMsim CAD tool, which is an electromagnetic simulator. As shown in Figure 10.26, S_{21} has a sharp null at 6 GHz that occurs because of the $\lambda/4$ short-circuited secondary coil at that frequency. In this case, maximum current flows through the grounded port 4 and negligible current flows through port 2. Thus, power flowing into the load at port 2 is negligible and results in a null in the S_{21} response. Below 6 GHz, the current in the spiral conductors is nearly constant and the power transfer is provided by magnetic coupling similar to that in classical coil transformers. Above 6 GHz, however, the spiral conductors are electrically long and the current distribution along the conductors has less of a standing wave nature. Their behavior becomes closer to that of coupled transmission lines supporting both magnetic and electric coupling.

When ports 3 and 4 are terminated in 50Ω loads, the measured two-port response of the transformer is shown in Figure 10.27, which shows a gradual increase of S_{11} and S_{21}. Because in this case there are no standing waves, the power transfer from primary to secondary occurs gradually from magnetic to magnetic and electric coupling. However, 50Ω terminating loads result in higher loss in the transformer.

The above examples show that efficient power transfer in a two-port transformer occurs at high frequencies through both magnetic and electric coupling.

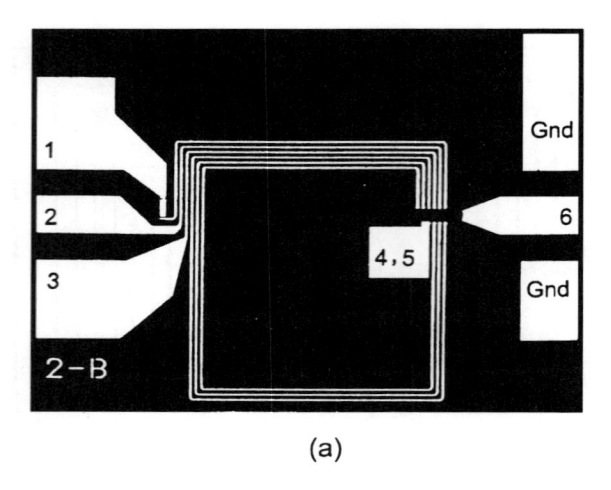

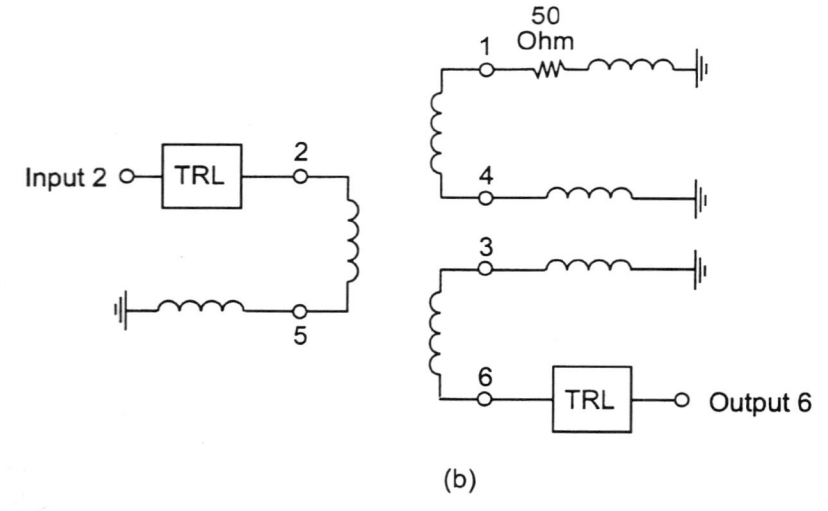

Figure 10.29 (a) Photograph of a 1.5-turn MMIC triformer; (b) Two-port equivalent circuit of the triformer. (From [37], © 1989 IEEE. Reprinted with permission.)

At such frequencies, the spiral conductors are longer than $\lambda/4$. This does not permit grounding of the center tap of the secondary spiral conductor to obtain balanced output at ports 2 and 3. On the other hand, at low frequencies, a center tap is possible; as in classical transformers, however, power transfer is inefficient as shown in Figure 10.27. The power transfer can be improved by making these transformers electrically small, minimizing the parasitic capacitances, and increasing the number of tightly coupled turns of the spirals, while maintaining the same spiral length.

Efficient power transfer can also be obtained by using more than two coils in a transformer. Figure 10.28 shows a layout of a 3-conductor coupled-line transformer. Figure 10.29(a) shows a photograph of this structure known as a *triformer* having 1.5 turns and fabricated on GaAs substrate using conductor width and spacing of 5 μm [37] and designed as a two-port matching network. An electrical equivalent circuit is shown in Figure 10.29(b). TRL denotes on wafer probe pads. The triformer structure may be used in the realization of wide-band baluns.

References

[1] Lacombe, D., and J. Cohen, "Octave-Band Microstrip DC Blocks," *IEEE Trans. Microwave Theory Tech.*, Vol. MTT-20, Aug. 1972, pp. 555–556.

[2] Ho, C. Y., "Analysis of DC Blocks Using Coupled Lines," *IEEE Trans. Microwave Theory Tech.*, Vol. MTT-23, Sept. 1975, pp. 773–774.

[3] Free, C. E., and C. S. Aitchison, "Excess Phase in Microstrip DC Blocks," *Electronics Letters*, Vol. 20, Oct. 1984, pp. 892–893.

[4] Garg, R., and I. J. Bahl, "Characteristics of Coupled Microstriplines," *IEEE Trans. Microwave Theory Tech.*, Vol. MTT-27, July 1979, pp. 700–705; also see correction in *IEEE Trans. Microwave Theory, Tech.*, Vol. MTT-28, Mar. 1980, p. 272.

[5] Gupta, K. C., et al., *Microstrip Lines and Slotlines*, Ch. 8. Sec. Ed., Artech House, Norwood, MA, 1996.

[6] Bahl, I. J., and P. Bhartia, *Microwave Solid-State Circuit Design*, Ch. 2. John Wiley, New York, 1988.

[7] Kajfez, D., and B. S. Vidula, "Design Equations for Symmetric Microstrip DC Blocks," *IEEE Trans. Microwave Theory Tech.*, Vol. MTT-28, Sept. 1980, pp. 974–981.

[8] Bahl, I. J., "Simulation Column," *Int. J. Microwave and Millimeter-Wave Computer-Aided Engineering*, Vol. 2, July 1992, pp. 204–206.

[9] Ho, T. Q., and Y. C. Shih, "Broadband Millimeter-Wave Edge-Coupled Microstrip DC Blocks," *Microwave Systems News and Communication Technology*, Vol. 17, Apr. 1987, pp. 74–78.

[10] Koscica, T. E., "Microstrip Quarter-Wave High-Voltage DC Block," *IEEE Trans. Microwave Theory Tech.*, Vol. 41, Jan. 1993, pp. 162–164.

[11] Easter, B., and B. S. Shivashankaran, "Some Results on the Edge-Coupled Microstrip Section as an Impedance Transformer," *IEE J. Microwaves*, Opt. Acoust., Vol. 3, Mar. 1979, pp. 63–66.

[12] Alley, G. D., "Interdigital Capacitors and Their Application to Lumped-Element Microwave Integrated Circuits," *IEEE Trans. Microwave Theory Tech.*, Vol. MTT-18, Dec. 1970, pp. 1,028–1,033.

[13] Hobdell, J. L., "Optimization of Interdigital Capacitors," *IEEE Trans. Microwave Theory Tech.*, Vol. MTT-27, Sept. 1979, pp. 788–791.

[14] Esfandiari, R., D. W. Maki, and M. Siracusa, "Design of Interdigitated Capacitors and Their Application to GaAs Filters," *IEEE Trans. Microwave Theory Tech.*, Vol. MTT-31, Jan. 1983, pp. 57–64.

[15] Joshi, J. S., J. R. Cockrill, and J. A. Turner, "Monolithic Microwave Gallium Arsenide FET Oscillators," *IEEE Trans. Electron Devices*, Vol. ED-28, Feb. 1981, pp. 158–162.

[16] Pettenpaul, E., et al., "CAD Models of Lumped Elements on GaAs up to 18 GHz," *IEEE Trans. Microwave Theory Tech.*, Vol. 36, Feb. 1988, pp. 294–304.

[17] Sadhir, V., I. Bahl, and D. Willems, "CAD-Compatible Accurate Models for Microwave Passive Lumped Elements for MMIC Applications," *Int. J. Microwave and Millimeter Wave Computer-Aided Engineering*, Vol. 4, Apr. 1994, pp. 148–162.

[18] EM, Sonnet Software, Liverpool, NY.

[19] Compact Explorer, Compact Software, Peterson, NJ.

[20] EM Sim, EEsof, Inc., Westlake Village, CA.

[21] High-Frequency Structure Simulator, HP, Santa Rosa, CA.

[22] LINMIC + Analysis Program, Jansen Microwave, Ratingen, Germany.

[23] MSC/EMAS, MacNeal Schwendler, Milwaukee, WI.

[24] IE3D, Zeland Software, San Francisco, CA.

[25] Kattapelli, K., J. Burke, and A. Hill, "Simulation Column," *Int. J. Microwave and Millimeter-Wave Computer-Aided Engineering*, Vol. 3, Jan. 1993, pp. 77–79.

[26] Rautio, J., "Simulation Column," *Int. J. Microwave and Millimeter-Wave Computer-Aided Engineering*, Vol. 3, Jan. 1993, pp. 80–81.

[27] Zhang, J. X., "Simulation Column," *Int. J. Microwave and Millimeter-Wave Computer-Aided Engineering*, Vol. 3, July 1993, pp. 299–300.

[28] Wheeler, H. A., "Simple Inductance Formulas for Radio Coils," *Proc. IRE*, Vol. 16, Oct. 1928, pp. 1,398–1,400.

[29] Grover, F. W., *Inductance Calculations*, Van Nostrand, Princeton. NJ, 1946, reprinted by Dover Publications, 1962, pp. 17–47.

[30] Daly, D. A., et al., "Lumped Elements in Microwave Integrated Circuits," *IEEE Trans. Microwave Theory Tech.*, Vol. MTT-15, Dec. 1967, pp. 713–721.

[31] Caulton, M., et al., "Status of Lumped Elements in Microwave Integrated Circuits—Present and Future," *IEEE Trans. Microwave Theory Tech.*, Vol. MTT-19, July 1971, pp. 588–599.

[32] Camp Jr., W. O., S. Tiwari, and D. Parson, "2-6 GHz Monolithic Microwave Amplifier," *IEEE MTT-S Int. Microwave Symp. Dig.*, 1983, pp. 46–49.

[33] Cahana, D., "A New Transmission Line Approach for Designing Spiral Microstrip Inductors for Microwave Integrated Circuits," *IEEE MTT-S Int. Microwave Symp. Dig.*, 1983, pp. 245–247.

[34] Zysman, G. I., and A. K. Johnson, "Coupled transmission line networks in an inhomogenous dielectric medium," *IEEE Trans. Microwave Theory Tech.*, Vol. MTT-17, Oct. 1969, pp. 753–759.

[35] Ferguson, D., et al., "Transformer Coupled High-Density Circuit Technique for MMIC," *IEEE GaAs IC Symp. Dig.*, 1984, pp. 34–36.

[36] Howard, G. E., et al., "The Power Transfer Mechanism of MMIC Spiral Transformers and Adjacent Spiral Inductors," *IEEE MTT-S Int. Microwave Symp. Dig.*, 1989, pp. 1,251–1,254.

[37] Boulouard, A., and M. LeRouzic, "Analysis of Rectangular Spiral Transformers for MMIC Applications," *IEEE Trans. Microwave Theory Tech.*, Vol. 37, Aug. 1989, pp. 1,257–1,260.

[38] Chow, Y. L., G. E. Howard, and M. G. Stubbs, "On the Interaction of the MMIC and its Packaging," *IEEE Trans. Microwave Theory Tech.*, Vol. 40, Aug. 1992, pp. 1,716–1,719.

[39] Chen, T. H., "Broadband Monolithic Passive Baluns and Monolithic Double-Balanced Mixer," *IEEE Trans. Microwave Theory Tech.*, Vol. 39, Dec. 1991, pp. 1,980–1,986.

[40] Marx, K. D., "Propagation Modes, Equivalent Circuits, and Characteristic Terminations for Multiconductor Transmission Lines with Inhomogeneous Dielectrics," *IEEE Trarns. Microwave Theory Tech.*, Vol. MTT-21, July 1973, pp. 450–457.

[41] Djordjevic, A., et al., *Matrix Parameters for Multiconductor Transmission Lines*, Artech House, Norwood, MA, 1989.

11

Baluns

11.1 Introduction

A balun (balanced-to-unbalanced) is a transformer used to connect balanced transmission line circuits to unbalanced transmission line circuits. Figures 11.1 and 11.2 show examples of balanced and unbalanced transmission lines, respectively. Two conductors having equal potential with 180-degrees phase difference constitute a balanced line. In this case, no current flows through a grounded shield (i.e., $I_1 = I_2$ and $I_s = I_g = 0$). When this condition is not satisfied, as shown in Figure 11.2, where I_s is finite, the transmission line is termed as *unbalanced*. In addition to providing a matched transition between a balanced and an unbalanced line, baluns also function as center-tap transformers for push-pull applications used in radio frequency applications.

Baluns are required for balanced mixers, push-pull amplifiers, balanced frequency multipliers, phase shifters, balanced modulators, dipole feeds, and numerous other applications. This transformation from a balanced medium to an unbalanced one requires special techniques, which are described in this chapter. Because the focus of this book is on coupled-line structures, only schemes based on the latter concept are discussed. Both narrowband and broadband circuits are described.

Over the past half-century, several different kinds of balun structures have been developed [1–41]. Examples of baluns are shown in Figures 11.3 to 11.6. Early coaxial baluns were used exclusively for feeding dipole antennas. Later, planar baluns using stripline techniques were developed for balanced mixers and printed antenna feeds. Current interest in transmission-line-type structures is focused toward making it planar, compact, and more suitable for mixers and push-pull power amplifiers. For high-efficiency broadband power amplifier

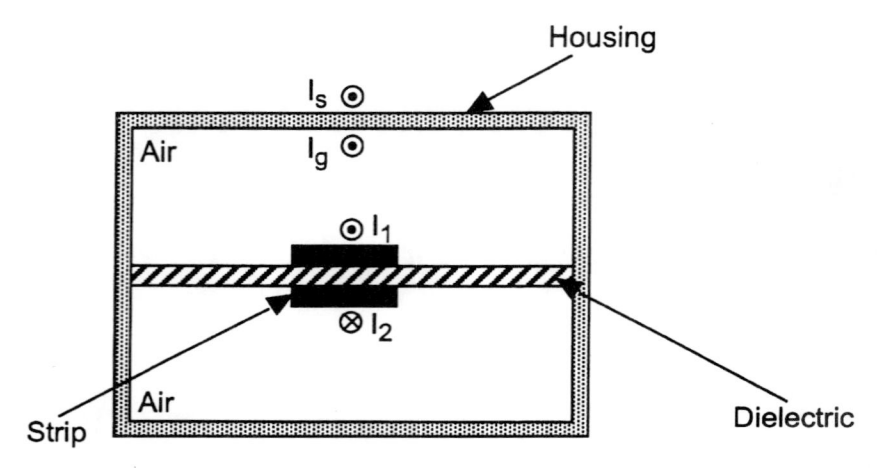

Figure 11.1 Shielded parallel strip balanced transmission line.

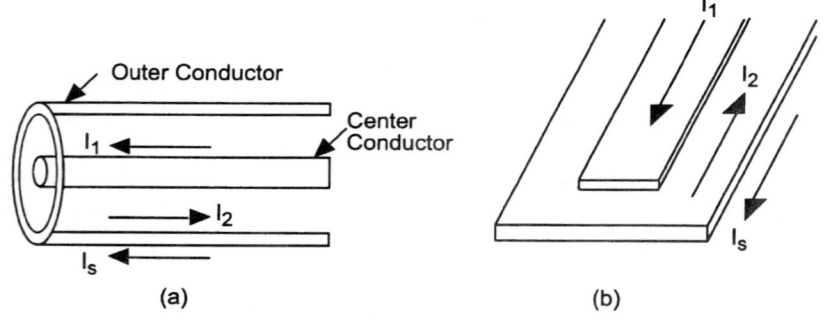

Figure 11.2 Unbalanced transmission lines (a) coaxial and (b) microstrip.

applications, these components are important for enhancing the power added efficiency (PAE) by about 10% or higher. However, the lack of a true RF center tap such as is available in the low-frequency transformer type baluns, is a problem in transmission-line-type baluns.

Planar baluns shown in Figure 11.5 provide greater flexibility and better performance in mixers and modulators. They are compatible with MIC and MMIC technologies, relatively small in size, can be fabricated on single-sided substrates, have wide bandwidths, and are very useful in surface-mounted packages for large-volume production. The realization of a decade bandwidth mixer in a microstrip-type medium, with excellent performance, is now possible with planar balun technology [23]. Marchand baluns of the planar and nonplanar variety are widely used at microwave frequencies, especially for mixers.

Baluns 393

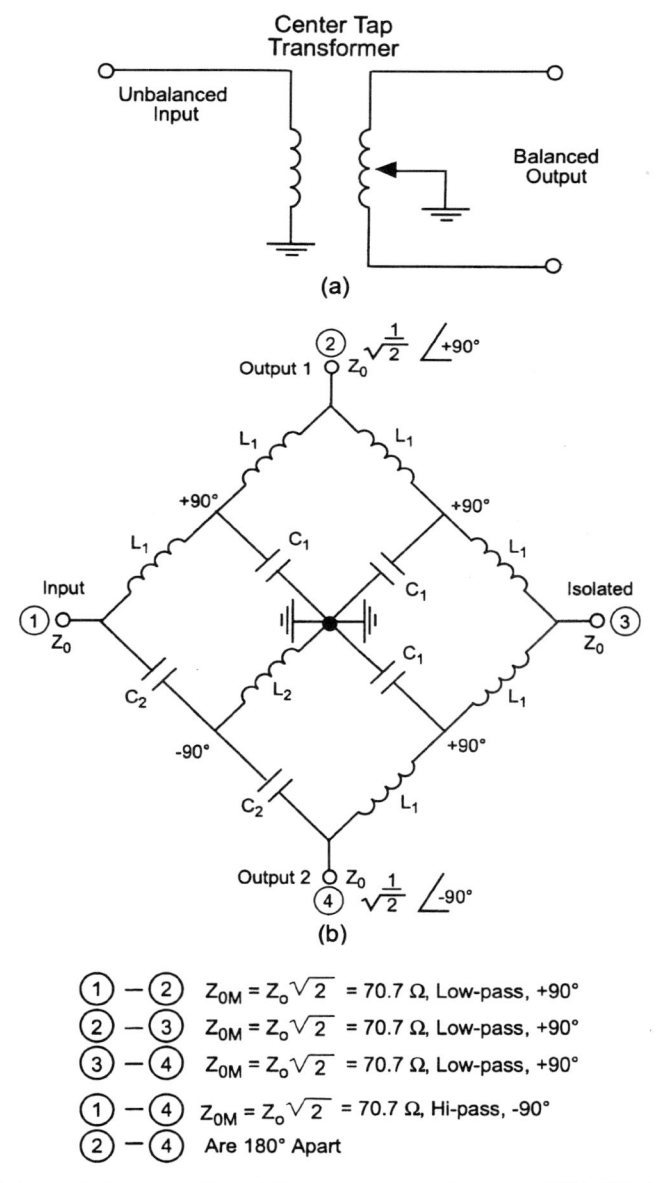

Figure 11.3 Lumped-element baluns: (a) center-tap transformer and (b) 180-degree rat-race coupler.

① — ② $Z_{0M} = Z_0\sqrt{2} = 70.7\ \Omega$, Low-pass, +90°
② — ③ $Z_{0M} = Z_0\sqrt{2} = 70.7\ \Omega$, Low-pass, +90°
③ — ④ $Z_{0M} = Z_0\sqrt{2} = 70.7\ \Omega$, Low-pass, +90°
① — ④ $Z_{0M} = Z_0\sqrt{2} = 70.7\ \Omega$, Hi-pass, -90°
② — ④ Are 180° Apart

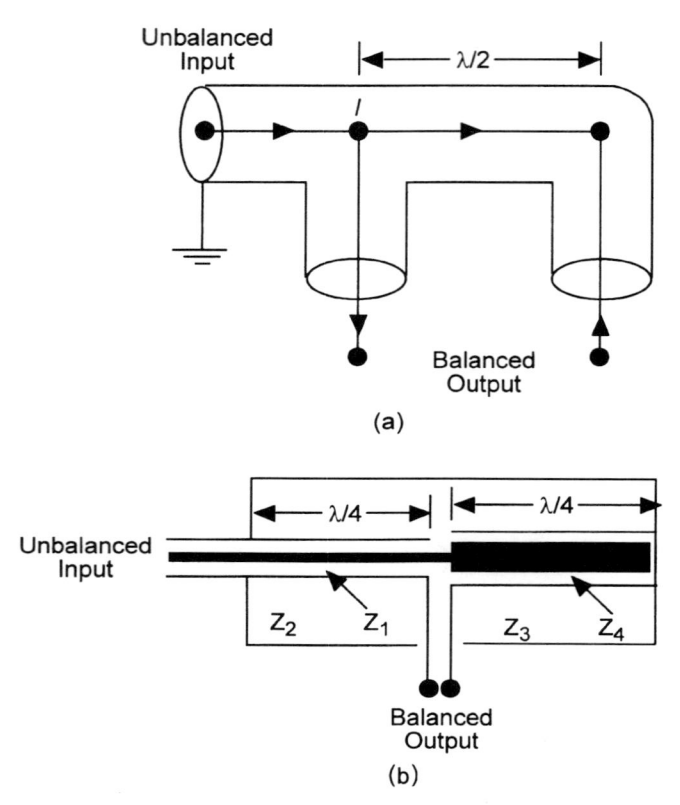

Figure 11.4 Coaxial baluns: (a) simple $\lambda/2$ line separation splitter, (b) Marchand. In (b), Z_1 and Z_4 are the characteristic impedances of the coaxial lines, and Z_2 and Z_3 are the characteristic impedances of the coaxial outer conductors with respect to the outer shield.

Four-port passive circuits, such as rat-race hybrids and waveguide magic tees, can also be used as baluns. Major limitations of these components are their narrow bandwidths and the lack of a method for center-tap grounding. Several kinds of baluns (e.g., ferrite, coaxial) have been successfully used below microwave frequencies. Now, baluns employing new coupled-line topologies are becoming more common at microwave frequencies. They provide a lower impedance-matching ratio, greater bandwidths, and reduced even-harmonic levels. The principal disadvantages include poor isolation between the two single-ended amplifiers in a push-pull configuration and poor VSWR at the input and output. In a push-pull amplifier, the reflected signals from the single-ended amplifiers do not cancel as they do in a balanced configuration.

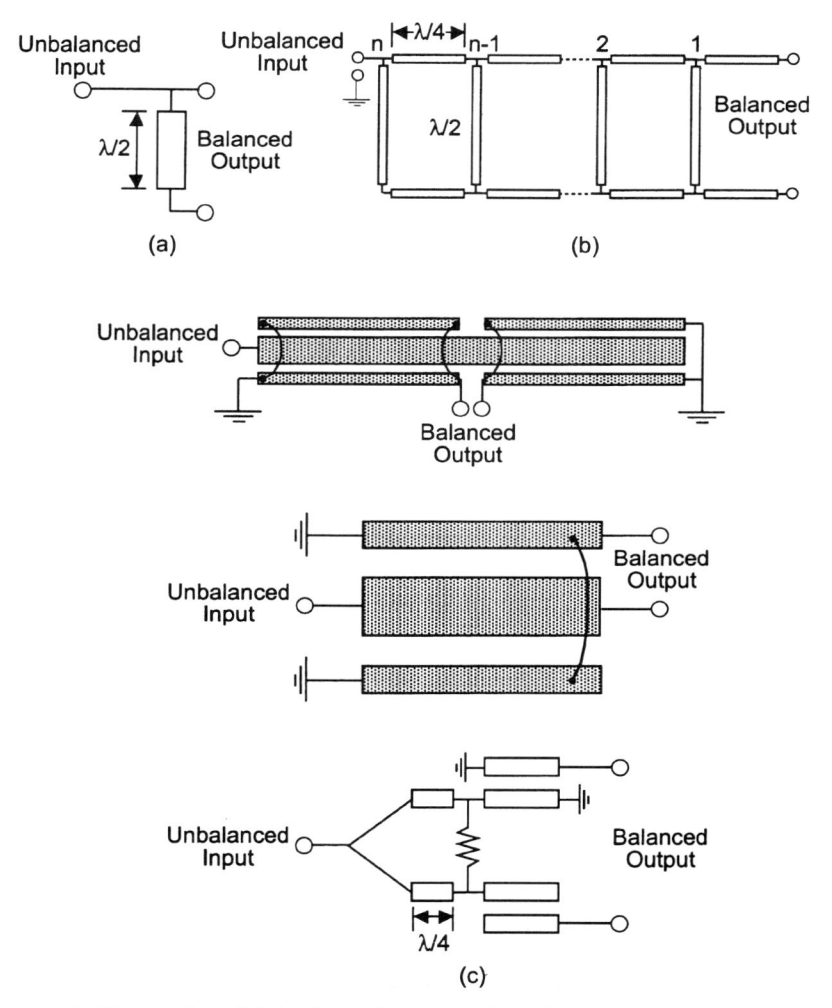

Figure 11.5 Planar baluns: (a) simple $\lambda/2$ line separation splitter, (b) multisection half-wave balun, and (c) coupled-line (coupled-line sections are $\lambda/4$ at center frequency).

11.2 Microstrip-to-Balanced Stripline Balun

A smooth transition from a microstrip to a balanced stripline as shown in Figure 11.6(a) works as a broadband balun. When a microstrip is joined to a balanced stripline, a step discontinuity between the ground plane of the microstrip line and the bottom conductor of the balanced stripline exists for the same characteristic impedances. A step discontinuity also exists between the top

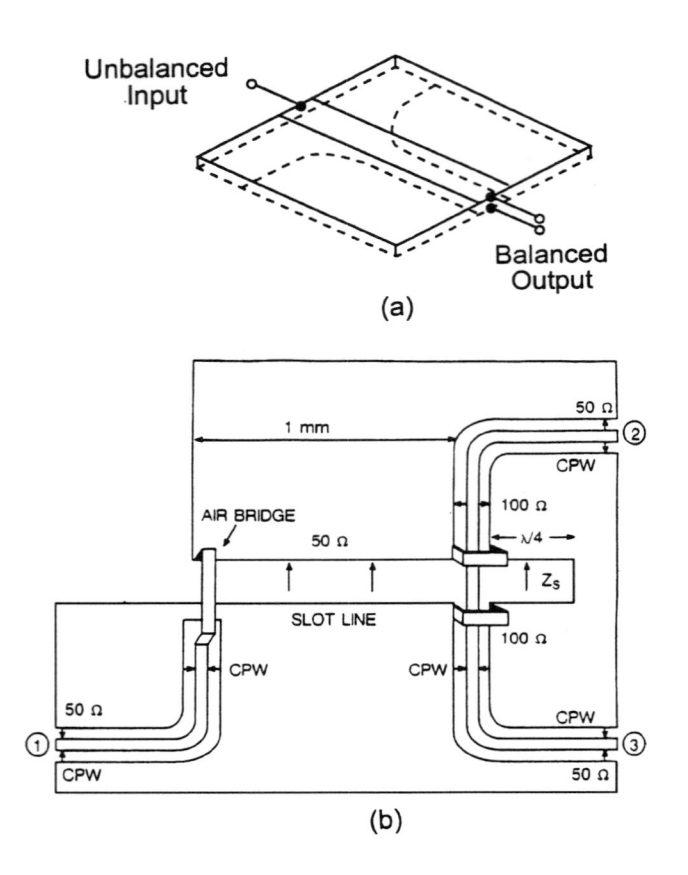

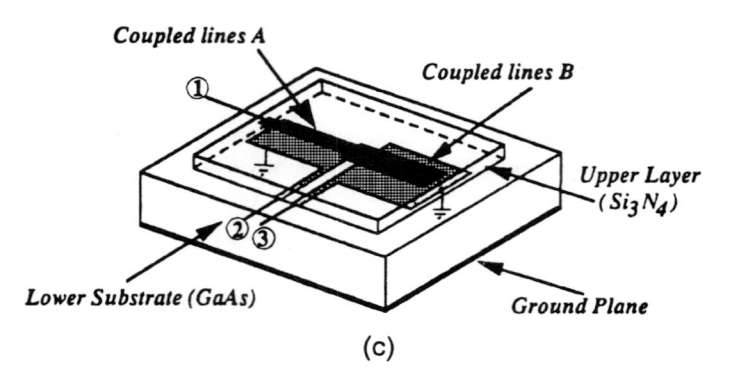

Figure 11.6 Nonplanar baluns: (a) simple broadside microstrip, (b) uniplanar, and (c) Marchand.

strip conductors of these two lines. But the step discontinuity in the former case is larger than in the later case. Transmission-line tapers are generally employed to achieve a good match between the two lines.

An example of a transition from microstrip to balanced stripline when connected back to back, to simplify measurements, is shown in Figure 11.7. For the same impedance, for example, 50 Ω, the stripwidth for the microstrip, W_m, is smaller than the stripwidth for the balanced stripline W_b. When these two lines are connected, both the stripline conductor and the ground plane are tapered to match the dimensions. Several tapering techniques are available, and we can select a taper shape based on the bandwidth requirement. A Chebyshev tapering contour along the balun's length yields excellent broadband performance [25]. A simple taper, such as shown in Figure 11.7, can easily achieve an octave bandwidth. For the fabricated example shown in the figure, $x = W_m$, $X = 3 W_m$, and $Y = 6 W_m$ and the transitions were characterized on 62.5-mil-thick polystyrene substrate ($\varepsilon_r = 2.55$) and on 50-mil-thick alumina substrate ($\varepsilon_r = 9.7$). The complete assembly was 1 inch long and the connectors used were of OSM-stripline-type, which have about 0.1-dB insertion loss per connector at S-band. Figure 11.8 shows typical measured VSWR for two transitions as well as for the through line. For the polystyrene transition the maximum VSWR over 2.4 to 4.5 GHz was 1.2, whereas for the alumina case the VSWR was less than 1.2 over

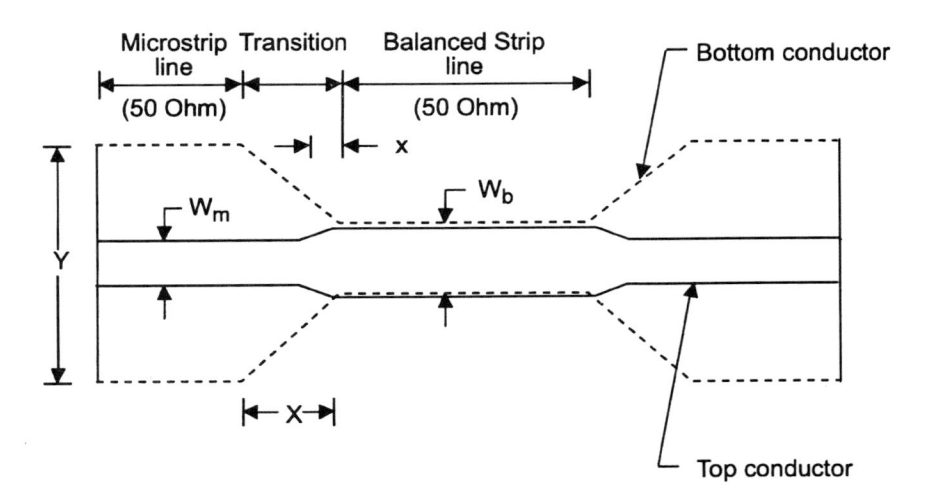

Figure 11.7 Unbalanced-to-balanced transitions connected back-to-back.

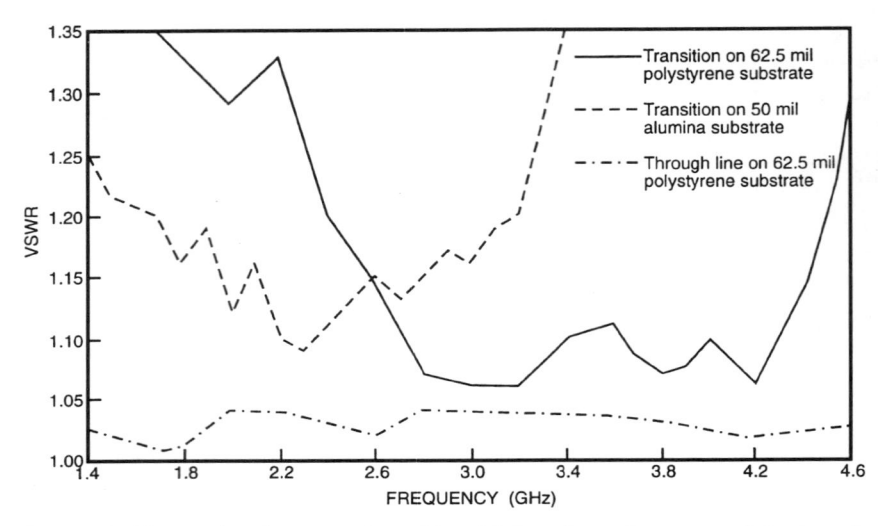

Figure 11.8 Measured performance of transitions fabricated on polystyrene and alumina substrates.

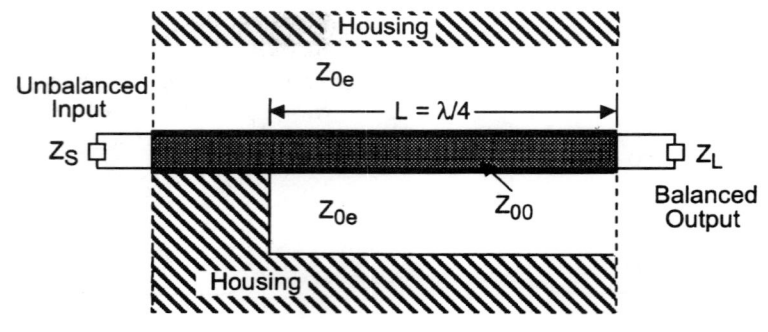

Figure 11.9 Side view of the parallel-plate balun. (From [34], © 1993 Applied Microwave. Reprinted with permission.)

the 1.7- to 3.2-GHz frequency range. Typical measured insertion loss was less than 0.6 dB and 0.4 dB, respectively, for these two transitions.

Another parallel-strip balun [34] is shown in Figure 11.9, in which the input is a microstrip and the output is a broadside-coupled microstrip [42]. One of the advantages of this structure is that it is possible to achieve a ground isolation at the balanced output port. That is, the balanced lines have their potential referenced to each other. The simple design equations given by [34]

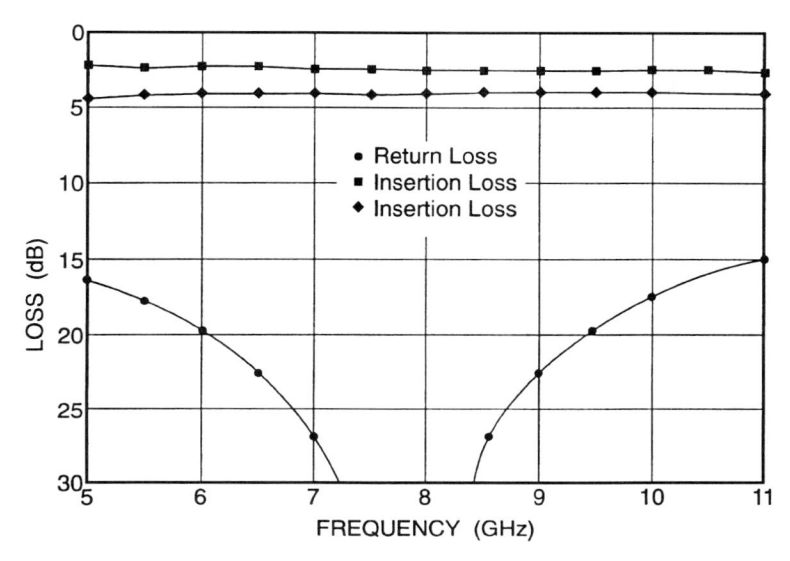

Figure 11.10 Frequency response of the parallel-plate balun. (From [34], © 1993 Applied Microwave. Reprinted with permission.)

are as follows:

$$Z_{0p} = \sqrt{Z_S Z_L} \tag{11.1}$$

$$Z_{0e} \geq 10 Z_{0o} \geq 5 Z_{0p} \tag{11.2}$$

$$Z_{0o} = Z_{0p}/2 \tag{11.3}$$

where Z_S and Z_L are the source and load impedances, respectively, and Z_{0p} is the parallel-plate impedance of the two conductors forming broadside-coupled lines with the housing. Z_{0e} and Z_{0o} are the even- and odd-mode characteristic impedances of the broadside-coupled lines, respectively.

Consider an example having source impedance $Z_S = 50\ \Omega$ and load impedance $Z_L = 100\ \Omega$. In this case, $Z_{0p} = 70.7\ \Omega$ and $Z_{0o} = 35.4\ \Omega$. Figure 11.10 shows a balun's performance when $Z_{0e} = 350\ \Omega$ and the structure is lossless. As can be seen, the power split is not the desired 3 dB. In this case, as the Z_{0e}/Z_{0o} ratio becomes higher and higher, the power split will approach 3 dB, and when Z_{0e} is infinite, the power split will be 3 dB.

11.3 Analysis of a Coupled-Line Balun

The operation of a balun can be easily explained by a pair of coupled microstrip lines of equal width, as shown in Figure 11.11. The length of the line is $\lambda/4$ at

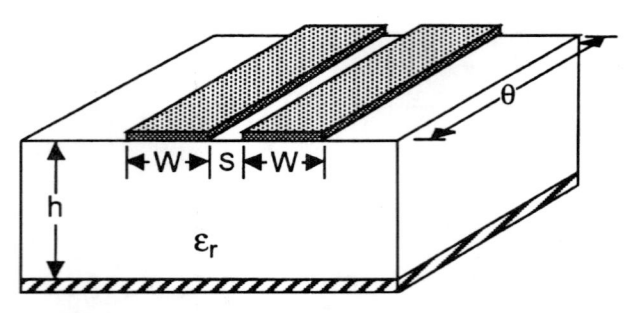

Figure 11.11 Coupled microstrip lines.

the center frequency. This structure can be analyzed by the even- and odd-mode excitation technique as shown in Figure 11.12. The admittance matrix for a four-port network consisting of transmission lines having the same propagation constants is given by [11, 43].

$$\begin{pmatrix} I_1 \\ I_2 \\ I_3 \\ I_4 \end{pmatrix} = \begin{pmatrix} Y_{11} & Y_{12} & Y_{13} & Y_{14} \\ Y_{21} & Y_{22} & Y_{23} & Y_{24} \\ Y_{31} & Y_{32} & Y_{33} & Y_{34} \\ Y_{41} & Y_{42} & Y_{43} & Y_{44} \end{pmatrix} \begin{pmatrix} V_1 \\ V_2 \\ V_3 \\ V_4 \end{pmatrix} \quad (11.4)$$

where for a homogenous dielectric medium

$$Y_{11} = Y_{22} = Y_{33} = Y_{44} = -j(Y_{0o} + Y_{0e})\frac{\cot\theta}{2} \quad (11.5a)$$

$$Y_{12} = Y_{21} = Y_{34} = Y_{43} = j(Y_{0o} - Y_{0e})\frac{\cot\theta}{2} \quad (11.5b)$$

$$Y_{13} = Y_{31} = Y_{24} = Y_{42} = -j(Y_{0o} - Y_{0e})\frac{\csc\theta}{2} \quad (11.5c)$$

$$Y_{14} = Y_{41} = Y_{23} = Y_{32} = j(Y_{0o} + Y_{0e})\frac{\csc\theta}{2} \quad (11.5d)$$

At $\theta = 90$ degrees

$$Y_{11} = Y_{22} = Y_{33} = Y_{44} = 0, \qquad Y_{12} = Y_{21} = Y_{34} = Y_{43} = 0; \text{ and}$$

$$Y_{13} = Y_{31} = Y_{24} = Y_{42} = -j(Y_{0o} + Y_{0e})/2 \quad (11.6)$$

$$Y_{14} = Y_{41} = Y_{23} = Y_{32} = j(Y_{0o} + Y_{0e})/2 \quad (11.7)$$

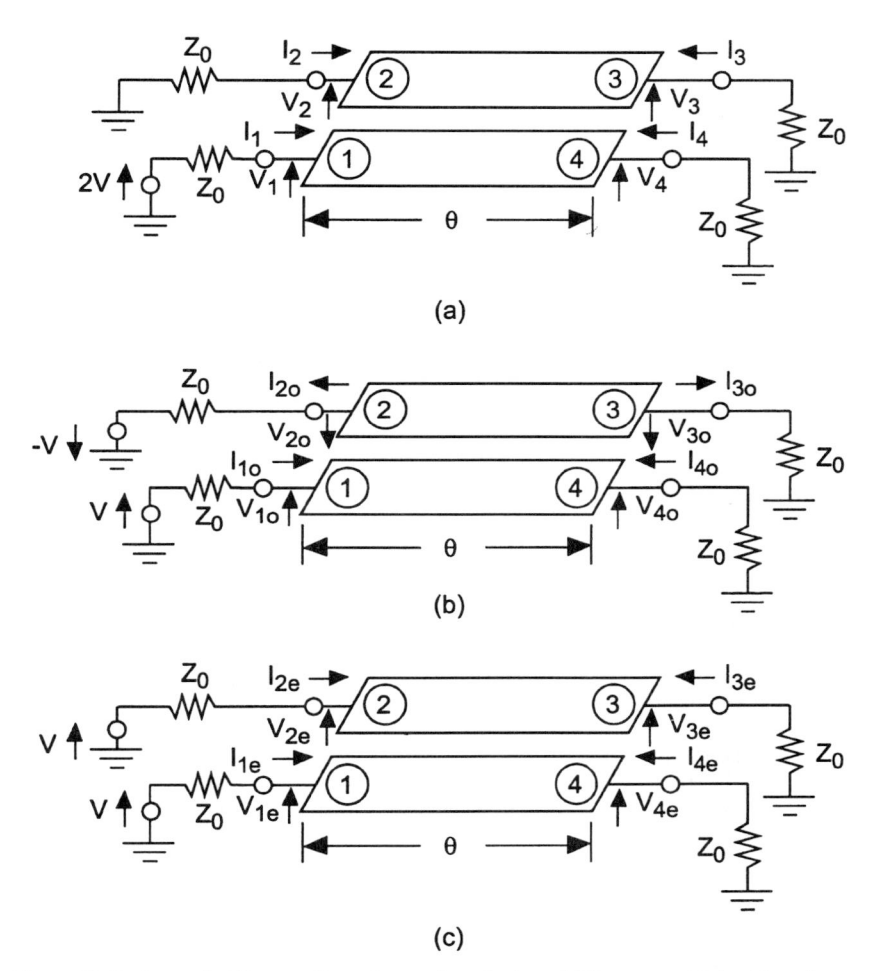

Figure 11.12 A coupled line configuration and excitation of (b) even and (c) odd modes on symmetrical coupled lines.

When port 2 is short-circuited (i.e., $V_2 = 0$), and the other ports are terminated in matched loads, the voltage transfer from port 1 to port 3 can be obtained by converting the Y-parameters into S-parameters, as follows [11]:

$$S_{31} = \frac{j2Y_0 Y_A}{Y_0^2 + Y_A^2 + Y_B^2} \tag{11.8}$$

where $Y_0 = 1/Z_0$ and $Y_A = (Y_{0o} - Y_{0e})/2$, $Y_B = (Y_{0o} + Y_{0e})/2$. Similarly, the voltage transfer from port 1 to port 4 can be obtained as

$$S_{41} = \frac{-j2Y_0 Y_B}{Y_0^2 + Y_A^2 + Y_B^2} \qquad (11.9)$$

and $S_{31}/S_{41} = -Y_A/Y_B = -(Y_{0o} - Y_{0e})/(Y_{0o} + Y_{0e})$. The above equation shows that S_{31} and S_{41} are 180 degree out of phase at $\theta = 90$ deg. A balun works over the required band if $|S_{31}/S_{41}| = 1$ and $|\angle S_{31} - \angle S_{41}| = 180$ degrees. These conditions can be achieved if $Y_{0e} = 0$ or $Z_{0e} = \infty$, that is, the effect of ground plane on the conductors is negligible. For a perfect match, $S_{11} = 0$, that is:

$$Y_0^2 = Y_A^2 + Y_B^2 \qquad (11.10)$$

or

$$Z_{0o} = Z_0/\sqrt{2} \qquad (11.11)$$

where Z_0 is the source or load impedance.

Figures 11.13 and 11.14 show the balun's performance where for $Z_{0o} = 35\,\Omega$ and $Z_{0e} = 10{,}000\,\Omega$, the amplitude response varied by less than 0.5 dB

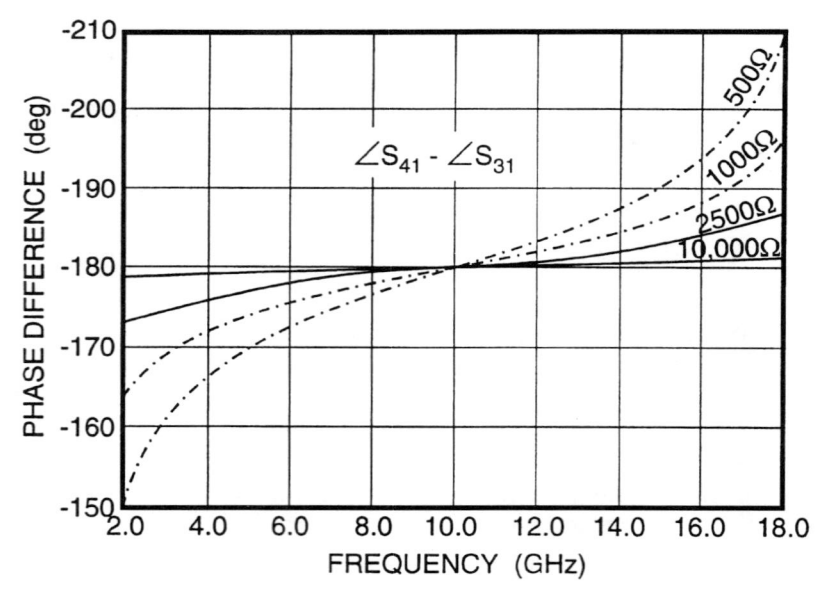

Figure 11.13 The effects of a ground plane can be analyzed by varying the value of the even-mode impedance to a relatively low level. (From [11], © 1985 Microwaves and RF. Reprinted with permission.)

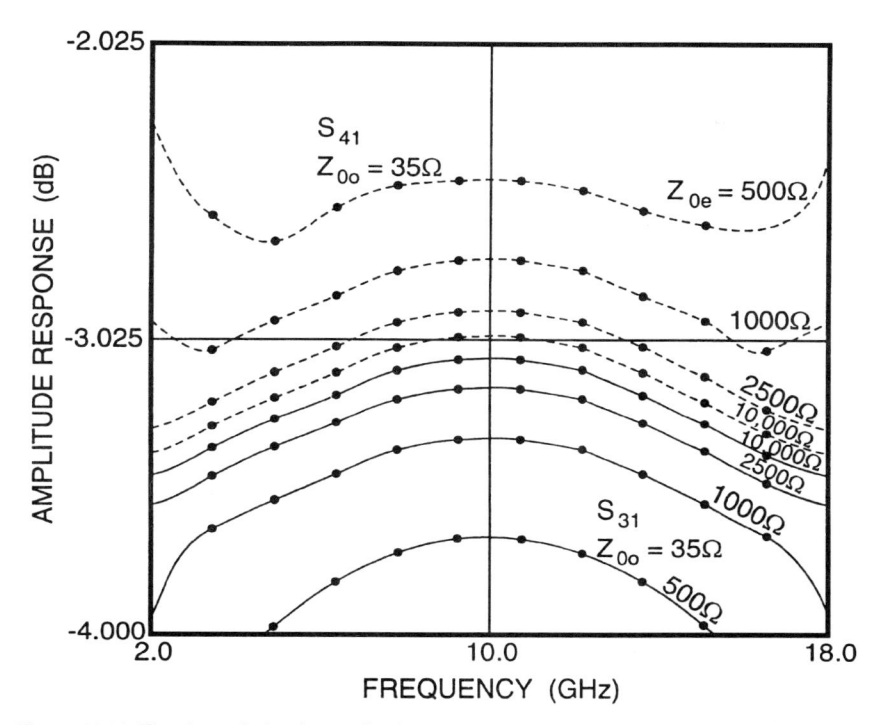

Figure 11.14 The degradation in amplitude response for a finite even-mode impedance indicates that a significant amplitude imbalance may result if ground plane effect is not considered. (From [11], © 1985 Microwave and RF. Reprinted with permission.)

and the phase difference is within ±3 degrees over the 2- to 18-GHz frequency range. This type of balun has a severe requirement for a very high even-mode impedance, which is not realizable in the microstrip configuration. Thicker and lower dielectric constant value substrates with narrower conductor width result in a higher even-mode impedance. Narrower conductors have also higher insertion loss. These kinds of baluns are therefore not very suitable for high-efficiency power amplifiers and low-loss mixer applications.

11.4 Planar Transmission Line Baluns

The planar transmission line baluns consist of two sections: the first section divides the signal into two signals having equal magnitude and phase over a broad frequency range and the second section provides −90 degrees and +90 degrees phase shift for these two signals, so that the balanced output signals have a

180-deg phase difference. The power divider section generally uses a Wilkinson in-phase power divider, which uses multisections for larger bandwidths. In general, short-circuited and open-circuited coupled lines are used for the phase-shifter sections. Basic configurations for these baluns are shown in Figure 11.15, where the divider uses only one section. A multisection divider has also been used [28] to obtain 6 to 18-GHz bandwidth performance from such baluns.

Figure 11.15(a) shows a basic balun where, in order to obtain tight coupling over a multioctave bandwidth, the edge-coupled line sections are commonly replaced with interdigital Lange couplers. Because of the inherent symmetry and broadband characteristics of coupled-line sections or Lange couplers, good

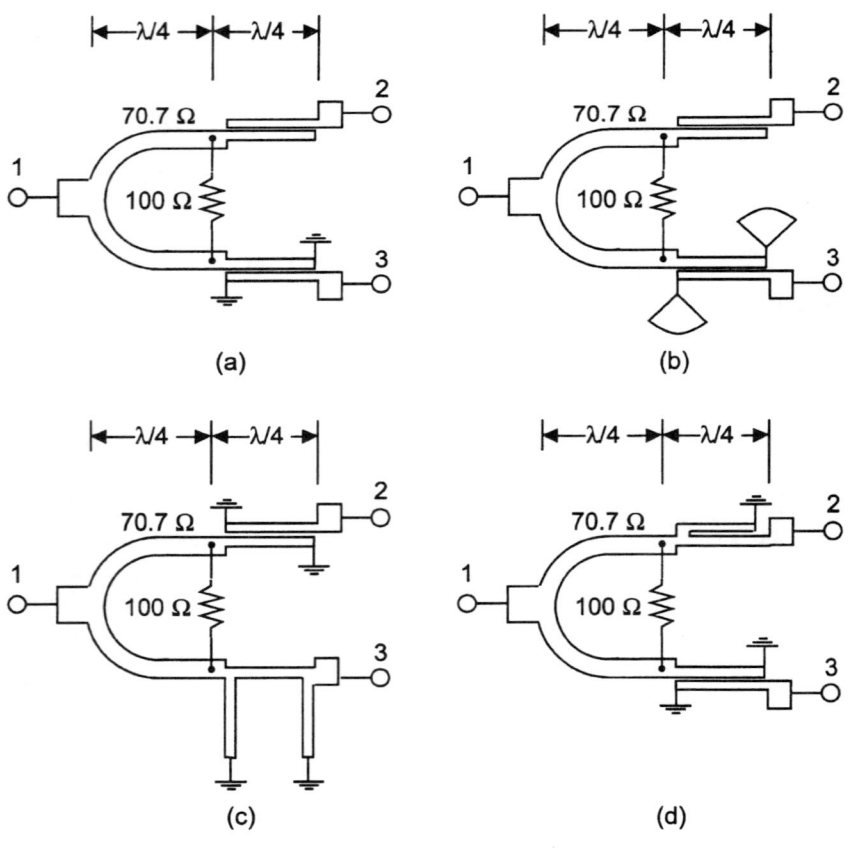

Figure 11.15 Planar balun configurations, where port 1 is the unbalanced, and ports 2 and 3 constitute balanced ports.

amplitude and phase balance performance are achievable. As the use of via holes or ground connections through alumina substrate in MICs require additional processing, broadband radial line stubs, [44] as shown in Figure 11.15(b), can also be used to simulate RF grounding without additional processing. When the open-circuit coupler is replaced by a "π" network of transmission lines, as shown in Figure 11.15(c), the 180-degree phase shift becomes independent of the electrical length of the two networks [21, 45]. Figure 11.15(d) shows another broadband balun topology [27] suitable for push-pull power amplifiers.

11.4.1 Analysis

An ultra broadband 180-degree phase shift can be realized by using the phase-reversal property of a tightly coupled (3-dB) four-port network. In a coupler when coupled and direct ports are switched from open circuit to short circuit, the transmission phase difference between the input and isolated ports changes from −90 degrees to +90 degrees as shown in Figure 11.16. In the case of the short-circuited condition, an additional −180-degrees phase is added, which brings the phase to −270 degrees or +90 degrees. Since tight coupling is required, a Lange coupler is generally used for this application. The S-parameters of a

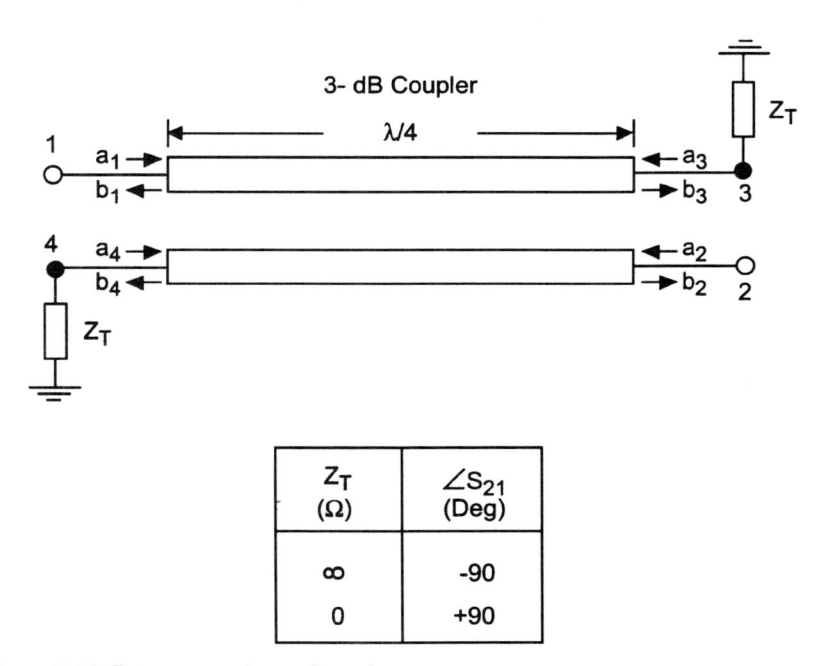

Z_T (Ω)	$\angle S_{21}$ (Deg)
∞	-90
0	+90

Figure 11.16 Two-port coupler configuration.

four-port network shown in Figure 11.16 are given by

$$\begin{pmatrix} b_1 \\ b_2 \\ b_3 \\ b_4 \end{pmatrix} = \begin{pmatrix} S_{11} & S_{12} & S_{13} & S_{14} \\ S_{21} & S_{22} & S_{23} & S_{24} \\ S_{31} & S_{32} & S_{33} & S_{34} \\ S_{41} & S_{42} & S_{43} & S_{44} \end{pmatrix} \begin{pmatrix} a_1 \\ a_2 \\ a_3 \\ a_4 \end{pmatrix} \quad (11.12)$$

or

$$b_1 = S_{11}a_1 + S_{12}a_2 + S_{13}b_3 + S_{14}b_4 \quad (11.13a)$$

$$b_2 = S_{21}a_1 + S_{22}a_2 + S_{23}b_3 + S_{24}b_4 \quad (11.13b)$$

$$b_3 = S_{31}a_1 + S_{32}a_2 + S_{33}b_3 + S_{34}b_4 \quad (11.13c)$$

$$b_4 = S_{41}a_1 + S_{42}a_2 + S_{43}b_3 + S_{44}b_4 \quad (11.13d)$$

In a matched 3-dB coupler:

$$S_{13} = S_{24} = S_{31} = S_{42} = -j/\sqrt{2} \quad (11.14)$$

$$S_{14} = S_{23} = S_{32} = S_{41} = 1/\sqrt{2} \quad (11.15)$$

$$S_{11} = S_{22} = S_{33} = S_{44} = 0 \quad (11.16)$$

$$S_{12} = S_{21} = S_{34} = S_{43} = 0 \quad (11.17)$$

Therefore, the four-port S-matrix becomes

$$[S] = \frac{1}{\sqrt{2}} \begin{bmatrix} 0 & 0 & -j & 1 \\ 0 & 0 & 1 & -j \\ -j & 1 & 0 & 0 \\ 1 & -j & 0 & 0 \end{bmatrix} \quad (11.18)$$

When ports 3 and 4 are terminated in open-circuit (i.e., $Z_T = \infty$) and other ports are matched, $a_3 = b_3$ and $a_4 = b_4$ and $a_1 = b_2$, in this case, (11.13) through (11.17) reduce to

$$b_1 = \frac{-j}{\sqrt{2}} b_3 + \frac{1}{\sqrt{2}} b_4 \quad (11.19a)$$

$$b_2 = \frac{1}{\sqrt{2}} b_3 - \frac{j}{\sqrt{2}} b_4 \quad (11.19b)$$

$$b_3 = \frac{-j}{\sqrt{2}} a_1 + \frac{j}{\sqrt{2}} a_2 \qquad (11.19c)$$

$$b_4 = \frac{1}{\sqrt{2}} a_1 - \frac{j}{\sqrt{2}} a_2 \qquad (11.19d)$$

By rearranging the equations above, the two-port S-parameters become,

$$[S]_{open} = \begin{bmatrix} 0 & -j \\ -j & 0 \end{bmatrix} \qquad (11.20)$$

Similarly, when ports 3 and 4 are terminated in short circuits, $a_3 = -b_3$ and $a_4 = -b_4$ and (11.18) simplifies to

$$[S]_{short} = \begin{bmatrix} 0 & j \\ j & 0 \end{bmatrix} \qquad (11.21)$$

(11.20) and (11.21) illustrate that open to short switching leads to a 180-degree phase shift, signals are combined, and ports 1 and 2 are still matched. This condition holds well over a wide frequency range.

When the open-circuited coupler is replaced by a "π" or equivalent network of transmission lines, the 180-degree phase shift becomes independent of the electrical length of the two networks; that is, "π" and short-circuited coupler, and this results in a wider bandwidth. Figure 11.17 shows these 180-degree phase-difference sections.

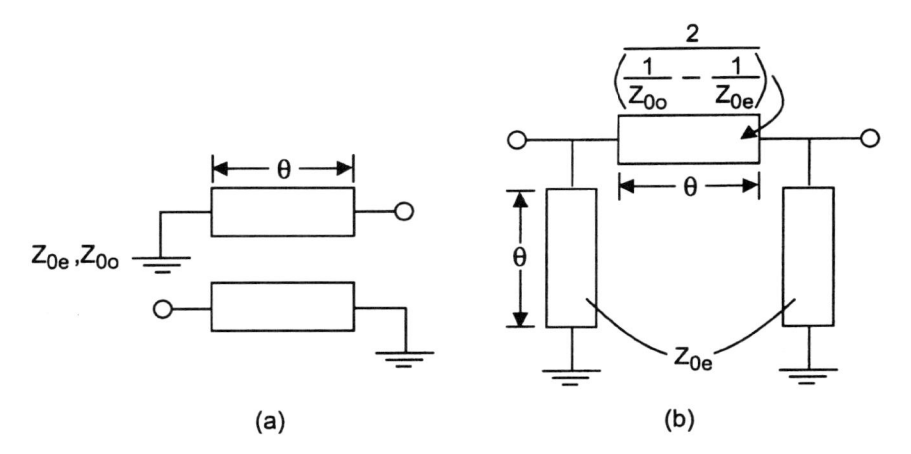

Figure 11.17 180-degree phase difference sections.

The two networks are exactly equivalent for all frequencies, except that the transmission phase difference between the two circuits is exactly 180 degrees. [45] This can be seen by developing ABCD matrices for both networks. For the "π" network:

$$\begin{bmatrix} A & B \\ C & D \end{bmatrix}_\pi = \begin{bmatrix} 1 & 0 \\ \frac{Y_{oe}}{j\tan\theta} & 1 \end{bmatrix}$$

$$\cdot \begin{bmatrix} \cos\theta & \frac{j2\sin\theta}{(Y_{0o}-Y_{0e})} \\ \frac{j(Y_{0o}-Y_{0e})}{2\csc\theta} & \cos\theta \end{bmatrix} \cdot \begin{bmatrix} 1 & 0 \\ \frac{Y_{oe}}{j\tan\theta} & 1 \end{bmatrix} \quad (11.22)$$

$$= \begin{bmatrix} \frac{(Y_{0o}+Y_{0e})}{(Y_{0o}-Y_{0e})}\cos\theta & \frac{j2\sin\theta}{(Y_{0o}-Y_{0e})} \\ \frac{(Y_{0o}+Y_{0e})^2\cos^2\theta-(Y_{0o}-Y_{0e})^2}{j2(Y_{0o}-Y_{0e})\sin\theta} & \frac{(Y_{0o}+Y_{0e})}{(Y_{0o}-Y_{0e})}\cos\theta \end{bmatrix} \quad (11.23)$$

where $Y_{0o} = 1/Z_{0o}$ and $Y_{0e} = 1/Z_{0e}$.

The ABCD matrix of the shorted coupled-line section is calculated from the y parameters as

$$\begin{bmatrix} A & B \\ C & D \end{bmatrix}_{coupled}$$

$$= \begin{bmatrix} -\frac{(Y_{0o}+Y_{0e})}{(Y_{0o}-Y_{0e})}\cos\theta & -\frac{j2\sin\theta}{(Y_{0o}-Y_{0e})} \\ -\frac{(Y_{0o}+Y_{0e})^2\cos^2\theta-(Y_{0o}-Y_{0e})^2}{j2(Y_{0o}-Y_{0e})\sin\theta} & -\frac{(Y_{0o}+Y_{0e})}{(Y_{0o}-Y_{0e})}\cos\theta \end{bmatrix}$$

$$(11.24)$$

From (11.23) and (11.24)

$$\begin{bmatrix} A & B \\ C & D \end{bmatrix}_\pi = \begin{bmatrix} -1 & 0 \\ 0 & -1 \end{bmatrix} \cdot \begin{bmatrix} A & B \\ C & D \end{bmatrix}_{coupled} \quad (11.25)$$

The above result is independent of the electrical length, and thus is independent of frequency. For tight coupling, however, $\theta = 90$ degree is required.

11.4.2 Examples

Topologies for the planar transmission line baluns shown in Figure 11.15 were analyzed and optimized using a commercial CAD tool and found that the configuration shown in Figure 11.15(c) gives the best results over an octave bandwidth. Physical dimensions and performance results are summarized in Table 11.1.

A broadband balun employing Lange couplers as shown schematically in Figure 11.15(a) was fabricated on a 0.635-mm-thick substrate using conventional photolighography and etching techniques. The coupler has nominal coupling of 2.8 dB and a 0.7-dB ripple. The Z_{0e} and Z_{0o} values were 126 Ω and 20 Ω, respectively [29]. The measured performance is plotted in Figure 11.18. Over 5 to 11 GHz, the two output signals have the magnitude of 3.6 ± 0.3 dB and 3.6 ± 0.5 dB. The VSWR and isolation were better than 10 and 14 dB, respectively. The phase difference between the output ports was 170 ± 5 degrees over the 5 to 11-GHz band.

Another planar balun was constructed on a 10-mil alumina substrate using short- and open-circuited Lange couplers. [28] The Lange couplers were realized using the ion-beam milling technique to achieve fine 1-mil-width and 0.5-mil gap dimensions. The photograph of the balun is shown in Figure 11.19. The

Table 11.1

Design and performance summary of a planar balun shown in Figure 11.15(c). Substrate dimensions: $\varepsilon = 9.9$, $h = 15$ mil, $t = 0.2$ mil, and conductors of gold
Wilkinson divider dimensions: 　Line width = 6.2 mil 　Line length = 96.2 mil 　Isolation resistor = 88.4 Ω Lange coupler dimensions: 　Line width = 2.6 mil 　Line spacing = 1.8 mil 　Coupler length = 98.0 mil Lowpass section dimensions: 　Shunt line width = 2.5 mil 　Shunt line length = 107.0 mil 　Series line width = 17.2 mil 　Series line length = 81.7 mil Performance over 8 to 16 GHz: 　Insertion loss = 3.15 ± 0.05 dB 　Phase difference = 180 ± 1 deg 　Amplitude difference = ± 0.01 dB 　Return loss >18 dB

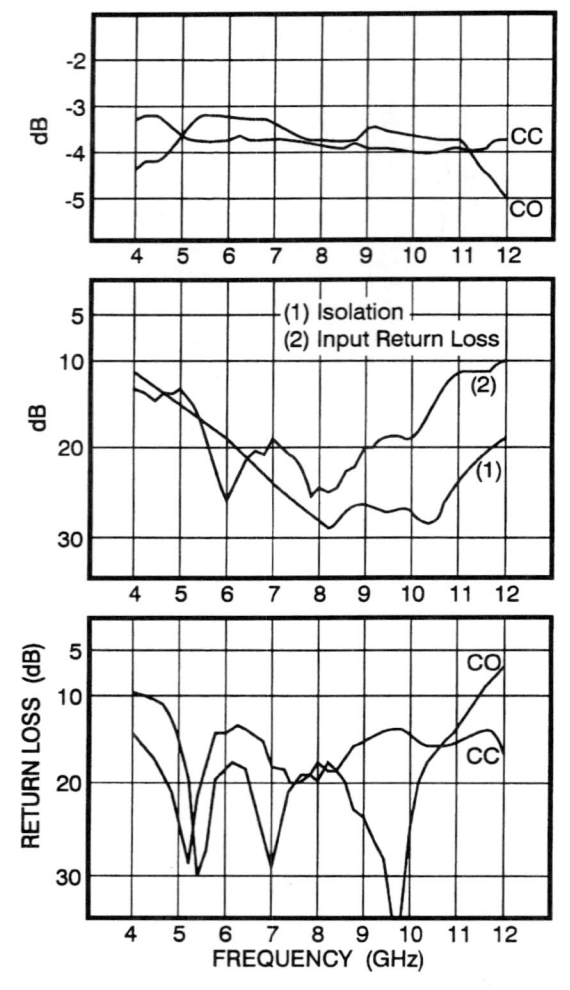

Figure 11.18 Measured performance of a broadband balun of Figure 11.15(a). (From [29], ©1991 European Microwave Conf. Reprinted with permission.)

balun was accurately characterized using thru-reflect-line (TRL) calibration techniques and the measured amplitude, phase, and return-loss response are shown in Figure 11.20. The maximal loss in each path was about 1.2 dB, and the amplitude and phase imbalance were ±0.6 dB and ±7 degree, respectively, over the 6 to 20-GHz frequency range. The VSWR and the isolation were better than 10 and 16 dB, respectively.

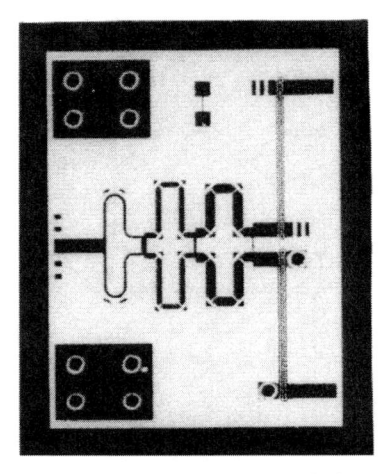

Figure 11.19 Photograph of fabricated balun chip. (From [28], © 1991 IEEE. Reprinted with permission.)

11.5 Marchand Balun

The Marchand balun, which has several versions, is the most commonly used component in broadband double-balanced mixers. As compared with a shorted coupled-line balun, this structure has less stringent requirements for Z_{0e}; generally $Z_{0e} \cong 3$ to 5 times of Z_{0o} is sufficient to obtain good performance for such baluns. Proper selection of balun parameters can achieve a bandwidth of more than 10:1.

The balun basically consists of an unbalanced, an open-circuited, two short-circuited, and balanced transmission line sections. Each section is about a quarter-wavelength long at the center frequency of operation. A coaxial version of a compensated Marchand balun is shown in Figure 11.21(a), while its equivalent circuit representation is shown in Figure 11.21(b). The compensation term is used in broadband baluns where the balanced output and reduced phase slope are maintained over a wide bandwidth. As shown in Figure 11.21(a), this structure basically consists of two coaxial lines, each $\lambda/4$ long at the center frequency. The left-hand line has the characteristic impedance of Z_1. The second line, which has a characteristic impedance of Z_2, is open-circuited. The outer conductors of these transmission line sections with housing make another two short-circuited $\lambda/4$ lines that are in series with each other and shunt the balanced lines, having a characteristic impedance of Z_B, at locations *a* and *b*. As shown in the equivalent circuit (Figure 11.21(b)), the stubs Z_{S1} and Z_{S2} are in series and shunt the balanced lines. Their characteristic impedance is made as large as

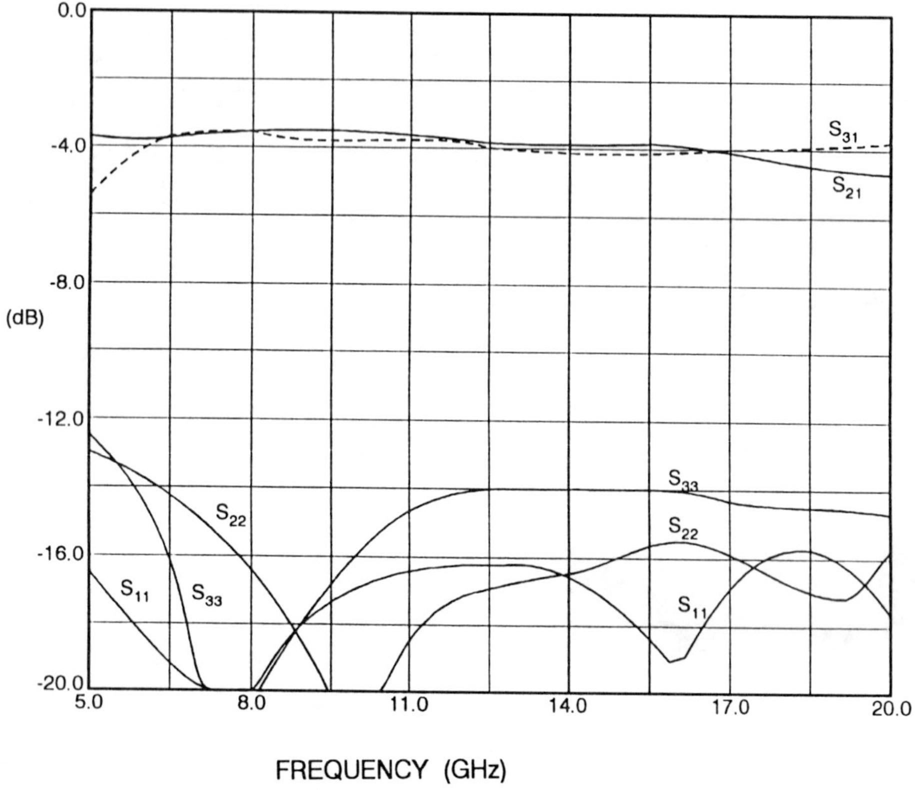

Figure 11.20 Measured (a) amplitude and (b) phase balance of planar balun. (From [28], © 1991 IEEE. Reprinted with permission.)

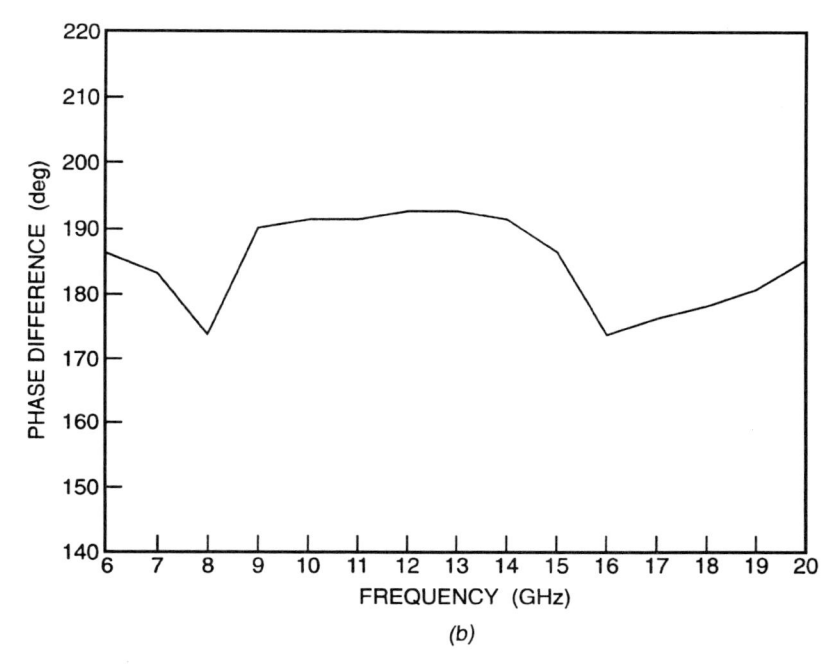

[Graph: Phase Difference (deg) vs Frequency (GHz), ranging 6-20 GHz on x-axis and 140-220 deg on y-axis]

(b)

Figure 11.20 *(continued.)*

possible. These impedances along with the other transmission line impedances determine the impedance transformation and bandwidth. Figure 11.22 shows a simplified equivalent circuit of the Marchand balun. Because of equal shunting effects on the balanced lines, these stubs provide greater bandwidths. The open-circuit stub Z_2 provides a low impedance at the junction of the four different lines and acts like a series resonant circuit. The series resonant circuit with a shunt resonant circuit as shown in Figure 11.23 reduces the phase variation over the designed bandwidth. The ratio of the characteristic impedances of short-circuited and open-circuited stubs determines the bandwidth; the higher the ratio, the wider the bandwidth. The transmission line Z_B can be designed with a characteristic impedance of the balanced line or can be used as an impedance transformer between the desired impedance and the balanced-line impedance.

A balun fabricated using printed-circuit topology is shown in Figure 11.24(a), and its simplified equivalent circuit is shown in Figure 11.24(b). In such baluns, the performance also depends upon the ground spacing and the housing in which the balun is placed. Like any other distributed balun (coaxial, microstrip, CPW, etc.), the tap center shown in Figure 11.24(b) only works at dc or low frequencies such as IF that are in the RF range. The center tap concept is not valid at microwave frequencies, however, as required for push-pull topology.

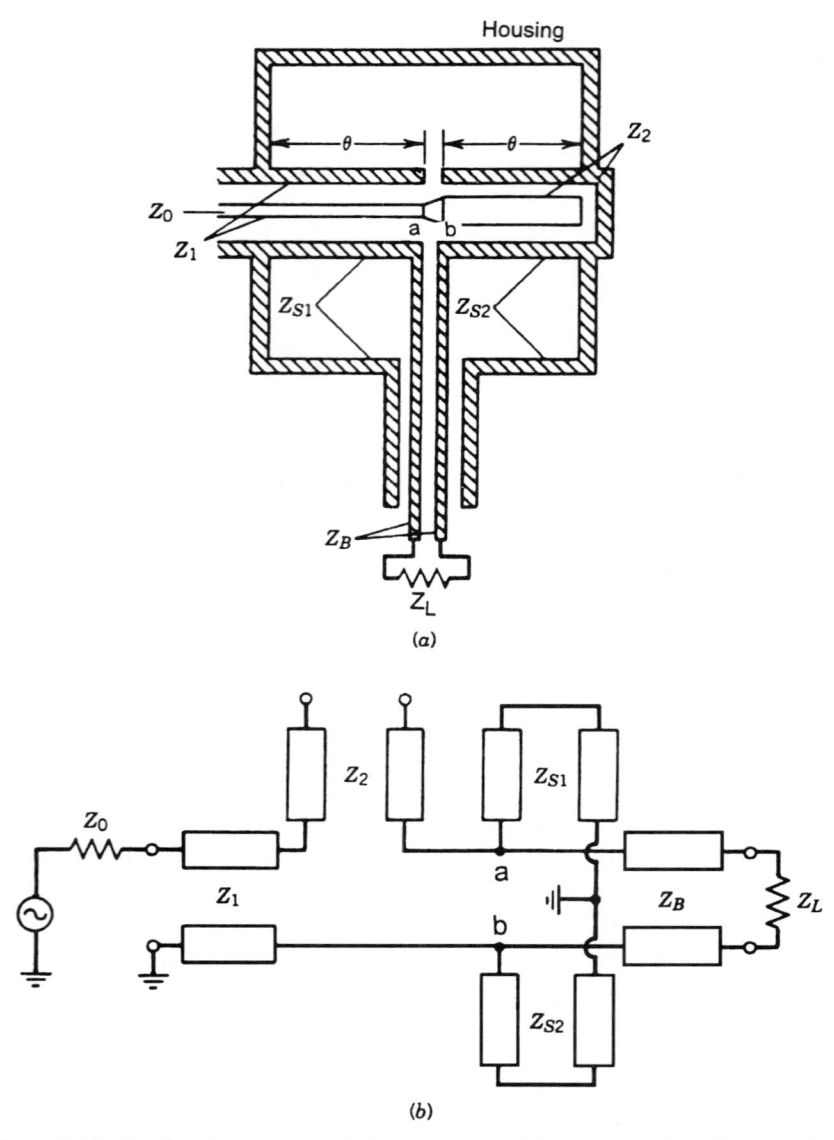

Figure 11.21 Marchand compensated balun; (a) coaxial cross section; (b) equivalent transmission-line model. (From [22], © 1990 IEEE. Reprinted with permission.)

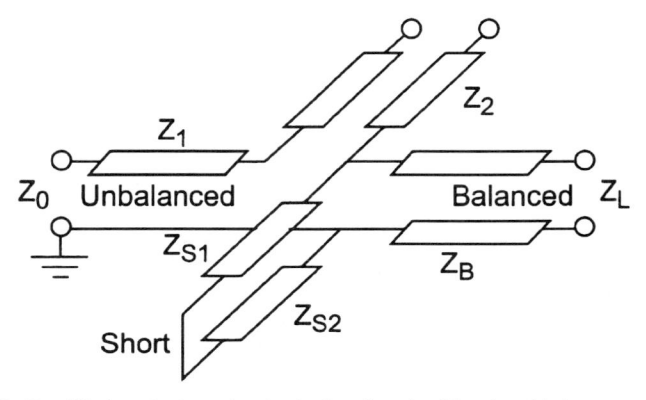

Figure 11.22 Simplified equivalent circuit of a fourth-order Marchand balun.

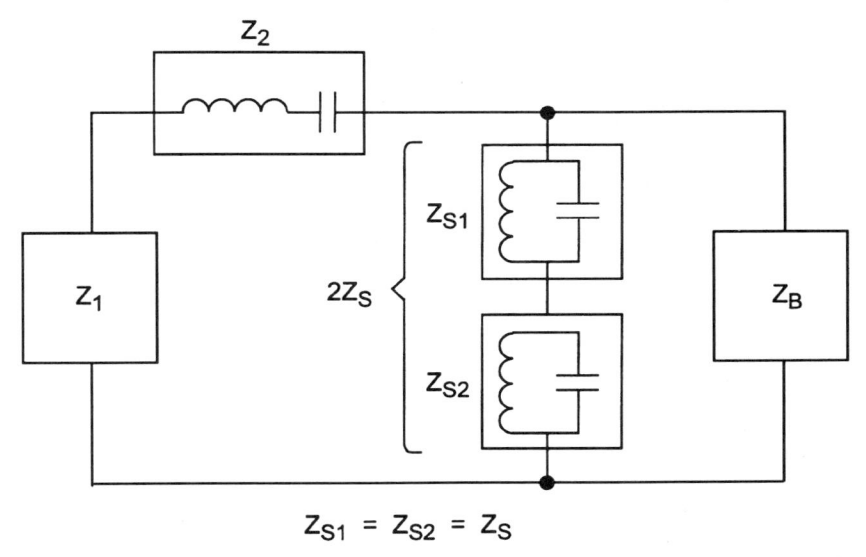

$$Z_{S1} = Z_{S2} = Z_S$$

Figure 11.23 Marchand balun's series and parallel-resonant compensating representation for wider bandwidth.

11.5.1 Coaxial Marchand Balun

Now we present a simple analysis of a coaxial Marchand balun shown in Figure 11.25(a). It consists of two coaxial lines, a and b, of characteristic impedances Z_a and Z_b, respectively. The signal is fed at the input point, and the balanced signal appears across the O and O' nodes. The center conductors of these lines are connected at nodes c and c', and the other end of the center conductor of

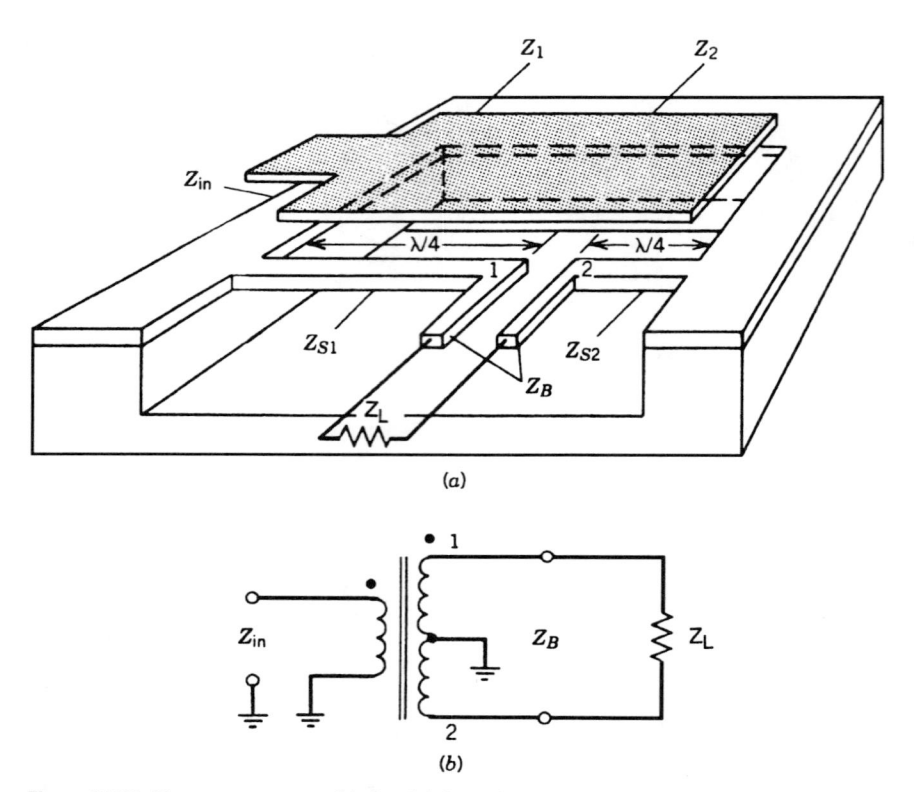

Figure 11.24 Planar compensated balun fabricated on a low-dielectric substrate: (a) metalization pattern, (b) lumped-element equivalent-circuit model. (From [25], © 1990 John Wiley. Reprinted with permission.)

line b is open-circuited. The outer conductor of these two lines are coupled to each other to form the balanced line of characteristic impedance Z_{ab}. The outer conductors of lines a and b are connected at d, and the electrical length θ at the center frequency is 90 degrees. The equivalent circuit of this balun is shown in Figure 11.25(b). Referring to this, the input impedance at nodes c and c', assuming a lossless structure, may be expressed as

$$Z_{in} = -jZ_b \cot \theta_b + \frac{jZ_L Z_{ab} \tan \theta_{ab}}{Z_L + jZ_{ab} \tan \theta_{ab}} \tag{11.26}$$

substituting $\theta_b = \theta_{ab} = \theta$ and after simplifying

$$Z_{in} = \frac{Z_L Z_{ab}^2 + j \cot \theta [Z_L^2(Z_{ab} - Z_b \cot^2 \theta) - Z_b Z_{ab}^2]}{Z_{ab}^2 + Z_L^2 \cot^2 \theta} \tag{11.27}$$

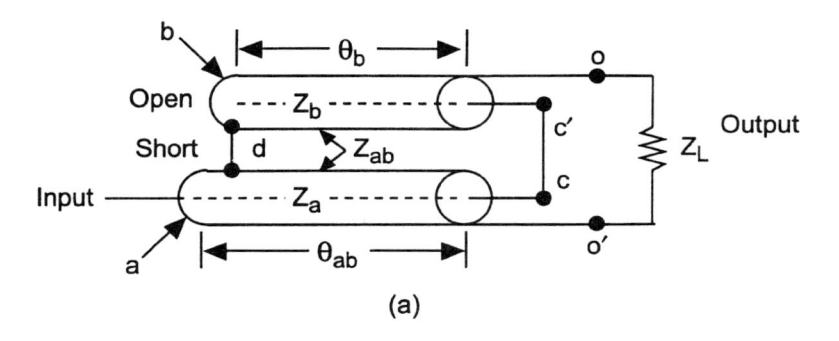

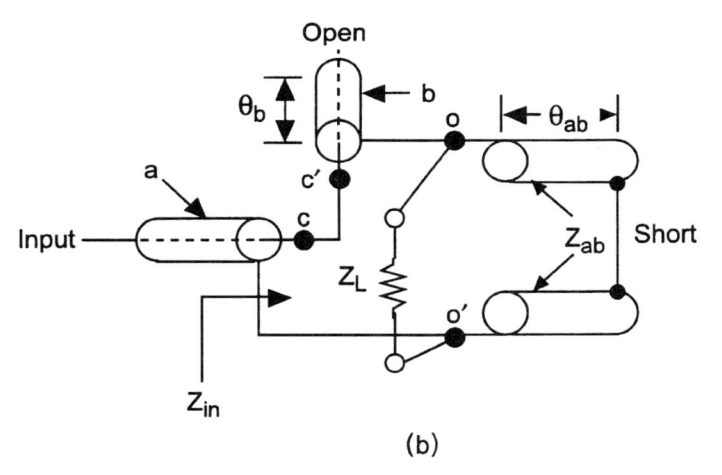

Figure 11.25 Wide-band balun: (a) schematic diagram; (b) equivalent circuit representation (from [4], © 1960 IEEE. Reprinted with permission.).

when $Z_b = Z_a$ and $Z_{ab} = Z_L$

$$Z_{in} = Z_L \sin^2 \theta + j \cot \theta (Z_L \sin^2 \theta - Z_a) \quad (11.28)$$

and the input impedance becomes perfectly matched to Z_a at two widely separated frequencies given by the solution of

$$\sin^2 \theta = Z_a/Z_L \quad (11.29)$$

These frequencies are symmetrically disposed about a center frequency corresponding to $\theta = 90$ degrees. A larger bandwidth is realized by choosing

$$Z_b = Z_L^2/Z_{ab} \quad (11.30)$$

and making Z_{ab} as large as possible [2, 4]. Figure 11.26 compares the input VSWR for two cases in which a 50 Ω unbalanced input was transformed to a 70 Ω

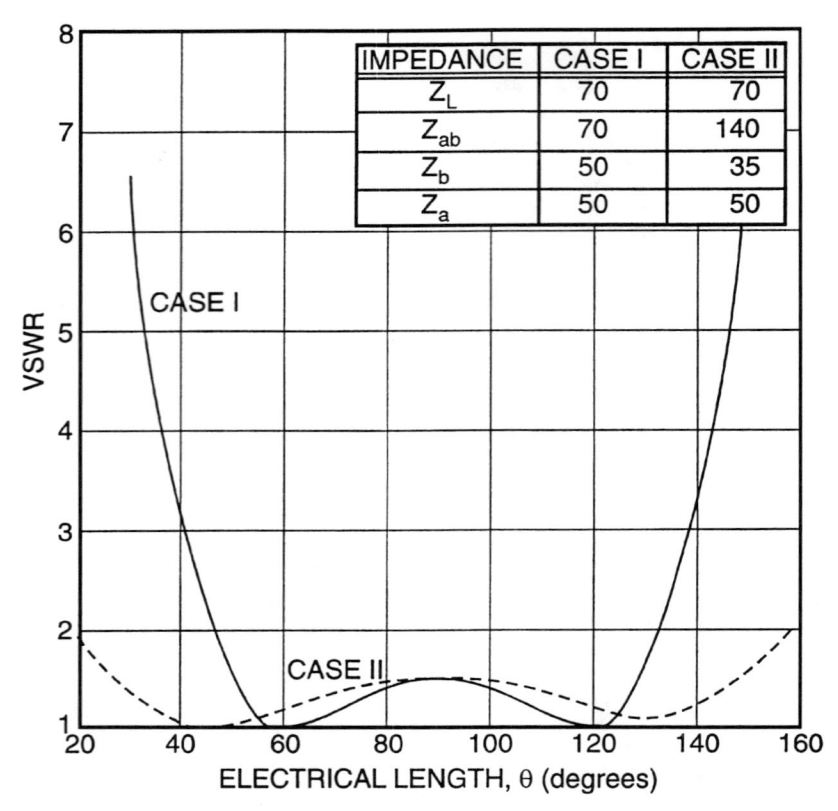

IMPEDANCE	CASE I	CASE II
Z_L	70	70
Z_{ab}	70	140
Z_b	50	35
Z_a	50	50

Figure 11.26 Calculated VSWR versus electrical length for two different balun design criteria. The impedance values are in ohms. (From [4], © 1960 IEEE. Reprinted with permission.)

balanced output. The improvement in the second case, where $Z_b = Z_L^2/Z_{ab}$ is quite obvious. The realization of this case in coaxial line form is difficult; however, it can be easily fabricated using coplanar striplines [4]. Figure 11.27 delineates a microstrip version of this balun. The load impedance Z_L is between O and O' and is 70 Ω in this case.

A printed balun with integrated dipole [12] is shown in Figure 11.28(a), where the unbalanced coaxial line is replaced by a microstrip line section and one end of the balanced line section is connected to the microstrip ground plane conductor, while the other end connects to the printed dipole. The widths of the balanced line conductors must be three times that of the microstrip line conductors in order to use microstrip design equations. Narrower widths can also be used if appropriate corrections are made. The characteristic impedance

Baluns

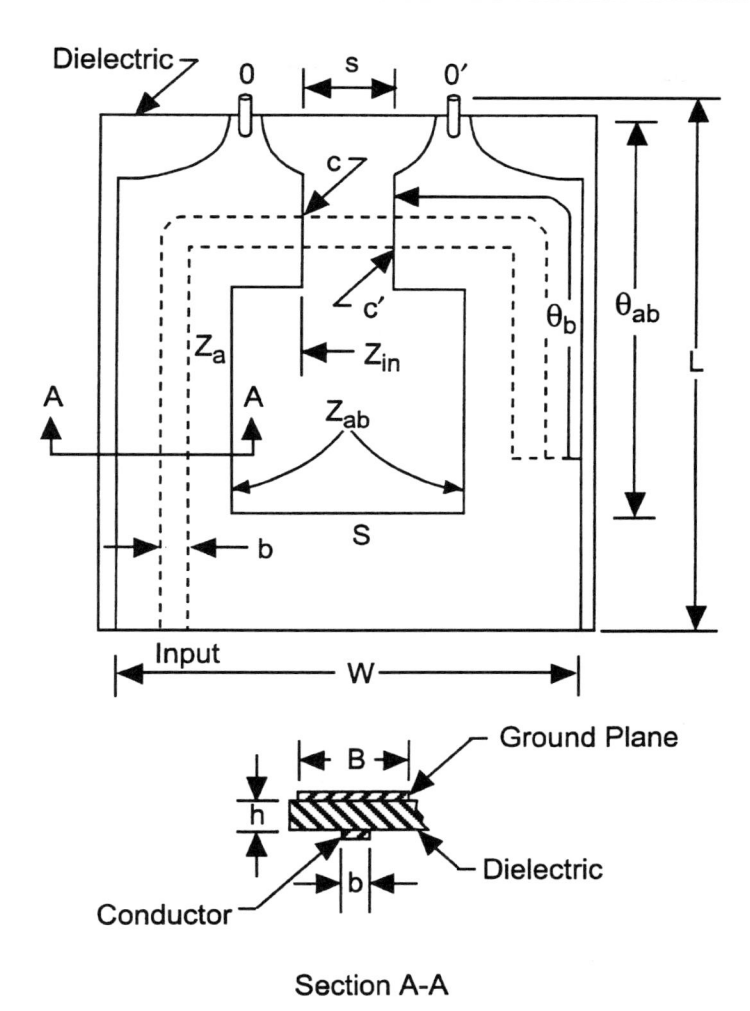

Figure 11.27 Illustration of the construction of a printed circuit balun. (From [4], © 1960 IEEE. Reprinted with permission.)

Z_{ab} of the balanced line may be calculated by treating it as a pair of coupled microstrip lines on a suspended substrate and excited in the odd mode. The distance between the balanced line conductors and the enclosure, which also acts as a ground plane, is kept large as compared with the spacing between the conductors. Under this assumption, the characteristic impedance of the balanced line is approximately twice the odd-mode impedance calculated for the coupled line. The effective dielectric constant for the balanced line is the same as that calculated for the coupled lines excited in the odd-mode.

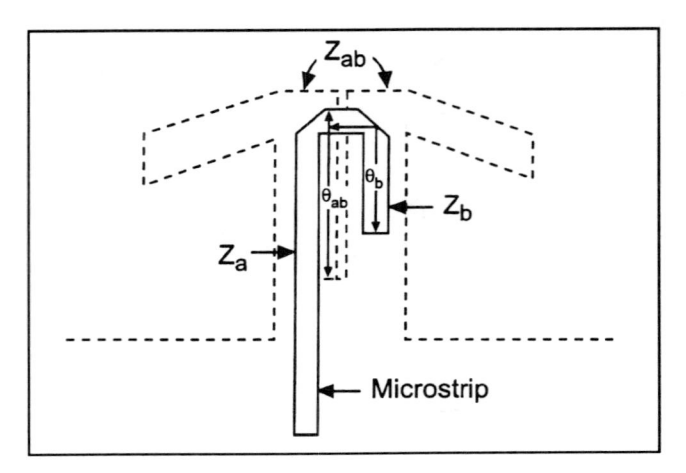

Figure 11.28(a) Printed circuit realization of balun structure with integrated dipole. (From [12], © 1987 *Microwave J.* Reprinted with permission.)

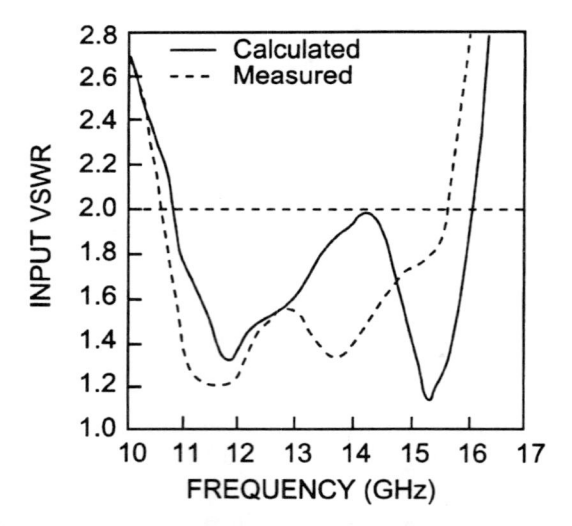

Figure 11.28(b) Measured and calculated input VSWR for the dipole/balun combination. (From [12], © 1987 *Microwave J.* Reprinted with permission.)

Lower values of Z_{ab} require narrower spacing and wider conductor widths. Fabrication techniques set the lower limit on spacing that can be realized. Thus a higher limit on Z_{ab} is set by the minimum line widths required to realize Z_a and Z_b. The effects of the enclosure, discontinuity, and coupling influence the performance of this balun in terms of bandwidth and frequency range. For applications in the antenna area, the radiating elements are connected to the balanced line by extending it and shaping it into the desired dipole configuration. In the final design, the dipole can be matched over a broadband by treating it as part of the balun design and by tuning out the reactances accordingly [12]. Figure 11.28(b) shows the simulated and measured input VSWR of a dipole/balun combination. In this example, $Z_b = Z_{ab} = 80\ \Omega$, which is equal to the dipole's radiation resistance at the resonant frequency at the 13 GHz center frequency. $\theta_b = 105$ degrees and $\theta_{ab} = 95$ degrees, and a $\lambda/4$ microstrip impedance of 63 Ω was used to transform 80 Ω to 50 Ω at the input. The structure was printed on a 0.64-mm-thick fused silica ($\varepsilon_r = 3.78$) and mounted a quarter-wavelength above-the-ground plane. Although such baluns have good bandwidth, they are not suitable for ultrabroad band applications.

11.5.2 Synthesis of Marchand Balun

Synthesis of coaxial Marchand baluns is available in the literature [8, 32]. Recent trends focus on planar technology baluns using single-layer and multilayer microstrip transmission media. A Marchand balun can easily be realized using a pair of coupled lines as shown in Figure 11.29(a). By properly selecting its parameters, we can design a planar balun to meet the desired response. Synthesis techniques of such baluns have been recently described [8, 32] and are summarized in this section.

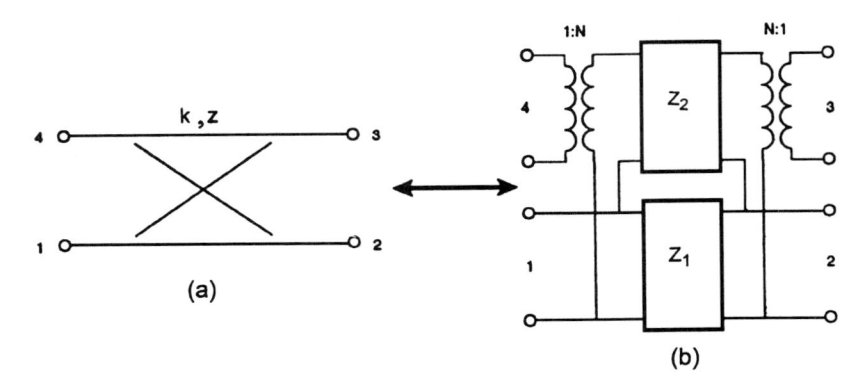

Figure 11.29 (a) A four-port coupled line and (b) its equivalent circuit.

An equivalent of a coupled-line of equal conductor widths is shown in Figure 11.29(b) where Z_1 and Z_2 are the characteristic impedances of distributed unit elements, and N is the transformation ratio. The characteristic impedance Z_{0c} and the coupling coefficient k of the coupled line are given by

$$Z_{0c} = \sqrt{Z_{0e} Z_{0o}} \tag{11.31a}$$

$$k = \frac{Z_{0e} - Z_{0o}}{Z_{0e} + Z_{0o}} \tag{11.31b}$$

and

$$Z_{0e} = Z_{0c}\sqrt{\frac{1+k}{1-k}} \tag{11.32a}$$

$$Z_{0o} = Z_{0c}\sqrt{\frac{1-k}{1+k}} \tag{11.32b}$$

are related to the network equivalent circuit elements as follows:

$$Z_1 = \frac{Z_{0c}}{\sqrt{1-k^2}} \tag{11.33}$$

$$Z_2 = Z_{0c}\frac{\sqrt{1-k^2}}{k^2} \tag{11.34}$$

and

$$N = \frac{1}{k}$$

where Z_{0e} and Z_{0o} are the even- and odd-mode impedances, respectively.

The balun in Figure 11.30(a) has four reactive elements and is known as a fourth-order Marchand balun. Matching section Z makes it a fourth-order balun with improved broadband performance. Figures 11.30(b) and 11.30(c) show equivalent circuit representations and Figure 11.30(d) shows further simplification when $N_a = N_b = N$. In this case, the middle transformers cancel each other.

The final equivalent circuit element values Z_1', Z_2', Z_3', Z_4', and R_5' can be expressed in terms of coupled line parameters Z_{ac}, Z_{bc}, and k as follows:

$$Z_1' = \frac{Z_{2a}}{N^2} = Z_{ac}\sqrt{1-k^2} \tag{11.35a}$$

$$Z_2' = \frac{Z_{2b}}{N^2} = Z_{bc}\sqrt{1-k^2} \tag{11.35b}$$

Baluns

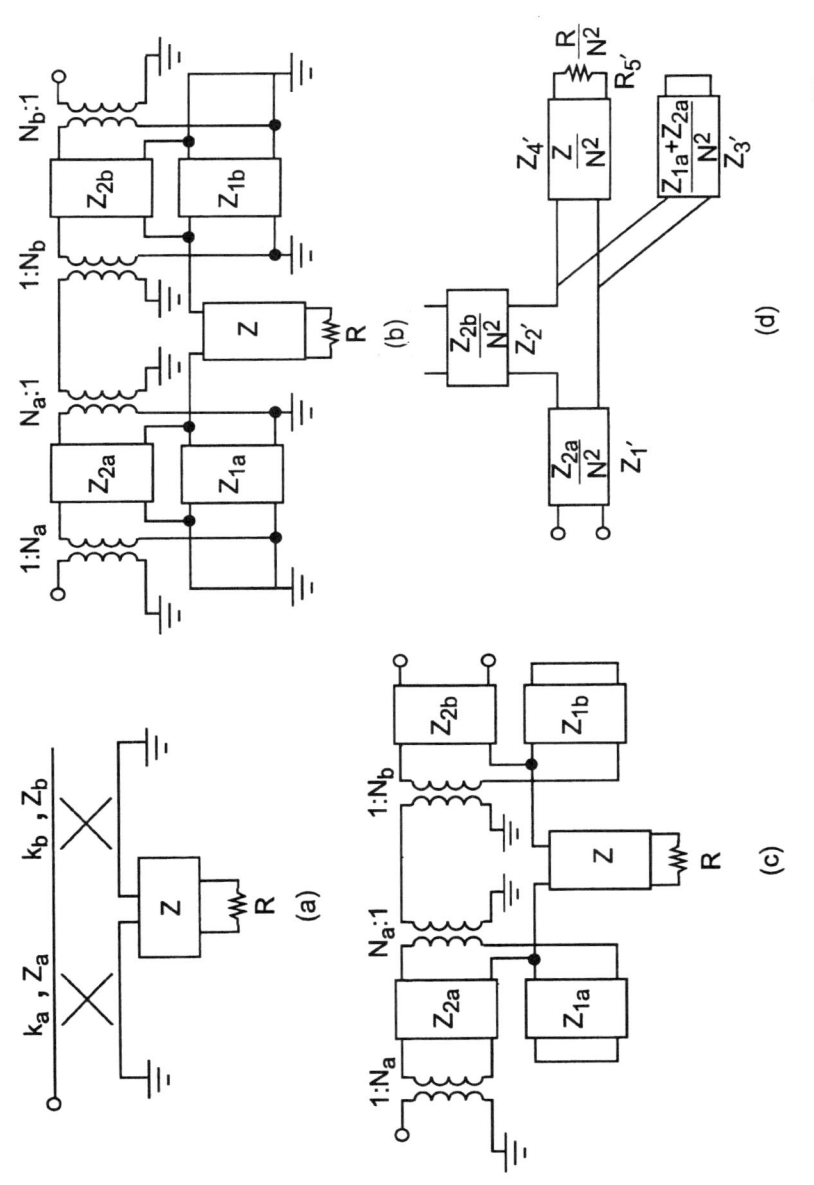

Figure 11.30 Coupled-line Marchand balun and its equivalent circuits. (From [32], © 1992 *Int. J. Microwave Millimeter Wave Comp. Aided Engineering*. Reprinted with permission.)

$$Z_3' = \frac{Z_{1a} + Z_{1b}}{N^2} = (Z_{ac} + Z_{bc})\frac{k^2}{\sqrt{1-k^2}} \quad (11.35c)$$

$$Z_4' = \frac{Z}{N^2} = Zk^2 \quad (11.35d)$$

$$R_5' = \frac{R}{N^2} = Rk^2 \quad (11.35e)$$

where R is the load impedance, and $Z_{ac}^2 = Z_{0e}^a Z_{0o}^a$ and $Z_{bc}^2 = Z_{0e}^b Z_{0o}^b$ are the characteristic impedances of the coupled lines a and b, respectively.

The equations above can be rearranged to solve for the coupled-line parameters in terms of the equivalent element values as follows:

$$k = \sqrt{\frac{Z_3'}{Z_1' + Z_2' + Z_3'}} \quad (11.36a)$$

$$Z_{ac} = \frac{Z_1'}{\sqrt{1-k^2}} \quad (11.36b)$$

$$Z_{bc} = \frac{Z_2'}{\sqrt{1-k^2}} \quad (11.36c)$$

Choosing element values for Z_1', Z_2', Z_3', Z_4', and R_5' for the final design of a balun to meet specifications, the coupled-line parameters can be calculated from (11.30) and (11.31). Finally, the physical dimensions for the balun are determined using suitable coupled-line structures; for example, edge-coupled microstrip [46, 47, 48], broadside-coupled striplines [48, 49], embedded microstrip [50], or any other topology.

The design of a balun starts with determining the network parameters given in Figure 11.30(d) to meet design specifications in terms of bandwidth, load impedance, and return loss. This can be achieved by either using nomographs [8] or using synthesis techniques [51], or by using commercial CAD tools. Balun equivalent circuit parameters as a function of return loss for fourth-order Chebyshev filters having 2:1, 3:1, and 5:1 bandwidth are given in Figure 11.31 [32] for a source impedance of 50 Ω. Coupled-line parameters are calculated using (11.30) and (11.31) and plotted in Figure 11.32. These figures also include the values of Z and the load impedance R.

In addition to the above synthesis analysis of planar Marchand baluns, several other analysis and design methods of such baluns have been reported. This includes equivalent circuit model for the two-coupled line sections presented

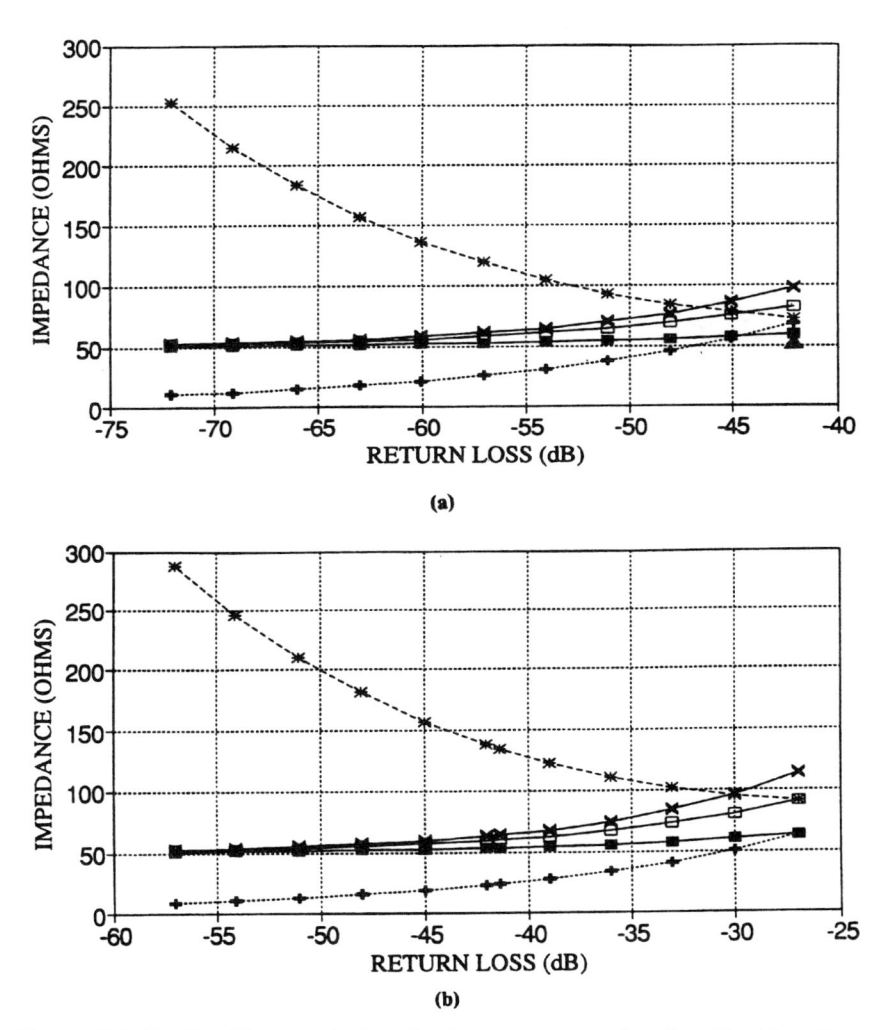

Figure 11.31 Marchand balun equivalent circuit parameters as a function of load impedance (a) 2:1 bandwidth, (b) 3:1 bandwidth, and (c) 5:1 bandwidth. (From [32], © 1992 *Int. J. Microwave Millimeter Wave Comp. Aided Engineering.* Reprinted with permission.)

by Tsai and Gupta [33], the method reported by Schwindt and Nguyen [37], that provides analytical expressions for the S-parameters for a planar multilayer Marchand balun based on broadside-coupled quasi-TEM normal-mode parameters, and the Engels and Jansen [38] design method based on the mode parameters for multiple coupled strip configurations. These methods provide a reasonably

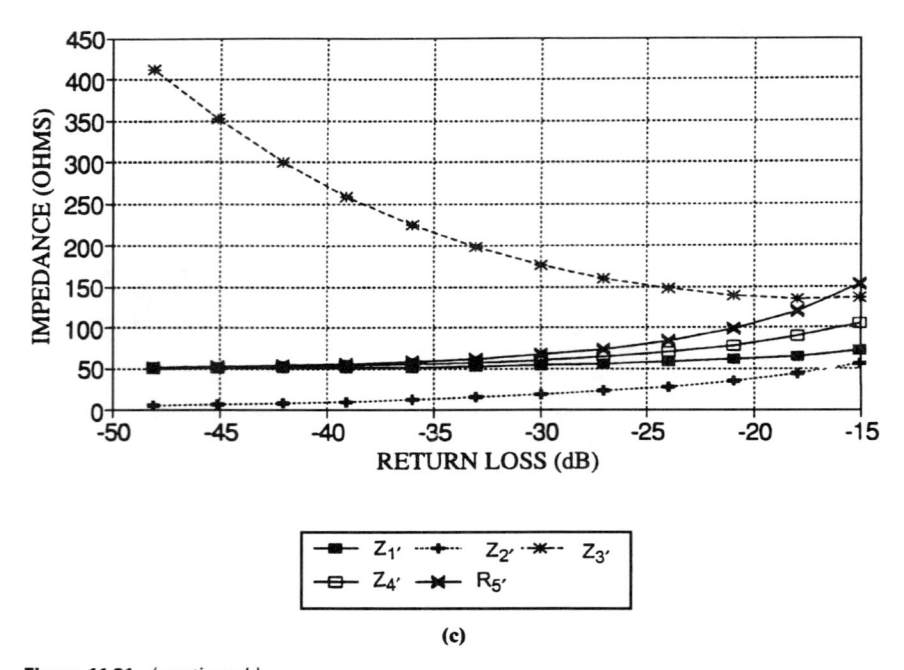

[Figure: Impedance (ohms) vs Return Loss (dB) plot with curves for $Z_{1'}$, $Z_{2'}$, $Z_{3'}$, $Z_{4'}$, $R_{5'}$]

(c)

Figure 11.31 *(continued.)*

good starting point for the design of complicated multilayer baluns. However, for accurate characterization of these baluns, three-dimensional electromagnetic simulators are essential and play a very important role in the development of such baluns.

11.5.3 Examples of Marchand Baluns

Let us consider an example for a balun design with 30-dB return loss, $R'_5 = 95\ \Omega$ and 3:1 bandwidth. From Figure 11.31(b), approximate values for filter network are $Z'_1 = 61.5\ \Omega$, $Z'_2 = 50\ \Omega$, $Z'_3 = 95\ \Omega$, and $Z'_4 = 80\ \Omega$. From (11.34) and (11.35), the coupled-line parameters are $k = 0.678$, $Z_{ac} = 83.7\ \Omega$, $Z_{bc} = 68\ \Omega$, $R = 206\ \Omega$, and $Z = 174\ \Omega$. From (11.30) and (11.31), $Z^a_{0e} = 191.1\ \Omega$, $Z^a_{0o} = 36.7\ \Omega$, $Z^b_{0e} = 155.3\ \Omega$, and $Z^b_{0o} = 29.8\ \Omega$. Figure 11.32 can also be used directly to obtain these parameters. These values are used to determine the physical parameters for a given substrate thickness and dielectric constant values for coupled-line balun [46–48] or broadside coupled-line [48, 49].

Tightly coupled sections of this balun are not possible with edge-coupled microstrip lines. The embedded microstrip technique [50] or Lange coupler

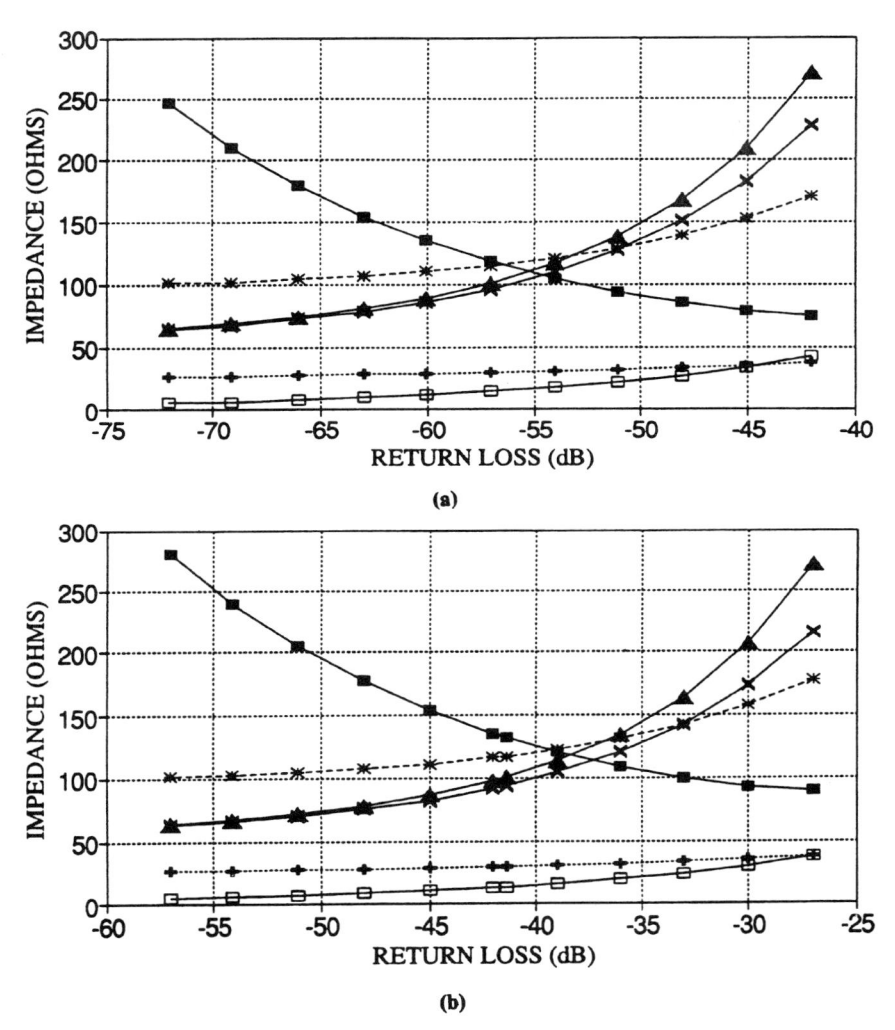

Figure 11.32 Marchand balun coupled-line parameters as a function of load impedance for (a) 2:1 bandwidth, (b) 3:1 bandwidth, and (c) 5:1 bandwidth. (From [32], © 1992 *Int. J. Microwave Millimeter Wave Comp. Aided Engineering.* Reprinted with permission.)

configuration, however, can be used to realize tightly coupled structures. Realization of such couplers with multilayer techniques on substrates such as alumina, quartz, or GaAs [31] have also been reported. A cross-sectional view of a broadside-type coupled line showing even- and odd-mode impedance realization is shown in Figure 11.33.

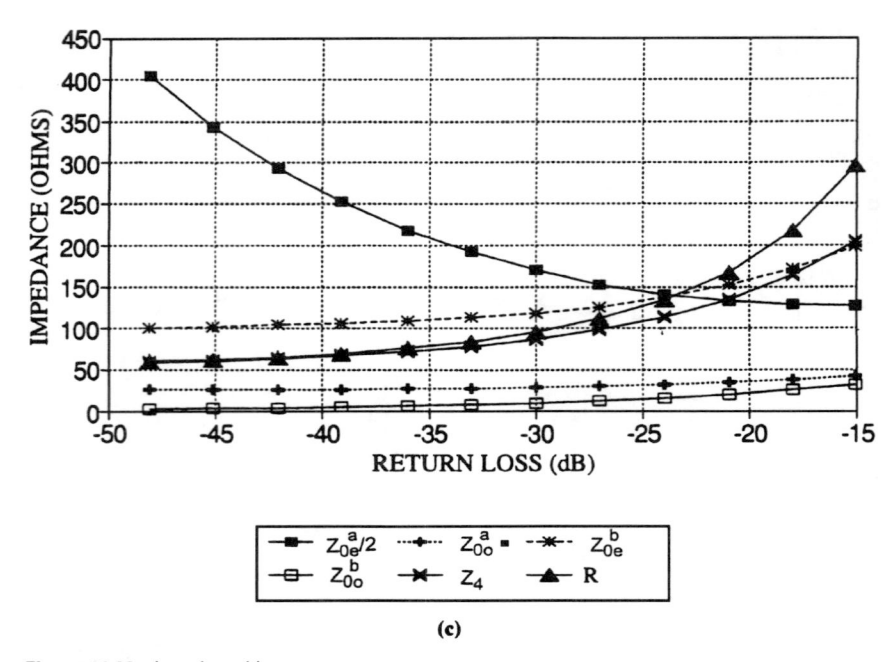

[Figure: Impedance vs Return Loss plot with curves for $Z_{0e}^a/2$, Z_{0o}^a, Z_{0e}^b, Z_{0o}^b, Z_4, and R]

Figure 11.32 *(continued.)*

Marchand Balun Using Lange Coupler

A Marchand balun using two tight couplers having connections as shown in Figure 11.34 has been described recently [52]. The couplers used are slightly under-coupled 3-dB Lange couplers, fabricated on 100-μm-thick GaAs substrate. The linewidth of the couplers is 10 μm and the spacings between the conductors are 12 μm and 10 μm for couplers 1 and 2, respectively. The length of the couplers is 1.6 mm. These dimensions are easily realized using MMIC processes. The ground connection was realized using a via hole. A photograph of the chip, which is 1 by 2 mm, is shown in Figure 11.35.

Simulated performance of the balun is shown in Figure 11.36. Measured insertion loss of the balun's output ports is shown in Figure 11.37, and Figure 11.38 shows the amplitude and phase differences between the two outputs over the 0.1 to 20-GHz frequency range. In an ideal case, the desired values of amplitude and phase differences are 0 dB and 180 degrees, respectively. This balun demonstrates good performance over the 6- to 20-GHz frequency range.

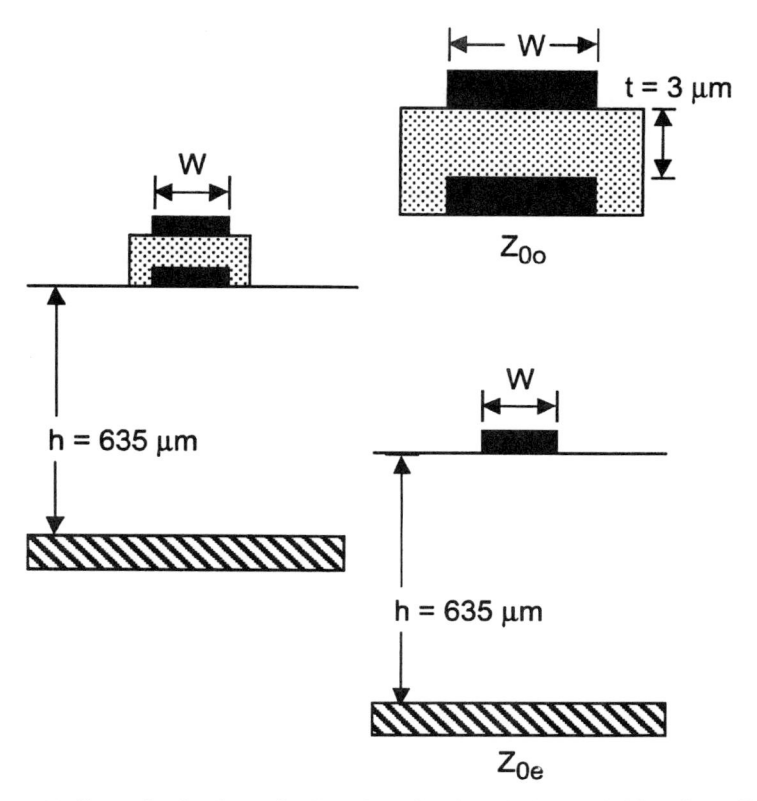

Figure 11.33 Illustration for the realization of a typical broadside-coupled line. (From [31], © 1991 IEEE. Reprinted with permission.)

Marchand Balun Using Re-Entrant Couplers

An alternative solution to broadside multilayer coupled lines and Lange couplers is to use re-entrant multilayer coupled lines [53–55] to realize the balun circuit. The later structure has the advantage of providing tight coupling without stringent dimensional requirements in circuit fabrication. Re-entrant couplers using multilayer microstrip lines have been reported, and a configuration appears in Figure 11.39. It consists of three conductors in which two conductors are deposited on the top surface of a two-layered dielectric substrate and a third conductor that is floating is sandwiched between the two dielectric layers. The floating conductor beneath the coupled lines helps in realizing higher even-mode impedance and tight coupling. Because there are a total of four conductors, the

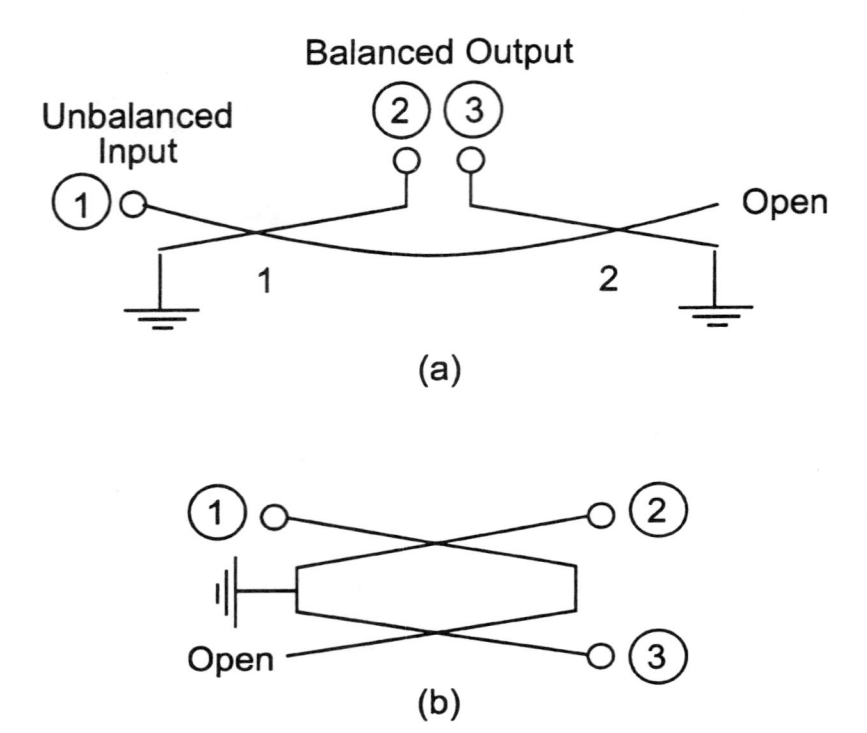

Figure 11.34 (a) Schematic using crossover couplers and (b) rearranged schematic further simplifying the circuit.

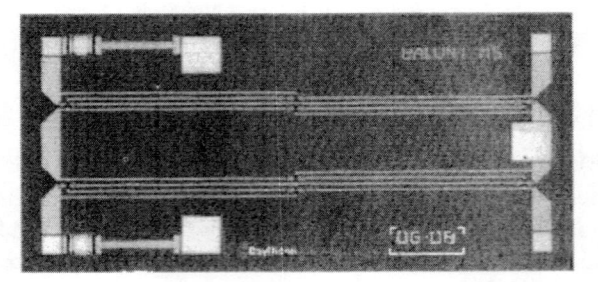

Figure 11.35 Photograph of a balun using Lange couplers (from [52], © 1993 IEEE. Reprinted with permission.)

structure can support three propagation modes instead of the two general even and odd nodes in the case of a three-conductor structure.

Although multilayer dielectric structures allow much greater freedom in designing the baluns in terms of process and size requirements, they give rise to other structurally related problems. A multilayer microstrip structure is

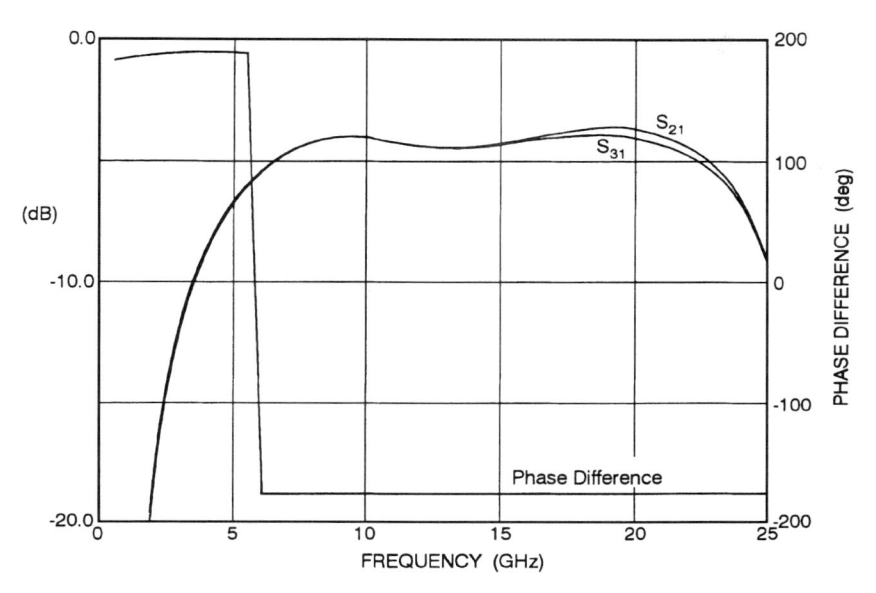

Figure 11.36 Simulated insertion loss and phase difference of the balun. (From [52], © 1993 IEEE. Reprinted with permission.)

inhomogeneous in character and there are no simple design expressions available for determining the physical dimensions. The commercially available EM simulators can be used to analyze the structure, but they usually require a considerable amount of computation time to optimize the design for best performance. Recently, a generalized network model for coupled multiconductor transmission lines has been reported [36] that can also be used to design re-entrant-type couplers. A simple analysis of this coupler has been discussed in Section 8.6.3, and the design procedure for the balun can be summarized as follows:

1. Calculate the even- (Z_{0e}) and odd- (Z_{0o}) mode impedances for each coupled section as described in the beginning of this section.
2. Because $Z_{02} = Z_{0o}$, calculate the physical dimensions of the top-layer structure such as ε_{r2}, d, and W_2 by keeping a large separation between the structure.
3. Calculate $Z_{01} = (Z_{0e} - Z_{02})/2$ and determine the physical dimensions of the bottom-layer structure such as ε_{r1}, h, and W_1.
4. Optimize the balun in terms of the structure's physical dimensions by fine-tuning the above calculated dimensions using EM simulators.

Table 11.2
Design parameters for a 3:1 bandwidth balun using re-entrant couplers calculated using EM simulator [32].

Design #	Z^a_{0e}	Z^a_{0o}	Z^b_{0e}	Z^b_{0o}	ε_{r2}	ε_{r1}	h (mil)	d (mil)	W_{1a} (mil)	W_{2a} (mil)	S_a (mil)	W_{1b} (mil)	W_{2b} (mil)	S_b (mil)
1	200	28	111	11	9.8	3.2	15	0.2	3.0	0.7	0.5	14.0	2.5	1.0
2	264	30	117	13	12.9	3.2	25	0.2	2.0	0.7	0.5	17.0	2.0	1.0

a and *b* denote coupler a and coupler b, respectively.

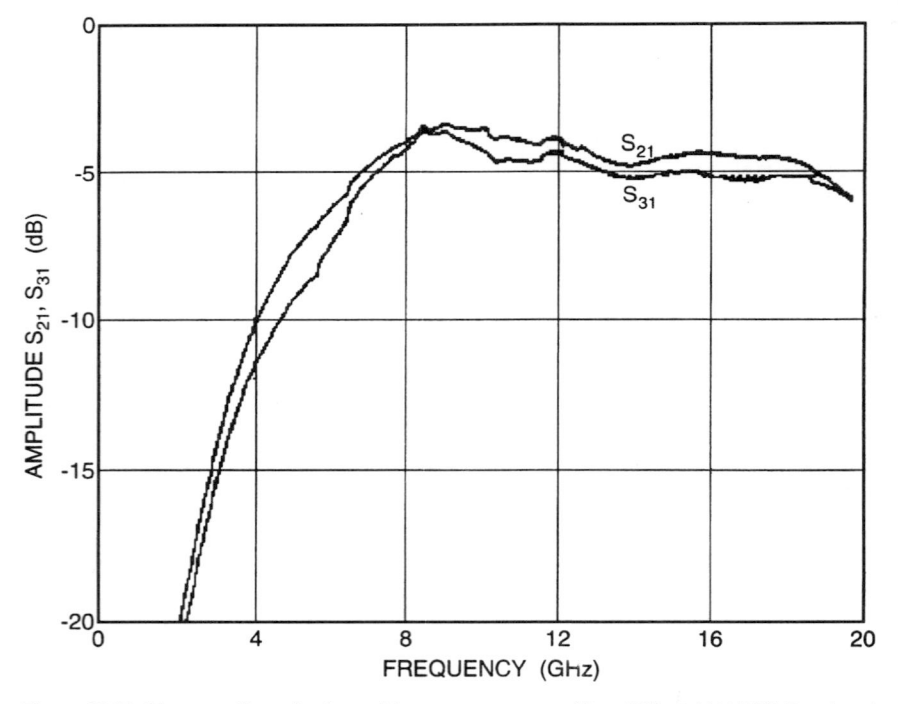

Figure 11.37 Measured insertion loss of the two output ports. (From [52], © 1993 IEEE. Reprinted with permission.)

For example, a Marchand balun covering 3:1 bandwidth with a 41.4-dB in-band return loss requires $Z^a_{0e} = 264\,\Omega$, $Z^a_{0o} = 30\,\Omega$, $Z^b_{0e} = 117\,\Omega$, and $Z^b_{0o} = 13\,\Omega$ [32]. The design dimensions are summarized in Table 11.2. Both these examples demonstrate the realization of Marchand baluns using multilayer microstrip structures. Figure 11.40 shows the simulated performance of a balun using re-entrant-type couplers.

Baluns 433

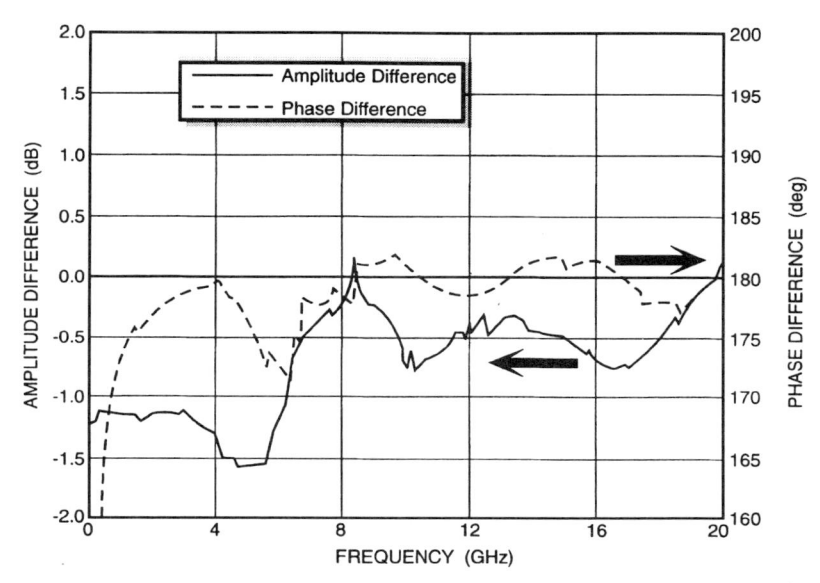

Figure 11.38 Measured amplitude and phase balance of the balun. (From [52], © 1993 IEEE. Reprinted with permission.)

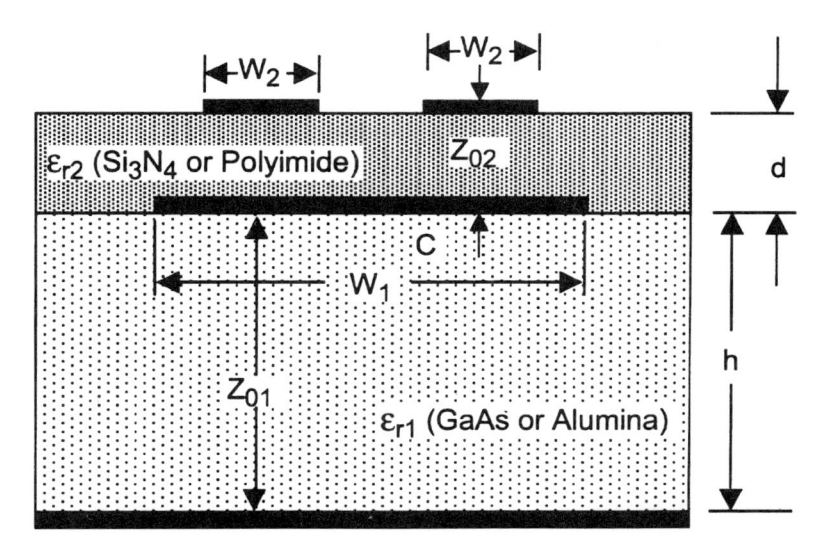

Figure 11.39 Microstrip re-entrant–type coupler.

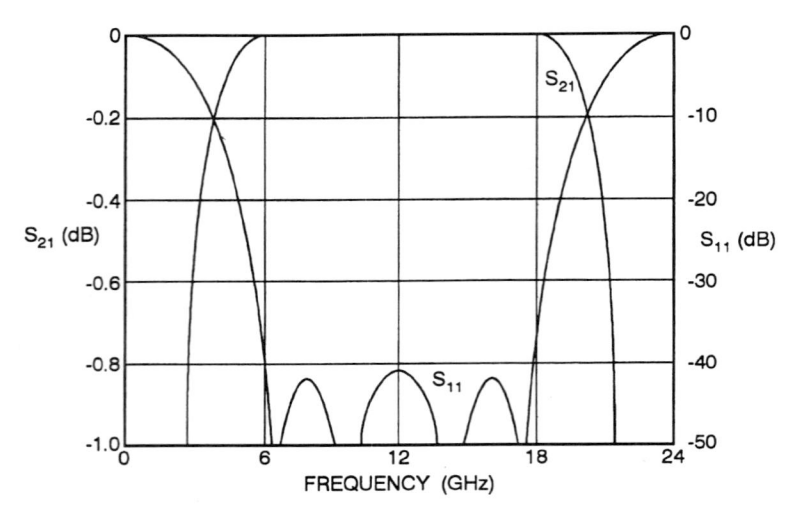

Figure 11.40 Linear circuit simulation of 3:1 bandwidth Marchand balun with ideal element values $Z^a_{oe} = 269.3\Omega$, $Z^a_{0o} = 29.7\Omega$, $Z^b_{oe} = 117.1\Omega$, $Z^b_{0o} = 13.2\Omega$, $Z'4 = 94\Omega$, and $R' = 100.4\,\Omega$. (From [32], © 1992 Int. J. Microwave Millimeter Wave. Comp. Aided Engineering. Reprinted with permission.)

Monolithic Marchand Balun

A monolothic Marchand balun has been developed [41] using multilayer dielectric monolithic technology, where polyimide ($\varepsilon_r = 3.2$) was used as the intermetal dielectric. Figure 11.41 shows the top view and side view of the monolithic Marchand balun. The dimensions for the various line sections are input line A has a 40-μm linewidth and is 1,440 μm long. The linewidth, length for the open-circuited line B and short-circuited bottom lines C are 140 μm, 1,420 μm, and 120 μm and 1,427 μm, respectively. The unbalanced input impedance is 25Ω and the balanced output impedance is 50Ω. The balun was designed using electromagnetic simulation. The input return shown in Figure 11.42(a) was measured by terminating the balanced port into a 50Ω load.

The insertion loss of the baluns when connected back to back is shown in Figure 11.42(b). For a single balun, the insertion loss is less than 0.7 dB from 6 to 21 GHz. The measured amplitude balance was within 0.5 dB from 7 to 21 GHz with a corresponding phase difference of 178 to 172 degrees.

Compact Marchand Balun

A novel compact lumped-element Marchand balun for wireless applications has been described by Jansen, et al. [39]. The structure used three metal levels and two polyimide intermetal layers to realize broadside-coupled sections to fabricate

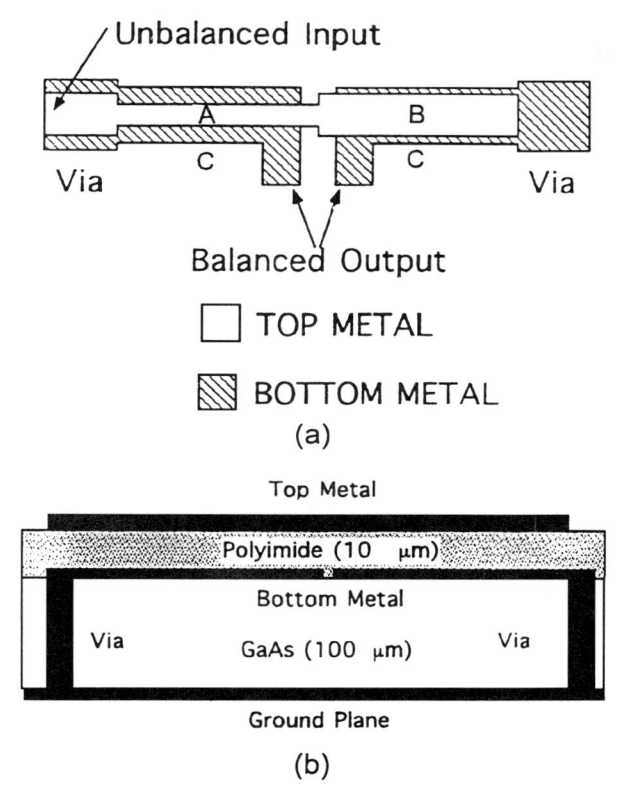

Figure 11.41 Monolithic Marchand balun (a) top view and (b) side view. (From [41], © 1997 IEEE. Reprinted with permission.)

a lumped-element Marchand balun on Si substrate. Figure 11.43 shows the top view of a 1.9-GHz balun and occupies only a 2.5-mm^2 area.

11.6 Other Baluns

11.6.1 Coplanar Waveguide Baluns

Apart from microstrip baluns, coplanar waveguide (CPW) baluns have also been reported [30, 40, 55]. An analysis of such baluns is given in [30], while a physical layout is shown in Figure 11.44. The physical length of the balun section is $\lambda/4$. This balun does not have as large a bandwidth as planar baluns or the Marchand baluns. Another CPW balun [55] is shown in Figure 11.45. This balun also has narrowband performance.

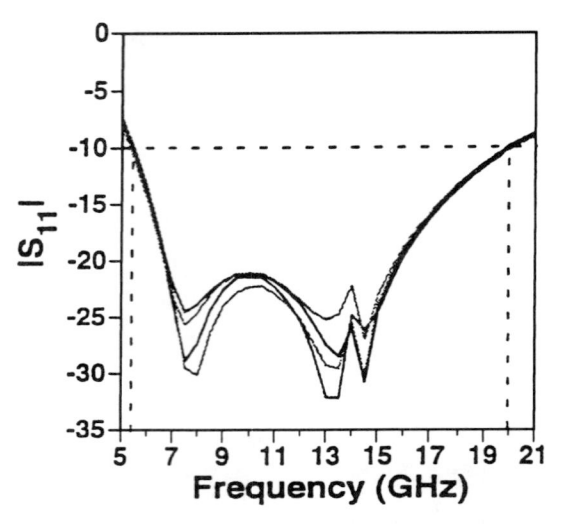

Figure 11.42(a) Measured return loss of four Marchand baluns terminated into a balanced 50 Ω load. (From [41], © 1997 IEEE. Reprinted with permission.)

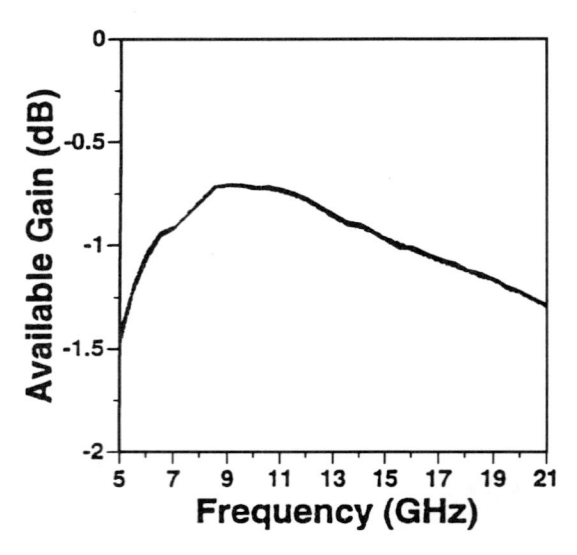

Figure 11.42(b) Measured loss of four pairs of back-to-back Marchand baluns. (From [41], © 1997 IEEE. Reprinted with permission.)

Baluns

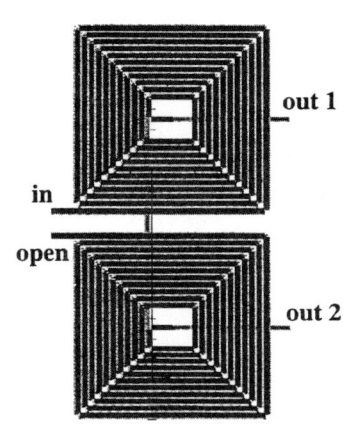

Figure 11.43 Top view of a lumped-element balun layout. (From [39], © 1997 IEEE. Reprinted with permission.)

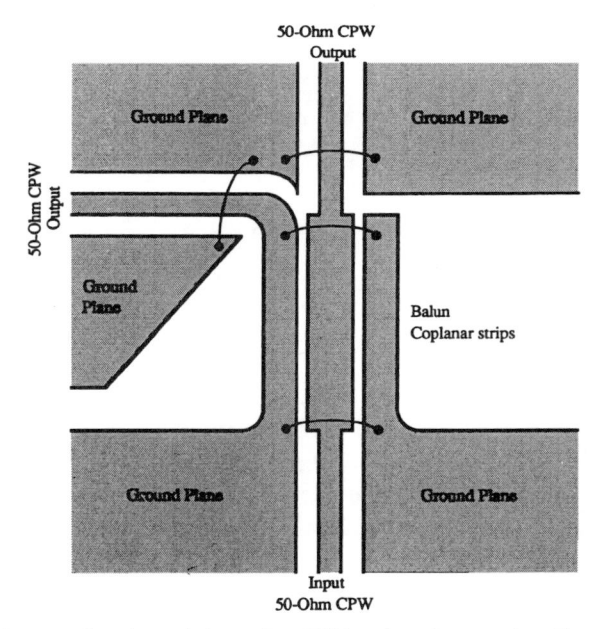

Figure 11.44 Layout of a planar balun using CPW and coplanar strips. (From [30], © 1991 *Optical Tech. Letts.* Reprinted with permission.)

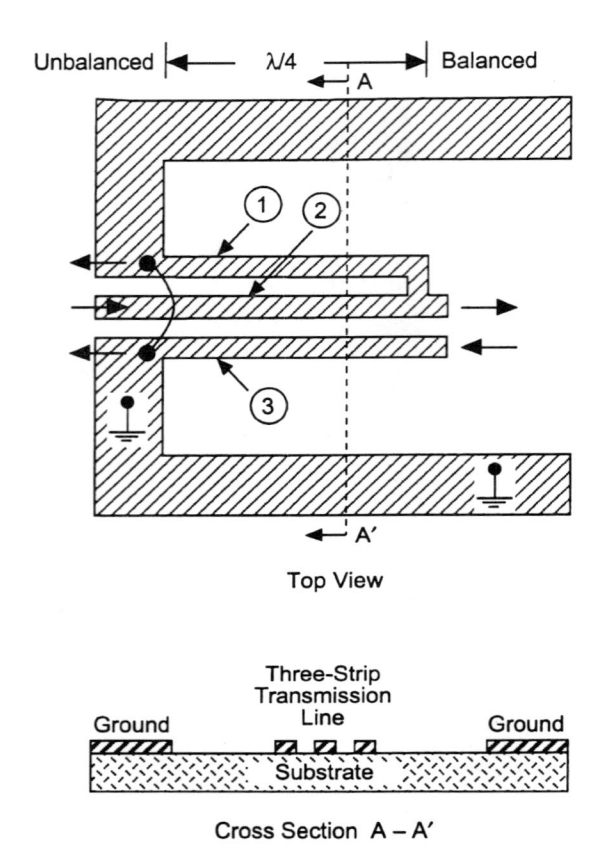

Figure 11.45 Top and cross sectional illustrations of a three-strip CPW balun.

11.6.2 Triformer Balun

A triformer balun that is a rectangular-spiral-shaped inductor using three coupled lines has been developed for microwave monolithic integrated circuit applications [19]. The balun was designed using the multiconductor coupled-line transformer chain matrix method. However, this structure can be accurately analyzed using EM simulators. The physical layout, equivalent circuit, and microphotograph of a one-turn triformer are shown in Figure 11.46. The structure was fabricated on GaAs ($\varepsilon_r = 12.9$, $\tan \delta = 0.0003$) substrate with a 2-μm conductor thickness, a 10-μm conductor width, and 5-μm spacing between the conductors. The gold conductors have metal resistivity of 0.03 Ω-μm, and the substrate thickness is 100 μm. Measured and simulated performance of the triformer is

Baluns 439

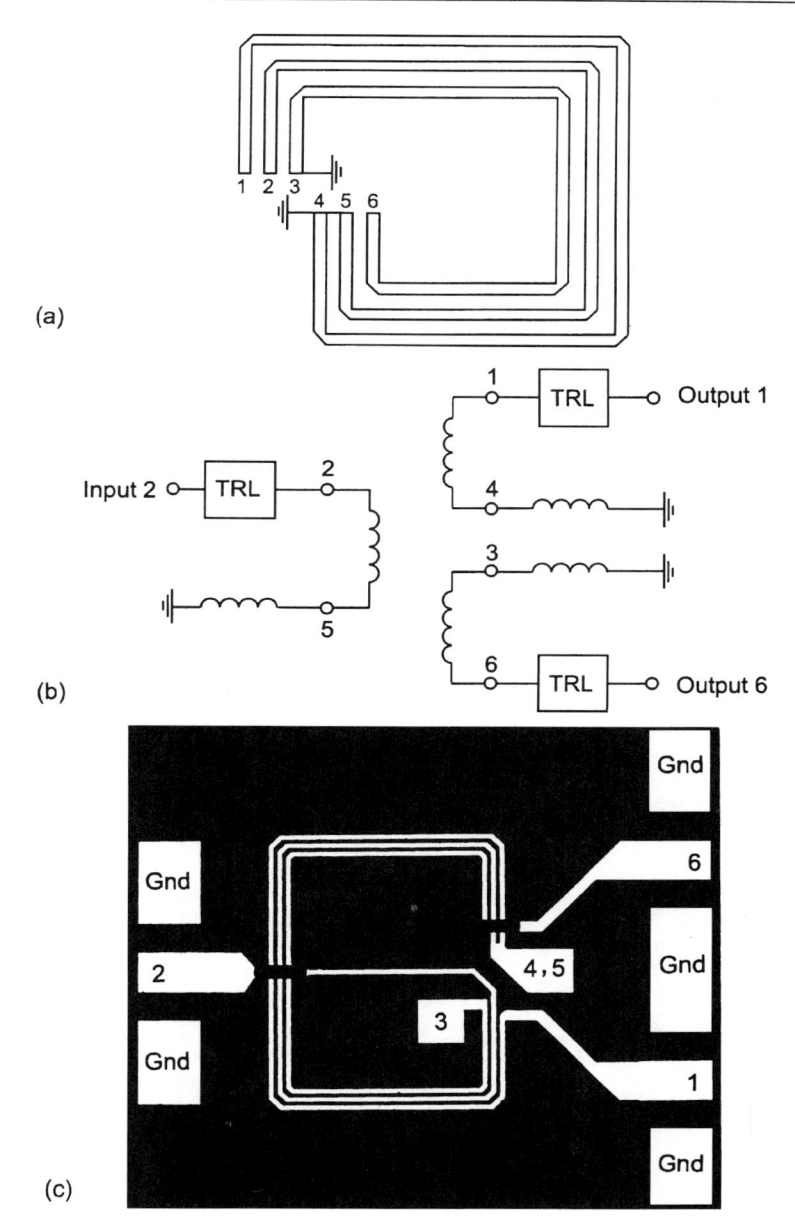

Figure 11.46 Multiconductor coupled line triformer (a) Physical layout, (b) equivalent circuit, and (c) Microphotograph. (From [19], © 1989 IEEE. Reprinted with permission.)

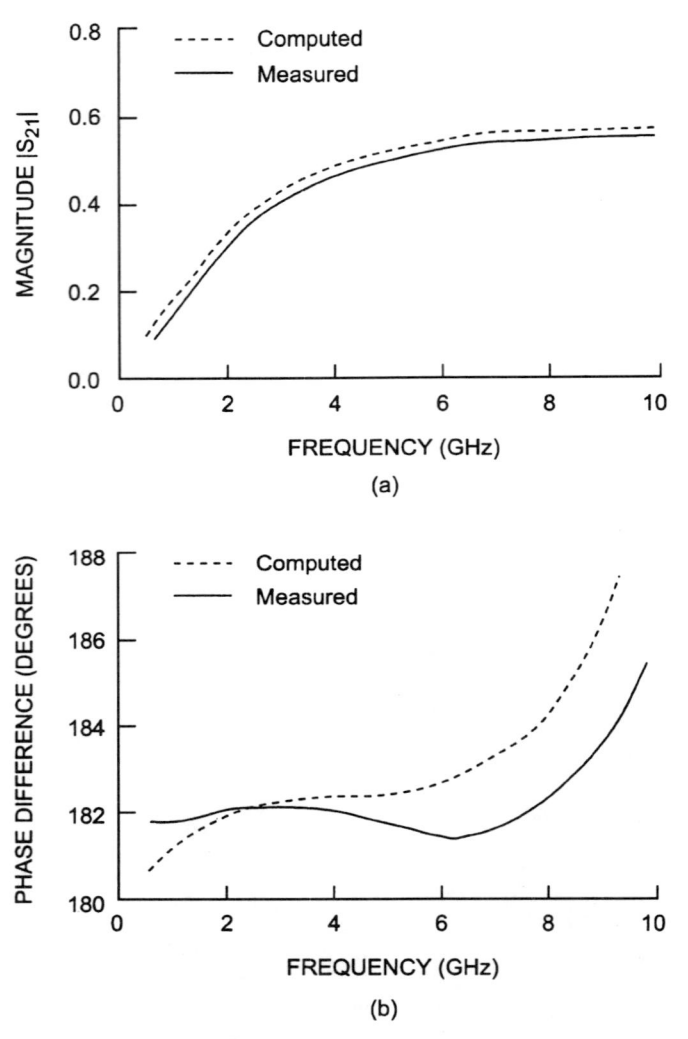

Figure 11.47 Comparison of computed and measured S-parameters. (a) Magnitude, (b) differential phase shift. (From [19], © 1989 IEEE. Reprinted with permission.)

shown in Figure 11.47. The differential phase-shift between the output ports 1 and 6 remains constant at about 182 degrees. A major shortcoming of this approach is the loss in the structure because of the thin substrate and narrow conductor width. This can be overcome to some extent by printing thick lines on thick alumina substrates.

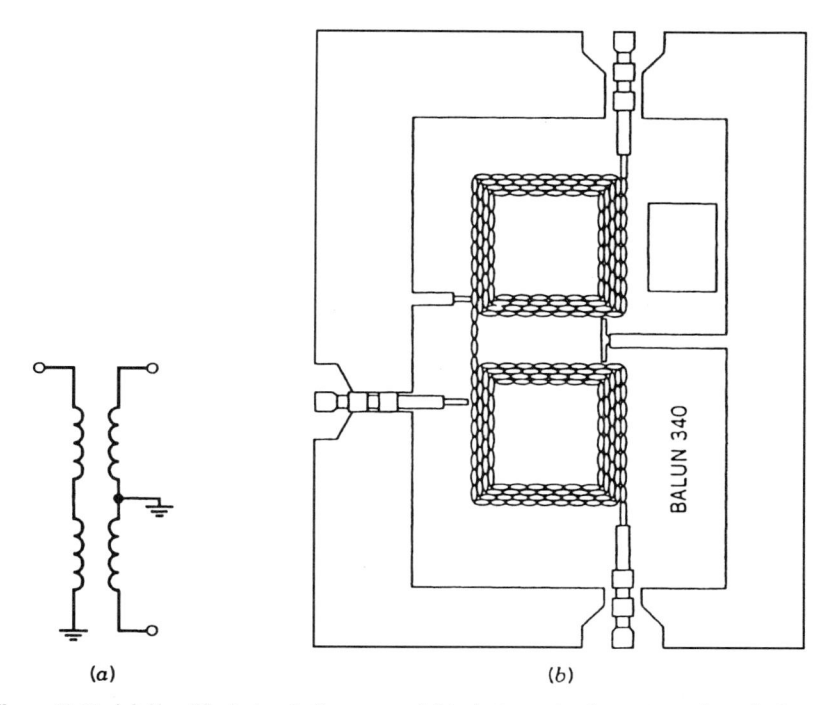

Figure 11.48 (a) Simplified circuit diagram and (b) photograph of a rectangular spiral transformer balun. (From [31], © 1991 IEEE. Reprinted with permission.)

11.6.3 Planar-Transformer Balun

The planar-transformer balun consists of two oppositely wrapped twin-coil transformers connected in series. In this configuration, one of the two outer nodes in the primary coil and the inner common node in the secondary coil are grounded as shown in Figure 11.48 [31]. Figure 11.48(a) shows the equivalent circuit, and Figure 11.48(b) is a microphotograph of the balun using rectangular spiral transformers. The chip measures about 1.5 mm^2, which demonstrates the compactness of this approach.

The resonant frequency of the coil transformers due to interturn and other parasitic capacitances with the ground limits the performance and bandwidth of this type of balun. Usually the balun is operated below the resonant frequency. The lower frequency bound of the bandwidth is set by the inductance, while the upper frequency bound is set by the reasonant frequency of the coil. The bandwidth can be increased either by increasing the resonant frequency while maintaining the same inductance or by increasing the inductance while maintaining the resonant frequency. The resonant frequency can be increased

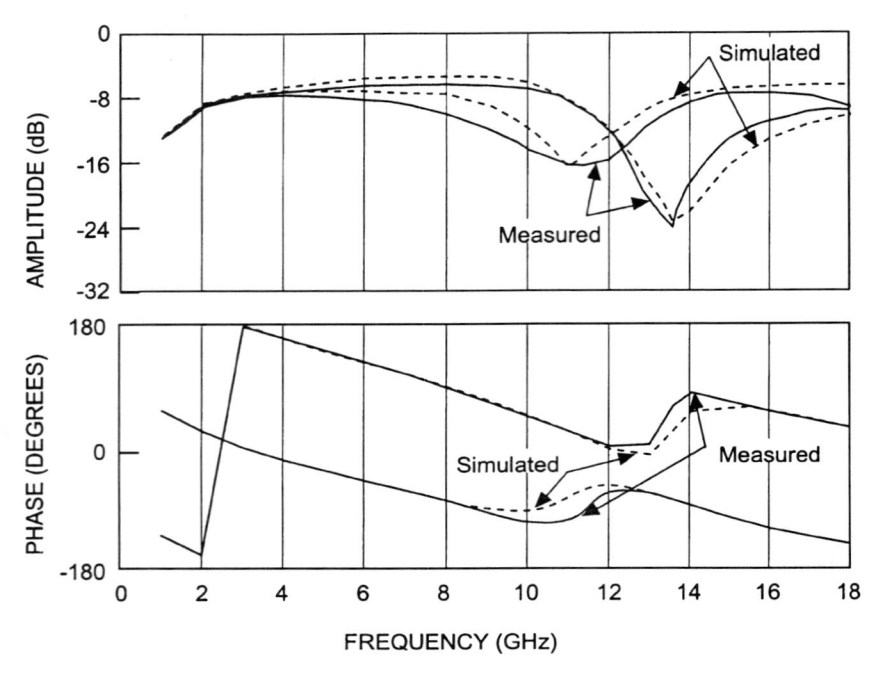

Figure 11.49 Comparison between simulated and measured performances of a planar-transformer balun. (From [31], © 1991 IEEE. Reprinted with permission.)

by reducing the interturn and parasitic capacitances by employing thick and low-dielectric constant substrates and air bridges in the coil. In addition, the inductance can be increased by optimizing the area-to-length ratio.

Figure 11.49 shows the simulated and measured performance of a planar-transformer balun. The amplitude and phase imbalances between the two balanced ports are less than 1.5 dB and 10 degrees, respectively, over the 1.5 to 6.5-GHz frequency band. The simulated results shown were obtained using EM analysis.

We described several kinds of baluns in this chapter. The selection of a particular type depends upon the application, performance, and cost limitations. Among all the baluns described, the Marchand balun achieves the maximum bandwidth and best performance.

References

[1] Marchand, N., "Transmission Line Conversion Transformers," *Electronics*, Vol. 17, Dec. 1944, pp. 142–145.

[2] Roberts, W. K., "A New Wide-Band Balun," *Proc. IRE*, Vol. 45, Dec. 1957, pp. 1,628–1,631.

[3] McLaughlin, J. W., D. A. Dunn, and R. W. Grow, "A Wide Band Balun," *IRE Trans. Microwave Theory Tech.*, Vol. MTT-6, July 1958, pp. 314–316.

[4] Bawer, R., and J. J. Wolfe, "A Printed-Circuit Balun for Use With Spiral Antennas," *IRE Trans. Microwave Theory Tech.*, Vol. MTT-8, May 1960, pp. 319–325.

[5] Oltman, H. G., "The Compensated Balun," *IEEE Trans. Microwave Theory Tech.*, Vol. MTT-14, Mar. 1966, pp. 112–119.

[6] Phelan, H. R., "A Wide-Band Parallel-Connected Balun," *IEEE Trans. Microwave Theory Tech.*, Vol. MTT-18, May 1970, pp. 259–263.

[7] Hallford, B. R., "A Designer's Guide to Planar Mixer Baluns," *Microwaves*, Vol. 18, Dec. 1979, pp. 52–57.

[8] Cloete, J. H., "Exact Design of the Marchand Balun," *Microwave J.*, Vol. 23, May 1980, pp. 99–110.

[9] Bassett, R., "Three Balun Designs for Push-Pull Amplifiers," *Microwaves*, Vol. 19, July 1980, pp. 47–52.

[10] Hallford, B. R., "Simple Balun Coupled Mixers," *IEEE MTT-S Int. Microwave Symp. Dig.*, 1981, pp. 304–306.

[11] Ho, C. Y., and R. Traynor, "New Analysis Technique Builds Better Baluns," *Microwaves and RF*, Vol. 24, Aug 1985, pp. 99–102.

[12] Edward, B., and D. Rees, "A Broadband Printed Dipole With Integrated Balun," *Microwave J.*, Vol. 30, May 1987, pp. 339–344.

[13] Seligman, J. M., E. S. Gillespie, and M. Ryan, "A Quasi-Lumped Planar Balun for Printed-Circuit Antennas," *IEEE AP-S International Symp. Dig.*, 1987, pp. 15–19.

[14] Shuhao, H., "The Balun Family," *Microwave J.*, Vol. 30, Sept. 1987, pp. 227–229.

[15] Baker, D. E., J. R. Nortier, and C. A. Van der Neut, "Design and Performance of a Microstrip Balun for Archimedes Spiral Antennas," *Trans. S. Afr. Inst. Electr. Eng.*, Vol. 78, Dec. 1987, pp. 34–37.

[16] Cripps, S. C., "Microstrip Balun Having Improved Bandwidth," U.S. Patent No. 4,739,289, Apr. 1988.

[17] Climer, B., "Analysis of Suspended Microstrip Taper Baluns," *Proc. IEE*, Vol. 135, Pt. H., Apr. 1988, pp. 65–69.

[18] Parisi, S. J., "180° Lumped Element Hybrid," *IEEE MTT-S Int. Microwave Symp. Dig.*, 1989, pp. 1,243–1,246.

[19] Boulouard, A., and M. LeRouzic, "Analysis of Rectangular Spiral Transformers for MMIC Applications," *IEEE Trans. Microwave Theory Tech.*, Vol. 37, Aug. 1989, pp. 1,257–1,260.

[20] Sevick, J., *Transmission Line Transformers*, published by The American Radio Relay League, Newington, CT, Ch. 9, 1989.

[21] Bharj, S. S., B. Thompson, and S. P. Tan, "Integrated Circuit 2-18 GHz Balun," *Applied Microwave*, Vol. 2, Nov./Dec. 1989, pp. 110–119.

[22] Pavio, A. M., and A. Kikel, "A Monolithic or Hybrid Broadband Compensated Balun," *IEEE MTT-S Int. Microwave Symp. Dig.*, 1990, pp. 483–486.

[23] Barber, R. G., "Enhanced Coupled, Even Mode Terminated Baluns and Mixers Constructed Therefrom," *ibid*, 1990, pp. 495–498.

[24] Pavio, A. M., and S. K. Sutton, "A Microstrip Re-Entrant Mode Quadrature Coupler for Hybrid and Monolithic Circuit Applications," *IEEE MTT-S Int. Microwave Symp. Dig.*, 1990, pp. 573–576.

[25] Vendelin, G. D., A. M. Pavio, and U. L. Rohde, *Microwave Circuit Design Using Linear and Nonlinear Techniques*, New York, John Wiley, 1990, Ch. 7.

[26] Lu, S., A. E. Riad, and S. M. Riad, "Design, Fabrication, and Characterization of a Wideband Hybrid Balun," *IEEE Trans. Instrumentation and Measurement*, Vol. 40, Apr. 1991, pp. 486–489.

[27] Minnis, B. J., and M. Healy, "New Broadband Balun Structures for Monolithic Microwave Integrated Circuits," *IEEE MTT-S Int. Microwave Symp. Dig.*, 1991, pp. 425–428.

[28] Rogers, J., and R. Bhatia, "A 6 to 20 GHz Planar Balun Using a Wilkinson Divider and Lange Couplers," *ibid*, 1991, pp. 865–868.

[29] Jacques, R., and D. Meignant, "Novel Wide Band Microstrip Balun," *Proc. of 11th European Microwave Conf.*, 1991, pp. 839–843.

[30] Nguyen, T., and J. B. Beyer, "Analysis and Design of a Planar Balun," *Microwave and Optical Technology Letters*, Vol. 4, Oct. 1991, pp. 451–455.

[31] Chen, T. H., et al., "Broadband Monolithic Passive Baluns and Monolithic Double-Balanced Mixer," *IEEE Trans. Microwave Theory Tech.*, Vol. 39, Dec. 1991, pp. 1,980–1,986.

[32] Goldsmith, C. L., A. Kikel, and N. L. Wilkens, "Synthesis of Marchand Baluns Using Multilayer Microstrip Structures," *Int. J. of Microwave and Millimeter-Wave Computer-Aided Engineering*, Vol. 2, July 1992, pp. 179–188.

[33] Tsai, C. M., and K. C. Gupta, "A Generalized Model for Coupled Lines and Its Applications to Two-Layer Planar Circuits," *IEEE Trans. Microwave Theory Tech.*, Vol. 40, Dec. 1992, pp. 2,190–2,199.

[34] Sturdivant, R., "Balun Designs for Wireless, Mixers, Amplifiers, and Antennas," *Applied Microwave*, Vol. 5, summer 1993, pp. 34–44.

[35] Maas, S. A., and K. W. Chang, "A Broadband, Planar, Doubly Balanced Monolithic Ka-Band Diode Mixer," *IEEE Microwave and Millimeter-Wave Monolithic Circuits Symp. Dig.*, 1993, pp. 53–56.

[36] Tsai, C. M., and K. C. Gupta, "CAD Procedures for Planar Re-Entrant Type Couplers and Three-Line Baluns," *IEEE MTT-S Int. Microwave Symp. Dig.*, 1993, pp. 1,013–1,016.

[37] Schwindt, R., and C. Nguyen, "Computer-Aided Analysis and Design of a Planar Multilayer Marchand Balun," *IEEE Trans. Microwave Theory Tech.*, Vol. 42., July 1994, pp. 1,429–1,434.

[38] Engels, M., and R. H. Jansen, "Design of Integrated Compensated Baluns," *Microwave and Opt. Tech. Letts*, Vol. 14, Feb. 1997, pp. 75–81.

[39] Jansen, R. H., J. Jotzo, and M. Engels, "Improved Compaction of Multilayer MMIC/MCM Baluns Using Lumped Element Compensation," *IEEE MTT-S Int. Microwave Symp. Dig.*, 1997, pp. 277–280.

[40] Gokdemir, T., et al., "Design and Performance of GaAs MMIC CPW Baluns Using Overlaid and Spiral Couplers," *ibid*, 1997, pp. 401–404.

[41] Tutt, M. N., H. Q. Tserng, and A. Ketterson, "A Low-Loss, 5.5 GHz–20 GHz Monolithic Balun," *IEEE MTT-S Int. Microwave Symp. Dig.*, 1997, pp. 933–936.

[42] Bahl, I. J., and P. Bhartia, "Characteristics of Inhomogeneous Broadside-Coupled Striplines," *IEEE Trans. Microwave Theory Tech.*, Vol. MTT-22, June 1980, pp. 529–535.

[43] Jones, E. M. T., and J. T. Bolljahn, "Coupled-Strip-Transmission-Line Filters and Directional Couplers," *IRE Trans. Microwave Theory Tech.*, Vol. MTT-4, Apr. 1956, pp. 75–81; also see correction in Vol. MTT-17, Oct. 1969, pp. 753–759.

[44] Sadhir, V., and I. Bahl, "Radial Line Structures for Broadband Microwave Circuit Applications," *Microwave J.*, Vol. 34, Aug. 1991, pp. 102–123.

[45] Boire, D. C., J. E. Degenford, and M. Cohn, "A 4.5 to 18 GHz Phase Shifter," *IEEE MTT-S Int. Microwave Symp. Dig.*, 1985, pp. 601–604.

[46] Gupta, K. C., R. Garg, I. J. Bahl, and P. Bhartia, *Microstrip Lines and Slotlines*, Sec. Ed., Norwood, MA: Artech House, 1996.

[47] Garg, R., and I. J. Bahl, "Characteristics of Coupled Microstriplines," *IEEE Trans. Microwave Theory Tech.*, Vol. MTT-27, July 1979, pp. 700–705; also see correction in *IEEE Trans. Microwave Theory, Tech.*, Vol. MTT-28, Mar. 1980, p. 272.

[48] Bahl, I. J., and P. Bhartia, *Microwave Solid State Circuit Design*, New York, John Wiley, 1988.

[49] Cohn, S. B., "Characteristic Impedances of Broadside-Coupled Strip Transmission Lines," *IRE Trans. Microwave Theory Tech.*, Vol. MTT-8, Nov. 1960, pp. 633–637.

[50] Willems, D., and I. Bahl, "A MMIC Compatible Tightly Coupled Line Structure Using Embedded Microstrip," *IEEE Trans. Microwave Theory Tech.*, Vol. 41, Dec. 1993, pp. 2303–2310.

[51] Horton, M., and R. Wenzel, "General Theory and Design of Optimum Quarter-Wave TEM Filters," *IEEE Trans. Microwave Theory Tech.*, Vol. MTT-13, May 1965, pp. 316–327.

[52] Tsai, M. C., "A New Compact Wideband Balun," *IEEE MTT-S Int. Microwave and Millimeter Wave Monolithic Circuit Symp. Dig.*, 1993, pp. 123–125.

[53] Cohn, S. B., "The Re-Entrant Cross Section and Wide-Band 3-dB Hybrid Couplers," *IEEE Trans. Microwave Theory and Tech.*, Vol. MTT-11, July 1963, pp. 254–258.

[54] Lavendol, L., and J. J. Taub, "Re-Entrant Directional Coupler Using Strip Transmission Line," *IEEE Trans. Microwave Theory and Tech.*, Vol. MTT-13, Sept. 1965, pp. 700–701.

[55] DeBrecht, R. E., "Coplanar Balun Circuits for GaAs FET High-Power Push-Pull Amplifiers," *IEEE MTT-S Int. Microwave Symp. Dig.*, 1973, pp. 309–312.

12

High-Speed Circuit Interconnects

The preceeding chapters have concentrated on the use of coupling between transmission lines to achieve certain functions and performance objectives in components and circuits. In some cases, however, unwanted coupling creates problems that a circuit designer needs to resolve. For example, in recent years, microwave techniques have been increasingly applied in the design of high-speed digital circuit interconnects [1–3]. With increasing speed of digital circuits (several hundred Mbps to Gbps), effects such as propagation delay, mismatch, dispersion, loss and electromagnetic coupling between lines have an impact on the performance. Therefore, the conducting lines interconnecting various elements of a digital circuit have to be treated as transmission line sections. An exhaustive discussion of all these effects is beyond the scope of this book. We have tried to limit our treatment to a brief coverage of the various effects discussed above. The most common technique used for the analysis of these effects is outlined. Some typical results published in the literature are included. Most of the discussion on interconnects in this chapter is devoted to the crosstalk effect as it is closely associated with the main theme of this book.

12.1 High-Speed Digital Circuit Interconnects

A typical high-speed digital circuit package consists of a number of individual chips attached to a circuitboard. Each chip has a number of input/output ports (pins). The connection between different chips is achieved by means of "interconnect" lines printed on the circuitboard. With the increasing number of input/output ports on a chip, the average length of the interconnect lines increases. Further, with increasing clock speeds, the interconnect length becomes of the order of a wavelength even at the lower order frequency components of the digital signal.

Because of different propagation delays in various paths, a pulse starting from a given port reaches other ports in the circuit at different times. This affects the synchronous operation of the circuit and therefore the maximal operating speed of the circuit. Typically, the effect of finite length of the transmission line should be taken into account if $T_r < 10\, T_d$, where T_r is the rise time of the pulse and T_d is the time taken by the signal to travel the finite length of the transmission line.

Proximity effects cause electromagnetic coupling between parallel-coupled lines. In general, a signal incident in the "active" line gets coupled to the "quiet" line in both the forward and backward directions. This effect is known as "crosstalk" and may result in false-switching if crosstalk levels are large enough.

Electrical oscillations (ringing) can occur in a digital circuit package if the interconnect lines are improperly terminated. A wave reflected successively from improper loads at the ends of a transmission line can lead to a resonance in the circuit, thereby causing failure of the circuit operation. The effect is more severe if the line is low loss, in which case the modes of oscillation are of relatively high Q and do not die quickly. When ringing is present, the output pulse is present for quite a long duration after the input pulse has ceased. Further in this case, near-end crosstalk levels may significantly increase.

A mismatch occurs if a transmission line is connected to a load impedance that is different from its characteristic impedance. The mismatch causes reflected waves to be set up on the transmission line, resulting in ringing and higher crosstalk levels.

A pulse traveling along a uniform transmission line is distorted because of the dispersive and lossy nature of the transmission line. The most commonly used transmission line for interconnects is the open microstrip, which is known to exhibit moderate dispersion. Further, in interconnects, it is not possible to assume that the lines are *lossless*. The width of the interconnect lines and substrate thickness are kept quite small in order to keep a high package density. Because the conductor loss of a microstrip line is nearly inversely proportional to the substrate thickness, it can be quite significant for interconnect lines. A short rise-time pulse may be assumed to contain high-frequency components of relatively large amplitudes. Because the loss of a line is greater at higher frequencies, it is expected that a short rise-time pulse is affected more by the loss in the circuit, and the distortion increases with an increase in the length of the line.

12.2 Analysis of Coupled Lossy Interconnect Lines

We now discuss the analysis of coupled lossy interconnect lines in a typical high-speed digital circuit package. Because a digital signal has a pulse form, it is preferable to solve the problem directly in the time domain. The analysis of lossy lines is convenient, however, only in the frequency domain [4]. The most common procedure adopted for the analysis of interconnects is therefore to first convert the input time-domain signal into the frequency domain using Fourier transforms. The response of the circuit is then determined in the frequency domain and converted back into the time domain using inverse Fourier transforms. The evaluation of Fourier and inverse Fourier transforms can be facilitated by using fast Fourier transform (FFT) procedures.

Figure 12.1 shows a two-port circuit where a time-domain signal $X(t)$ is applied at the input. If $X(\omega)$ denotes the Fourier transform of the input signal $X(t)$, then

$$Y(\omega) = H(\omega) X(\omega) \tag{12.1}$$

where $Y(\omega)$ denotes the Fourier transform of the output signal and $H(\omega)$ denotes the transfer function of the two-port network. The output signal in the

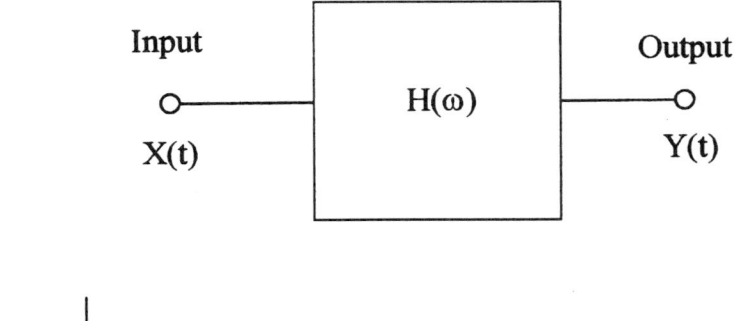

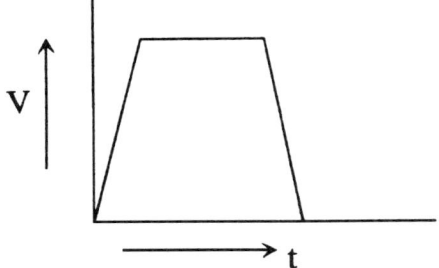

Figure 12.1 A two-port network of known transfer function $H(\omega)$.

time domain is then given by

$$Y(t) = F^{-1}[Y(\omega)] = F^{-1}[H(\omega)X(\omega)] \qquad (12.2)$$

where F^{-1} denotes the inverse Fourier transform. The input pulse is assumed to have finite rise and fall times. This is a good approximation of a typical digital pulse encountered in practice. For such an input signal, the Fourier transform is easily determined. Therefore, the main task in finding the response is to determine the transfer function $H(\omega)$.

In a typical interconnect, a large number of transmission lines run parallel to each other. The rigorous analysis of interconnect lines should be carried out using multiconductor transmission line theory [4]. The use of multiconductor transmission line theory, however, may prove to be quite complicated. As a first-order approximation, it is sufficient to consider the model of two parallel-coupled lines to determine the transfer function $H(\omega)$.

An interconnect coupled-line circuit differs from a coupled-line microwave circuit in its manner of termination. In a microwave circuit, the source generally has a fixed impedance of 50Ω. The impedances of the ports of the network and load impedances are also chosen to be 50Ω to ensure maximum power transfer from the source. High-speed interconnects, however, require that the voltage output of the driver (source) be available at the receiver (load) connected at the other end of the transmission line with minimal attenuation. Therefore, terminations of high-speed interconnect lines are generally different from those used in microwave circuits.

It is much simpler to analyze the performance of interconnects using their Z- or Y-parameters rather than S-parameters. This is so because the ports of an interconnect circuit are generally not match-terminated. Figure 12.2 shows a four-port network formed using two parallel-coupled asymmetrical lines. It is assumed that a voltage source V_g in series with an impedance Z_1 is connected to port 1. Ports 2, 3, and 4 of the network are assumed to be terminated in load impedances Z_2, Z_3, and Z_4, respectively. It is required to determine the terminal currents and voltages at the various ports of the circuit. The terminal voltages and currents are related by the Z-parameters of the circuit as follows:

$$\begin{aligned} V_1 &= Z_{11}I_1 + Z_{12}I_2 + Z_{13}I_3 + Z_{14}I_4 \\ V_2 &= Z_{21}I_1 + Z_{22}I_2 + Z_{23}I_3 + Z_{24}I_4 \\ V_3 &= Z_{31}I_1 + Z_{32}I_2 + Z_{33}I_3 + Z_{34}I_4 \\ V_4 &= Z_{41}I_1 + Z_{42}I_2 + Z_{43}I_3 + Z_{44}I_4 \end{aligned} \qquad (12.3)$$

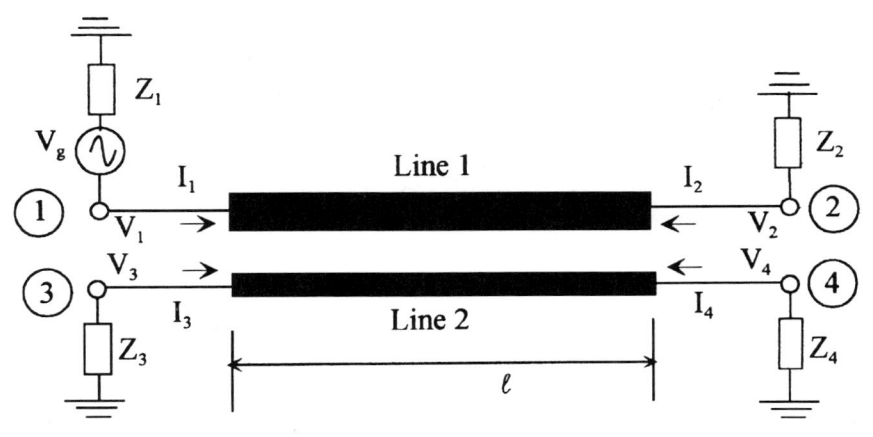

Figure 12.2 A four-port network composed of parallel coupled asymmetrical lines.

Further, the terminal voltages and currents are related by

$$V_1 = V_g - I_1 Z_1$$
$$V_2 = -I_2 Z_2$$
$$V_3 = -I_3 Z_3$$
$$V_4 = -I_4 Z_4$$
(12.4)

Substituting the boundary conditions of (12.4) in (12.3) gives four simultaneous equations in four unknowns I_1, I_2, I_3, and I_4, which can be solved. The values of V_1, V_2, V_3, and V_4 can then be found using (12.4). Depending on the input/output quantities of interest, the appropriate transfer function can then be found (e.g., the transfer function between the output voltage at port 3 and the source voltage is given by $H(\omega) = V_3/V_g$).

The Z- and Y-parameters of a four-port network composed of asymmetrical parallel-coupled lines were given in terms of characteristic impedances and propagation constants of normal (c- and π-) modes in Chapter 4 (Section 4.3.1). In a lossless case, the characteristic impedance of a transmission line is real and its propagation constant γ purely imaginary (i.e., $\gamma = j\beta$, where β is a real quantity). However, the characteristic impedance and propagation constant become complex (with both non-zero real and imaginary parts) if the losses of the transmission lines are also considered. When the frequency-dependent complex characteristic impedances and propagation constants of the normal modes are known, these can be substituted directly in (4.72) to yield the complex Z-parameters. In most situations, the effect of loss on the characteristic impedance is relatively small, and only a small error is incurred if the characteristic impedance is assumed to be real, even in the lossy case. Both the attenuation and phase constant, however, should be taken into account for the complex propagation constant γ. When the line parameters (i.e., [R], [L], [C] and [G] matrices) of lossy coupled lines are known, the complex characteristic impedances and propagation constants can be determined using (4.86) through (4.98) where the self- and mutual impedance parameters z_1, z_2, and z_m are given by,

$$\begin{bmatrix} z_1 & z_m \\ z_m & z_2 \end{bmatrix} = \begin{bmatrix} R_{11} & R_{12} \\ R_{21} & R_{22} \end{bmatrix} + j\omega \begin{bmatrix} L_{11} & L_{12} \\ L_{12} & L_{22} \end{bmatrix} \qquad (12.5)$$

In (12.5), R_{ij} and L_{ij} denote the elements of [R] and [L] matrices, respectively. Similarly, the self- and mutual admittance parameters y_1, y_2, and y_m are given by

$$\begin{bmatrix} y_1 & y_m \\ y_m & y_2 \end{bmatrix} = \begin{bmatrix} G_{11} & G_{12} \\ G_{21} & G_{22} \end{bmatrix} + j\omega \begin{bmatrix} C_{11} & C_{12} \\ C_{12} & C_{22} \end{bmatrix} \qquad (12.6)$$

where G_{ij} and C_{ij} denote the elements of [G] and [C] matrices, respectively. The capacitance matrix of coupled lines was defined in Chapter 3 (Section 3.2). Note that in general, C_{12} is a negative quantity and is usually expressed as $C_{12} = -C_m$, where C_m is a positive quantity.

12.2.1 Frequency Dependence of Line Parameters

A pulse signal may be assumed to contain a broad range of frequencies. It is therefore important to characterize the interconnect lines over a broad frequency range to determine the time-domain response. We now discuss the frequency dependence of line parameters of a single transmission line. Similar rules of frequency dependence can apply to coupled lines.

The series resistance R of a transmission line is representative of the losses in the imperfect conductors of the transmission line (the relation between R and conductor loss is given by (3.11). For normal metals at low frequencies, the current in the conductor can be assumed to be uniform in the cross section of the conductor. Therefore, at these frequencies, the resistance R can be determined from a knowledge of the dc conductivity and cross section of the conductor. At high frequencies where the skin effect is completely onset (when the skin depth becomes much smaller than the thickness of the conductors), the resistance R increases as the square root of the frequency. The resistance R can then be determined at any frequency if its value at a reference frequency is known. In the intermediate frequency region, however, the frequency dependence of the resistance R becomes very complicated and depends on many factors, such as the cross section and shape factor of the conductor. It can only be determined by using numerical methods [5–9]. Figure 12.3 shows a conductor of rectangular cross section. The dc resistance per-unit length of the rectangular bar is given by

$$R_{dc} = \frac{1}{\sigma Wt} \qquad (12.7)$$

where σ denotes the conductivity of the bar material in S/m. As the frequency increases, the resistance increases due to the skin effect. The ratio of the ac resistance to the dc resistance of the bar has been plotted in Figure 12.4 for various aspect ratios ($AR = W/t$) as a function of a parameter p, which is defined as follows [7]:

$$p = \frac{1}{\delta}\sqrt{\frac{2}{\pi}Wt} \qquad (12.8)$$

In the (12.8), δ is the skin depth given by (3.29). The ac resistance of a conductor of circular cross section is also shown in Figure 12.4 (curve V) for comparison.

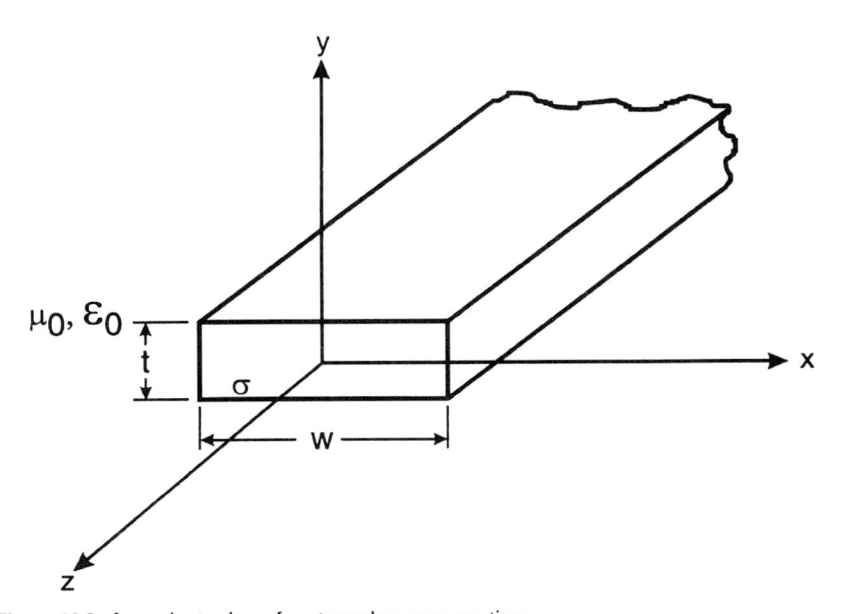

Figure 12.3 A conductor bar of rectangular cross section.

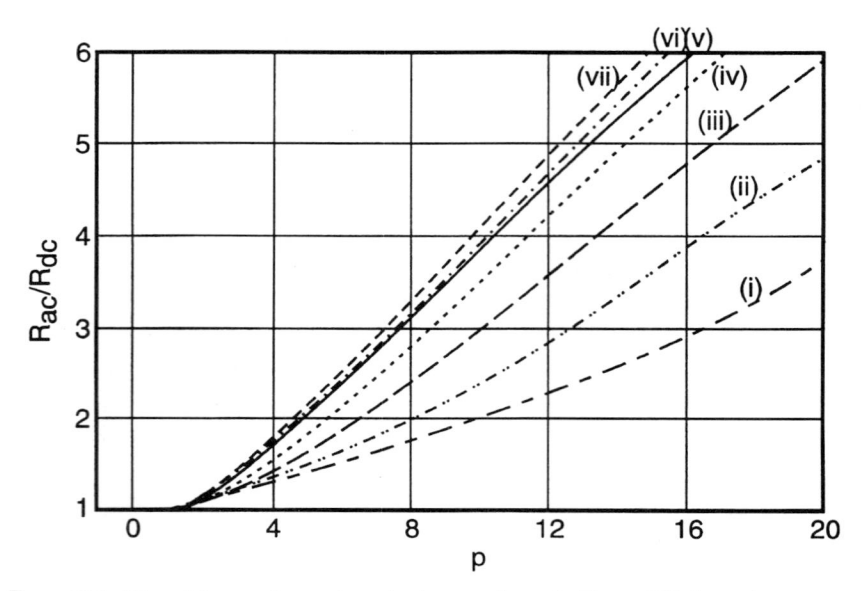

Figure 12.4 AC resistance of a rectangular bar as shown in Figure 12.3 for various aspect ratios. (i) $W/t = 32$, (ii) $W/t = 16$, (iii) $W/t = 8$, (iv) $W/t = 4$, (v) Circle, (vi) $W/t = 2$, and (vii) $W/t = 1$. (From [7], © 1997 IEE. Reprinted with permission.)

The inductance of a transmission line is a consequence of the magnetic field linked with the conductor when the current flows through it and can be assumed to consist of two parts: one resulting from the magnetic field that exists inside the conductor and the other resulting from the magnetic field outside the conductor [5, 6]. These two inductances can be termed as L_{int} and L_{ext}, respectively. The total inductance can then be expressed as $L = L_{int} + L_{ext}$. At low frequencies, the contribution of the internal inductance L_{int} to the total inductance L is generally not negligible. As the operating frequency increases, the current in the conductor gets increasingly confined near the surface of the conductor and causes the magnetic field inside the conductor to decrease. Therefore, the internal inductance decreases as the frequency increases. On the other hand, the external inductance of a transmission line is nearly independent of frequency. At sufficiently high frequencies, the internal inductance becomes very small compared with the external inductance, such that the total inductance is nearly equal to the external inductance. Note that the external inductance of a transmission line is given by

$$L_{ext} = \frac{1}{c^2 C_0} \tag{12.9}$$

where c denotes the velocity of light in free space and C_0 denotes the free-space capacitance of the transmission line (i.e., when the dielectric constant of the material surrounding the transmission line is unity). Because at very high frequencies, the total inductance is nearly the same as the external inductance, the quantity L_{ext} can be replaced by the total inductance L in (12.9). When the skin effect is completely onset, the reactance from the internal inductance has the same value in magnitude as the series resistance. Because the series resistance R increases as the square root of the frequency in the skin effect region, the internal inductance varies as the inverse of the square root of frequency (inductance = reactance/ω). Note that for the lossless case, the skin depth is zero at any frequency. In that case, the internal inductance is zero, and the total inductance is the same as the external inductance.

The variation of the capacitance of a transmission line with frequency depends on the particulars of the structure. For a transmission line in a homogeneous dielectric material, the capacitance varies with frequency only if the dielectric constant of the material varies with frequency. For a transmission line placed in an inhomogeneous medium, the capacitance varies with frequency even if the dielectric constant is independent of frequency. In an interconnect environment, however, the variation of the capacitance with frequency can generally be neglected.

There appears to be no fixed rule to describe the frequency dependence of the conductance G of a transmission line. The conductance G results from losses in the dielectric material in which the transmission line is placed. The relation between G and the dielectric loss is given by (3.12). The frequency dependence of G varies from material to material (even for the same material, G may vary widely depending on the processing). Usually in an interconnect environment, the dielectric loss will be much smaller than the conductor loss if the loss tangent of the dielectric material is 0.005 or smaller. In that case, the effect of G on the total loss can be neglected.

It may be useful to remark here that the conductance G can be readily computed with the same technique used to determine the capacitance C [10]. A lossy dielectric material can be characterized by a complex dielectric constant $\epsilon_r = \epsilon'_r(1 - j\tan\delta)$. With the capacitance matrix of a lossy dielectric denoted by $\hat{C}$:

$$G = Re(j\omega\hat{C}) \tag{12.10}$$

where $Re(j\omega\hat{C})$ denotes the real part of the complex quantity $j\omega\hat{C}$.

Equation 12.10 can also be used for multiconductor transmission lines if the conductance and capacitance are replaced by conductance and capacitance matrices respectively.

It has been reported that if the conductor and dielectric losses of a transmission line are assumed to be independent of frequency, the inductance and capacitance of the transmission line can also be assumed to be independent of frequency when determining the time-domain response [11, 12]. If the frequency dependence of the loss is considered, however, it is essential to consider the proper frequency dependence of other line parameters as well. For example, when the skin effect is completely onset, the series resistance R is proportional to the square root of the frequency. In that case, although the internal inductance is very small compared with the external inductance, the internal inductance and its proper frequency dependence should be taken into account to determine the total inductance. A noncausal response results (the output appears before the input is switched on) if the proper frequency dependence is not taken into account. The capacitance of a transmission line can be assumed to be independent of frequency. If the frequency dependence of the dielectric loss tangent is considered, however, its proper frequency variation should be taken into account. A causal response results if it is assumed that the loss tangent increases linearly with frequency.

12.2.2 Results

We now discuss some typical numerical results reported in the literature of the various effects discussed above [2]. For the sake of simplicity, coupling between two symmetrical lines is considered as shown in Figure 12.5(a). A voltage source is connected to one of the lines. The line to which the voltage source is connected is called the "active" line, while the other line is called the "quiet" line. The shape of the input pulse is depicted in Figure 12.5(b).

We assume that the lines are either of microstrip or stripline type. The coupled microstrip and striplines are shown in Figures 12.6(a) and (b), respectively. Computations were made for the following parameters of microstrip and striplines:

Case A:

Coupled microstrip lines $W = 10$ μm, $h = 10$ μm, $S = 20$ μm, $t = 5$ μm, $\epsilon_{r1} = 1.0$, and $\epsilon_{r2} = 3.5$.

Case B:

All the parameters are the same as in Case A except that $S = 10$ μm.

Case C:

Coupled striplines $W = 10$ μm, $h_1 = h_2 = 10$ μm, $S = 20$ μm, $t = 5$ μm, $\epsilon_{r1} = 1.0$, and $\epsilon_{r2} = 3.5$.

For all cases, the inductance and capacitance matrices of the coupled lines (assuming no loss) were computed using a numerical method based on the

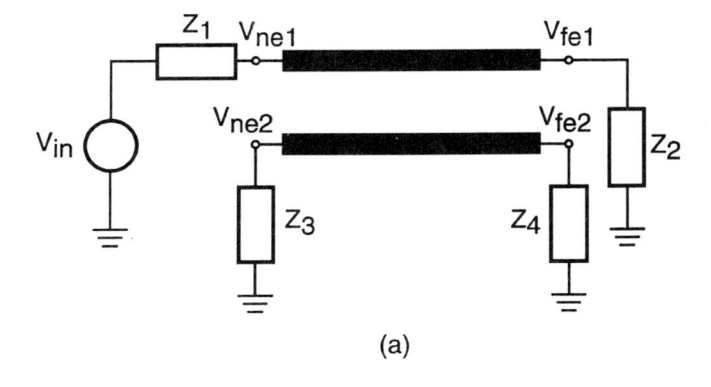

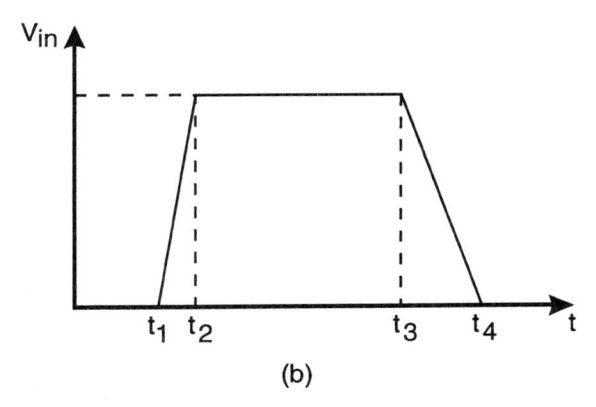

Figure 12.5 (a) Coupled interconnect lines with one active and one quiet line, (b) input pulse waveform. (From [2], © IEEE. Reprinted with permission.)

method of moments and the TEM mode assumption. For case A, the inductance and capacitance matrices are given by

$$[\mathbf{L}] = \begin{bmatrix} L_{11} & L_{12} \\ L_{12} & L_{22} \end{bmatrix} = \begin{bmatrix} 3.4714 & 0.49054 \\ 0.49054 & 3.4714 \end{bmatrix} \text{ nH/cm} \quad (12.11)$$

$$[\mathbf{C}] = \begin{bmatrix} C_{11} & C_{12} \\ C_{12} & C_{22} \end{bmatrix} = \begin{bmatrix} 0.72715 & -0.04814 \\ -0.04814 & 0.72715 \end{bmatrix} \text{ pF/cm} \quad (12.12)$$

As already explained, the inductance and resistance parameters of transmission lines are frequency dependent because of the skin effect. The inductance and resistance parameters of the coupled microstrip lines for Case A are shown

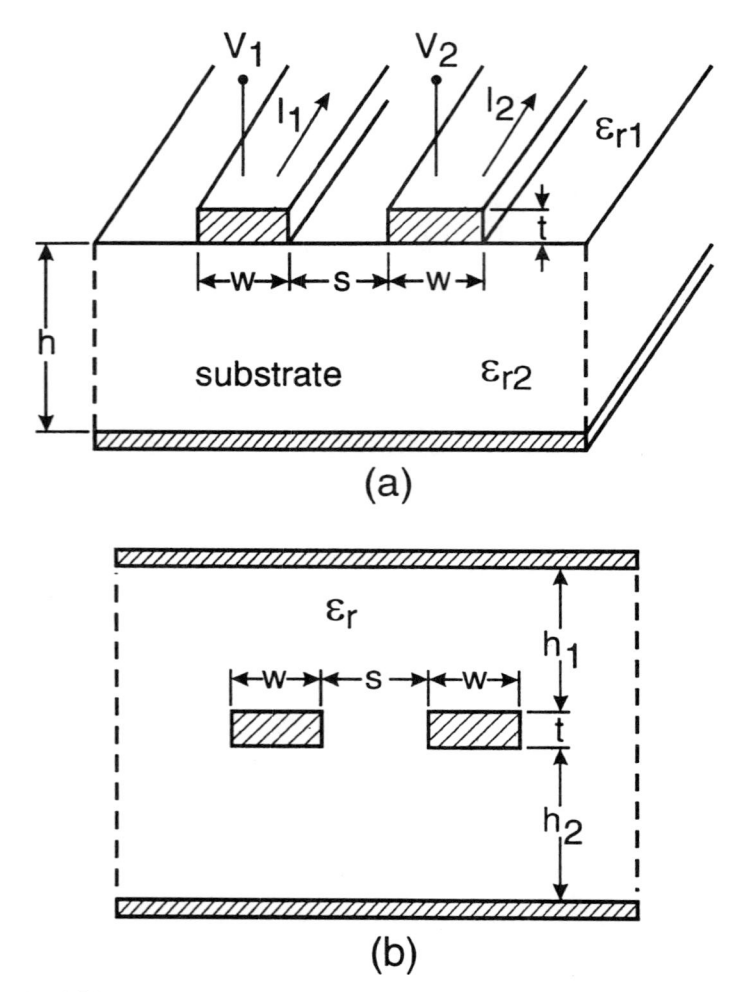

Figure 12.6 (a) Coupled microstrip lines and (b) Cross section of coupled strip lines. (From [2], © IEEE. Reprinted with permission.)

in Table 12.1 at various frequencies assuming that the conductors are made of copper. We see that the total inductance decreases as the frequency increases. This is due to a decrease in the internal inductance with an increase in frequency. Further, as the frequency becomes high, the inductance for the lossy case approaches the lossless value. The dc resistance of the lines (R_{dc}) is 3.4 Ω/cm. This also represents the resistance of the line at very low frequencies. The resistance increases significantly at microwave frequencies because of the skin effect. At high frequencies, the resistance increases as the square root of frequency as expected.

Table 12.1
L(f) and R(f) for microstrip line structure of case A
W = 10 μm, t = 5 μm, S = 20 μm, h = 10 μm,
ϵ_{r1} = 1.0, ϵ_{r2} = 3.5, C_{11} = C_{22} = 0.72715 pF/cm,
C_{12} = C_{21} = −0.04814 pF/cm

Freq (GHz)	$L_{11} = L_{22}$ (μH/m)	$L_{12} = L_{21}$ (μH/m)	$R_{11} = R_{22}$ (kΩ/m)	Skin depth (μm)
0.0	0.429 35	0.054 004	0.340 00	—
0.1	0.416 30	0.053 826	0.340 11	6.5637
0.5	0.407 39	0.053 668	0.354.83	2.9354
1.0	0.404 01	0.053 221	0.385 70	2.0756
3.0	0.387 58	0.052 430	0.563 94	1.1984
5.0	0.378 76	0.052 231	0.716 19	0.9283
7.0	0.372 72	0.052 229	0.845 76	0.7845
9.0	0.368 63	0.052 120	0.979 89	0.6919
10.0	0.36717	0.052198	1.04120	0.6564

For Case B, the inductance and capacitance matrices of the coupled lines (assuming no loss) are given by

$$[\mathbf{L}] = \begin{bmatrix} L_{11} & L_{12} \\ L_{12} & L_{22} \end{bmatrix} = \begin{bmatrix} 3.4171 & 0.91166 \\ 0.91166 & 3.4171 \end{bmatrix} \text{nH/cm} \quad (12.13)$$

$$[\mathbf{C}] = \begin{bmatrix} C_{11} & C_{12} \\ C_{12} & C_{22} \end{bmatrix} = \begin{bmatrix} 0.75465 & -0.12147 \\ -0.12147 & 0.75465 \end{bmatrix} \text{pF/cm} \quad (12.14)$$

For Case C, the inductance and capacitance matrices of the coupled striplines (assuming no loss) are

$$[\mathbf{L}] = \begin{bmatrix} L_{11} & L_{12} \\ L_{12} & L_{22} \end{bmatrix} = \begin{bmatrix} 2.6468 & 0.10338 \\ 0.10338 & 2.6468 \end{bmatrix} \text{nH/cm} \quad (12.15)$$

$$[\mathbf{C}] = \begin{bmatrix} C_{11} & C_{12} \\ C_{12} & C_{22} \end{bmatrix} = \begin{bmatrix} 1.4718 & -0.06051 \\ -0.06051 & 1.4718 \end{bmatrix} \text{pF/cm} \quad (12.16)$$

We first discuss the effect of conductor loss on the voltages at the ends of the active and quiet lines of the structure of Case A. The load impedances are assumed to be $Z_1 = 0$, $Z_2 = R_0 \parallel C_L$ (R_0 in parallel with C_L), and $Z_3 = Z_4 = R_0$. C_L denotes the typical capacitance of a receiver and its value was assumed to be 0.1 pF. R_0 is approximately equal to the characteristic impedance Z_0 of either line (assuming lossless lines) and its value was determined as follows:

$$R_0 = \sqrt{\frac{L_{11}}{C_{11}}} \qquad (12.17)$$

The characteristic impedance for cases A, B, and C were determined to be 69.1 Ω, 67.3 Ω, and 42.3 Ω, respectively.

The transient response at the near and far ends of the quiet and active lines are shown in Figures 12.7(a) to (c). The length of the microstrip lines was assumed to be 6 cm. V_{ne1} and V_{fe1} denote the voltages at the near end (end nearest to the voltage source) and far end, respectively, of the active line. Similarly, V_{ne2} and V_{fe2} denote the voltages at the near and far ends, respectively, of the quiet line. The results correspond to three different cases of loss. Those in Figure 12.7(a) are based on the assumption that the transmission line loss is zero; in Figure 12.7(b) on the assumption that the loss resistance of the line is independent of frequency and is the same as its dc value; while Figure 12.7(c) shows the results for the case of skin effect loss of lines having $R_{dc} = 3.4$ Ω/cm. We see that the line loss has a significant effect on the amplitude and shape of the voltage waveform at the far end of the active line. The rising edge of the input pulse is rounded when the skin effect losses are considered. This is due to the fact that the sharp rising edge of the pulse is associated with high-frequency components of relatively large amplitude and the high frequency components are increasingly attenuated because of the skin effect.

Figures 12.8(a) and (b) show the maximum near- and far-end crosstalk voltages (V_{ne2} and V_{fe2}, respectively) as a function of line length for various cases of loss for the structure and dimensions of Case A. The terminations of the lines are assumed to be $Z_1 = 0$, $Z_2 = R_0 \parallel C_L$ (R_0 in parallel with C_L), and $Z_3 = Z_4 = R_0$. The peak amplitude of the input pulse is assumed to be 1V and the rise and fall times are 50 ps. The results are shown for the no-loss case, for a frequency-independent line resistance of $R_{dc} = 3.4$ Ω/cm, $R_{dc} = 8.6$ Ω/cm, and frequency-dependent loss of lines having a dc resistance of $R_{dc} = 3.4$ Ω/cm. The value of $R_{dc} = 8.6$ Ω/cm assumed corresponds to the per-unit length dc resistance of aluminum lines. These results show that the maximum voltage induced at the near end of a quiet line (near-end crosstalk voltage) is quite small and will probably not cause false switching. The far-end crosstalk voltage, however, can be quite large in amplitude. For the lossless case, the far-end crosstalk voltage increases with an increase in the length of the line. For lossy lines, which is the more practical case, the far-end crosstalk voltage increases with an increase in line length up to a certain length, beyond which it starts to reduce.

High-Speed Circuit Interconnects 461

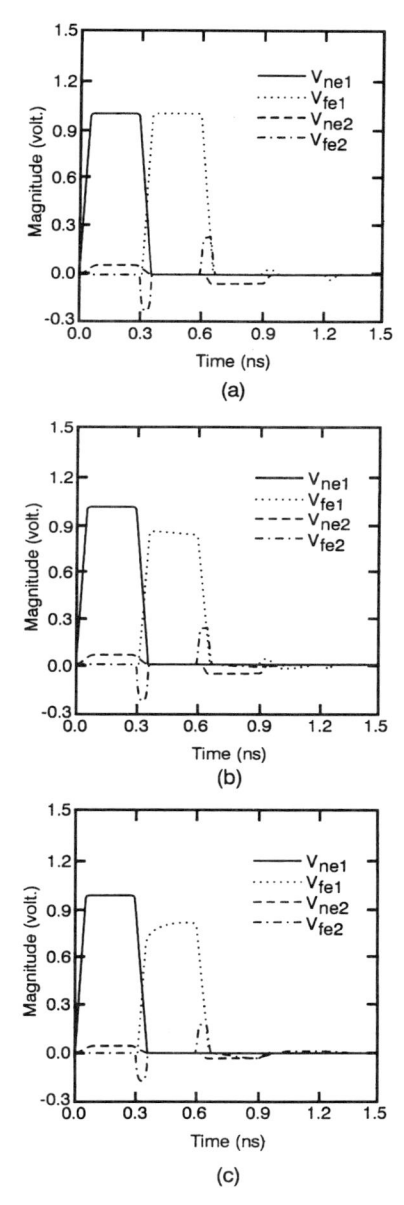

Figure 12.7 (a) Transient response of lossless coupled lines with a length of 6 cm. (b) Transient response of coupled lines with a dc loss of $R_{dc} = 3.4\ \Omega/\text{cm}$. (c) Transient response of coupled lines with skin effect loss (for copper conductors with $R_{dc} = 3.4\ \Omega/\text{cm}$). (From [2], © IEEE. Reprinted with permission.)

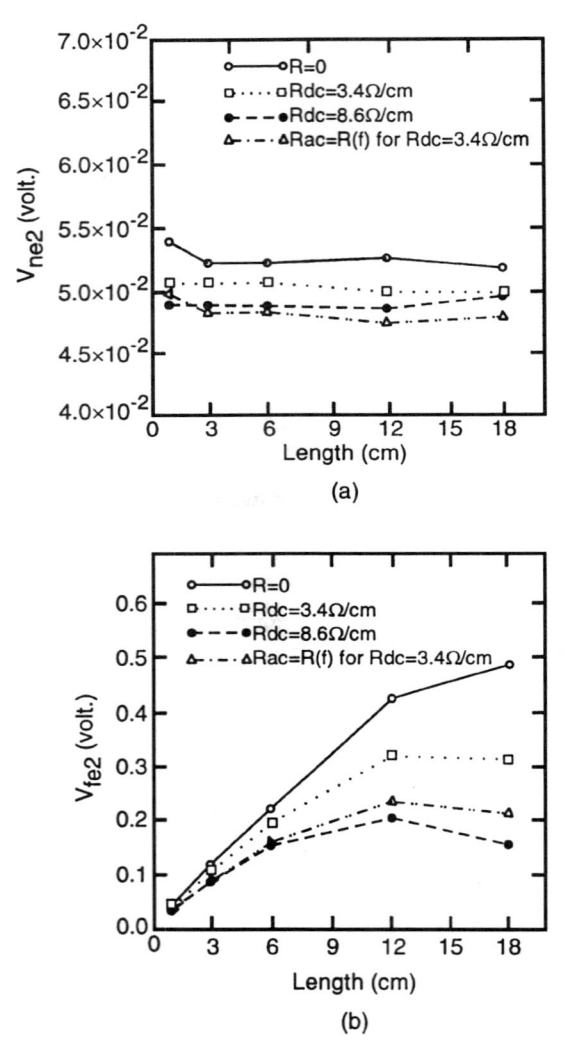

Figure 12.8 (a) Near-end (V_{ne2}) and (b) far-end crosstalk voltage (V_{fe2}) of structure of case A for various line losses. $Z_1 = 0$, $Z_2 = R_0 \parallel C_L$ (R_0 in parallel with C_L) and $Z_3 = Z_4 = R_0$. (From [2], © 1993 IEEE. Reprinted with permission.)

Figures 12.9(a) and (b) show the effect of spacing between coupled microstrip lines on the crosstalk (Case A versus Case B). The results show that, as expected, the crosstalk increases as the spacing between the lines decreases. The near-end crosstalk is nearly doubled when the spacing between the microstrip lines is reduced from $S = 20\ \mu$m to $S = 10\ \mu$m. We see from Figure 12.9(b), however, that the effect of loss on the far-end crosstalk is more dominant than the effect of spacing between the lines. For example, for a line length of greater than 9 cm, V_{fe2} of $R_{dc} = 3.4\ \Omega$/cm and $S = 10\ \mu$m is smaller than V_{fe2} of $R_{dc} = 0$ and $S = 20\ \mu$m.

Figures 12.10(a) and (b) show the effect of using striplines (Case C) in place of microstrip lines (Case A). We find that both the near- and far-end voltages are reduced by using the stripline geometry. The far-end voltage is nearly reduced to zero. (A small voltage is probably due to a numerical error.) This is an expected result because as discussed in Chapter 4 (Section 4.2) in TEM coupled lines such as striplines, there is no wave coupled in the forward direction because of equal even- and odd-mode phase velocities.

The crosstalk voltages shown in Figures 12.8 to 12.10 were determined for a specific case of termination in which the far end of the active line was terminated by a parallel load of R_0 and C_L. Because R_0 is nearly equal to the characteristic impedance of either line and the value of C_L is quite small, any wave reflected at the far end of the active line is quite small in magnitude. The effect of terminating the far end of an active line by a capacitance C_L only gives the results in Figures 12.11 to 12.13. Here, all the parameters are assumed to be the same as those considered in deriving the results shown in Figures 12.8 to 12.10, respectively, except for the load impedance at the far end of the active line. Because the load impedance at the far end of the active line is now very different from the characteristic impedance of the line, a reflected wave of large amplitude is set up on the active line. The reflected wave on the active line causes the near-end voltage on the quiet line (V_{ne2}) to increase significantly. This is mainly due to the secondary far-end crosstalk voltage created by the reflected wave. The near-end crosstalk voltage is maximum when the length of the line is about 3 cm. For this line length, the line delay is approximately equal to one-half of the pulsewidth. Results show, however, that the magnitude of the near-end voltage is still considerably smaller than that of the far-end crosstalk. For line lengths greater than 6 cm, it does not make much difference, in terms of far-end crosstalk voltage, whether the active line is properly terminated or not. For line lengths of less than about 3 cm, the termination of the far end of the active line by R_0 reduces the far-end crosstalk voltage V_{fe2}. The response of the voltage at the far end of the active line is shown in Figure 12.14 as a function of time. It is seen that ringing exists on the active line when it is about 1 cm long.

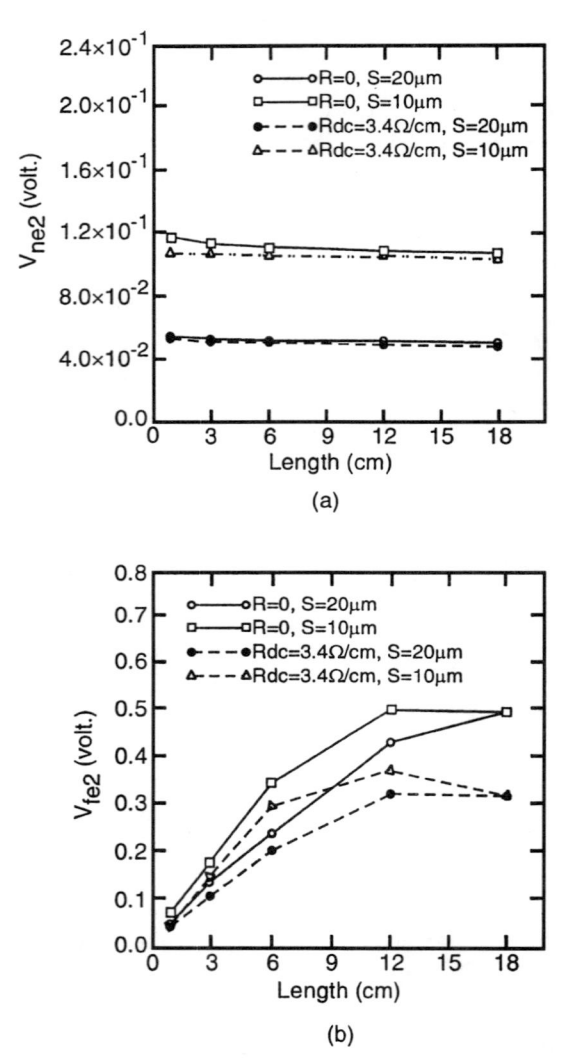

Figure 12.9 (a) Near-end (V_{ne2}) and (b) far-end crosstalk voltage (V_{fe2}) for coupled microstrip lines of different spacings (corresponding to case A and case B respectively). $Z_1 = 0$, $Z_2 = R_0 \parallel C_L$ (R_0 in parallel with C_L) and $Z_3 = Z_4 = R_0$. (From [2], © 1993 IEEE. Reprinted with permission.)

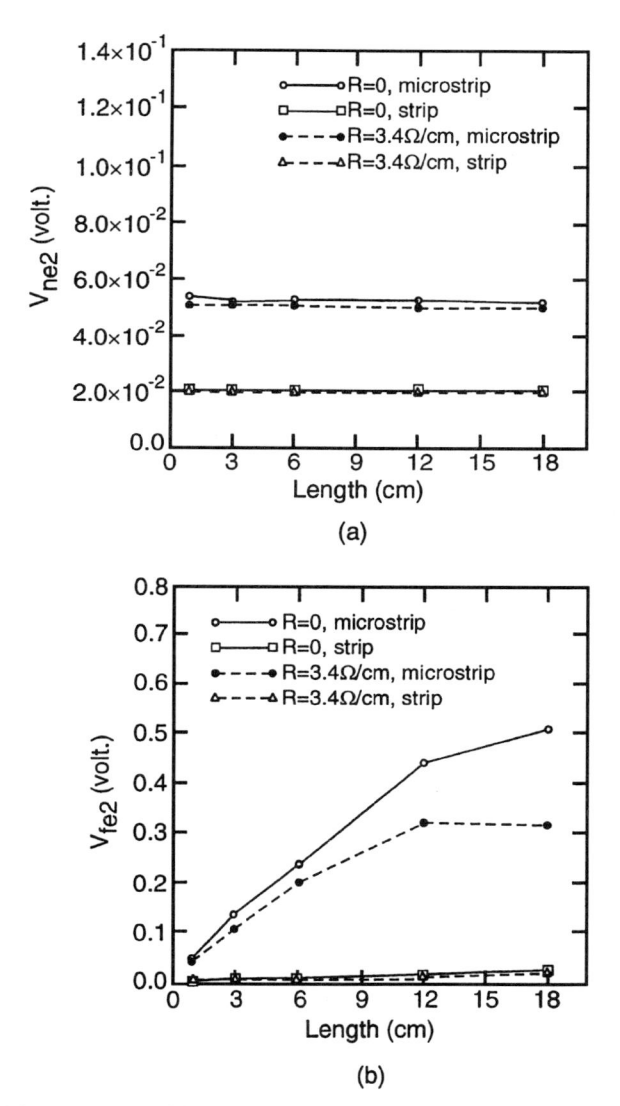

Figure 12.10 (a) Near-end and (b) far-end crosstalk voltages for microstrip and strip line configurations (corresponding to case A and case C respectively). $Z_1 = 0$, $Z_2 = R_0 \parallel C_L$ (R_0 in parallel with C_L) and $Z_3 = Z_4 = R_0$. (From [2], © 1993 IEEE. Reprinted with permission.)

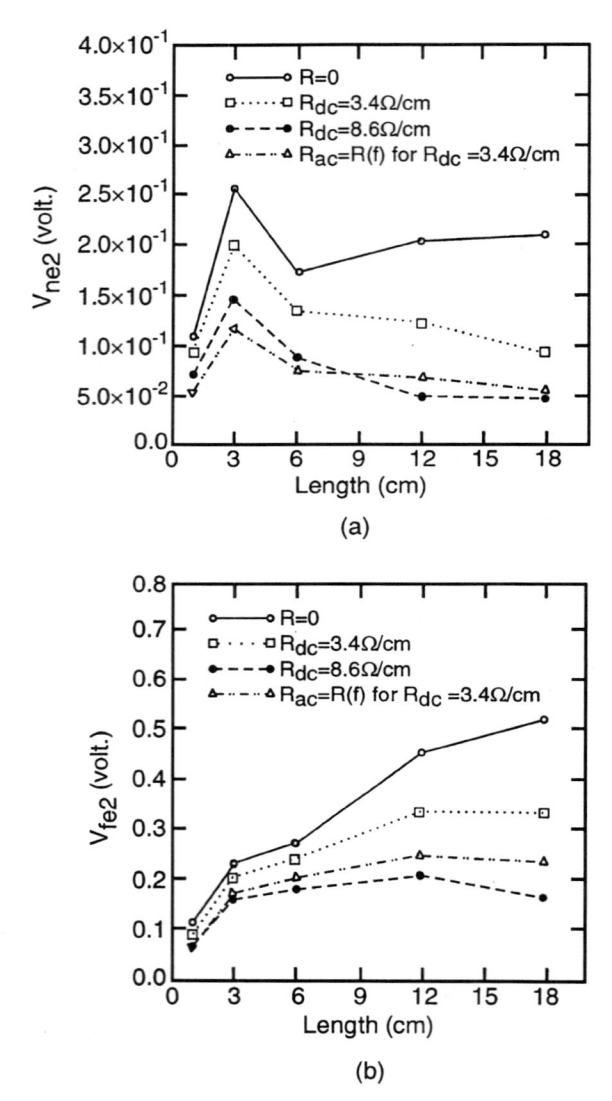

Figure 12.11 (a) Near-end and (b) far-end crosstalk voltage for structure of case A for various line losses. $Z_1 = 0$, $Z_2 = C_L$, and $Z_3 = Z_4 = R_0$. (From [2], © 1993 IEEE. Reprinted with permission.)

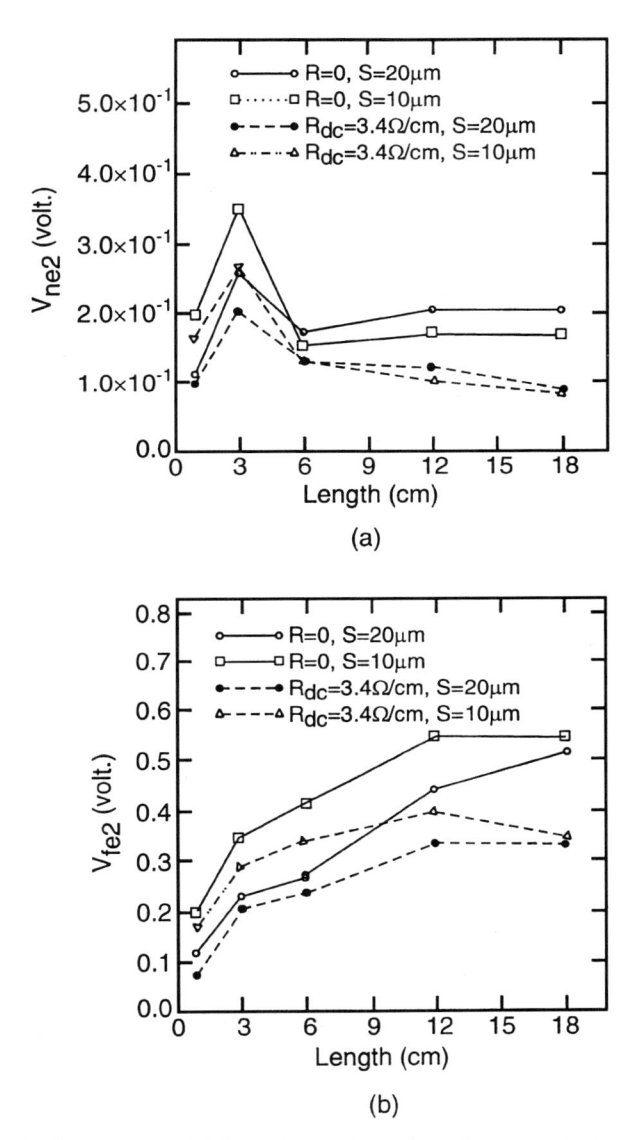

Figure 12.12 (a) Near-end and (b) far-end crosstalk voltage for coupled microstrip lines of different spacings (corresponding to case A and case B respectively). $Z_1 = 0$, $Z_2 = C_L$, and $Z_3 = Z_4 = R_0$. (From [2], © 1993 IEEE. Reprinted with permission.)

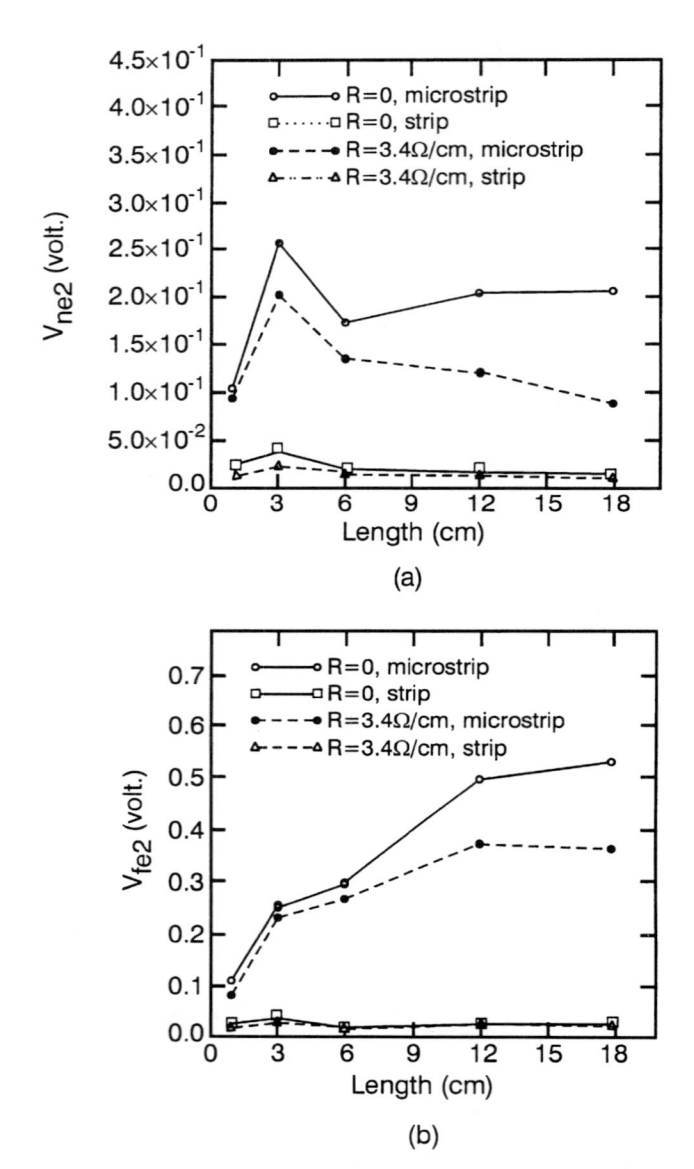

Figure 12.13 (a) Near-end and (b) far-end crosstalk voltages for microstrip and strip line configurations (corresponding to case A and case C respectively). $Z_1 = 0$, $Z_2 = C_L$ and $Z_3 = Z_4 = R_0$. (From [2], © 1993 IEEE. Reprinted with permission.)

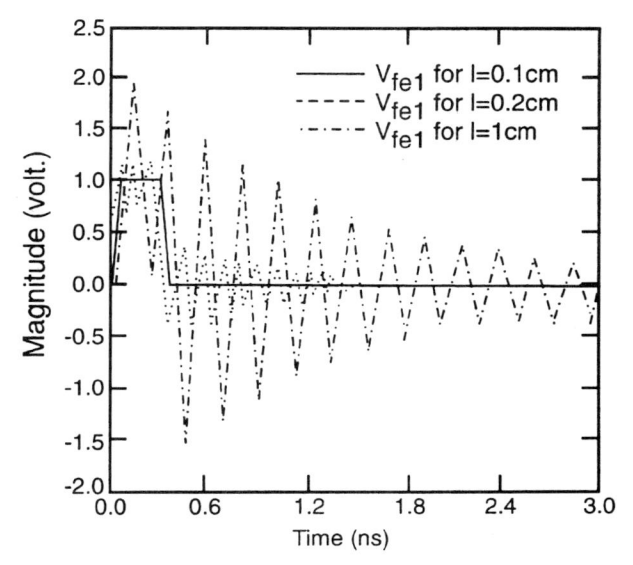

Figure 12.14 Far-end voltage (V_{fe1}) on the active line for various line lengths of the structure of case A. $Z_1 = 0$, $Z_2 = C_L$ and $Z_3 = Z_4 = R_0$. (From [2], © 1993 IEEE. Reprinted with permission.)

In Figures 12.15(a) and (b), the effect of terminating the quiet line in nonmatching impedances is compared with the case when the line is nearly match terminated. Both the active and quiet lines are assumed to have a loss resistance of 3.4 Ω/cm $Z_1 = 0$, $Z_2 = R_0 \parallel C_L$. We see from Figure 12.15(a) that the near-end crosstalk voltage V_{ne2} is largest for the case when $Z_3 = C_L$ and $Z_4 = 10\Omega$. The near-end crosstalk voltage is large because the far-end crosstalk is reflected back and adds to the near-end crosstalk. From Figure 12.15(b), we see that the far-end crosstalk voltage V_{fe2} is maximum for the case when $Z_3 = 10\Omega$ and $Z_4 = C_L$. V_{fe2} could be as large as the signal on the far end of the active line for line lengths longer than about 12 cm. Therefore, when the interconnect lines are relatively long, proper care should be taken in their termination.

Figure 12.16 compares the effect of matching the active line at its near end compared with when it is match-terminated at its far end. More specifically, the comparison is made between two terminal conditions: $Z_1 = 0$, $Z_2 = R_0 \parallel C_L$ (R_0 in parallel with C_L), and $Z_3 = Z_4 = R_0$ versus $Z_1 = R_0$, $Z_2 = C_L$, and $Z_3 = Z_4 = R_0$. In both cases, both lines are assumed to be lossy with $R_{dc} = 3.4\Omega$. Figure 12.16(b) shows that the far-end crosstalk (which is the predominant component of the crosstalk), is smaller in the latter case (i.e., when a series impedance is used at the near end of the active line). We also see from Figure 12.16(a), however, that the near-end crosstalk also increases when the line length is more than

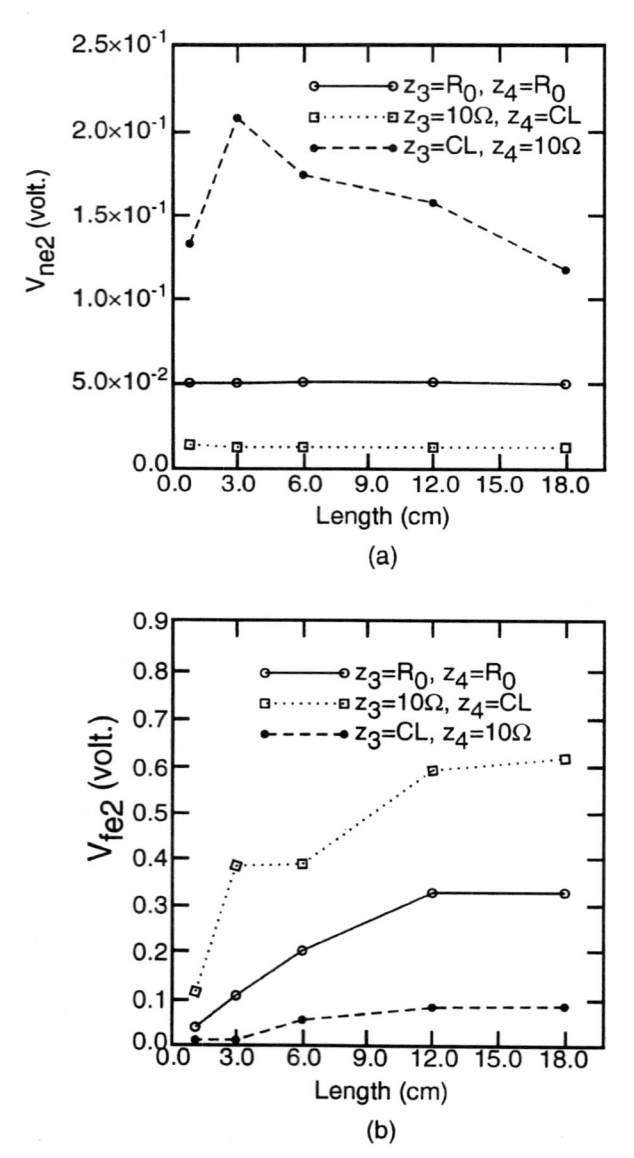

Figure 12.15 (a) Near-end and (b) far-end crosstalk voltage for the structure of case A for various terminations of the quiet line. $Z_1 = 0$ and $Z_2 = R_0 \parallel C_L$ (R_0 in parallel with C_L). (From [2], © 1993 IEEE. Reprinted with permission.)

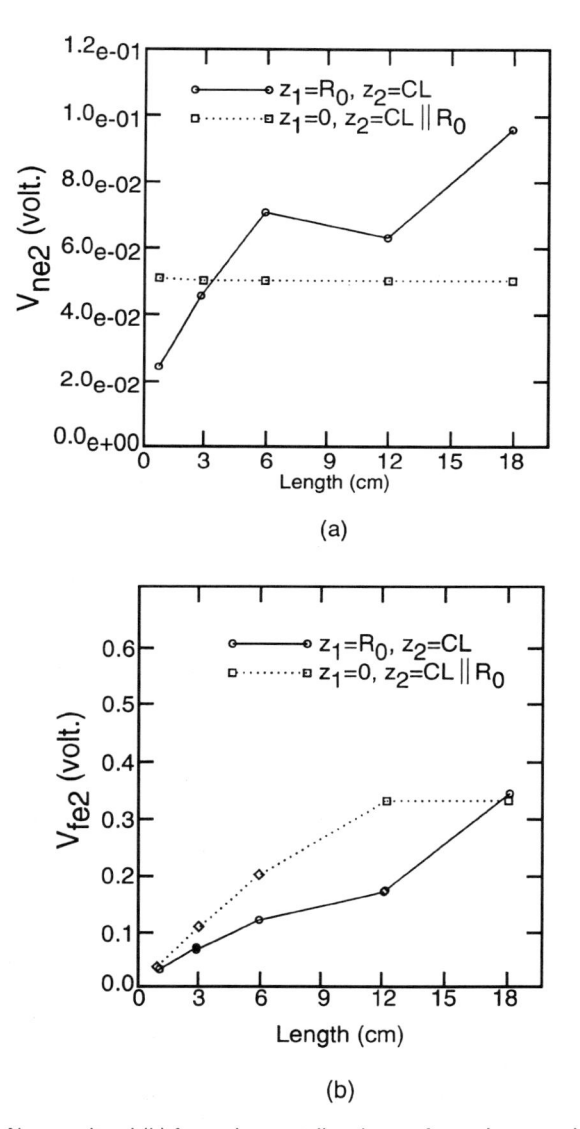

Figure 12.16 (a) Near-end and (b) far-end crosstalk voltages for various terminations of the active line for the structure of case A. $Z_3 = Z_4 = R_0$. (From [2], © 1993 IEEE. Reprinted with permission.)

about 3 cm. This is due to the far-end crosstalk voltage created by the reflected wave on the active line.

The results discussed above amply show that the interconnect performance depends on many factors, such as the type of transmission line, spacing between lines, terminations, and loss. In general, the following remarks can be made for the crosstalk effect:

- The far-end (forward) crosstalk is generally the predominant component of the crosstalk and can cause false switching. However, a large far-end crosstalk can lead to a large near-end crosstalk if the lines are not properly terminated. Forward crosstalk exists because of unequal even- and odd-mode phase velocities and is always present for the basic microstrip configuration. Forward crosstalk can be reduced by choosing transmission lines that support a pure TEM mode of propagation, such as striplines.
- The effect of conductor loss on the interconnect performance is significant. The peak voltage of the output pulse at the far end can be significantly reduced and its shape affected because of line losses. The loss, however, has a positive impact on the crosstalk. Both the near- and far-end crosstalk are reduced by losses in the transmission line. For a matched lossy line, the far-end crosstalk increases with increase in length until a certain distance beyond which it starts decreasing.
- The far-end crosstalk is reduced by placing a matched load in series with the voltage source compared with a matched load in parallel with the receiver capacitance connected at the far end of the active line.
- The active line should be match-terminated for moderate line lengths to avoid large crosstalk voltages and ringing.

References

[1] Hwang, L. T., D. Nayak, I. Turlik, and A. Reisman, "Thin-Film Pulse Propagation Analysis Using Frequency Techniques," *IEEE Trans. Comp., Hybrids and Manufac. Tech.*, Vol. 14, Mar. 1991, pp. 192–198.

[2] Voranantakul, S., J. L. Prince, and P. Hsu, "Crosstalk Analysis for High-Speed Pulse Propagation in Lossy Electrical Interconnections," *IEEE Trans. Comp., Hybrids and Manufac. Tech.*, Vol. 16, Feb. 1993, pp. 127–136.

[3] Sainati, R. A., and T. J. Moravec, "Estimating High-Speed Circuit Interconnect Performance," *IEEE Trans. Circuits Systems*, Vol. 3, Apr. 1989, pp. 533–541.

[4] Djordjevic, A. R., T. K. Sarkar, and R. F. Harrington, "Time-Domain Response of Multi-conductor Transmission Lines," *Proc. IEEE*, Vol. 75, June 1987, pp. 743–764.

[5] Ruheli, A. E., "Equivalent Circuit Models for Three-Dimensional Multiconductor Systems," *IEEE Trans.*, Vol. MTT-22, 1974, pp. 216–221.

[6] Weeks, W. T., L. L. Wu, M. F. McAllister, and A. Singh, "Resistive and Inductive Skin Effect in Rectangular Conductors," *IBM J. Res. Develop.*, Vol. 23, Nov. 1979, pp. 652–660.

[7] Guo, J., D. Kajfez, and A. W. Glisson, "Skin-Effect Resistance of Rectangular Strips," *Electronic Letters*, Vol. 33, May 1997, pp. 966–967.

[8] Gardiol, F. E., *LOSLIN: Lossy Line Calculation Software and Users' Manual*, Norwood, MA: Artech House.

[9] Gardiol, F. E., *Lossy Transmission Lines*. Norwood, MA, Artech House.

[10] Harrington, R. F., and C. Wei, "Losses in Multiconductor Transmission Lines in Multilayered Dielectric Media," *IEEE Trans.*, Vol. MTT-32, July 1984, pp. 705–710.

[11] Arabi, T. R., et. al., "On the Modelling of Conductor and Substrate Losses in Multiconductor, Multidielectric Transmission Line Systems," *IEEE Trans.*, Vol. MTT-39, July 1991, pp. 1,090–1,097.

[12] Djordjevic, A. R., "SPICE-Compatible Models for Multiconductor Transmission Lines in Laplace-Transform Domain," *IEEE Trans.*, Vol. MTT-45, May 1997, pp. 569–579.

13

Multiconductor Transmission Lines

Multiconductor transmission lines find many practical applications in microwave circuits and high-speed digital circuit interconnects. Multiconductor lines are frequently used in the design of Lange couplers and coupled-line filters (hairpin, combline, and interdigital). Multiconductor transmission lines are also used for carrying power over large distances at low frequencies. In this chapter, we first discuss the theory and design of multiconductor transmission lines [1–18]. It is shown that once the line parameters (per-unit length inductances and capacitances) of multiconductor transmission lines are known, the propagation constants and the voltage and current eigenvectors of various modes can be easily obtained using matrix theory. The effect of conductor and dielectric loss is then discussed. The equivalent circuit of a section of multiconductor transmission line that is terminated in arbitrary networks (containing loads and sources) at its end is derived. Finally, the theory is illustrated by means of examples. The multiconductor transmission line theory described in this chapter is based largely on the work of Marx [2, 3].

Figure 13.1 shows a $(N+1)$ conductor transmission line. We assume that of the $(N+1)$ conductors, N conductors are signal conductors and the $(N+1)^{th}$ conductor is the reference conductor at ground potential. Earlier in the introduction of Chapter 3 we discussed that a two-conductor line, shown in Figure 13.2(a), supports a TEM (or quasi-TEM) mode of propagation. Further, a three-conductor line (of which one conductor is assumed to be a reference conductor), as shown in Figure 13.2(b), can support two independent TEM (or quasi-TEM) modes of propagation. For symmetrical coupled lines, these two modes are known as the even and the odd modes. For asymmetrical coupled lines, these two modes are called the c- and π-modes, respectively. Similarly, it can be shown that the $(N+1)$ conductor line of Figure 13.1,

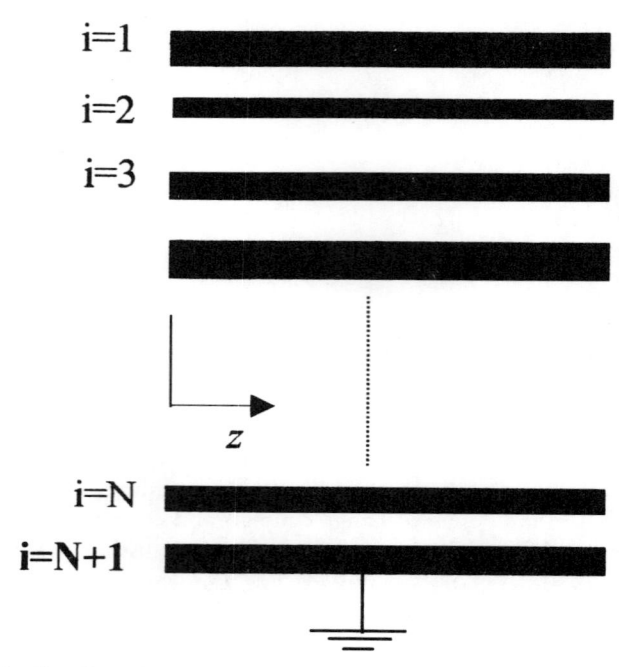

Figure 13.1 A ($N + 1$) conductor transmission line. Conductor ($N + 1$) is assumed to be the reference conductor at ground potential.

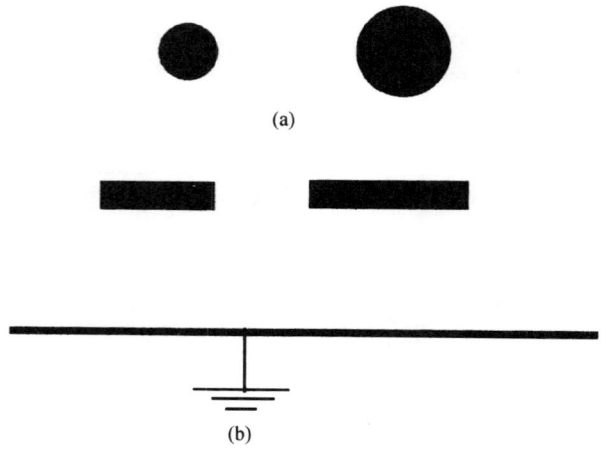

Figure 13.2 (a) A two-conductor transmission line. (b) A three-conductor transmission line.

which represents that N coupled transmission lines can support N independent TEM (or quasi-TEM) modes of propagation, also called N *normal* or *decoupled* modes. Each of these N modes can propagate independently. For quasi-TEM modes, each mode has, in general, a unique propagation constant. (For TEM modes, all the modes have the same propagation constant.) Further, for each normal or decoupled mode, non-zero voltages and currents exist on each conductor. For a given mode, the ratio of voltages on any two conductors is fixed. The same is true of the currents. For example, we discussed in Chapter 4 (Section 4.3) that for two asymmetrical coupled lines (having a common ground conductor), the ratio of the voltages on the two lines is R_c for the c-mode and is R_π for the π-mode. On the other hand, the ratio of the currents on the two lines is $-1/R_\pi$ for the c-mode and is $1/R_c$ for the π-mode. For each mode of a multiconductor transmission line, the ratio of the voltages on different conductors can be represented by a voltage eigenvector. Similarly the ratio of the currents on different conductors can be represented by a current eigenvector.

When a signal is launched at the input of one of the N lines, voltage and current waves are also excited on other lines because of electromagnetic coupling. The voltage and current on each line can be expanded as a sum of N voltage and N current waves, respectively (corresponding to N normal modes of the structure). The coefficients of the N normal modes can be determined by applying the proper boundary conditions.

13.1 Theory

The telegraphist equations for two coupled transmission lines are given by (4.75) to (4.78). By analogy, the voltages and currents on N coupled lines as shown in Figure 13.1 are governed by the following equations:

$$-\frac{\partial V_1(z,t)}{\partial z} = L_{11}\frac{\partial I_1(z,t)}{\partial t} + L_{12}\frac{\partial I_2(z,t)}{\partial t} \cdots + L_{1N}\frac{\partial I_N(z,t)}{\partial t}$$

$$-\frac{\partial V_2(z,t)}{\partial z} = L_{21}\frac{\partial I_1(z,t)}{\partial t} + L_{22}\frac{\partial I_2(z,t)}{\partial t} \cdots + L_{2N}\frac{\partial I_N(z,t)}{\partial t}$$

$$\vdots = \vdots \qquad \vdots \qquad \vdots \qquad \vdots \qquad \vdots \qquad (13.1)$$

$$-\frac{\partial V_N(z,t)}{\partial z} = L_{N1}\frac{\partial I_1(z,t)}{\partial t} + L_{N2}\frac{\partial I_2(z,t)}{\partial t} \cdots + L_{NN}\frac{\partial I_N(z,t)}{\partial t}$$

and

$$-\frac{\partial \mathcal{I}_1(z,t)}{\partial z} = C_{11}\frac{\partial \mathcal{V}_1(z,t)}{\partial t} + C_{12}\frac{\partial \mathcal{V}_2(z,t)}{\partial t} \cdots + C_{1N}\frac{\partial \mathcal{V}_N(z,t)}{\partial t}$$

$$-\frac{\partial \mathcal{I}_2(z,t)}{\partial z} = C_{21}\frac{\partial \mathcal{V}_1(z,t)}{\partial t} + C_{22}\frac{\partial \mathcal{V}_2(z,t)}{\partial t} \cdots + C_{2N}\frac{\partial \mathcal{V}_N(z,t)}{\partial t}$$

$$\vdots = \vdots \qquad \vdots \qquad \vdots \qquad \vdots \qquad (13.2)$$

$$-\frac{\partial \mathcal{I}_N(z,t)}{\partial z} = C_{N1}\frac{\partial \mathcal{V}_1(z,t)}{\partial t} + C_{N2}\frac{\partial \mathcal{V}_2(z,t)}{\partial t} \cdots + C_{NN}\frac{\partial \mathcal{V}_N(z,t)}{\partial t}$$

where we assume that the lines are lossless. $\mathcal{V}_i(z,t)$ and $\mathcal{I}_i(z,t)$ denote the voltage and current, respectively, on i^{th} conductor at distance z along the transmission line at time t. L_{ij} and C_{ij} denote, respectively, the ith element of the inductance and capacitance matrices of the multiconductor transmission line. In matrix notation, (13.1) and (13.2) can be stated respectively as

$$-\frac{\partial}{\partial z}[\mathcal{V}(z,t)] = [\mathbf{L}]\frac{\partial}{\partial t}[\mathcal{I}(z,t)] \qquad (13.3)$$

$$-\frac{\partial}{\partial z}[\mathcal{I}(z,t)] = [\mathbf{C}]\frac{\partial}{\partial t}[\mathcal{V}(z,t)] \qquad (13.4)$$

where

$$[\mathcal{V}(z,t)] = \begin{bmatrix} \mathcal{V}_1(z,t) \\ \mathcal{V}_2(z,t) \\ \vdots \\ \mathcal{V}_N(z,t) \end{bmatrix}, \quad [\mathcal{I}(z,t)] = \begin{bmatrix} \mathcal{I}_1(z,t) \\ \mathcal{I}_2(z,t) \\ \vdots \\ \mathcal{I}_N(z,t) \end{bmatrix}$$

$$[\mathbf{L}] = \begin{bmatrix} L_{11} & L_{12} & \cdots & L_{1N} \\ L_{21} & L_{22} & \cdots & L_{2N} \\ \vdots & \vdots & \vdots & \vdots \\ L_{N1} & L_{N2} & \cdots & L_{NN} \end{bmatrix}$$

and

$$[\mathbf{C}] = \begin{bmatrix} C_{11} & C_{12} & \cdots & C_{1N} \\ C_{21} & C_{22} & \cdots & C_{2N} \\ \vdots & \vdots & \vdots & \vdots \\ C_{N1} & C_{N2} & \cdots & C_{NN} \end{bmatrix}$$

At high frequencies (where almost all the stored magnetic energy can be assumed to lie outside the conductors), the inductance matrix [L] of a multiconductor transmission line is given by

$$[L] = \epsilon_0 \mu_0 [C_0]^{-1} \tag{13.5}$$

where $[C_0]$ denotes the free-space capacitance matrix of the multiconductor transmission line (i.e., the capacitance matrix when the dielectric constant of the material surrounding the transmission lines is unity[1]).

A method to determine the inductance and capacitance matrices of lossless multiconductor transmission lines of arbitrary cross section is described in [5]. A method to determine the resistance and conductance matrices of lossy lines has also been reported [6]. Computer programs based on the above methods are also commercially available [7]–[10].

13.1.1 Eigenvalues and Eigenvectors

For a mode propagating in the positive z-direction, the voltage and current on the kth line can be expressed as follows:

$$\mathcal{V}_k(z, t) = V_k e^{j(\omega t - \beta z)} \tag{13.6}$$

$$\mathcal{I}_k(z, t) = I_k e^{j(\omega t - \beta z)} \tag{13.7}$$

where $k = 1, 2, \ldots, N$. Here, ω and β denote, respectively, the radian frequency of the signal and propagation constant of the mode and hence (13.3) and (13.4) reduce, respectively, to

$$[V] = v[L][I] \tag{13.8}$$

$$[I] = v[C][V] \tag{13.9}$$

where $v = \omega/\beta$ denotes the phase velocity of the mode. In (13.8) and (13.9), [V] and [I] are the column vectors of voltages and currents (independent of z and t), respectively, on N lines, that is:

$$[V] = \begin{bmatrix} V_1 \\ V_2 \\ \vdots \\ V_N \end{bmatrix} \quad \text{and} \quad [I] = \begin{bmatrix} I_1 \\ I_2 \\ \vdots \\ I_N \end{bmatrix}$$

---

1. At low frequencies where the magnetic fields can penetrate into conductors made of normal metals, the inductance matrix [L] would be somewhat different from that given by (13.5). Strictly speaking, the inductance matrix [L] should be evaluated by independent means at these frequencies [4].

Eliminating [I] from (13.8) and (13.9) results in

$$([L][C])[V] = \frac{1}{v^2}[V] \qquad (13.10)$$

or

$$\left([L][C] - \frac{1}{v^2}[U]\right)[V] = 0 \qquad (13.11)$$

where [U] is an identity matrix of order N (all the diagonal elements of an identity matrix are unity while all its other elements are zero).

Equation 13.11 represents an eigenvalue equation. For this equation to have a nontrivial solution, the determinant of the matrix on the left-hand side should vanish, that is:

$$det\left([L][C] - \frac{1}{v^2}[U]\right) = 0 \qquad (13.12)$$

where *det* denotes the determinant. If the inductance and capacitance matrices ([L] and [C]) are known, the values of v^2 can be found by solving (13.12). Because [L] and [C] are matrices of order $N \times N$, the highest power of v^2 in the determinant of (13.12) is N. Therefore, (13.12) results in N values of v^2, which are also known as N eigenvalues. Each eigenvalue corresponds to a particular mode of the structure. Therefore, N eigenvalues result in N modes of the structure. Corresponding to each eigenvalue, there is a voltage eigenvector that can be found using (13.11). The voltage eigenvector denotes to within an arbitrary multiplication constant, the voltages on the N transmission lines corresponding to that mode. Once a voltage eigenvector has been found, the corresponding current eigenvector can be determined using (13.8) or (13.9). For each forward-traveling mode, there is a corresponding backward-traveling mode with the same phase velocity. In the discussion to follow, the voltage and current eigenvectors of the ith mode will generally be denoted as follows:

$$[V]_i = \begin{bmatrix} V_{1i} \\ V_{2i} \\ \vdots \\ V_{Ni} \end{bmatrix} \quad \text{and} \quad [I]_i = \begin{bmatrix} I_{1i} \\ I_{2i} \\ \vdots \\ I_{Ni} \end{bmatrix}$$

Orthogonal Properties of Voltage and Current Eigenvectors

The *current* eigenvector of a mode is orthogonal to the *voltage* eigenvector of any other mode, that is:

$$[I]_j \cdot [V]_i = 0 \quad \text{for} \quad i \neq j \qquad (13.13)$$

or
$$I_{1j}V_{1i} + I_{2j}V_{2i} + \ldots + I_{Nj}V_{Ni} = 0 \quad \text{for } i \neq j$$

The orthogonal property given by (13.13) is easily proved when the eigenvalues corresponding to the two eigenvectors are different (i.e., when $v_i \neq v_j$). In case of degeneracy (i.e., when $v_i = v_j$ for some i and j) there is arbitrariness in the choice of eigenvectors. In this case, the eigenvectors can be orthogonalized by the Gram-Schmidt procedure [11]. Note that in general, the *voltage* eigenvectors of two different modes are not orthogonal. The same is true of current eigenvectors.

Normalization of Voltage and Current Eigenvectors

As already remarked, an eigenvector denotes to within an arbitrary multiplication constant the values of its different components. Therefore, eigenvectors are generally normalized. A suitable normalization of current and voltage eigenvectors is as follows:

$$[\mathbf{I}]_i \cdot [\mathbf{V}]_i = 1 \tag{13.14}$$

Because of the properties given by (13.13) and (13.14), the voltage and current eigenvectors are said to be *bi-orthonormal*. When the voltage and current eigenvectors are normalized according to (13.14), some of the physical properties of multiconductor lines can be represented using elegant matrix expressions. For the sake of clarity, the normalized eigenvectors will be denoted with a bar symbol, such that $[\bar{\mathbf{I}}]_i$ and $[\bar{\mathbf{V}}]_i$ are used to denote the ith normalized current and voltage eigenvectors, respectively.

Voltage and Current Eigenvector Matrices

Let $[\mathbf{M}_V]$ and $[\mathbf{M}_I]$ denote $N \times N$ matrices whose columns are voltage and current eigenvectors, respectively. Similarly, let $[\bar{\mathbf{M}}_V]$ and $[\bar{\mathbf{M}}_I]$ denote $N \times N$ matrices whose columns are normalized voltage and current eigenvectors, respectively. Because the normalized eigenvectors are defined according to (13.14), we obtain

$$[\bar{\mathbf{M}}_V][\bar{\mathbf{M}}_I]_t = [\bar{\mathbf{M}}_I][\bar{\mathbf{M}}_V]_t = [\mathbf{U}] \tag{13.15}$$

where $_t$ denotes the transpose.

13.1.2 Decomposition in Terms of Normal Modes

An arbitrary signal on a $(N+1)$ conductor multiconductor transmission line can be expressed as a sum of N normal modes of the structure. Let $[\mathbf{V}^+(0)]$ denote the voltage vector of an arbitrary forward-travelling wave at the location

$z = 0$. Because, in general, the voltage vector $[\mathbf{V}^+(0)]$ is not identical to the eigenvector of one of the normal modes of the structure, the voltage wave on each line can then be expanded as a sum of voltages of various modes of the multiconductor-transmission line. For example, the forward-traveling voltage wave on line j can be expressed as

$$V_j^+(z) = A_1^+ V_{j1} e^{-j\beta_1 z} + A_2^+ V_{j2} e^{-j\beta_2 z} + \cdots + A_N^+ V_{jN} e^{-j\beta_N z} \quad (13.16)$$

where $V_{j1}, V_{j2}, \ldots, V_{jN}$ denote the elements of the jth row of matrix $[\mathbf{M}_V]$. Further, A_i^+ denotes the coefficient of the ith mode. The coefficient vector $[\mathbf{A}^+]$ can be determined using

$$[\mathbf{A}^+] = [\mathbf{M}_V]^{-1} [\mathbf{V}^+(0)] \quad (13.17)$$

where $[\mathbf{M}_V]$ denotes the voltage eigenvector matrix defined in the last section.

Premultiplying both sides of (13.17) by $[\mathbf{M}_V]$, we obtain

$$[\mathbf{V}^+(0)] = [\mathbf{M}_V][\mathbf{A}^+] \quad (13.18)$$

In compact notation, the voltage waves on different lines can be expressed using a voltage vector $[\mathbf{V}^+(z)]$:

$$[\mathbf{V}^+(z)] = \sum_{i=1}^{N} A_i^+ [\mathbf{V}]_i e^{-j\beta_i z} \quad (13.19)$$

The current waves can then be expressed as

$$[\mathbf{I}^+(z)] = \sum_{i=1}^{N} A_i^+ [\mathbf{I}]_i e^{-j\beta_i z} \quad (13.20)$$

The voltage wave vector $[\mathbf{V}^+(z)]$ can also be expanded in terms of normalized voltage eigenvectors as follows:

$$[\mathbf{V}^+(z)] = \sum_{i=1}^{N} \bar{A}_i^+ [\bar{\mathbf{V}}]_i e^{-j\beta_i z} \quad (13.21)$$

where $\bar{A}_i^+$ denotes the element in the ith row of column vector $[\bar{\mathbf{A}}^+]$ given by

$$[\bar{\mathbf{A}}^+] = [\bar{\mathbf{M}}_I]_t [\mathbf{V}^+(0)] \quad (13.22)$$

In the above equation, $[\bar{\mathbf{M}}_I]_t$ denotes the transpose of the normalized current eigenvector matrix $[\bar{\mathbf{M}}_I]$.

13.1.3 Admittance and Impedance Matrices

The admittance matrix of a multiconductor transmission line can be defined as follows:

$$[\mathbf{I}^+] = [\mathbf{Y}_0][\mathbf{V}^+] \qquad (13.23)$$

where $[\mathbf{I}^+]$ and $[\mathbf{V}^+]$ denote the forward-traveling current and voltage wave vectors, respectively, at any location along the transmission line.

The admittance matrix $[\mathbf{Y}_0]$ is then given by

$$[\mathbf{Y}_0] = [\mathbf{M}_I][\mathbf{M}_V]^{-1} \qquad (13.24)$$
$$= [\bar{\mathbf{M}}_I][\bar{\mathbf{M}}_I]_t \qquad (13.25)$$

The impedance matrix of a multiconductor transmission line is defined as

$$[\mathbf{V}^+] = [\mathbf{Z}_0][\mathbf{I}^+] \qquad (13.26)$$

The impedance matrix is related to the admittance matrix by

$$[\mathbf{Z}_0] = [\mathbf{Y}_0]^{-1} \qquad (13.27)$$

If a ($N + 1$) terminal (including a common ground terminal) resistive network is connected at one of the ends of a section of a ($N + 1$) conductor multiconductor transmission line, the network will completely absorb any arbitrary wave incident upon it if the impedance matrix of the network is the same as the characteristic impedance matrix of the multiconductor transmission line. There are, therefore, no resulting reflections on the lines from this end. This is analogous to a single transmission line where a line terminated in a load impedance equal to its characteristic impedance does not suffer any reflections.

Note that in the case of multiconductor transmission lines, characteristic impedances based on alternative definitions are also possible [12]. Different definitions lead to different results and have different physical meaning.

Reflection and Transmission Coefficient Matrices

It is interesting to note that the equations for the reflection and transmission coefficient matrices of multiconductor transmission lines are analogous to those of a single transmission line. The reflection coefficient matrix Γ of a multiconductor transmission line connected to a network can be defined as

$$[V^-] = \Gamma[V^+] \tag{13.28}$$

where $[V^-]$ and $[V^+]$ denote the reflected and incident voltage wave vectors respectively. If $[Z_L]$ denotes the impedance matrix of the terminal network, then the reflection matrix Γ is given by

$$\Gamma = ([Z_L] - [Z_0])([Z_L] + [Z_0])^{-1} \tag{13.29}$$

Similarly, the transmitted voltage vector $[V]$ which denotes the sum of incident and reflected voltage vectors can be expressed as

$$[V] = T[V^+] \tag{13.30}$$

where T denotes the transmission coefficient vector and is given by

$$T = 2Z_L(Z_L + Z_0)^{-1} \tag{13.31}$$

13.1.4 Eigenvalues and Eigenvectors of Lossy Lines

When the losses of a multiconductor transmission loss are also taken into account, the propagation constants of the various modes become complex ($\gamma = \alpha + j\beta$). The eigenvalue equation for the complex propagation constants of lossy lines is as follows:

$$([R] + j\omega[L])([G] + j\omega[C])[V] = \gamma^2[V] \tag{13.32}$$

where $[R]$ and $[G]$ denote the resistance and conductance matrices of the multiconductor transmission line. Once the eigenvalues are known, the voltage eigenvectors associated with different eigenvalues can be determined. Further, for each voltage eigenvector, the corresponding current eigenvector can be determined using one of the following two telegraphist's equations:

$$\gamma[I] = ([G] + j\omega[C])[V] \tag{13.33}$$
$$\gamma[V] = ([R] + j\omega[L])[I] \tag{13.34}$$

13.1.5 Equivalent Circuit of Multiconductor Transmission Line

We now discuss one of the possible equivalent circuits of a section of a multiconductor transmission line terminated by arbitrary linear networks at its ends [2]. More details on this and some other forms of equivalent circuits can be found in [13–18]. In the equivalent circuit discussed in this section, the voltage and current vectors at one end of the section are related to those at the other end by multiconductor transmission line parameters. By solving the resulting equations, the voltages and currents at either end can be determined.

Figure 13.3 shows a $(N+1)$ conductor multiconductor transmission line that is connected to arbitrary linear networks at its near and far ends. The length of the multiconductor transmission line is assumed to be D. Let the near and far ends of the multiconductor transmission line section be located at $z = 0$ and $z = D$, respectively. The voltage and current vectors at $z = 0$ can be expressed as a sum of forward and backward waves on the multiconductor transmission line as follows:

$$[\mathbf{V}(0)] = [\mathbf{V}^+(0)] + [\mathbf{V}^-(0)] \tag{13.35}$$

$$[\mathbf{I}(0)] = [\mathbf{I}^+(0)] - [\mathbf{I}^-(0)] \tag{13.36}$$

where $[\mathbf{V}^+(0)]$ and $[\mathbf{V}^-(0)]$ denote the forward and backward voltage wave vectors respectively at $z = 0$. Similarly $[\mathbf{I}^+(0)]$ and $[\mathbf{I}^-(0)]$ denote the forward and

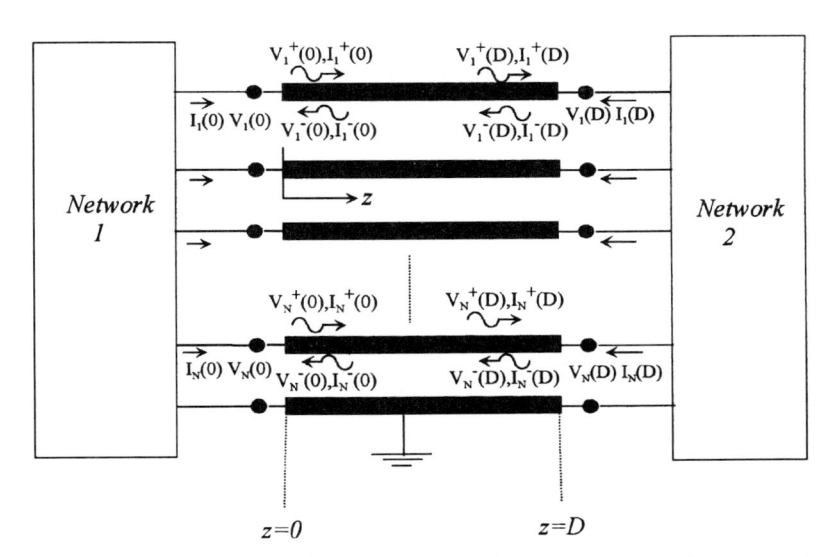

Figure 13.3 A section of multiconductor transmission line connected to arbitrary networks at either end.

backward current wave vectors, respectively, at $z = 0$. Premultiplying (13.36) by $[\mathbf{Z}_0]$ and subtracting the resulting equation from (13.35), we obtain

$$[\mathbf{V}(0)] - [\mathbf{Z}_0][\mathbf{I}(0)] = 2[\mathbf{V}^-(0)] \tag{13.37}$$

Therefore, at the near end, the multiconductor transmission line can be replaced by an equivalent circuit as shown in Figure 13.4 where $[\mathbf{Z}_0]$ denotes the characteristic impedance matrix of the multiconductor transmission line. The reflected voltage wave vector $[\mathbf{V}^-(0)]$ can be determined in terms of voltage and current vectors at the far end as described below:

At the far end, the voltage and current vectors can be expressed as

$$[\mathbf{V}(D)] = [\mathbf{V}^+(D)] + [\mathbf{V}^-(D)] \tag{13.38}$$

$$[\mathbf{I}(D)] = -[\mathbf{I}^+(D)] + [\mathbf{I}^-(D)] \tag{13.39}$$

In writing (13.39), we have followed the convention that the current flowing

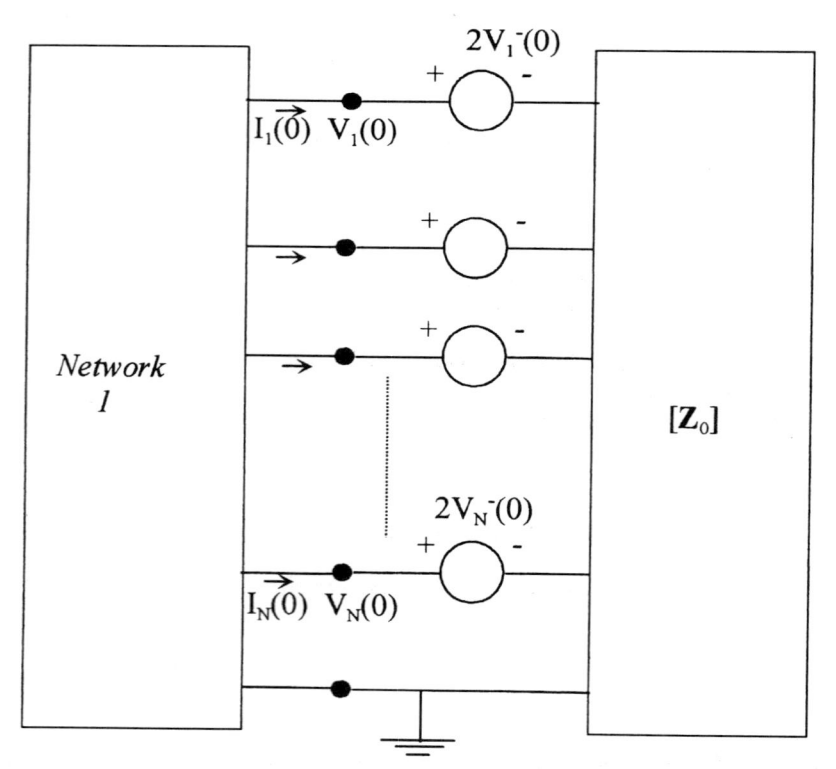

Figure 13.4 Equivalent circuit at the near-end of the network shown in Figure 13.3.

into the network is positive. Premultiplying (13.39) by $[\mathbf{Z}_0]$ and adding the resulting equation to (13.38):

$$2[\mathbf{V}^-(D)] = [\mathbf{V}(D)] + [\mathbf{Z}_0][\mathbf{I}(D)] \qquad (13.40)$$

Further, using (13.18), we can write for the reflected waves

$$[\mathbf{V}^-(0)] = [\mathbf{M}_V][\mathbf{A}^-(0)] \qquad (13.41)$$

$$[\mathbf{V}^-(D)] = [\mathbf{M}_V][\mathbf{A}^-(D)] \qquad (13.42)$$

where $[\mathbf{A}^-(0)]$ and $[\mathbf{A}^-(D)]$ denote vectors of the modal coefficients at $z = 0$ and $z = D$, respectively. Furthermore:

$$[\mathbf{A}^-(0)] = [\mathbf{E}][\mathbf{A}^-(D)] \qquad (13.43)$$

where $[\mathbf{E}]$ is a diagonal $N \times N$ matrix given by

$$[\mathbf{E}] = \begin{bmatrix} e^{-j\beta_1 D} & 0 & \cdots & 0 \\ 0 & e^{-j\beta_2 D} & \cdots & 0 \\ \vdots & \vdots & \vdots & \vdots \\ 0 & 0 & \cdots & e^{-j\beta_N D} \end{bmatrix} \qquad (13.44)$$

Using (13.41)–(13.43), we obtain

$$[\mathbf{V}^-(0)] = [\mathbf{M}_V][\mathbf{E}][\mathbf{M}_V]^{-1}[\mathbf{V}^-(D)] \qquad (13.45)$$

By the substitution of $[\mathbf{V}^-(D)]$ from (13.40) in (13.45), the expression for the voltage source $2[\mathbf{V}^-(0)]$ shown in the equivalent circuit of Figure 13.4 is found to be

$$2[\mathbf{V}^-(0)] = [\mathbf{M}_V][\mathbf{E}][\mathbf{M}_V]^{-1} \left([\mathbf{V}(D)] + [\mathbf{Z}_0][\mathbf{I}(D)] \right) \qquad (13.46)$$

The equivalent circuit at the far end ($z = D$) of the multiconductor transmission line can be built in a similar fashion. At the load end, the voltages and currents are related by

$$[\mathbf{V}(D)] - [\mathbf{Z}_0][\mathbf{I}(D)] = 2[\mathbf{V}^+(D)] \qquad (13.47)$$

The equivalent circuit at the far end ($z = D$) is shown in Figure 13.5. The voltage source vector $2[\mathbf{V}^+(D)]$ is given by

$$2[\mathbf{V}^+(D)] = [\mathbf{M}_V][\mathbf{E}][\mathbf{M}_V]^{-1} \left([\mathbf{V}(0)] + [\mathbf{Z}_0][\mathbf{I}(0)] \right) \qquad (13.48)$$

Once the equations are obtained in terms of terminal currents and voltages, these can be solved for the unknown quantities.

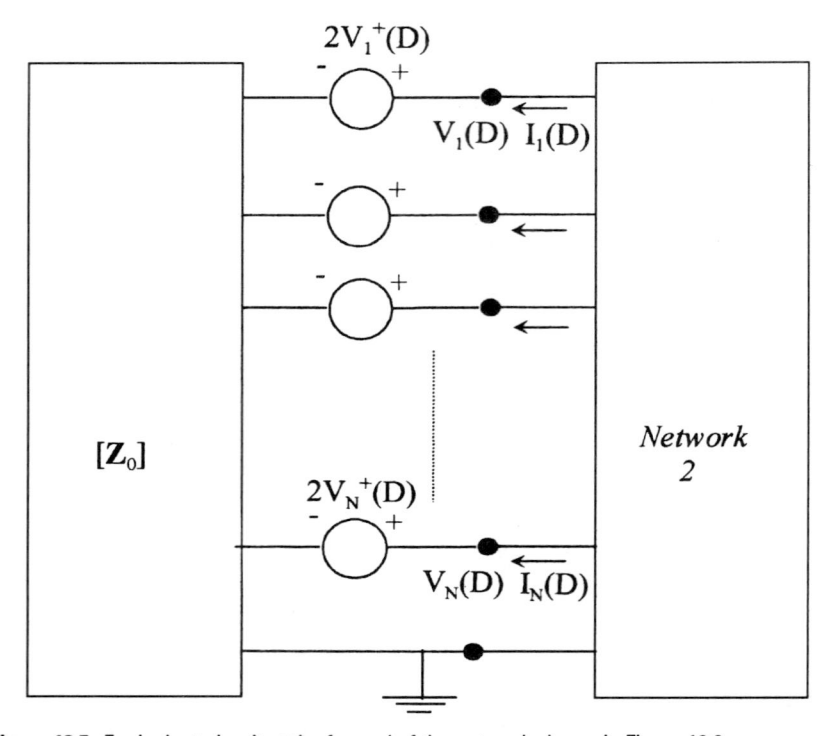

Figure 13.5 Equivalent circuit at the far-end of the network shown in Figure 13.3.

13.2 Examples

The multiconductor transmission line theory is best demonstrated by means of examples. A few illustrative examples are given below.

13.2.1 Eigenvalues and Eigenvectors of Symmetrical Coupled Lines

A three-conductor transmission line of which one conductor is a reference conductor at ground potential is shown in Figure 13.6. The two signal conductors of the transmission line are assumed to be symmetrical. The symmetrical coupled lines can support even and odd-modes of propagation. The equations for the phase velocities and characteristic impedances of the even and odd-modes of symmetrical coupled lines in terms of inductance and capacitance parameters were given in Section 4.3.3. We now derive those equations using the multiconductor transmission line theory. Let the capacitance and inductance matrices of

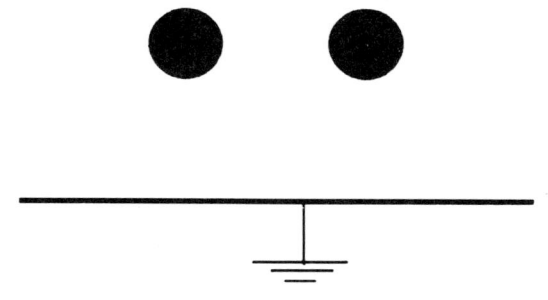

Figure 13.6 Symmetrical coupled lines.

symmetrical coupled lines be given by

$$[\mathbf{C}] = \begin{bmatrix} C_1 & -C_m \\ -C_m & C_1 \end{bmatrix} \quad (13.49)$$

and

$$[\mathbf{L}] = \begin{bmatrix} L_1 & L_m \\ L_m & L_1 \end{bmatrix} \quad (13.50)$$

Further, let the voltages on the two lines be represented by a voltage vector [**V**] as follows:

$$[\mathbf{V}] = \begin{bmatrix} V_1 \\ V_2 \end{bmatrix} \quad (13.51)$$

(13.11) then becomes,

$$\begin{bmatrix} (L_1 C_1 - L_m C_m - \frac{1}{v^2}) & (L_m C_1 - L_1 C_m) \\ (L_m C_1 - L_1 C_m) & (L_1 C_1 - L_m C_m - \frac{1}{v^2}) \end{bmatrix} \begin{bmatrix} V_1 \\ V_2 \end{bmatrix} = 0 \quad (13.52)$$

where v denotes the phase velocity. For a nontrivial solution, the determinant of the matrix on the left-hand side of (13.52) should vanish, that is:

$$\left(L_1 C_1 - L_m C_m - \frac{1}{v^2} \right) = \pm (L_m C_1 - L_1 C_m) \quad (13.53)$$

The solution of (13.53) leads to two possible values of v^2, which means that the structure can support *two* independent modes of propagation. The two values of v^2 (which are known as the two eigenvalues and are denoted by v_1^2 and v_2^2, respectively) are given by

$$v_1^2 = \frac{1}{(L_1 - L_m)(C_1 + C_m)}$$

or

$$v_1 = \frac{1}{\sqrt{(L_1 - L_m)(C_1 + C_m)}} \tag{13.54}$$

and

$$v_2^2 = \frac{1}{(L_1 + L_m)(C_1 - C_m)}$$

or

$$v_2 = \frac{1}{\sqrt{(L_1 + L_m)(C_1 - C_m)}} \tag{13.55}$$

When the possible values of v^2 are substituted in (13.52), two simultaneous equations are obtained in variables V_1 and V_2. By solving these two simultaneous equations, the values of V_1 and V_2 can be found. For example, by substituting the value of v_1 from (13.54) in (13.52), the following equations are obtained:

$$(L_m C_1 - L_1 C_m) V_1 + (L_m C_1 - L_1 C_m) V_2 = 0 \tag{13.56}$$
$$(L_m C_1 - L_1 C_m) V_1 + (L_m C_1 - L_1 C_m) V_2 = 0 \tag{13.57}$$

The solution of (13.56) and (13.57) is $V_2 = -V_1$, or the voltage eigenvector **V** (corresponding to the eigenvalue $v = v_1$) is given by

$$[\mathbf{V}] = [\mathbf{V}]_1 = \begin{bmatrix} V_1 \\ V_2 \end{bmatrix} = \begin{bmatrix} a_{11} \\ -a_{11} \end{bmatrix} \tag{13.58}$$

where a_{11} is an arbitrary constant. From (13.58), we conclude that for the mode whose velocity is given by (13.54), the voltages on the two lines are equal in magnitude but out of phase. This mode can therefore be termed as the odd mode.

Similarly, by substituting the value of v_2 from (13.55) in (13.52), two simultaneous equations are obtained whose solution leads to $V_2 = V_1$. Therefore, in this case the voltage eigenvector [V] becomes

$$[\mathbf{V}] = [\mathbf{V}]_2 = \begin{bmatrix} V_1 \\ V_2 \end{bmatrix} = \begin{bmatrix} a_{12} \\ a_{12} \end{bmatrix} \quad (13.59)$$

where a_{12} is an arbitrary constant. We conclude from (13.59), that for the mode whose velocity is given by (13.55), the voltages on the two lines are equal in magnitude and in phase. This mode can therefore be called the even mode.

The unnormalized voltage eigenvector matrix $[\mathbf{M}_V]$ (whose columns are the voltage eigenvectors) is then given by

$$[\mathbf{M}_V] = [[\mathbf{V}]_1 \ [\mathbf{V}]_2] = \begin{bmatrix} a_{11} & a_{12} \\ -a_{11} & a_{12} \end{bmatrix} \quad (13.60)$$

Once the voltage eigenvectors have been determined, the current eigenvectors can be found using either (13.8) or (13.9). For example, substitution of velocity v from (13.54) and the corresponding voltage eigenvector from (13.58) in (13.9) leads to

$$[\mathbf{I}] = \frac{1}{\sqrt{(L_1 - L_m)(C_1 + C_m)}} \begin{bmatrix} C_1 & -C_m \\ -C_m & C_1 \end{bmatrix} \begin{bmatrix} a_{11} \\ -a_{11} \end{bmatrix}$$

or

$$[\mathbf{I}] = \begin{bmatrix} I_1 \\ I_2 \end{bmatrix} = \begin{bmatrix} a_{11}\sqrt{\frac{(C_1+C_m)}{(L_1-L_m)}} \\ -a_{11}\sqrt{\frac{(C_1+C_m)}{(L_1-L_m)}} \end{bmatrix} \quad (13.61)$$

It is therefore seen that for the odd mode, the currents on the two lines are also equal in magnitude but out of phase. Further, for this mode, let the ratio of voltage and current on either line be denoted by Z_{0o}. From (13.58) and (13.61)

$$Z_{0o} = \frac{V_1}{I_1} = \frac{V_2}{I_2} = \sqrt{\frac{(L_1 - L_m)}{(C_1 + C_m)}} \quad (13.62)$$

The current eigenvector of the odd mode can therefore be expressed as,

$$[\mathbf{I}] = \begin{bmatrix} I_1 \\ I_2 \end{bmatrix} = \begin{bmatrix} \frac{a_{11}}{Z_{0o}} \\ -\frac{a_{11}}{Z_{0o}} \end{bmatrix} \quad (13.63)$$

Similarly, by the substitution of velocity v of the even mode from (13.55) and the corresponding voltage eigenvector from (13.59) in (13.9) leads to

$$[\mathbf{I}] = \begin{bmatrix} I_1 \\ I_2 \end{bmatrix} = \begin{bmatrix} \frac{a_{12}}{Z_{0e}} \\ \frac{a_{12}}{Z_{0e}} \end{bmatrix} \qquad (13.64)$$

where

$$Z_{0e} = \sqrt{\frac{(L_1 + L_m)}{(C_1 - C_m)}} \qquad (13.65)$$

The unnormalized current eigenvector matrix whose columns are the current eigenvectors is therefore given by

$$[\mathbf{M}_I] = \begin{bmatrix} \frac{a_{11}}{Z_{0o}} & \frac{a_{12}}{Z_{0e}} \\ -\frac{a_{11}}{Z_{0o}} & \frac{a_{12}}{Z_{0e}} \end{bmatrix} \qquad (13.66)$$

It is easily seen that the voltage and current eigenvectors satisfy the orthogonality condition given by (13.13). Further, if the voltage and current eigenvectors are normalized according to (13.14), we obtain

$$a_{11} = \sqrt{\frac{Z_{0o}}{2}} \qquad (13.67)$$

$$a_{12} = \sqrt{\frac{Z_{0e}}{2}} \qquad (13.68)$$

The normalized voltage and current eigenvector matrices are therefore given by

$$[\bar{\mathbf{M}}_V] = \frac{1}{\sqrt{2}} \begin{bmatrix} \sqrt{Z_{0o}} & \sqrt{Z_{0e}} \\ -\sqrt{Z_{0o}} & \sqrt{Z_{0e}} \end{bmatrix} \qquad (13.69)$$

and

$$[\bar{\mathbf{M}}_I] = \frac{1}{\sqrt{2}} \begin{bmatrix} \frac{1}{\sqrt{Z_{0o}}} & \frac{1}{\sqrt{Z_{0e}}} \\ -\frac{1}{\sqrt{Z_{0o}}} & \frac{1}{\sqrt{Z_{0e}}} \end{bmatrix} \qquad (13.70)$$

The characteristic admittance matrix of symmetrical coupled lines can be found using (13.24) as follows:

$$[\mathbf{Y}_0] = [\mathbf{M}_I][\mathbf{M}_V]^{-1} = \frac{1}{2}\begin{bmatrix} \left(\frac{1}{Z_{0e}} + \frac{1}{Z_{0o}}\right) & \left(\frac{1}{Z_{0e}} - \frac{1}{Z_{0o}}\right) \\ \left(\frac{1}{Z_{0e}} - \frac{1}{Z_{0o}}\right) & \left(\frac{1}{Z_{0e}} + \frac{1}{Z_{0o}}\right) \end{bmatrix} \qquad (13.71)$$

Further, the characteristic impedance matrix of the coupled lines is found as

$$[\mathbf{Z}_0] = [\mathbf{Y}_0]^{-1} = \frac{1}{2}\begin{bmatrix}(Z_{0e} + Z_{0o}) & (Z_{0e} - Z_{0o})\\(Z_{0e} - Z_{0o}) & (Z_{0e} + Z_{0o})\end{bmatrix} \quad (13.72)$$

We may easily see that if (13.25) is used to find the characteristic admittance matrix, exactly the same result as given by (13.71) is obtained. This shows that the characteristic admittance or impedance matrix of a multiconductor transmission line is *unique* and does not depend on the normalization of eigenvectors as long as (13.8) and (13.9) are satisfied.

When the [**Y**] matrix of a network is known, a π network can be constructed as discussed in Section 2.5.3. For example, a network whose [**Y**] matrix is given by (13.71) is shown in Figure 13.7(a) where the different elements of the network are given by

$$R_p = Z_{0e} \quad (13.73)$$

$$R_m = \frac{2Z_{0e}Z_{0o}}{Z_{0e} - Z_{0o}} \quad (13.74)$$

If the load network of Figure 13.7(a) is used to terminate symmetrical coupled lines as seen in Figure 13.7(b), any arbitrary wave traveling on

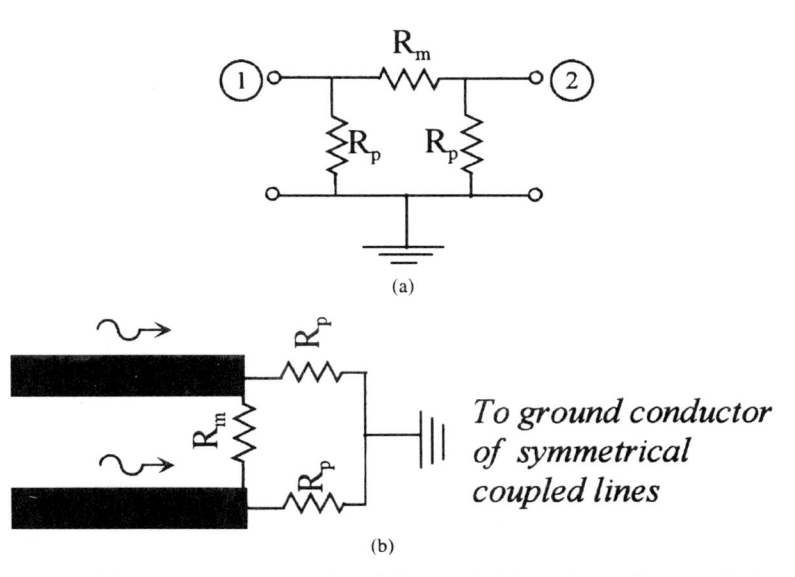

Figure 13.7 (a) π network representation of characteristic impedance of symmetrical coupled lines. (b) Symmetrical coupled lines terminated by a matched network.

symmetrical coupled lines and incident on the load network will be completely absorbed.

13.2.2 Equivalent Circuit of Single Transmission Line Network

We now discuss an example to illustrate the technique discussed in Section 13.1.5, which is useful for the analysis of multiconductor transmission line circuits. For the sake of illustration, the simple example of a single transmission line connected to a load and source impedance is chosen. The technique however, can be easily extended to multiconductor transmission lines. For the circuit of Figure 13.8, the characteristic impedance of the line is assumed to be Z_0 and its length is assumed to be D. Further, it is assumed that a voltage source of impedance Z_g is connected at the near end of the transmission line and its far end is terminated in a load impedance Z_L. It is required to find the terminal voltages, V_1 and V_2, and terminal currents, I_1 and I_2. In a general case, both forward and backward waves exist on the line. Let the forward- and backward traveling voltage waves at $z = 0$ be denoted by $V^+(0)$ and $V^-(0)$, respectively. Similarly, let forward- and backward-traveling current waves at $z = 0$ be denoted by $I^+(0)$ and $I^-(0)$, respectively. We then have

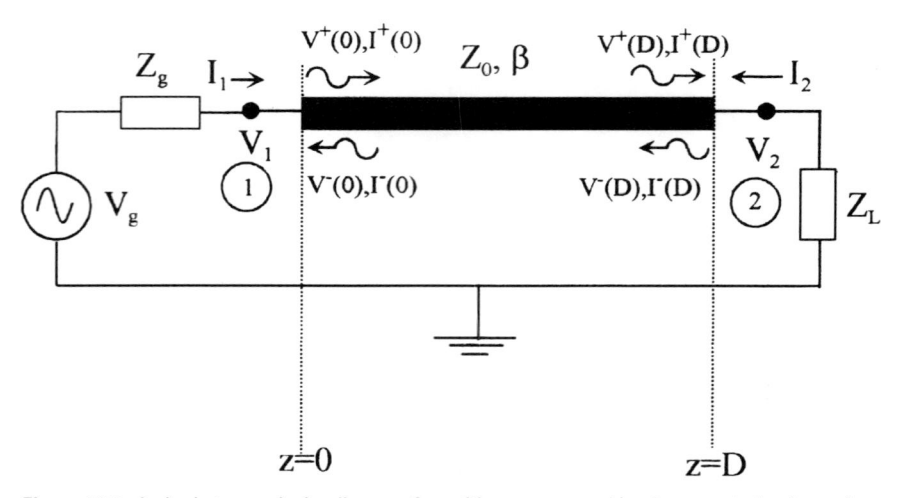

$$V_1 = V^+(0) + V^-(0) \tag{13.75}$$

$$I_1 = I^+(0) - I^-(0) \tag{13.76}$$

Figure 13.8 A single transmission line section with a source and load connected at its ends.

Multiplying (13.76) by Z_0 and subtracting the resulting equation from (13.75), we obtain

$$V_1 - Z_0 I_1 = 2V^-(0) \qquad (13.77)$$

Equation 13.77 shows that the circuit at the right-hand side of terminal 1, including the transmission line, can be replaced by a Thevenin equivalent circuit as shown in Figure 13.9(a). The voltage source $V^-(0)$ can be expressed in terms of a terminal current and voltage at the load end (I_2 and V_2, respectively) and transmission line parameters. At the load end:

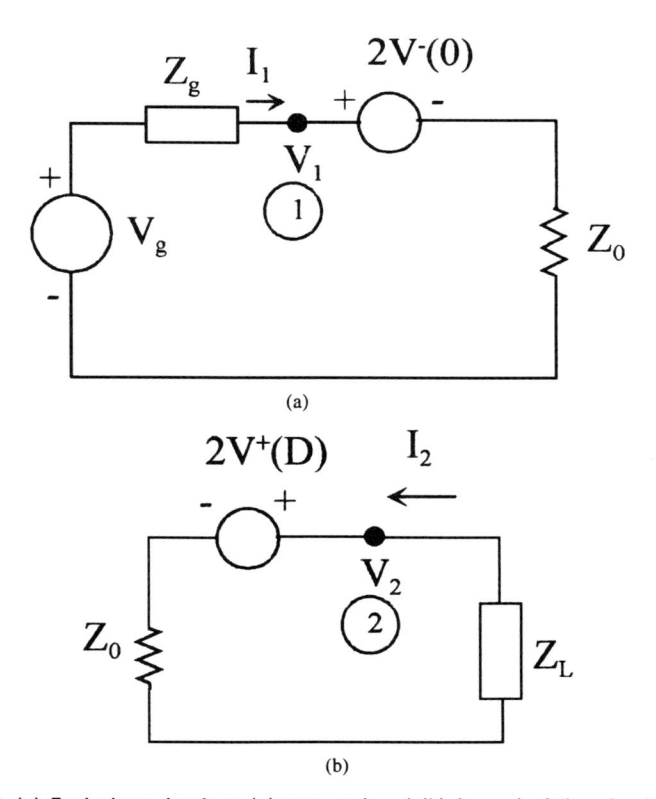

$$V_2 = V^+(D) + V^-(D) \qquad (13.78)$$
$$I_2 = -I^+(D) + I^-(D) \qquad (13.79)$$

Figure 13.9 (a) Equivalent circuit at (a) near-end and (b) far-end of the circuit shown in Figure 13.8.

Multiplying (13.79) by Z_0 and adding the resulting equation to (13.78), we obtain

$$2V^-(D) = V_2 + Z_0 I_2 \qquad (13.80)$$

Further:

$$2V^-(0) = 2V^-(D)e^{-j\beta D} = (V_2 + Z_0 I_2)e^{-j\beta D} \qquad (13.81)$$

where β denotes the propagation constant of the transmission line. (13.77) then becomes

$$V_1 - Z_0 I_1 = (V_2 + Z_0 I_2)e^{-j\beta D} \qquad (13.82)$$

(13.82) represents one relationship between the terminal quantities at the input and output. To obtain another, we proceed in a similar manner as above and obtain an equivalent circuit at the load end as shown in Figure 13.9(b). The terminal voltage and current at the load end are related to the forward-traveling voltage wave at $z = D$ as follows:

$$V_2 - Z_0 I_2 = 2V^+(D) \qquad (13.83)$$

After a few steps similar to those used in deriving (13.82), we obtain

$$V_2 - Z_0 I_2 = (V_1 + Z_0 I_1)e^{-j\beta D} \qquad (13.84)$$

Further, substituting

$$V_1 = V_g - I_1 Z_g \qquad (13.85)$$
$$V_2 = -I_2 Z_L \qquad (13.86)$$

in (13.82) and (13.84), two simultaneous equations in the unknowns I_1 and I_2 are obtained. The solution of these two equations gives I_1 and I_2. Further, by substituting the values of I_1 and I_2 in (13.85) and (13.86), respectively, the terminal voltages V_1 and V_2 can be determined.

References

[1] Chang, F. Y., "Transient Analysis of Lossless Coupled Transmission Lines in a Nonhomogeneous Dielectric Medium," *IEEE Trans.*, Vol. MTT-18, Sept. 1970, pp. 616–626.

[2] Marx, K. D., "Propagation Modes, Equivalent Circuits, and Characteristic Terminations for Multiconductor Transmission Lines With Inhomogeneous Dielectrics," *IEEE Trans.*, Vol. MTT-21, July 1973, pp. 450–457.

[3] Sun, Y. Y., and K. D. Marx, "Comments and Reply on 'Propagation Modes, Equivalent Circuits, and Characteristic Terminations for Multiconductor Transmission Lines With Inhomogeneous Dielectrics,'" *IEEE Trans.*, Vol. MTT-26, Nov. 1978, pp. 915–918.

[4] Weeks, W. T., L. L. Wu, M. F. McAllister, and A. Singh, "Resistive and Inductive Skin Effect in Rectangular Conductors," *IBM J. Res. Develop.* Vol. 23, Nov. 1979, pp. 652–660.

[5] Wei, C., R. F. Harrington, J. R. Mautz and T. K. Sarkar, "Multiconductor Transmission Lines in Multilayered Dielectric Media," *IEEE Trans.*, Vol. MTT-32, Apr. 1984, pp. 439–450.

[6] Harrington, R. F., and C. Wei, "Losses in Multiconductor Transmission Lines in Multilayered Dielectric Media," *IEEE Trans.*, Vol. MTT-32, July 1984, pp. 705–710.

[7] Djordjevic, A. R., D. D. Cvetkovic, G. M. Cujic, T. K. Sarkar, and M. B. Bazdar, *MULTLIN for Windows: Circuit Analysis Models for Multiconductor Transmission Lines, Software and User's Manual*, Norwood, MA: Artech House, 1996.

[8] Djordjevic, A. R., T. K. Sarkar, R. F. Harrington, and M. B. Bazdar, *Time-Domain Response of Multiconductor Transmission Lines, Software and User's Manual*, Norwood, MA: Artech House, 1989.

[9] Djordjevic, A. R., et al., *Scattering Parameters of Microwave Networks With Multiconductor Transmission Lines, Software and User's Manual*, Norwood, MA: Artech House, 1989.

[10] Djordjevic, A. R., M. B. Bazdar, R. F. Harrington, and T. K. Sarkar, *LINPAR for Windows: Matrix Parameters for Multiconductor Transmission Lines, Software and User's Manual*, Norwood, MA: Artech House, 1995.

[11] Mathews, J., and R. L. Walker, *Mathematical Methods of Physics*, CA: The Benjamin Cummings Publishing Co., 1970.

[12] Gentili, G. G., and M. Salazar-Palma, "The Definition and Computation of Modal Characteristic Impedances in Quasi-TEM Coupled Transmission Lines," *IEEE Trans.*, Vol. MTT-43, Feb. 1995, pp. 338–343.

[13] Tripathi, V. K., and J. B. Rettig, "A SPICE Model for Multiple Coupled Microstrip and Other Transmission Lines," *IEEE Trans.*, Vol. MTT-33, Dec. 1985, pp. 1,513–1,518.

[14] Tripathi, V. K., and A. Hill, "Equivalent Circuit Modelling of Losses and Dispersion in Single and Coupled Lines for Microwave and Millimeter-Wave Integrated Circuits," *IEEE Trans.*, Vol. MTT-36, Feb. 1988, pp. 256–262.

[15] Carin, L., and K. J. Webb, "An Equivalent Circuit Model for Terminated Hybrid Mode Multiconductor Transmission Lines," *IEEE Trans.*, Vol. MTT-37, Nov. 1989, pp. 1,784–1,793.

[16] Djordjevic, A. R., T. K. Sarkar, and R. F. Harrington, "Time-Domain Response of Multiconductor Transmission Lines," *Proc. IEEE*, Vol. 75, No. 6, June 1987, pp. 743–764.

[17] Arabi, T. R., et al., "On the Modelling of Conductor and Substrate Losses in Multiconductor, Multidielectric Transmission Line Systems," *IEEE Trans.*, Vol. MTT-39, July 1991, pp. 1,090–1,097.

[18] Djordjevic, A. R., "SPICE-Compatible Models for Multiconductor Transmission Lines in Laplace-Transform Domain," *IEEE Trans.*, Vol. MTT-45, May 1997, pp. 569–579.

Appendix

Table A.1
Physical Constants

Permittivity of vacuum, $\varepsilon_0 = 8.854 \times 10^{-12} \cong (1/36\pi) \times 10^{-9}$ F/m
Permeability of vacuum, $\mu_0 = 4\pi \times 10^{-7}$ H/m
Impedance of free space, $\eta_0 = 376.7 \cong 120\pi \, \Omega$
Velocity of light, $c = 2.998 \times 10^8$ m/sec
Charge of electron, $e = 1.602 \times 10^{-19}$ C
Mass of electron, $m = 9.107 \times 10^{-31}$ kg
Boltzmann's constant, $k = 1.380 \times 10^{-23}$ J/K
Planck's constant, $h = 6.547 \times 10^{-34}$ J-sec

Table A.2
Properties of Various Dielectric Materials @ 10 GHz

Material	Dielectric Constant, ε_r	Loss Tangent $\tan \delta \, (\times 10^{-4})$	Thermal Conductivity K(W/m.°C)
Alumina	9.8	2	37.0
Sapphire	11.7	1	46.0
Quartz	3.8	1	1.0
Si($\rho = 10^3 \Omega$ cm)	11.7	40	145.0
GaAs($\rho = 10^8 \Omega$ cm)	12.9	5	46.0
InP	14.0	5	68.0
AlN	8.8	10	230.0
BeO	6.4	3	250.0
SiC	40.0	50	270.0
Polyimide	3.0	10	0.2
Teflon	2.1	5	0.1
Duroid	2.2	9	0.26
Air	1.0	0	0.024

Table A.3
Properties of Various Conductor Materials

Metal	Melting Point (°C)	Electrical Resistivity ($10^{-6}\Omega \cdot$ cm)	Thermal Expansion Coefficient ($10^{-6}/°$C)	Thermal Conductivity K(W/m·°C)
Copper	1093	1.7	17.0	393
Silver	960	1.6	19.7	418
Gold	1063	2.2	14.2	297
Tungsten	3415	5.5	4.5	200
Molybdenum	2625	5.2	5.0	146
Platinum	1774	10.6	9.0	071
Palladium	1552	10.8	11.0	070
Nickel	1455	6.8	13.3	092
Chromium	1900	20.0	6.3	066
Kovar	1450	50.0	5.3	017
Aluminum	660	4.3	23.0	240
Au-20% Sn	280	16.0	15.9	057
Pb-5% Sn	310	19.0	29.0	063
Cu-W(20% Cu)	1083	2.5	7.0	248
Cu-Mo(20% Cu)	1083	2.4	7.2	197

About the Authors

Dr. Rajesh Mongia obtained a Ph.D. in electrical Engineering from Indian Institute of Technology (IIT), New Delhi, India in 1989. From 1981–1989 he was employed at the Center for Research in Electronics at IIT where he worked on various projects in the area of microwave and millimeter-wave circuits. Dr. Mongia also held post-doctoral positions at Florida State University, Tallahassee Florida, University of Ottawa, and Ottawa Communications Research Centre. He also worked as an Advanced Member of the technical staff at COMDEV, in Cambridge, Ontario, Canada where he was responsible for the design of dielectric resonator filters and mutiplexers for communications statellites. Presently, Dr. Mongia is employed at Bosch Telecom Inc., in Dallas, Texas where his work involves the design of RF circuits for LMDS applications.

Dr. Inder Bahl received a Ph.D. in electrical engineering from the Indian Institute of Technology, Kanpur, India in 1975. From 1969 to 1981, Dr. Bahl held positions at ITT Kanpur and the Ottawa University and Defense Research Establishment, Ottawa, Canada. He joined the ITT Gallium Arsenide Technology Center in 1981 and has been working on microwave- and millimeter-wave GaAs Ics. In his present capacity as an executive scientist at ITT GaAstek, Dr. Bahl's interests are in the areas of device modeling, high-efficiency high-power amplifiers, 3-D MMICs, and the development of MMIC products for commercial and military applications.

Dr. Bahl is the author/co-author of over 120 research papers. He authored/co-authored *Microstrip Lines* (Artech House), *Microstrip Antennas* (Artech House), *Millimeter Wave Engineering and Applications* (John Wiley), *Microwave Solid State Circuit Design* (John Wiley), *Microwave and Millimeter-Wave Heterostructure Transistors and Their Applications* (Artech House), and the *Gallium Arsenide IC Applications Handbook* (Academic Press). Dr. Bahl also contributed two chapters to the *Handbook of Electrical Engineering* (CRC Press), and three chapters to the *Gallium Arsenide IC Applications Handbook*. He holds 14 patents in the areas of microstrip antennas and microwave circuits.

Dr. Bahl is an IEEE fellow and a member of the Electro-magnetic Academy.

Dr. Prakash Bhartia obtained his MSc and Ph.D. degrees in Electrical Engineering from the University of Manitoba in Winnipeg. He served as an Associate Professor and Assistant Dean of Engineering at the University of Regina until 1977. In that year he joined the Department of National Defense in Ottawa, and served in a number of Director-level positions. In 1992 he was appointed Director General of the Defense Research Establishment Atlantic in Halifax and in 1998 as Director General of the Defense Research Establishment Ottawa, in which position he currently serves.

Dr. Bhartia has had considerable consulting experience with many companies while serving at the University and is the author of over 150 papers in the areas of radar, microwave, and millimeter-wave circuits, components and transmission lines. He is also the co-author of a number of books including *Microstrip Antennas* (Artech House), *Millimeter Wave Engineering and Applications* (John Wiley), *E-Plane Integrated Circuits* (Artech House), *Microwave Solid State Circuit Design* (John Wiley), *Millimeter Wave Microstrip and Printed Circuit Antennas* (Artech House) and *Microstrip Lines and Slotlines* (Artech House). He has also contribued chapters to other texts and holds a number of patents. Dr. Bhartia is a Fellow of the Royal Society of Canada, a Fellow of the IEEE, and a Fellow of the Institution of Electrical and Telecommunications Engineers.

Index

π/4 quadrature phase-shift keying (QPSK) modulation, 18
180-degree hybrid, 262

ABCD matrices, 49, 51
    branch-line coupler, 252
    lumped-element coupler, 253, 256
    parallel-coupled line, 324
    planar balun, 408
ABCD parameters, 47–49
    advantages, 48
    determining, 51
    of elementary two-port networks, 54
    normalized, 49, 51, 53
    properties, 49
    reflection/transmission coefficients, 49–53
    representation illustration, 47
    unnormalized, 49, 51
Admittance
    asymmetrical coupled line, 149–50
    characteristic, 35
Admittance matrix, 34
    coupled-line balun, 400
    multiconductor transmission lines, 483
    spiral inductor, 377
    symmetrical coupled lines, 492

Amplitudes
    broadside microstrip coupler, 287
    ideal coupler, 224
    instantaneous, 177
    Marchand balun, 433
    planar balun, 412–13
    time-varying, 177
Asymmetrical couplers, 166–68
    defined, 181
    design, 166–67
    fabrication, 166
    multisection, 194
    output ports, phase difference, 167–68
    performance, 167–68
    strip pattern, 166
    theoretical/experimental response, 167
    See also Directional couplers
Asymmetrical nonuniform TEM couplers, 237–42
    design, 242
    equivalent circuit, 240
    even-model equivalent circuit, 239
    illustrated, 239
    reflection coefficient, 239
    scattering matrix, 241–42
    scattering parameters, 238
    tandem connection, 241

Asymmetrical nonuniform TEM
    couplers (continued)
    See also Nonuniform TEM directional
        couplers
Asymmetric coupled lines, 140–59
    admittances, 149–50
    π mode, 141–42
    backward-wave coupler, 157–59
    characteristic impedances, 149–50
    c mode, 140–41
    coupling coefficients, 150
    current wave, 141
    directional couplers using, 154–59
    distributed equivalent circuit, 145–50
    forward-wave coupler, 154–57
    four-port network composed of, 143
    maximum power between, 176
    normal mode vs. distributed line
        parameters, 151–52
    parameters, 140–45
    propagation constants, 148
    uniformly coupled, 170
    voltage wave, 141
    Y-parameters, 145
    Z-parameters, 144, 145
Attenuation
    constant, 67, 69
    expression, 69
    Q-factor, 70
    total, 69–71

Backward-wave couplers, 123
    ABCD matrices, 136
    defined, 10, 181
    illustrated, 158
    remarks, 139
    scattering matrix, 138
    scattering parameters, 137
    using asymmetrical coupled
        lines, 157–59
    See also Directional couplers
Baluns, 391–442
    analysis, 399–403
    below microwave frequencies, 394
    broadband, 409, 410
    coaxial, 394
    coplanar waveguide, 435–38

coupled-line, 399–403
defined, 391
distributed, 413
introduction to, 391–95
lumped-element, 393
Marchand, 392, 411–35
microstrip-to-balanced stripline, 395–99
multilayer, 425
nonplanar, 396
parallel-strip, 398–99
planar, 392, 395, 403–11
planar-transformer, 441–42
printed circuit, 418–19
triformer, 438–40
usage requirement, 391
wide-band, 417
Bandpass filters
    coplanar waveguide, 341
    frequency response, 317
    high-K ceramic back combline, 342
    response, 307
    series inductors, 316
    structure, 317
    transformation, 316–18
    See also Filters
Bandstop filters
    frequency response, 318
    parallel-tuned circuit element values, 319
    response, 307
    series elements, 319
    series-tuned circuit element values, 319
    shunt element values, 318
    structure, 318
    transformation, 318–19
    See also Filters
Bandwidth
    CPW balun, 435
    fractional, 185
    frequency, ratio, 184
    lumped-element coupler, 260
    Marchand balun, 411, 413, 425,
        427, 432
    rat-race coupler, 264
    useful operating, 185
Biasing circuits, 357–62
    illustrated, 359
    insertion loss, 362

network structure parameters, 361
return loss, 362
schemes, 357–59
simulated response, 361
using mulisections shunt stubs, 360
Bi-orthonormal eigenvectors, 481
Branch-line couplers, 243, 244
   broadband, 260
   characteristic impedance, 245–46
   coupling variation, 248
   defined, 244
   frequency response, 251
   isolation, 247
   isolation variation, 248
   layout, 245
   for loose coupling, 247
   lumped-element, 255–60
   modified, 247–51
   perfectly matched, 246
   physical implementation, 247
   port impedance, 244
   properties, 246–47
   reduced-size, 251–55
   scattering matrix, 245
   scattering parameters, 244–45
   VSWR coupling, 247
   VSWR variation, 248
   See also Directional couplers; Tight couplers
Broadband baluns, 409, 410
   measured performance, 410
   photograph, 411
   See also Baluns
Broadband branch-line coupler, 260
Broadband coupling, 168–69
Broadband dc block, 357
Broadband forward-wave couplers, 161–78
Broadband rat-race coupler, 265
Broadside-coupled striplines, 7, 109–12
   characteristic impedance, 110–12
   loose coupling, 114, 117
   offset, 114–17
   tight coupling, 114, 116–17
   See also Striplines
Broadside coupled structures, 1, 3, 109–17
   inverted microstrip lines, 7
   offset striplines, 114–17

   striplines, 7, 109–12
   suspended microstrip lines, 7, 8, 112–14
Broadside-coupled suspended microstrip lines, 7, 8, 112–14
   characteristic impedance, 111
   even-/odd-mode characteristic impedance, 112
   even-/odd-mode effective dielectric constants, 112–14
   phase velocity, 111
   See also Suspended microstrip lines
Broadside couplers, 285–87
   amplitude characteristics, 287
   cross section, 286
   defined, 285
   measured performance, 286
   photograph, 287
   structure, 286
   See also Multilayer couplers
Butterworth filter, 309–10
   element values, 310
   insertion loss, 309
   lowpass prototype, 309
   See also Filters

Capacitance
   characteristic impedance and, 71
   even-mode, 76
   fringing, 80–81, 216
   lumped, 209
   mutual, 271–73
   odd-mode, 76–79, 217
   parallel-plate, 80–81
   parameters, 139
   representation, 75
   series, 371–72
   shunt, 381
   total, breakup, 83
   wiggly-coupled line, 214
Capacitance matrix, 75–76, 157
   of lossy dielectric material, 455
   of multiconductor transmission lines, 479
Capacitive coupling coefficient, 150
Capacitors
   interdigital, 367–73
   lumped, 209–12

Capacitors (continued)
    MIM, 357, 362
    series, 353
    shunting, 159
Ceramic filters, 339–48
    combline bandpass, 342
    coupling coefficient, 343, 346, 347
    cross-sectional view, 345
    effective dielectric constant, 345
    frequency response, 347
    low-profile, 348
    materials, 342
    resonator length, 343
    specifications, 346
    See also Filters
Characteristic impedances, 26, 30, 35, 40, 67
    asymmetrical coupled lines, 149–50
    branch-line coupler, 245–46
    broadside-coupled stripline, 110–12
    capacitance and, 71
    CPW, 103
    determining, 69
    edge-coupled microstrip line, 99–101
    of edge-coupled striplines, 88, 89
    even-mode excitation, 79
    finite strip thickness effect on, 84–88
    interconnect, 451–52, 460
    inverted microstrip line, 105, 111
    Marchand balun, 411–13, 422
    microstrip line, 91, 93
    microstrip line of finite thickness, 92
    phase velocity and, 69
    quasi-static, 91, 93, 94
    quasi-TEM mode, 72
    reduced-size branch-line coupler, 253
    slot-coupled microstrip line, 118–20
    stripline, 85
    suspended microstrip line, 105, 111
Chebyshev response filter, 310–12
    element values, 313–14
    insertion loss, 310
    nomograph, 311
    ripple value, 312
    See also Filters
Circuit components, 353–87
    coupled-line transformers, 364–67

DC blocks, 353–64
interdigital capacitor, 367–73
spiral inductors, 373–83
spiral transformers, 383–87
Circulator, scattering matrix, 43
Coaxial line, 66
Coaxial Marchand balun, 415–21
    calculated VSWR vs. electrical length for, 418
    defined, 415–16
    equivalent circuit representation, 417
    illustrated, 394
    performance influences, 421
    schematic, 417
    See also Baluns; Marchand balun
Codirectional couplers. *See* Forward-wave couplers
Combline filters, 14, 328–30
    capacitances, 328
    coaxial, 342
    defined, 328
    design procedures, 330
    ground connections, 330
    illustrated, 16, 17
    See also Filters
Compact couplers, 192
    lumped-element, 298–99
    meander line, 300–301
    spiral, 299–300
    tight, 296–301
Compact Marchand balun, 434–35
Computer-aided engineering (CAE) tools, 372
Computer-assisted design (CAD), 319–20
Conductor loss
    factor, 71
    microstrip line, 94–95
    normalized, 86
    of quasi-TEM mode, 74
    of TEM mode, 74
Conductor materials, properties, 500
Conversion relations, 55–56
Coplanar waveguide (CPW) balun, 435–38
    analysis, 435
    bandwidth, 435
    cross section, 438
    physical layout, 435, 437

top view, 438
See also Baluns
Coplanar waveguides, 66, 101–5
   bandpass filter, 341
   characteristic impedance, 103
   conductor-backed, 102, 104–5
   configurations, 102
   coupled, 105
   effective dielectric constant, 103
   illustrated, 6, 67
   on finite thickness dielectric substrate, 102
   with upper shielding, 102, 103–5
Coupled-line balun, 399–403
   admittance matrix, 400
   amplitude response, 403
   configuration and excitation, 401
   disadvantages of, 403
   for even-mode impedance, 403
   ground plane effects, 402
   illustrated, 400
   Marchand, 423–24
   performance, 402–3
   See also Baluns
Coupled-line circuit components, 353–87
Coupled-line filters. *See* Filters
Coupled-line transformers, 364–67
   bandwidth, 367
   broadband impedance-matching, 364
   illustrated, 365
   spiral, 382
   symmetrical, 367
   See also Circuit components
Coupled lossy interconnect lines, 449–72
Coupled-mode theory, 161–62, 169–77
   asymmetrical lines and, 175–77
   defined, 161
   even-/odd-mode analysis and, 174
   symmetrical lines and, 173–74
   for weakly coupled resonators, 177–78
Coupled structures, 1–9
   broadside, 1, 3
   components based on, 9–18
   cross-sectional view, 4
   edge, 1
   forms, 1–2
   symmetric and uniformly coupled, 4

types of, 4–6
Coupling
   abrupt discontinuities in, 11
   broadband, 168–69
   capacitance parameters and, 139
   desirable, 2–3
   dielectric guides, 8
   mechanism, 6–9
   parasitic, 3
   response, 225–26, 227
   between symmetrical lines, 173–74
   in terms of even-mode characteristic impedance, 221–23
   tight, 243–302
Coupling coefficients, 7
   capacitive, 150
   ceramic filter, 343, 346, 347
   inductive, 150
   Marchand balun, 422
   nature of, 171–72
   re-entrant mode coupler, 296
   structure specifics and, 173
   voltage, 178
Crosstalk voltages, 460, 462–69
   for coupled microstrip lines of different spacings, 464, 467
   for microstrip/stripline configurations, 465, 468
   on active line for various line lengths, 469
   for various line losses, 462, 466
   for various terminations of active lien, 471
   for various terminations of quiet line, 470
   See also Interconnects
Current eigenvectors, 480–81
   bi-orthonormal, 481
   matrix, 481
   normalization, 481
   normalized, matrix, 492
   properties, 480–81
Currents
   equivalent, 24
   interconnect terminal, 451
   normalized, 24–28
   on asymmetrical coupled lines, 141
   total, 31
   unnormalized, 28–29

Dc blocks, 353–64
    analysis, 353–57
    biasing circuits, 357–62
    broadband, 357
    designed vs. measured dimensions, 363
    fabricated on RT/duroid substrate, 364
    high-voltage, 362–64
    millimeter-wave, 362
    quarter-wave coupled-line section as, 354
    simulated VSWR vs. fractional bandwidth of, 356
    See also Circuit components
Decoupled modes, 477
Desirable coupling, 2–3
Dielectric loss
    determination of, 71
    edge-coupled stripline, 89–90
    microstrip line, 94–95
    quasi-TEM mode, 72
    transmission lines, 455
Dielectric materials, properties, 499
Dielectric overlays, 212–13
    defined, 212
    design curves, 214
    fabrication and, 213
Directional couplers, 9–14
    90-degree hybrid, 14
    asymmetrical, 166–68, 181
    asymmetrical nonuniform TEM, 237–42
    backward-wave, 10, 123, 136–39
    branch-line, 243, 244–60
    compact, 192, 296–301
    crossover, 430
    forward-wave, 123, 132, 133–36, 161–78
    four-port, 182, 183
    ideal, 224–28
    isolation, 182–83
    length of, 164
    meander line, 300–301
    measured directivity of, 165
    microstrip, 209–17
    multiconductor, 270–81
    multilayer, 284–96
    multisection, 193–209
    nonuniform, 169, 219–42
    overlay, 212–13
    parallel-coupled TEM, 181–217
    power loss ratio, 196–99
    quarter-wave, 192–93
    rat-race, 243, 244, 260–70
    scattering matrix, 44, 46
    with shunting capacitors, 159
    single-section, 182, 183–93
    slot-coupled microstrip lines and, 117–18
    spiral, 299–300
    symmetrical, 181
    symmetrical nonuniform TEM, 219–37
    tandem, 11, 243, 281–84
    TEM, 181–217
    tight, 15–16, 243–302
    uses, 9–10
    using uniform coupled lines, 130–39
    wiggly two-line, 13
Directivity, 209–17
    branch-line coupler, 250
    improving, 209–17
Dispersion, microstrip line, 95–97
Distributed equivalent circuits, 145–50
Distributed line parameters
    of asymmetrical coupled lines, 152–53
    determining, 151
    normal mode vs., 150–52

Edge-coupled microstrip lines, 98–101
    characteristic impedances, 99–101
    cross section, 98
    effective dielectric constant, 98
    frequency-dependent characteristics, 100
    See also Microstrip lines
Edge-coupled striplines, 88–90
    characteristic impedances, 88, 89
    cross section, 83
    defined, 83–84
    dielectric loss, 89–90
    propagation constant, 89
    synthesis equations, 88
    See also Striplines
Edge-coupled structures
    defined, 1
    tight-coupling between, 244
Effective dielectric constant, 71–72
    broadside-coupled suspended microstrip line, 112–14

CPW, 103
  of even-/odd-modes, 109
  inverted microstrip line, 107, 111
  microstrip line, 92
  quasistatic, 104
  slot-coupled effective dielectric
       constant, 118–20
  suspended microstrip line, 105–7
Eigenvalues, 479–81
  of lossy lines, 484
  of symmetrical coupled lines, 488–94
Eigenvectors
  current, 480–81
  of lossy lines, 484
  of odd mode, 491
  of symmetrical coupled lines, 488–94
  voltage, 480–81, 491
Electromagnetic (EM) simulators,
       372, 373, 431
Embedded microstrip couplers, 287–93
  cross section, 289
  defined, 287–88
  insertion phase difference, 293
  measured performance, 291–92
  phase difference, 289
  physical layout, 289
  top view, 290
  See also Multilayer couplers
Equal ripple function, 232
Equivalent circuits
  asymmetric nonuniform coupler, 240
  coaxial Marchand balun, 417
  distributed, 145–50
  even-mode scattering parameters,
       131, 132
  even-/odd-mode excitation, 210
  at far-end of network, 488
  lumped-element coupler, 259, 298
  Marchand balun, 415
  multiconductor transmission lines,
       475, 485–88
  multisection directional coupler, 194
  at near-end of network, 486
  odd-mode scattering matrix, 132
  quarter-wave couplers, 192–93
  single transmission line network, 494–96
  spiral inductor, 378, 380

symmetrical multisection coupler, 196
  triformer balun, 439
Equivalent RLC networks, 320, 321–22
Even-mode excitation, 76, 126–28
  capacitance representation, 78
  characteristic impedance, 79
  defined, 126
  equivalent circuit, 210
  illustrated, 77
  phase constant, 79
  phase velocity, 79

False switching, 472
Far-end crosstalk voltage, 460
  for coupled microstrip lines of different
       spacings, 464, 467
  false switching and, 472
  for microstrip/stripline configurations,
       465, 468
  on active line for various line lengths, 469
  reduction, 472
  for various line losses, 462, 466
  for various terminations of active
       line, 471
  for various terminations of quiet line, 470
  See also Crosstalk voltage
Fast Fourier transform (FFT)
       procedures, 449
Filter analysis, 319
Filters, 14–18, 305–48
  applications, 308
  bandpass, 307
  bandstop, 307
  Butterworth, 309–10
  CAD and, 319–20
  Chebyshev response, 310–12
  coaxial-line, 16
  combline, 14, 16, 17, 328–30
  configuration, 306
  construction considerations, 320
  coupled-line, 16–18
  design, 308–23
  first use of, 14
  group delay, 306, 323
  hairpin-line, 14, 17, 330–34
  highpass, 307
  insertion loss, 305

Filters (continued)
  interdigital, 14, 16, 17, 326–28
  introduction to, 305–8
  lowpass, 307
  microwave passive, 14
  miniature, 337–48
  Minnis-type, 337–39, 340
  parallel-coupled line, 14, 17, 323–26
  parallel-coupled stepped-impedance, 335–36
  parameters, 305
  practical considerations, 320–23
  printed circuit, 322–23
  realization of, 19
  response-type, 312–14
  return loss, 305
  theory, 308–23
  transformation, 314–19
  types of, 307, 323–36
Finite Q, 320, 322
Finite-thickness microstrip, 91–93
Forward-wave couplers, 123
  with asymmetrical directional couplers, 154–57
  bandwidth of, 161
  broadband, 161–78
  defined, 132
  illustrated, 154
  realization of, 161
  remarks, 135–36
  scattering parameters, 134
  with symmetrical coupled lines, 162
  types of, 161
  ultra-broadband, 168–69
  See also Directional couplers
Four-port directional coupler, 182, 183
Four-port networks, 43–46
  of asymmetrical coupled lines, 143
  composed of parallel coupled asymmetrical lines, 451
  reciprocal, 45
  reduction into two-port network, 62–63
  reflection coefficients, 62
  scattering matrix, 62
  symmetrical, 124, 127
  with uniform coupled symmetrical lines, 131

Z-parameters of, 152
  See also N-port networks
Fourth-order Marchand balun, 422–24
Fringing capacitance, 80–81
  defined, 80
  determining, 83
  parallel capacitances, 81–82
Fringing fields, 80

Gibb's phenomenon, 226
Gotte function, 231
Group delay, 306
  constant, 323
  defined, 323

Hairpin-line filters, 14, 330–34
  defined, 330
  design, 330–31
  ground connections, 330
  illustrated, 17
  inter-resonator coupling, 331
  layout, 333
  measured coupling coefficients, 331
  measured response, 334
  M/FILTER software for, 334
  physical layout, 331
  substrate material, 334
  See also Filters
High-dielectric constant ceramics, 339–48
Highpass filters
  frequency response, 315
  response, 307
  schematic, 315
  transformation, 314–16
  See also Filters
High-speed circuit interconnects.
  See Interconnects
High-voltage dc block, 362–64
  defined, 362–64
  illustrated, 364
  See also Dc blocks
Hyperbolic scent, 84

Ideal directional coupler, 224–28
  amplitude, 224
  coupling response, 225
  See also Directional couplers
Impedance

Index

even-/odd-mode, 190–91
normalized, 24, 30
ratio, 278, 335
surface, 73
transformation of, 30–31
See also Characteristic impedance
Impedance matrix, 33–34
multiconductor transmission lines, 483
normalized, 34
symmetrical coupled lines, 493
unnormalized, 34, 35
Imperfect conductors, 73
Incident wave, 36, 124
Incremental inductance rule, 74
Inductive coupling coefficient, 150
Insertion loss, 51
biasing circuits, 362
Butterworth filter, 309
Chebyshev response, 310
filter, 305
function, 196
Marchand balun, 431, 432
monolithic Marchand balun, 434
Interconnects, 447–72
"active" line, 456
capacitance variation, 455
cases, 456
characteristic impedance, 451–52, 460
chips, 447
conductor bar, 453
conductor loss, 459, 472
coupled lines, 457
electrical oscillations, 448
external inductance, 454–55
far-end crosstalk voltage, 460, 462–71
frequency dependence of line parameters, 452–56
inductance/capacitance matrices, 457, 459
input pulse waveform, 457
line length, 447–48
near-end crosstalk voltage, 460, 462–68, 470–71
performance analysis, 450
propagation delays, 448
proximity effects, 448
"quiet" line, 456

results, 456–72
series resistance, 452–53
terminal voltages/currents, 451
transient response, 460, 461
transmission line inductance, 454
Interdigital capacitors, 367–73
approximate analysis, 370–72
approximate expression, 370
configuration, 370
coupled microstrip line, 371
defined, 367
illustrated, 370, 371
responses, 374
series capacitance, 371–72
series resistance, 372
single microstrip line, 371
structure parameters, 373
subcomponents, 371
T-junction, 371
unsymmetrical bend, 371
unsymmetrical gap, 371
See also Circuit components
Interdigital couplers, 271–81
coaxial, 342
design, 274–81
impedance parameters, 278
impedance ratio, 278
length of, 277
N-conductor, 276
results, 277
theory, 271–74
top/side views, 274
See also Directional couplers
Interdigital filters, 14, 326–28
design bandwidths, 326
design equations, 327
illustrated, 16, 17
popularity, 326
realization, 328
singly-loaded Q, 327
See also Filters
Inverted microstrip lines, 105–9
broadside-coupled, 7
characteristic impedance, 105, 111
cross section, 110
effective dielectric constant, 107, 111
illustrated, 5

Inverted microstrip lines (continued)
  See also Microstrip lines
Isolation, 182–83
  branch-line coupler, 247, 248
  rat-race coupler, 263

Lange couplers, 243, 274–81
  3-dB coupling, 279
  design data, 279–81
  design of, 274–81
  dimensional ratios, 280
  gap dimensions, 281
  Marchand balun using, 428
  metalization thickness, 281
  planar baluns and, 409
  realization, 279
  unfolded, 275
  See also Directional couplers; Tight couplers
LIBRA, 211
Loaded Q, 322, 327
Lowpass filters
  prototype, 308
  response, 307
  See also Filters
Lumped capacitors, 209–12
  compensated microstrip coupler, 210
  directivity improvement and, 212
  example, 211
  fabrication and, 213
  physical length of, 211
  See also Capacitors
Lumped-element baluns
  illustrated, 393
  layout, 437
  See also Baluns
Lumped-element-based miniaturization, 337–39
Lumped-element branch-line coupler, 255–60
  ABCD matrix, 253, 256
  bandwidth, 260
  defined, 255
  element values, 259, 260
  equivalent circuit model, 259
  implementation, 256
  quarter-wave section, 258

See also Branch-line couplers
Lumped-element rat-race coupler, 268–70
  design, 268
  illustrated, 272
  lumped elements, 270
  See also Rat-race couplers

Marchand balun, 392, 411–35
  amplitude, 433
  bandwidth, 411, 413, 425, 427, 432
  characteristic impedance, 411–13, 422
  coaxial, 415–21
  coaxial cross section, 414
  compact, 434–35
  coupled-line, 423–24
  coupled-line parameters, 424, 427
  coupling coefficient, 422
  with crossover couplers, 430
  defined, 411
  design parameters, 432
  elements, 411
  equivalent circuit, 415
  equivalent circuit parameters, 425
  equivalent-line model, 414, 424
  examples, 426–35
  fourth-order, 422
  illustrated, 414
  insertion loss, 431, 432
  with Lange coupler, 428
  linear circuit simulation, 434
  measured loss of, 436
  monolithic, 434
  with multilayer microstrip structures, 432
  phase balances, 433
  phase difference, 431
  physical dimensions, 424
  printed-circuit fabrication, 413
  realization, 421
  with re-entrant coupler, 429–33
  return loss, 436
  series/parallel-resonant compensating representation, 415
  synthesis of, 421–26
  tightly-coupled sections, 426
  See also Baluns
Meander line directional coupler, 300–301
  defined, 300–301

illustrated, 300
measured performance, 301
physical dimensions, 301
See also Directional couplers
Metal-insulator-metal (MIM) shunt
    capacitors, 255, 265
M/FILTER software, 334
Microstrip lines, 90–101
    broadside-coupled, 110
    characteristic impedance, 91, 92, 93, 94
    characteristics of, 90
    conductor loss, 94–95
    coupled, 458
    cross section, 90
    dielectric loss, 94–95
    dispersion, 95–97
    edge-coupled, 98–101
    effective dielectric constant, 92
    even-/odd-mode field configurations, 10
    finite-thickness, 91–93
    illustrated, 5, 67
    inverted, 5, 7, 105–9
    length of, 165
    normalized conductor Q-factor, 96
    single, 90–97
    slot-coupled, 117–20
    striplines vs., 90
    suspended, 5, 7, 8, 105–9, 112–14
    symmetrical, 164–65
    synthesis equations, 92–93
    with unequal impedances, 9
Microstrip-to-balanced stripline
    balun, 395–99
    defined, 395–97
    example, 397–98
    parallel-plate, 398–99
Microwave integrated circuits (MICs), 90
Microwave network
    ports, 37
    theory, 23–63
Millimeter-wave dc block, 362
MIM capacitors, 357, 362
Miniature filters, 337–48
    high-dielectric constant ceramics, 339–48
    lumped-element, 337–39
    MMIC, 339
    See also Filters

Minnis-type filter, 337–39
    defined, 337
    illustrated, 340
    See also Filters
Modified branch-line coupler, 247–51
    directivity, 250
    frequency response, 250
    illustrated, 250
    simulated response of, 250
    See also Branch-line couplers
Modified rat-race coupler, 264–65
Monolithic Marchand balun, 434
    defined, 434
    illustrated, 435
    insertion loss, 434
    See also Baluns
Monolithic microwave integrated circuits
    (MMICs), 18
    filters, 339
    triformer, 386
Multiconductor couplers, 270–81
    coupling factor, 287
    defined, 271
    design, 274–81
    theory, 271–74
    See also Directional couplers
Multiconductor transmission lines, 475–96
    ($N+1$) conductor, 476, 481, 485
    admittance matrix, 483
    applications, 475
    capacitance matrix, 479
    connected to arbitrary networks, 485
    decomposition in terms of normal
        modes, 481–83
    eigenvalues/eigenvectors, 479–81
    equivalent circuit, 475, 485–88
    examples, 488–96
    impedance matrix, 483
    inductance matrix, 479
    reflection matrix, 484
    theory, 477–88
    three-conductor, 476
    transmission coefficient matrix, 484
    two-conductor, 476
    See also Transmission lines
Multilayer couplers, 284–96
    broadside, 285–87

Multilayer couplers (continued)
    defined, 284
    embedded microstrip, 287–93
    re-entrant mode, 294–96
    See also Directional couplers; Tight couplers
Multioctave bandwidth, 11
Multiplexers, 308
Multisection directional couplers, 193–209
    asymmetrical, 194
    branch, 260
    equivalent circuit, 194
    example, 199–208
    ideal, power loss ratio, 198–99
    illustrated, 194
    length of, 234
    limitations, 208
    physical layout, 209
    power loss ratio, 196–97
    symmetrical, 194
    synthesis, 199
    theory, 193–208
    See also TEM directional couplers
Mutual capacitance, 271–73

Near-end crosstalk voltage, 460
    for coupled microstrip lines of different spacings, 464, 467
    for microstrip/stripline configurations, 465, 468
    reduction, 472
    for various line losses, 462, 466
    for various terminations in active line, 471
    for various terminations in quiet line, 470
    See also Crosstalk voltage
N normal modes, 477
Nomographs, 309
Nonplanar baluns, 396
Nonreciprocal networks
    three-port, 43
    two-port, 41
Nonuniformly coupled symmetric lines
    defined, 6
    illustrated, 9
    See also Symmetrical coupled lines

Nonuniform TEM directional couplers, 219–42
    asymmetrical, 237–42
    design procedure, 234–37, 238
    symmetrical, 219–37
    weighting function, 229
    See also TEM directional couplers
Normalized currents, 24–28
    determining, 25
    incident, 26, 27
    total, 27
    See also Currents
Normalized scattering matrix, 40–41
    two-port network, 40
    unitary property, 41
Normalized voltages, 24–28
    amplitude, 36
    determining, 25
    incident, 26, 27, 36
    total, 27
    See also Voltages
N-port networks, 23
    four-port, 43–46
    illustrated, 32
    P ports, 24
    representative matrix elements, 39
    three-port, 42–43
    two-port, 40–42
Odd-mode excitation, 76–79, 128–30
    capacitance representation, 78
    defined, 126
    equivalent circuit, 210
    illustrated, 77
Overlay couplers, 212–13

Parallel-coupled filters, 14, 323–26
    ABCD matrix, 324
    bandpass characteristics, 325
    design, 323
    illustrated, 17, 324
    performance, 326
    physical dimensions, 325
    physical lengths, 325–26
    See also Filters
Parallel-coupled stepped-impedance filters, 335–36

design, 335
illustrated, 336
impedance ratio, 335
spurious resonance frequencies, 335
See also Filters
Parallel-coupled TEM couplers, 181–217
Parallel-plate baluns, 398–99
advantages, 398–99
frequency response, 399
side view, 398
See also Baluns
Parallel-plate capacitance, 80–81
Parasitic coupling, 3
Per-unit wavelength, 170, 171, 177
Phase balance
Marchand balun, 433
planar balun, 412–13
Phase constant, 67
determining, 71
even-mode excitation, 79
Phase-reversal property, 405
Phase velocity, 71
characteristic impedance and, 69
determining, 69
even-mode excitation, 79
suspended microstrip line, 111
Physical constants, 499
Planar baluns, 403–11
180-degree phase difference sections, 407
ABCD matrices, 408
amplitude, 412–13
analysis, 405–8
broadband, 409
configurations, 404
defined, 392
examples, 408–11
four-port S-matrix, 406
illustrated, 395
Lange couplers, 409
multisection divider, 404
performance results, 409
phase difference, 412–13
physical dimensions, 409
return loss, 412–13
sections, 403–4
two-port S-parameters, 407
See also Baluns

Planar-transformer balun, 441–42
circuit diagram, 441
defined, 441
resonant frequency, 441–42
simulated/measured performances comparison, 442
See also Baluns
Planar transmission lines, 65–120
Power coupling coefficient, 157
Power loss ratio, 196–97
of ideal directional coupler, 198–99
of single-section coupler, 197
of symmetrical multisection coupler, 197
Printed circuit
filters, 322–23
realization of balun structure, 420
Propagation
constants, 89, 148
delays, 448
lossless, 31
velocities, 6, 70
Properties
branch-line coupler, 246–47
conductor material, 500
current eigenvector, 480–81
dielectric material, 499
rat-race coupler, 262–64
voltage eigenvector, 480–81

Q-factors, 69–70
microstrip line, 96
normalized, 87
quasi-TEM mode, 74
stripline, 87
TEM mode, 74
Quantities, 31–33
normalized, 31
normalized/unnormalized relationship between, 33
unnormalized, 32
Quarter-wave couplers, 192–93
Quasi-TEM mode, 71–73
characteristic impedance, 72
conductance, 72
defined, 66
dielectric loss, 72
effective dielectric constant, 71–72

Quasi-TEM mode (continued)
  effective filling fraction, 73
  equivalent circuit, 68
  general characteristics, 65–71
  parameters, 66
  Q-factor, 74
  transmission lines, 67
  See also TEM mode

Rat-race couplers, 243, 244, 260–70
  bandwidth, 264
  broadband, 265
  coupling variation, 263
  design, 262
  example, 262
  isolation variation, 263
  junction discontinuities, 264
  layout, 261
  lumped-element, 268–70
  modified, 264–65
  phase difference, 263
  properties of, 262–64
  reduced-size, 265–68
  scattering matrix, 261
  scattering parameters, 260
  VSWR variation, 263
  See also Directional couplers; Tight couplers

Reciprocal networks, 39
  four-port, 45
  three-port, 42–43
  two-port, 42

Reduced networks
  four-port into two-port, 62–63
  scattering parameters, 58–59
  three-port into two-port, 59–61
  two-port into one-port, 61–62

Reduced-size branch-line coupler, 251–55
  ABCD matrix, 252
  bandwidth, 254
  characteristic impedance, 253
  defined, 252
  illustrated, 252, 255
  measured performance, 257
  photograph, 257
  See also Branch-line couplers

Reduced-size rat-race coupler, 265–68
  advantages, 265–67
  illustrated, 268
  measured performance, 271
  MMIC, 268
  parallel LC elements, 265
  photograph, 270
  port interchange, 269
  uses, 265
  See also Rat-race couplers

Re-entrant mode couplers, 294–96
  coupling coefficient, 296
  cross section, 294
  defined, 294
  even-/odd-mode impedances, 294, 295
  magnetic wall, 296
  Marchand balun using, 429–33
  measured performance, 296, 297
  microstrip, 295, 433
  schematic view, 294
  typical dimensions, 296
  See also Multilayer couplers

Reflection coefficient, 29–30, 61
  asymmetrical nonuniform TEM coupler, 239
  differential, 221
  even-mode, 131, 239
  four-port network, 62
  illustrated, 50
  normalized input impedance, 30
  odd-mode, 131
  at output port, 51
  two-port network, 49–53, 61
  unnormalized input impedance, 30

Reflection matrix, 484

Reflective wave, 24, 36, 124

Resonators, 326
  coupled-mode theory, 177–78
  length, 343
  weakly coupled, 177–78

Response-type filters, 312–14

Return loss, 51
  biasing circuits, 362
  defined, 30
  Marchand balun, 436
  planar balun, 412–13

Richards transformations, 320

Index

Scattering matrix, 35–40
  asymmetrical nonuniform TEM coupler, 241–42
  backward-wave coupler, 138
  branch-line coupler, 245
  circulator, 43
  directional coupler, 44, 46
  even-/odd-mode, determining, 130
  four-port network, 62
  interconnected networks, 56–63
  normalized, 40, 125
  overall network, 57–58
  port location, 38
  rat-race couplers, 261
  reciprocal networks, 39
  reduced networks, 58–59
  representation, 35
  symmetrical network, 125
  tandem couplers, 282
  terminal plane position, 38–39
  three-port network, 60
  two-port network, 61
  unitary property, 37–38, 46
  unnormalized, 125
Scattering parameters, 36
  amplitudes, 46
  asymmetrical nonuniform TEM coupler, 238
  backward-wave coupler, 137
  branch-line coupler, 244–45
  forward-wave coupler, 134
  rat-race coupler, 260
  reduced networks, 58–59
  symmetrical nonuniform TEM coupler, 221
Schwartz-Christoffel transformation, 84
Series capacitors, 353
Shunt capacitors, 255
Single microstrip, 91–97
Single-section directional couplers, 183–93
  compact, 192
  design, 185–90
  dimensions, calculation of, 190
  equivalent circuit, 192–93
  examples, 190–92
  fractional bandwidth, 185
  frequency bandwidth ratio, 184
  frequency response, 183–85
  illustrated, 182
  power loss ratio, 197
  quarter-wave, 192–93
  useful operating bandwidth, 185
  See also TEM directional couplers
Single stripline, 84–88
Skin depth, 73
Slot-coupled microstrip lines, 117–20
  characteristic impedances, 118–20
  directional couplers and, 117–18
  effective dielectric constant, 118–20
  illustrated, 118
  See also Microstrip lines
Spiral directional couplers, 299–300
  defined, 299
  illustrated, 300
  measured coupled power, 300
  top conductor layout, 299
  See also Directional couplers
Spiral inductors, 373–83
  π equivalent circuit representation, 380
  admittance matrix, 377
  admittance parameters, 380
  design, 374–76
  electrical characteristics, 377
  equivalent circuit model, 378
  equivalent LC circuit representation, 381
  four-port representation, 379
  illustrated, 376
  inductance calculation, 379
  low-loss, 373
  lumped-element design, 375–76
  in MICs, 373
  physical layout, 378
  reflection responses, 382
  transmission responses, 382
  uses, 373
  See also Circuit components
Spiral transformers, 383–87
  1.5-turn rectangular, 385
  3-port, 383
  efficient power transformer, 387
  example illustration, 382
  MMIC, 386
  rectangular, 384
  twin-coil four-port rectangular, 384–85

Spiral transformers (continued)
    See also Circuit components
Striplines
    broadside-coupled, 7, 109–12
    characteristic impedance, 85
    coupled, 458
    cross section, 83
    cutoff frequency, 88
    edge-coupled, 83, 88–90
    effective width of, 84
    microstrip lines vs., 90
    normalized conductor Q-factor, 87
    offset, broadside-coupled, 114–17
    shielded, 66
    single, 84–88
    synthesis, 85
SUPER COMPACT, 211
Suspended microstrip lines, 105–9
    broadside-coupled, 7, 8
    characteristic impedance, 105
    cross section, 110
    effective dielectric constant, 105–7
    illustrated, 5
    See also Microstrip lines
Symmetrical coupled lines, 76, 81–82
    admittance matrix, 492
    characteristic impedance matrix, 493
    coupling, 173–74
    cross section, 82
    eigenvalues/eigenvectors of, 488–94
    forward-wave, 162
    illustrated, 489
    microstrip, 164–65
    nonuniform, 6, 9
    normal mode vs. distributed line parameters, 150–51
    strip pattern, 165
Symmetrical couplers
    defined, 181
    equivalent circuit, 196
    multisection, 194, 195
    physical layout, 209
    schematic, 195, 199
    tables of parameters, 200–207
    See also Directional couplers
Symmetrical networks
    even-/odd-mode analysis of, 124–30

    four-port, 124, 127
    illustrated, 124, 127
    scattering matrix, 125
Symmetrical nonuniform TEM couplers, 219–37
    coupling determination, 223
    coupling response, 226–28
    design procedure, 234–37
    electrical/physical length, 233–34
    equal-ripple response, 233
    equivalent transmission line circuit, 222
    even-mode characteristic impedance, 221–23
    example, 235–37
    ideal, 224–28
    illustrated, 220
    multisection, length of, 234
    ripples, number of, 233
    scattering parameters, 221
    synthesis, 224–30
    total electrical length, 233
    weighting functions, 230–33
    See also Nonuniform TEM directional couplers
Tandem couplers, 243, 281–84
    defined, 281–82
    fractional power, 284
    output, 283
    physical configuration of, 285
    realization of, 284
    scattering matrix, 282
    schematic, 282
    See also Directional couplers; Tight couplers
Telegraphist equations, 477
TEM directional couplers, 181–217
    coupling variation, 186–89
    directivity, improving, 209–17
    interdigital configuration, 275
    introduction, 181–82
    isolation, 182–83
    multisection, 193–209
    nonuniform, 219–42
    parameters, 182–83
    single-section, 183–93
    two parallel-coupled lines, 275

Index 519

See also Directional couplers
TEM mode, 70–71
   equivalent circuit, 68
   general characteristics, 65–71
   lossless, 151
   Q-factor, 74
   transmission lines, 66
Three-conductor transmission line, 476
Three-port networks, 42–43
   illustrated, 44
   nonreciprocal, 43
   port three terminated in arbitrary load, 60
   reciprocal, 42–43
   reduction into two-port network, 59–61
   scattering matrix, 60
   unitary property, 43
   See also N-port networks
Thru-Reflect-Line (TRL) de-embedding techniques, 288
Tight couplers, 243–302
   braided microstrip, 301
   branch-line, 243, 244–60
   combline, 301
   compact, 296–301
   connections, 15–16
   coplanar waveguide, 301
   dielectric waveguide, 302
   finline, 301
   introduction to, 243–44
   Lange, 143, 274–81
   multiconductor, 270–81
   multilayer, 284–96
   rat-race, 243, 244, 260–70
   slot-coupled, 301
   tandem, 243, 281–84
   uses, 243
   vertically installed, 301
   wiggly two-line, 302
   See also Directional couplers
Transformation ratios, 368
Transmission coefficient
   illustrated, 50
   two-port network, 49–53
   vector, 484
Transmission Line Design Handbook, 120
Transmission lines

capacitance variation, 455, 456
dielectric losses of, 455
external inductance, 454–55
homogeneous, 66
impedance Z in, 53
inductance, 454–55, 456
multiconductor, 455, 475–96
parameters, 67
planar, 65–120
quasi-TEM mode, 67
TEM mode, 66
three-conductor, 476
two-conductor, 476
uniformly-coupled, 123–59
Triformer, 386–87
   defined, 387
   illustrated, 386
Triformer balun, 438–40
   computed/measured S-parameters comparison, 440
   defined, 438
   equivalent circuit, 439
   photograph, 439
   physical layout, 439
   See also Baluns
Twin wire line, 66
Two-conductor transmission line, 476
Two-port networks, 40–42
   ABCD parameters, 47–49, 54
   cascade, 48
   equivalent, 53–55
   four-port network reduction into, 62–63
   illustrated, 41
   interdigital, 145
   of known transfer function, 449
   lossless, 49
   matrice conversion relationships, 52–53
   nonreciprocal, 41
   normalized scattering matrix, 40
   passive, 49
   with port two terminated in arbitrary load, 62
   reciprocal, 42
   reduction into one-port network, 61–62
   reflection/transmission coefficients, 49–53
   scattering matrix, 61

Two-port networks (continued)
    special representation of, 47–55
    T and Π network representation, 55
    three-port network reduction into, 59–61
    See also N-port networks

Ultra-broadband forward-wave
    couplers, 168–69
Uncoupled resonators, 177
Uniformly-coupled lines, 123–59
    asymmetric, 4, 140–53
    directional couplers using, 130–39
Unitary matrix, 37
    reciprocal three-port network, 43
    two-port network, 41
Unnormalized currents, 28–29
Unnormalized voltages, 28–29

Very-large-scale integrated (VLSI) chips, 3
Voltage coupling coefficient, 178
Voltage eigenvectors, 480–81, 491
    bi-orthonormal, 481
    matrix, 481
    normalization, 481
    normalized, matrix, 492
    orthogonal properties, 480–81
    unnormalized, matrix, 491, 492

Voltages
    equivalent, 24
    far-end crosstalk, 460, 462–71
    forward-traveling wave, 482
    interconnect terminal, 451
    near-end crosstalk, 460, 462–68, 470–71
    normalized, 24–28
    on asymmetrical coupled lines, 141
    reflection coefficient, 29–30
    total, 31
    unnormalized, 28–29
Voltage standing wave radio (VSWR), 3, 30
    branch-line coupler, 247, 248
    rat-race coupler, 263
Voltage vector, 482

Weakly coupled resonators, 177–78
Weighting function, 226–28
    determining, 230–33
    example, 232–33
    of nonuniform coupler, 229
    terms, 233
Wiggly lines, 213–17
    capacitance parameters, 216
    effective length, 214
    geometrical parameters of, 213
    illustrated, 215
    odd-mode capacitances, 214

Recent Titles in the Artech House Microwave Library

Behavioral Modeling of Nonlinear RF and Microwave Devices,
Thomas R. Turlington

Computer-Aided Analysis of Nonlinear Microwave Circuits,
Paulo J. C. Rodrigues

Design of FET Frequency Multipliers and Harmonic Oscillators,
Edmar Camargo

Design of RF and Microwave Amplifiers and Oscillators,
Pieter L. D. Abrie

EMPLAN: Electromagnetic Analysis of Printed Structures in Planarly Layered Media, Software and User's Manual, Noyan Kinayman and M. I. Aksun

Feedforward Linear Power Amplifiers, Nick Pothecary

Generalized Filter Design by Computer Optimization,
Djuradj Budimir

High-Linearity RF Amplifier Design, Peter B. Kenington

Introduction to Microelectromechanical (MEM) Microwave Systems,
Hector J. De Los Santos

Microwave Engineers' Handbook, Two Volumes,
Theodore Saad, editor

Microwave Filters, Impedance-Matching Networks, and Coupling Structures, George L. Matthaei, Leo Young, and E.M.T. Jones

Microwave Materials and Fabrication Techniques, Third Edition,
Thomas S. Laverghetta

Microwave Mixers, Second Edition, Stephen Maas

Microwave Radio Transmission Design Guide, Trevor Manning

Microwaves and Wireless Simplified, Thomas S. Laverghetta

Neural Networks for RF and Microwave Design, Q. J. Zhang and K. C. Gupta

RF Design Guide: Systems, Circuits, and Equations, Peter Vizmuller

The RF and Microwave Circuit Design Handbook, Stephen A. Maas

RF and Microwave Coupled-Line Circuits, Rajesh Mongia, Inder Bahl, and Prakash Bhartia

RF Power Amplifiers for Wireless Communications, Steve C. Cripps

RF Systems, Components, and Circuits Handbook, Ferril Losee

Understanding Microwave Heating Cavities, Tse V. Chow Ting Chan and Howard C. Reader

For further information on these and other Artech House titles, including previously considered out-of-print books now available through our In-Print-Forever® (IPF®) program, contact:

Artech House
685 Canton Street
Norwood, MA 02062
Phone: 781-769-9750
Fax: 781-769-6334
e-mail: artech@artechhouse.com

Artech House
46 Gillingham Street
London SW1V 1AH UK
Phone: +44 (0)20 7596-8750
Fax: +44 (0)20 7630 0166
e-mail: artech-uk@artechhouse.com

Find us on the World Wide Web at:
www.artechhouse.com